THE MARINE AND FRESH-WATER PLANKTON

THE MARINE AND FRESH-WATER PLANKTON

by

Charles C. Davis

Associate Professor of Biology
Western Reserve University

1955

MICHIGAN STATE UNIVERSITY PRESS

Respectfully dedicated

to

TAKAMOCHI MORI

lost to planktology at Hiroshima, Japan

on

August 6, 1945

LIBRARY OF CONGRESS CATALOGUE CARD NUMBER 54-8843
MANUFACTURED IN THE UNITED STATES OF AMERICA
BY THE ADVANCE LITHOGRAPHING COMPANY OF CHICAGO

Table of Contents

Preface

IN THE COURSE of any broad study within the biological sciences, the student inevitably obtains at least a passing glimpse of the plankton—the world of drifting or weakly swimming aquatic beings that so long remained unknown to science because of the minute size of most of its members. Yet the plankton is very intimately involved ecologically with other aquatic organisms, and with terrestrial animals, and even with man himself. The most important of these relationships lies in the fact that the plankton, both in marine and fresh-water situations, is the most important ultimate or direct source of food for all the larger forms. Hence, in any serious program of study, especially in the field of aquatic biology (including ichthyology and fisheries biology), the desirability of obtaining a much more intimate knowledge of the plankton becomes obvious.

The literature on the plankton is both voluminous and scattered. The bulk of it is purely taxonomic, but much of the more recent literature is concerned with ecological and commercial problems, and it is in these two closely-related fields that the greatest future research advances will be made. In fact, the term "plankton," referring as it does to a group of organisms with a common habitat, is in itself an ecological term. The evolutionary history, physiology, or anatomy of an individual plankton species might well be studied with profit, but the primary interest of the general student of the plankton must remain either a taxonomic interest or an ecological one. To the modern plankton student (unless he be a dilettante), the major reason for a taxonomic study of the plankton, except in areas previously unexplored or inadequately explored, would be to obtain a better and more precise basis for ecological studies.

It was the author's privilege to offer a course in plankton for graduate students and seniors at the University of Miami in the summer session of 1948, and it is upon the lectures and laboratory work for the course that this book is based. Despite the volume of the literature on plankton, there is no book in print that deals adequately with either the fresh-water or the marine plankton, and it is hoped that the present book will help, at least to some degree, to fill the gap for those teachers and students who in the future may offer or take similar courses. In addition, it is hoped that the book will be of value to all those who are concerned in any way with general or specific studies of aquatic biology.

Inasmuch as most courses in biology are incomplete without a thorough laboratory program, an attempt is made in the final section of the present book to lay the basis for the study by students and research workers of the major types of plankton organisms that might commonly be met with in samples of fresh-water or marine plankton. The keys are not intended to be complete, nor are the discussions of the various subdivisions of the plant and animal kingdoms intended to be exhaustive. Such a coverage of the material would occupy many volumes. An attempt to compensate in part for the omissions is made by referring interested readers to some of the more important literature dealing with each field.

It was planned to add an Appendix dealing with methods of plankton research, but the timely appearance of the valuable book *Limnological Methods* by Paul S. Welch (1948), with its thorough Chapter XV on plankton methods (pp. 231-297), makes such a task superfluous so far as fresh-water techniques—and most marine techniques—are concerned. Additional equipment for marine problems is described in Chapter X (pp. 376-385) of *The Oceans,* by Sverdrup, Johnson, and Fleming (1946).

It is a pleasure as well as a duty to acknowledge my indebtedness to others in the preparation of this work. I am especially grateful to my students for their criticisms and suggestions concerning the material covered, and to Dr. E. Morton Miller, Chairman of the Department of Zoology, and Dr. F. G. Walton Smith, Director of the Marine Laboratory, of the University of Miami for the continued interest and encouragement that made it possible to offer the course in plankton described above. In addition, my debt is great to Dr. A. H. Banner, Mrs. Sally M. Davis, Dr. E. M. Miller, Dr. G. W. Prescott, Dr. F. G. W. Smith, and Dr. R. A. Woodmansee for their critical reading of the manuscript in part or in whole, as well as to the following specialists who made many valuable suggestions towards the improvement of the taxonomic keys for the various groups of planktonic organisms: Dr. G. W. Prescott for the Cyanophyta, Chlorophyta, Bacillariaceae, and Dinoflagellata; Dr. J. H. Hoskins for the Cyanophyta; Dr. Paul S. Conger for the Bacillariaceae; Dr. A. S. Campbell for the Radiolaria and Tintinnidiidae; Dr. Mary Sears and Dr. P. L. Kramp for the Coelenterata; Dr. W. T. Edmondson for the Rotifera; Dr. Harold A. Rehder for the Gastropoda; Mr. Rufus W. Kiser for the Cladocera; Dr. Clarence R. Shoemaker for the Amphipoda; and Dr. A. H. Banner for the Mysidacea and the Euphausiacea. Western Reserve University furnished equipment for making many of the illustrations.

Necessarily, such a work as this is based largely upon the sincere and selfless studies of generations of prominent and obscure investigators, and though their names perforce cannot be mentioned here, my debt to them is beyond measure.

CHARLES C. DAVIS

March, 1955
Western Reserve University
Cleveland, Ohio

I. *Introductory*

General Ecological Considerations

THE EARTH has solid, gaseous, and aqueous constituents. The solid portion (the lithosphere) is covered by water (the hydrosphere) on the major portion of its surface, while both are overlaid by an extensive mixture of gases (the atmosphere). Each is characterized by a certain faunal and floral population, though it is true that the atmosphere has a much more limited population than the other two, and it is for the most part a temporary population (birds, flying insects, etc.) coming from the surface of the lithosphere. The inhabitants of the lithosphere consist mostly of more highly specialized types of animals and plants, while those of the hydrosphere, in general, are less specialized. This is because of evolutionary adaptations to the more rigorous and variable environmental conditions that are to be found in terrestrial locations.

Because of the enormous number of insect species, most of the existing kinds of organisms are terrestrial, but if we were to disregard the insects, most of the remaining species would be found to be aquatic, and hence members of the population of the hydrosphere. For our purpose here, of course, the populations of the lithosphere and the atmosphere need not be considered further, because plankton by definition is aquatic.

Aquatic organisms can be classified, on an artificial basis, according to the broad conditions of their existence. Thus they may be considered as benthos, nekton, plankton, or neuston. The organisms classified as benthos are those animals and plants that are attached to, crawl over, or that burrow into the bottom in aquatic situations, whereas the other three categories are divorced in their habitats from the bottom. The nekton consists of animals only, inasmuch as by definition the nekton consists of those organisms which can move around in the water "purposively," and independently of the movements of the water itself. The plankton and the neuston consist of free-floating organisms, both animals and plants, whose intrinsic movements, if existent, are so feeble that they remain essentially at the mercy of every water current or other water movement. The neuston is distinguished from the plankton by the fact that it floats on the surface of the water, either under or above the surface film. Most plankton organisms seldom approach the surface and never have more than a temporary and accidental association with the surface film.

There are two main environmental subdivisions of the hydrosphere, namely marine and fresh-water, the two, however, merging into each other in certain situations, as in the estuaries of rivers, forming intermediate brackish-water environments. Estuarine waters have been discussed by Hedgpeth (1951), who extends the classification of Redeke and Välikangas. Estuarine waters are classified as follows: *fresh waters* are those having a chlorinity (see below, p 18) of less than 0.1 ‰; *oligohaline brackish waters* are those having a chlorinity between 0.1 and 1.0 ‰; there are two degrees of *mesohaline brackish waters,* namely *alpha* with chlorinities lying between 1.0 and 5.5 ‰, and *beta* with chlorinities lying between 5.5 and 10.0 ‰; *polyhaline brackish waters* are those with chlorinities lying between 10.0 and 17.0 ‰; and *marine waters* have chlorinities above 17.0 ‰. *Metahaline waters* are those of shallow lagoons with high evaporation and low recruitment of fresh water, hence with chlorinities higher than those of the open sea.

Of the two subdivisions of aquatic environments, the marine is by far the more extensive, for the seas occupy some 71 per cent of the surface of the earth, and their average depth is 3,500 m. (2¼ mi.), with depths known up to 10,550 m. (34,612 ft.), as in the Mindanao Trench in the Pacific Ocean close off the Philippine Islands, and the Ramapo Depth in the Japan Trench off the coast of the Japanese Archipelago.

Nonmarine aquatic environments.—Nonmarine (or fresh) waters are not nearly so extensive, either in area or, especially, in volume as the marine. In fact, both in area and volume, fresh-water is comparatively infinitesimal in extent. However, the influence of fresh-water environments is important not only to the aquatic organisms living therein, but also to terrestrial organisms. Because of his physiological need for fresh water to drink and his need of it for agricultural and industrial purposes, as well as his use of fresh water as a source of fish and other food, mankind as a whole undoubtedly finds fresh water to be economically of greater importance to himself than salt water. On the other hand, the potential direct economic importance of marine environments as a source of food for man has hardly yet even been contemplated more than superficially, and much less has it been acted upon (see, for example, pages 134 ff).

In North America, the exact extent of fresh-water areas is difficult to estimate accurately. It has been estimated that lakes alone in North America occupy about 0.2 per cent of the area of the land. In addition, much more area is occupied by flowing waters, swamps, etc. In the state of Florida the data on the official map of the state's Department of Agriculture indicates 3,805 sq. mi. of water in a total area of 58,666 sq. mi. Thus the water area would constitute 6.5 per cent of the total. However in addition to ponds, extensive swamps, and flowing rivers, this includes the extensive estuaries of the St. John's River, the Indian River, etc., which in some parts are more nearly akin to marine environmental situations than to fresh-water.

Fresh waters may be either lentic (standing water) or lotic (running water). Lentic waters include, in addition to ground water (which need not concern us here inasmuch as its plankton content is almost always negligible), lakes, ponds, and swamps. A lake can be defined as a body of standing water isolated from the sea. It is of sufficient depth so that there is a large area of open water devoid of rooted vegetation. A pond is simply a shallow lake with rooted submerged vegetation, and a swamp is a pond so shallow that its whole expanse is occupied by emergent vegetation, rooted in the bottom. In the evolutionary history of lentic waters, the general tendency has been for lakes to be transformed into ponds, ponds into swamps, and swamps into dry land. In general it can be said that, with very few exceptions, lentic environments are (speaking geologically) temporary environments, and the average age of a lake perhaps does not exceed about 25,000 years. Many small bodies of water may change appreciably within the life span of individual humans. On the other hand, a very few lakes, because of great depth and large size, seem to be much older than the average. Lake Tanganyika in Africa and Lake Baikal in Siberia are such lakes. The fauna of each includes large numbers of forms that are distinctly reminiscent of marine species. In Lake Baikal, for example, a species of seal occurs, as well as tubiculous polychaete worms and large sponges related to marine forms. In Lake Tanganyika fresh-water medusae, crabs, and marine-type Bryozoa have been found. The faunas, however, are thought to be primitive fresh-water faunas rather than relicts of marine faunas, though the distinction between those two categories is tenuous.

The evolution of lakes.—Lakes may have their origin in numerous ways. Perhaps the majority of lakes have originated through glaciation, the glaciers either scooping out depressions or piling up debris in such a manner as to dam up small or even large valleys. Other lakes have originated by the folding of the surface layers of rock, which produces depressions or ridges. Still others may be produced by the damming of valleys, by landslides or lava flows, by the changing of the course of rivers (e.g., ox-bow lakes, the lakes of the Grand Coulee in Washington, etc.), or by the solution of underlying limestone rocks through the action of natural waters. Most of the many lakes of Florida have evidently been formed by the last-mentioned means. Natural waters have dissolved and subsequently carried away the underlying limestone formations, resulting in the formation of caverns and depressions of varying extent. Frequently the roofs of caverns have collapsed to form sinkholes, which, upon becoming filled with water, form smaller or larger lakes. Areas with many small lakes having this type of origin are called karst areas, after the Karst Plain, an area of this type in Jugoslavia. The lake region of Florida is a karst area.

Once formed, a lake begins its (geologically) rapid evolution. Vegetation grows richly in the shallow areas close to shore, with emergent vegetation (cattails, etc.) in the shallowest parts, rooted vegetation (water lilies, etc.) in somewhat deeper portions, and rooted submerged vegetation yet farther from the shore. The bottom in the deepest portion of lakes, however, is devoid of rooted vegetation (see fig. 1-A), and any vegetation occuring in such places in the water above the bottom is planktonic or otherwise floating.

As time goes on, the lake vegetation grows, dies, and falls or sinks to the bottom, where it undergoes partial decomposition (the degree depending upon the temperature, the amount of oxygen present, the rapidity of accumulation, etc.). Undecomposed portions of the dead plants, added to by dust and sand deposited in the water through the agency of the wind, and by material washing in from the inlets of the lake, accumulate over the years. This results in the gradual shoaling of the lake. A portion of the lake nearest the shore accumulates sufficient material so that it becomes dry land and its emergent aquatic vegetation is replaced by terrestrial vegetation. That portion somewhat farther from shore becomes sufficiently shallow so that its floating plants are replaced by emergent plants. That portion even farther from shore becomes sufficiently shallow so that submerged plants are replaced by floating rooted plants. Meanwhile, the margin of the center of the lake, which formerly did not support any rooted vegetation, now becomes sufficiently shallow to support the growth of rooted submerged vegetation. Beyond this, the center of the lake becomes somewhat more shallow, but yet not perhaps sufficiently so to support any rooted vegetation. In this manner, as time goes on, the lake is gradually transformed into a pond (see fig. 1-B, C), and in a similar manner a pond will be transformed gradually into a swamp (see fig. 1-D), where the dominant vegetation is all emergent. In turn, the swamp is transformed by similar processes into dry land (see fig. 1-E). In other cases, lakes may disappear by means of rim-cutting (see fig.2), or by the aid of rim-cutting superimposed upon the process of filling described above.

Among lentic fresh waters the lakes and ponds (especially the lakes) concern us the most.

Variations among lakes.—There is a very great diversity among lakes, though at first sight there might not appear to be; there is diversity, e.g., in the historical origin of lakes as well as in their size, depth, elevation, latitude, chemical and physical conditions, etc.

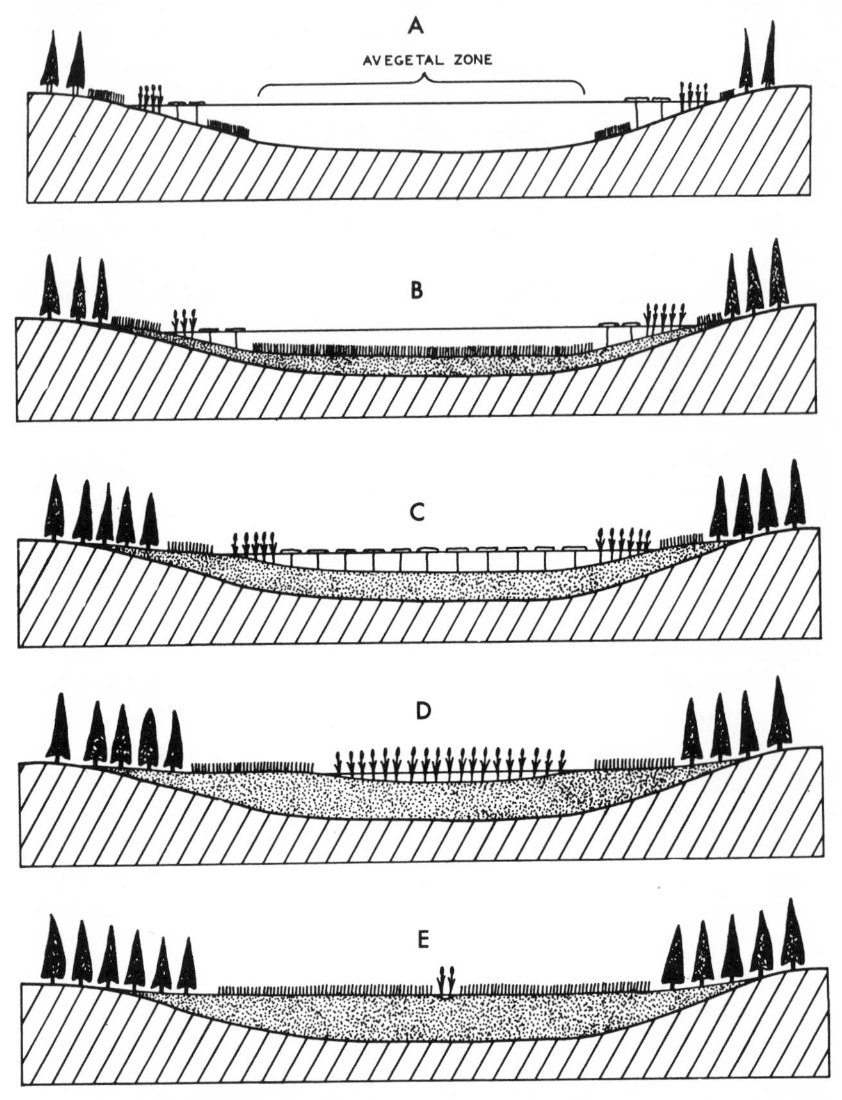

Fig. 1. Idealized diagrams to show the evolution of a lake with an extensive avegetal zone into a pond, into a marsh, and finally into dry land. The symbols signify respectively trees, grasses on land, emergent vegetation, floating vegetation, and submerged vegetation. The original basin of the lake is indicated by crosshatching. Sediments are stippled. (A) Original condition of the lake, with extensive avegetal zone. (B) Pond conditions, with extensive open water, but no avegetal zone. (C) Advanced pond conditions with little or no open water. (D) Marsh conditions, with emergent vegetation and no open water. (E) Dry land conditions, with only a small, marshy, central creek.

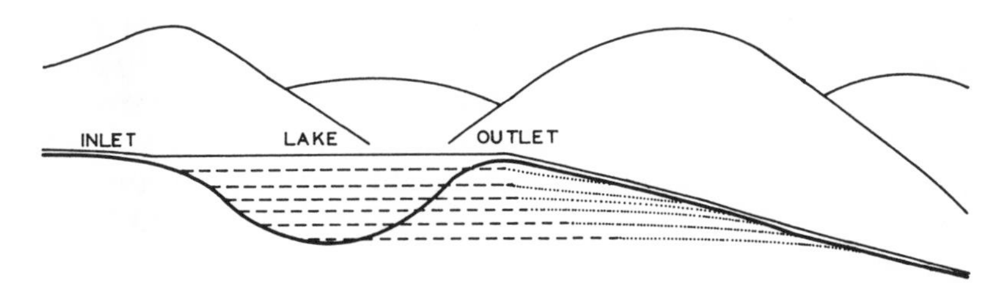

Fig. 2. Idealized diagram to show reduction of the level of a lake and its final disappearance by rim-cutting. The solid lines indicate the original levels of land and water. Dotted lines indicate new levels of the outlet. Dashed lines indicate consequent new water levels in the lake. Coincident lowering of the inlet by the cutting action of the water, and possible filling of the lake basin by sedimentation, are not shown.

The historical development of a lake has a great influence upon its fauna and flora, regardless of what environmental conditions are present. Thus Lake Tanganyika and Lake Baikal, being, relatively, extremely ancient lakes, have entirely different faunas from those that would be found in other lakes in which general environmental conditions might be similar. Less spectacularly, but equally validly, two ox-bow lakes of equal age and size, within a few miles of each other along the Mississippi River, may contain very different plankton populations because of the fact that one of them (because, say, of the chance lack of browsing fish within its confines) developed a rich growth of rooted submerged plants during its first year of existence, while the other did not do so.

The largest lake in the world is, of course, the salty lake called the Caspian Sea, with an area of some 170,000 sq. mi. The Caspian Sea is so large that it is difficult to think of it as a lake. However, in its historical development it has not been an arm of the ocean since the Eocene and Oligocene, and salts are not present in it in the same proportions as are the salts in the sea. Like most salt lakes, the Caspian Sea has many inlets, by means of which much-diluted salts enter the lake basin, but no outlets through which they can leave. Evaporation during the course of time has resulted in a considerable concentration of these salts.

The largest strictly fresh-water lake in the world is Lake Superior in North America, with an area of 31,000 sq. mi. A few other large lakes are Lake Michigan, with 22,400 sq. mi., Lake Erie with 4,990 sq. mi., Great Bear Lake in Canada with

11,500 sq. mi., Lake Baikal with 13,530 sq. mi., and Lake Chad in Africa with 23,000 sq. mi. (the last-mentioned is an average area, for Lake Chad may occupy 40,000 sq. mi. in the wet season and decrease to only 6,000 sq. mi. in the dry season). Such large lakes, however are unusual. Most lakes are relatively small. For example, aside from the Great Lakes, the state of Michigan has some 5,000 lakes, of which the largest has an area of only 30 sq. mi. Florida has an estimated 30,000 lakes and ponds, of which by far the largest is Lake Okeechobee with an area of 717 sq. m. And, Lake Okeechobee, with the exception of the Great Lakes, is the second largest lake in the United States, only Great Salt Lake in Utah being larger.

Lakes also vary greatly in depth, though no lakes any place in the world approach the sea in depth. Lake Baikal is the deepest known lake, with a recorded depth of 5,600 ft. (1,706 m.). Other deep lakes include Crater Lake in Oregon with a depth of 1,996 ft. and Lake Superior with a depth of 1,000 ft. Most lakes, however, are much shallower than these, being less than 90 ft. deep. For example, it has been stated that of the numerous lakes in Florida, few exceed 27 ft. in depth (American Guide Series, 1939). Of course, with regard to depth, many lakes are at the border line between lakes and ponds.

Depth has many influences (some only vaguely understood) upon the organisms living in a lake. A deeper lake, other conditions being similar, will have a longer evolutionary history than a shallower one, and this will influence the type of organism that lives there. Some deep lakes, such as Lake Baikal, have a distinct abyssal fauna. Heavier growth of planktonic diatoms has been observed over the deeper waters of deep lakes than nearer shore in shallower water. On the other hand, nutrients salts such as nitrates and phosphates would have a tendency to sink below the lighted surface waters in deep lakes, and thus be lost for the growth of plants.

Elevation and latitude have somewhat similar effects upon lakes with respect to temperature, and lakes are found at all levels from considerably below sea level (Caspian Sea, Dead Sea in Israel, Salton Sea in California, etc.) to considerably more than 12,000 ft. in the case of mountain and high plateau lakes, and they are found from the Equator to the frigid zones.

Tropical lakes and lakes at low altitude may never become sufficiently cold to form ice, whereas some lakes at high altitudes or at high latitudes may never be completely free of ice. All intermediate conditions also exist.

Altitude and latitude, in certain respects other than temperature, have differing effects, notably upon the length of the day in various seasons. A lake high in the Rocky Mountains of Colorado near the timber line may have conditions of temperature and length of season similar to those of a lake in northern Canada at a low altitude and near the border line between the forests and the tundra, but the northern lake will have longer periods of sunshine in the summer and shorter periods during the winter than the lake in Colorado, and this one condition would be of extreme importance to those plants and animals living in the lakes.

Lakes have the widest variety in their chemical and physical conditions. Some "lakes," such as West Lake and Seven Palm Lake in southern Florida, Laguna de Leche in Cuba, Lake Maracaibo in Venezuela, Lakes Piratininga and Itaipú in Brazil (see Oliveira, 1948), etc., are nothing more than enclosed arms of the sea. Though their salts are often either less or more concentrated than in sea water, the proportion of the various salts in the water is similar to that found in the ocean. Other lakes, and these constitute the vast majority, are completely cut off from the sea, and if saline the various salts are present in entirely different proportions from those obtaining in the sea.

The salinity of those lakes that are completely cut off from the sea depends upon several factors, the most important of which are: (1) whether there is any runoff from the lake toward some lower basin or towards the sea (i.e., whether there is an outlet as well as inlets); (2) the rate of evaporation characteristic of the region in which the lake lies; (3) the rate of precipitation in the region; and (4) the nature of the terrain composing the drainage area of the lake (e.g., what soluble materials are present in the soil and rocks, etc.). Lakes such as we are considering range from those so fresh and pure that their waters can be used by motorists as a substitute for distilled water in storage batteries (e.g., most of the lakes in the western portion of the state of Washington) to those in which the water is nearly saturated with dissolved salts (e.g., Great Salt Lake in Utah). In some cases, a lake as such may exist only during the wet season, losing all of its water by evaporation and leaving nothing but a thicker or thinner layer of salt or borax during the dry season (e.g., the Humboldt Salt Marsh of Nevada, the numerous "dry lakes" of eastern California, etc.). The nature of the salts to be found in the saline lakes varies greatly, depending upon the chemistry of the surrounding terrain.

The water of most lakes has very little buffering action, because of relatively small concentration of buffering salts in solution, and as a result lakes vary greatly in their hydrogen and hydroxide ion concentrations (pH). Lakes that are surrounded by swamps and bogs, or that are otherwise situated where there is considerable organic decay, tend to have an acid pH, because of the liberation of organic acids by the process of decay. Such is the case in the bog lakes of northern Wisconsin and in the mangrove lakes of Florida. On the other hand, lakes in desert regions tend to have a higher or alkaline pH, because of the presence of sodium and potassium salts of such weak acids as boric acid, carbonic acid, etc. Many other lakes, especially in regions of high rainfall, tend to be nearly or quite neutral in reaction.

Other variations could be cited, such as lakes with hard waters versus lakes with soft waters (due, of course, to the presence or absence of small quantities of calcium salts, etc., in solution), turbid lakes versus clear lakes, colored lakes (e.g., the reddish waters of bog lakes and mangrove lakes, or the green of algae-filled lakes) versus "colorless" lakes, etc. In fact, no two lakes are exactly alike biologically, but each has its own peculiar characteristics and its own fauna and flora.

Thus lake environments are characterized by two things: great variation of living conditions, and temporary (geologically speaking) existence. These two characteristics are of most profound importance in understanding the life of lakes, including, of course, their planktonic life.

Ecological subdivisions and seasonal history of lakes.—Ecologically, lakes may be subdivided into horizontal and vertical portions. Horizontally, the relatively shallow area close to shore, characterized by rooted emergent or floating vegetation, is called the littoral region, while the region of open water is known as the limnetic (or pelagic) region. Life conditions, of course, are very different in the two regions. Plankton occurs in both, but the limnetic region is much more extensive than the littoral in most lakes, and it is here that the plankton dominates. In the littoral region it is overshadowed by attached forms among the rooted vegetation. For all practical purposes, our subsequent discussion of the plankton of lakes will be restricted to the plankton of the limnetic region.

The vertical ecological subdivisions of a lake are perhaps best approached through a discussion of the seasonal history of lakes in temperate regions. The discussion does not necessarily hold for tropical or frigid zone lakes. The conditions in such lakes are, comparatively, very little known.

The water of a temperate lake in the spring of the year (providing the previous winter has been sufficiently severe) eventually will have a rather uniform temperature throughout, and this temperature will be approximately that of water at its greatest density (around 4° C.). Under these conditions the water of the lake will be free to mix thoroughly throughout. Persistent winds will serve to pile the water up on the lee side of the lake (see fig. 3), and this displaced water will sink in order to compensate for its unstable position at a higher level than the water at the windward end of the lake. Thus a system of currents will be established. Surface currents will be roughly with the wind, the water piled up thereby will sink towards the leeward end of the lake, a current will pass along near the bottom in a direction counter to the wind, and there will be an upwelling of water near the windward end of the lake. In this way the water at all levels in the lake, having recently been in contact with overlying atmosphere, will be well supplied with dissolved oxygen, and deep-water accumulations of plant nutrient salts will invade surface waters. Furthermore, all levels will tend to take on similar temperatures, while all other chemical and physical characteristics will also tend to be similar throughout the water. In a lake of sufficient depth these relationships may not hold in such a simple manner, but in essence the discussion applies to all temperate zone lakes.

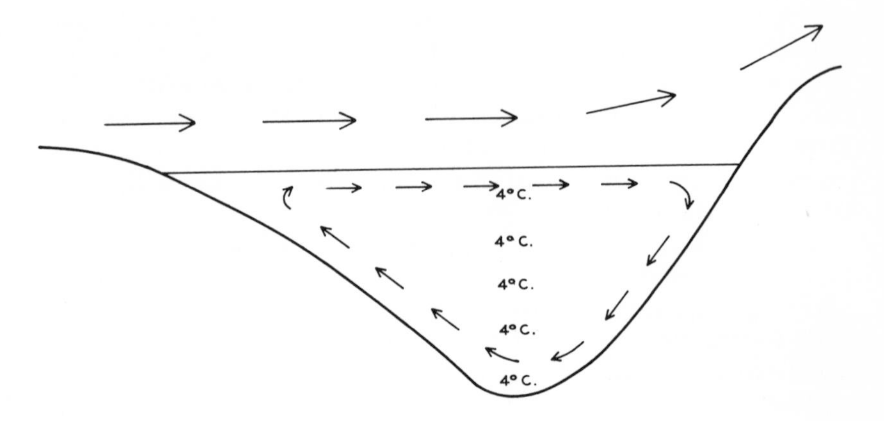

Fig. 3. Idealized diagram of spring conditions in a temperate lake. The water throughout has risen to the temperature of its greatest density, and therefore winds can cause a complete circulation of the water.

As the season progresses, the warmer winds blowing across the lake from the land, and the direct rays of the sun beating down upon the lake itself, tend to increase the temperature of the surface layers of the water. Water expands as its temperature rises above 4° C., and thus these warmer surface waters are lighter per unit volume than the underlying colder waters. Under proper conditions, the upper water becomes relatively so warm that mixture with lower layers becomes difficult. If persistent winds occur, they will produce surface currents as before, and these currents will result in the piling up of the water at the leeward end of the lake. This water will then sink, and the return current will no longer be at the bottom of the lake, but at some intermediate level, just above the colder bottom water (see fig. 4). The countercurrent in the upper warmer layer will induce a current of somewhat lesser magnitude in the lower colder layer, and the water piled up at the windward end of the lake by this process will sink and return as a countercurrent, in much the same direction as the original wind, along the bottom. In this way, both the upper warmer layer and the lower colder layer have a circulation of water, but the two circulations, and hence the two water masses, are isolated from each other.

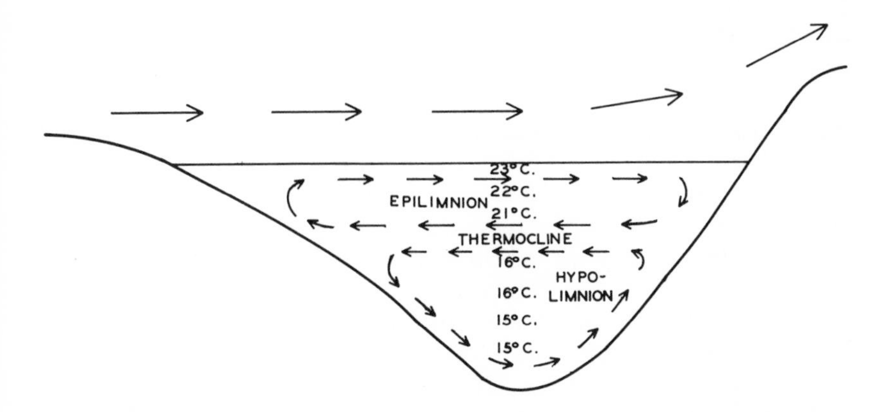

Fig. 4. Idealized diagram of summer conditions in a temperate lake. The lighter, warmer layer of superficial water (epilimnion) is unable to mix with the heavier, colder layer of deep water (hypolimnion). Temperatures within the epilimnion and the hypolimnion are relatively uniform, but between them lies a region of rapid temperature drop, the thermocline. Winds cause a circulation of water within the epilimnion, and the currents in the epilimnion, by friction, cause the circulation of the water of the hypolimnion.

Between the upper and the lower water masses, in typical, well-developed cases, the temperature drops suddenly, for the upper layer is in contact with the warm winds and warming radiation, and its circulation tends to make the whole water mass relatively uniform in temperature, whereas all increase of temperature of the lower layer must come from the upper by conduction and convection, which are very slow processes.

The line of demarcation between the two layers is known as the thermocline, which may be defined as a region wherein the temperature drops at least 1° C. per meter of depth. The water mass above the thermocline is the epilimnion, and that below is the hypolimnion. In addition to relatively high and relatively uniform temperatures, the epilimnion is characterized by high oxygen content, good illumination, and relatively uniform and favorable chemical and physical conditions in general. In this zone, both planktonic animals and plants reach a high degree of development. The hypolimnion, on the other hand, is characterized, in addition to relatively low temperature, by low oxygen content and stagnant conditions. The large amount of decay taking place in the hypolimnion not only frequently reduces the oxygen content to extinction, but also results in the production of extremely toxic hydrogen sulfide (H_2S). Most animals require the presence of free oxygen for their life processes, and only a very few can withstand the presence of any considerable concentration of H_2S, though it has been found that a number of Protozoa, worms and molluscs are very tolerant of the lack of oxygen and the presence of H_2S (see Galadziev and Malm, 1929, and Jakubowa and Malm, 1931). As a consequence, when it is well developed, the hypolimnion contains practically no life other than its anaerobic population of bacteria. A notable exception among planktonic organisms is the aquatic larva of the dipteran genus *Corethra*, whose larvae appear to be able to withstand the complete lack of oxygen, at least for appreciable periods.

The vertical division of lake water into epilimnion, thermocline, and hypolimnion is primarily an estival phenomenon. During the fall, cold winds blow over the lake, cooling the surface waters more or less quickly. As the water cools, it becomes more dense, and thus, if the immediately lower layers are warmer, it sinks until it finds water of the same density as itself. At first this will be in the thermocline. Thus the eplimnion gradually becomes cooler and cooler, while the hypolimnion maintains much the same temperature. Eventually the point is reached where the epilimnion and the hypolimnion come to have similar temperatures; at this time the thermocline ceases to exist, and the upper and lower waters are free to mix without barrier (see fig. 5). Such mixing of the water is known as the fall overturn, and is a regular phenomenon of lakes in temperate regions. The fall overturn results in the oxygenation of the water that was formerly in the hypolimnion, and the removal of the toxic H_2S. Thus all the waters of the lake become habitable for animals, and for plants as well, if there is sufficient light available for photosynthesis.

Mixing continues as the water cools more and more, until the temperature throughout reaches 4° C. Above 4° C., the cool water produced at the surface is heavier than the warmer underlying water, and sinks, allowing more of the warmer water to come in contact with the cold atmosphere. However, water expands upon being cooled below 4° C., the temperature

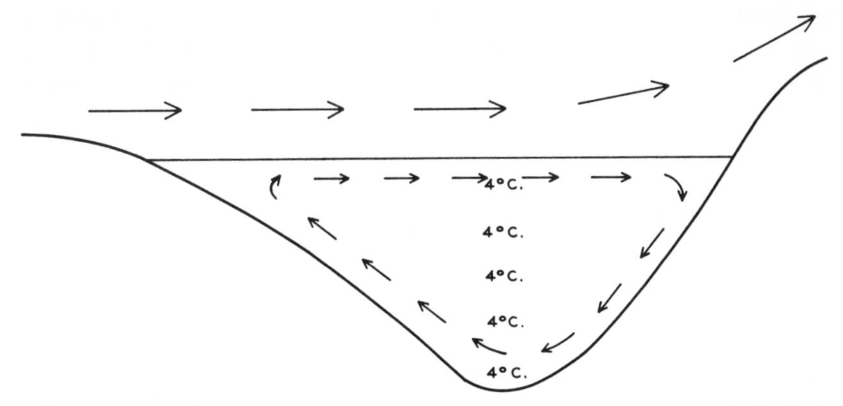

Fig. 5. Idealized diagram of fall conditions in a temperate lake. The water throughout has fallen to the temperature of its greatest density, and therefore winds again can cause a complete circulation of the water.

of its greatest density. After this temperature has been reached throughout the water of a lake, any subsequent cooling results in lighter water at the surface, and this floats, rather than sinks. The cooling of the surface water will continue, without undue mixing with the heavier underlying water (the amount and depth of the mixing depends upon the strength of the winds prevailing in that season), until finally ice forms on the surface (at 0° C.). The formation of a layer of surface ice effectively removes the effect of the wind, and also to a considerable degree cuts off contact of the water with the oxygen of the atmosphere (see fig. 6). Thus a period of stagnation sets in, relieved

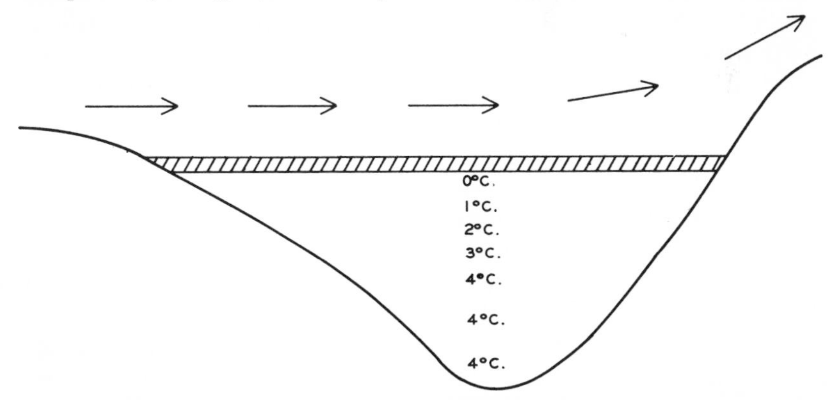

Fig. 6. Idealized diagram of winter conditions in a temperate lake. An ice cover prevents circulation of the water due to winds. Bottom water remains at temperature of its greatest density. More superficial water layers are colder, and hence lighter.

perhaps by photosynthesis occurring under the ice, which would supply oxygen to the upper layers of the water. Hydrogen sulfide may again be produced, especially in the lower layers of the water, and thus animal life again may be depleted during the winter stagnation.

Finally, when spring arrives, the warmer atmosphere and stronger solar radiation serve first to melt the ice and re-establish wind-derived circulation in the upper layers of water, and then finally to warm the water to a uniform temperature of 4° C. throughout the body of the lake. At this time there are again no barriers to mixture of all of the water of the lake. Thus the spring overturn is accomplished, oxygenating all of the water, and removing any accumulated H_2S. The spring overturn continues until a new thermocline can be established as described above.

Lotic environments.—Lentic fresh-water environments are of the greatest importance to us in the study of the plankton, but some plankton may also exist in lotic environments, and occasionally the quantity of plankton produced is very large. Lotic environments include brooks, creeks, and rivers. In a geological sense brooks tend, at a given geographical point, to erode their beds and widen their valleys to become creeks, while creeks have the same tendency to become rivers. Meanwhile, as brooks evolve into creeks, the headwaters recede geographically, so that in effect the brook has migrated (eaten its way) inland, and actually has not been transformed into a creek in the same sense that a lake may be transformed into a pond. Similarly, as a creek evolves into a river at any given geographical point, in actual fact the creek has migrated inland, leaving a river in its place.

Brooks and creeks, although their pools may develop a sparse plankton population (which to some degree may become the captive of the currents and thus be carried downstream), are not of especial interest to us here. However, certain rivers may develop a rather heavy plankton population. Rivers may be classified, according to the velocity of their current, as (geologically) young or old rivers. The former have a swift current and a reduced plankton content, while the latter are sluggish and meandering and may have a rich plankton content. Lotic environments, although not as variable as the lentic environments, are nevertheless subject to considerable variation with respect to their chemical and physical conditions.

Marine environments.—Compared with the fresh-water environments discussed above, marine environments are very stable. This is not to say that the sea is static, for exactly the opposite is the case; but whereas each fresh-water lake is different from all other lakes in chemical, physical, and biological characteristics, these factors vary to a much lesser extent over broad regions in the sea.

Dissolved salts.—In the open ocean the total concentration of salts is remarkably constant through all the seas. The salinity averages about 35.5 ‰, but fluctuates around this figure from about 34 ‰ to about 37 ‰. In general, of course, the salinity is higher in regions of high temperatures and low rainfall, and is lower in regions near the mouths of rivers, while in estuaries, bays, etc., the water of the open ocean becomes considerably diluted, and thus brackish, ranging all the way to nearly fresh. For example, in Chesapeake Bay (Cowles, 1930) the water near the mouth of the Bay approaches the open ocean in salinity, while it becomes progressively more brackish away from the mouth, and at the head of the bay, near the mouth of the Susquehanna River, the water is completely fresh. In other cases of marine waters, enclosed or semi-enclosed by land, the water may become much higher in salinity than the open sea through a high rate of evaporation uncompensated for by a runoff of fresh water. Such, for example, are the northern portion of the Red Sea, where salinities may reach 41 ‰, and some of the enclosed salt-water lakes and bays in southern Florida, where during the wet season the water may become very dilute, yet in the dry season evaporation may cause the salinity to rise as high as 43.6 ‰, as reported by J. H. Davis (1940) from Lake Surprise on Key Largo. Such great variations of salinity have a considerable effect upon the organisms living in the water, including the plankton organisms. All typically open-sea organisms are killed by such variable conditions. However, considering the sea as a whole, these variations affect a very insignificant portion of the total.

Not only is the total concentration of salts remarkably constant in the sea, but the proportion of the various kinds of salts is also constant. For example, it has been found by Dittmar (1884) and others that samples of sea water from all the seas and from all depths have definite and constant relationships among the amounts of the major kinds of salts present, though they may vary somewhat in the total concentration of salts. This being the case, the tedious process of analysis necessary to determine the total salt content of a sample of water need not be attempted during routine analysis. It is sufficient to analyze the amount of chloride ion (plus the bromide) present in the water by titration with a solution of silver nitrate of known strength. This quantity can be expressed as chlorinity (grams of chloride plus bromide per 1,000 grams of sea water, or parts per thousand— ‰). Salinity (grams of total soluble solids per 1,000 grams of sea water—‰) can easily be calculated by the simple equation:

$$\text{Salinity} = 0.03 + (1.805)\,(\text{Chlorinity}).$$

Not only do the chlorinity and the total salinity bear a constant ratio to each other, but also most of the other constituent salts do likewise, as shown in Table 1.

It is true, however, that in certain special circumstances the ratio may differ from the normal. In Baltimore Harbor, for example, the ratio of iron to other chemical elements is much higher than in seawater, because of the discarding of ferrous sulfate wastes into the waters of the harbor area by industrial concerns (see Olson, Brust, and Tressler, 1941). In other cases, natural waters entering the sea carry relatively large quantities of particular minerals, and these alter the proportions to a certain extent. However, such conditions as these are exceptional, and for all practical purposes it may be stated that the proportions of nearly all of the chemical ingredients of sea water remain constant.

A few salts, namely some of the salts constituting the plant nutrients, are normally present in small quantities only, and when plants are actively growing in a particular locality they use up large quantities, or even all, of the available material, thus altering the proportions of these salts to the total. The

TABLE 1*

VALUES IN GRAMS PER KILOGRAM (‰)
for the major constituents of sea water

Ion	Symbol	1940 values	
		Chlorinity = 19‰	Percentage
Chloride.	. . . Cl^-	18.980	55.04
Bromide.	. . . Br^-	0.065	0.19
Sulfate.	. . . $SO_4^=$	2.649	7.68
Bicarbonate. . . .	. . . HCO_3^-	0.140	0.41
Fluoride.	. . . F^-	0.001	0.00
Boric Acid.	. . . H_3BO_3	0.026	0.07
Magnesium. . . .	. . . Mg^{++}	1.272	3.69
Calcium.	. . . Ca^{++}	0.400	1.16
Strontium.	. . . Sr^{++}	0.013	0.04
Potassium.	. . . K^+	0.380	1.10
Sodium.	. . . Na^+	10.556	30.61
TOTAL.	. . . .	. . . . 34.482	99.99

major plant nutrients involved are the phosphates, the nitrates, the nitrites, and the silicates.

Dissolved gases.—Those dissolved gases which are of biological significance may fluctuate considerably in sea water, but those such as nitrogen and argon, which are not used nor produced to any extent by organisms, remain very constant. Oxygen and carbon dioxide, however, entering as they do into the metabolic activities of both animals and plants, vary in quantity from time to time and from place to place. The oxygen content of surface waters is usually high, because of the solution of this gas from the atmosphere. Indeed, the oxygen dissolved in surface waters frequently exceeds its saturation point, for oxygen may be produced in large quantities by the photosynthetic activities of plants. This is clearly shown by Table 2.

*Adapted by permission from Sverdrup, Johnson, and Fleming, *The Oceans*, p. 166. Table 33 Copyright 1942 by Prentice-Hall, Inc.

TABLE 2 *

PER CENT SATURATION OF OXYGEN AT VARIOUS DEPTHS—NORTH PACIFIC OCEAN AND BERING SEA

Station	Depth (meters)	Per Cent Saturation 0_2
	0	117.8
G2	25	64.7
	50	79.0
	0	82.4
	25	103.8
G35	50	78.7
	75	95.0
	100	89.7
	0	156.2
	10	105.6
C45b	25	77.5
	40	74.4
	0	105.6
	10	104.9
C7	25	100.2
	50	94.4
	75	77.9

Oxygen, however, is used up by the metabolic activities of animals, and thus in deeper waters there may be a relative, or even in a few instances a complete, depletion of the oxygen supply. On the other hand, most of the deepest waters of the oceans have had an arctic or an antarctic origin. Because gases are more soluble in the colder waters of these regions, and because there is a more rapid circulation near the bottom than in intermediate layers, animal activities do not result in as great a depletion as elsewhere. Thus in intermediate latitudes the surface waters and the deepest waters may be relatively rich in oxygen, while layers of water between the two are poor, forming what is known as an oxygen-minimum layer. In the North Pacific Ocean the oxygen content of the oxygen-minimum layer approaches zero, over a wide area and through a considerable depth, yet animals live apparently in good health as part of the plankton of the region (see fig. 7).

* Adapted from Barnes and Thompson, 1938, pp. 67-88, Tables 16-17.

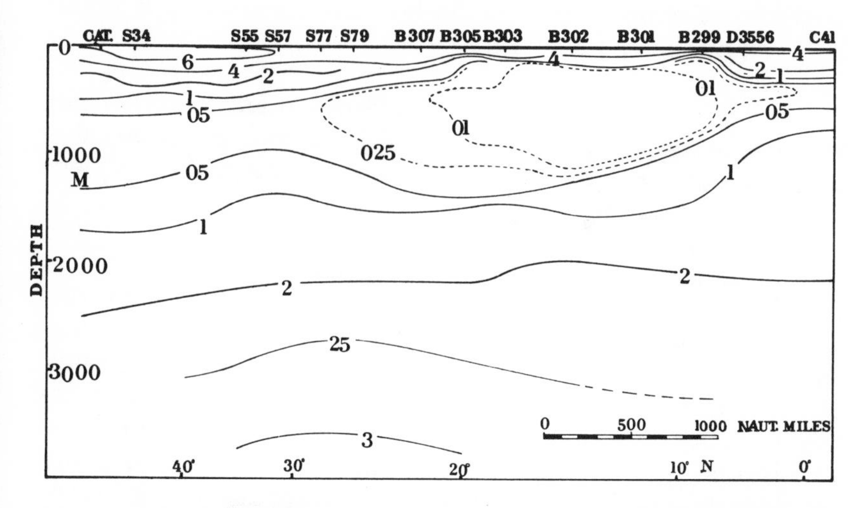

Fig. 7. Vertical distribution of oxygen in the North Pacific Ocean (a vertical section a few hundred miles from the west coast of North and Central America). Mainly based on observations of E. W. Scripps and Bushnell. From Sverdrup, Johnson, and Fleming. Reproduced by permission from the *The Oceans* by H. U. Sverdrup, Martin W. Johnson, and Richard H. Fleming. Copyright, 1942, by Prentice-Hall, Inc.

Carbon dioxide is rather the opposite of oxygen in its distribution. Carbon dioxide is utilized by plants as one of the raw materials for photosynthesis, while it is produced by animals as a waste product of metabolism. Thus where the oxygen content of sea water is high, the carbon dioxide content is low, and vice versa. However, because of a large reserve of bicarbonates, extremely low concentrations of carbon dioxide are unknown in the sea. For a discussion of the equilibria existing in the sea among carbon dioxide, carbonates, and bicarbonates, the interested reader is referred to Sverdrup, Johnson, and Fleming (1946), pp. 192-210.

Other chemical conditions.—Other chemical characteristics of sea water are remarkably constant. Among these, perhaps the most important is the pH of the water. Sea water in general is slightly on the alkaline side of neutrality (a pH of around 8.2), except in the immediate vicinity of the shore, where the entrance of fresh water, the presence of salt-water swamps adjacent to the sea, and the decay of organic detritus may change the condition somewhat. Marine environments on the acid side of neutrality are virtually unknown. The constancy of pH of sea water is, of course, due to the fact that sea water is highly buffered by the dissolved salts.

Physical conditions.—Physical conditions at any locality (particularly in the great depths of the sea) also may be remarkably uniform, but they vary more from time to time and from place to place than do the chemical conditions.

The temperature of the ocean water varies with latitude and with depth. As would be expected, surface waters in high latitudes have a lower temperature than those in low latitudes, because of the lesser angle of incidence of the rays of the sun as one approaches the poles. However, the great ocean currents, carrying as they do large quantities of relatively cold or relatively warm water (as the case may be) beyond the latitudes at which such temperatures are produced, serve to disturb any simple system of distribution of temperature according to latitude. Thus warm waters derived from the great Gulf Stream in the Atlantic Ocean serve to keep the arctic port of Murmansk ice-free in winter, while places much farther south along the Atlantic Coast of North America become completely frozen over, and thus inaccessible every winter, due to the influence of the cold Labrador Current.

Temperatures at the surface in the salt water of the ocean are almost always warmer than those at lower levels, for water of sufficiently high salinity ($> 24.7‰$) continues to contract all the way down to its freezing point (which is approximately at $-2°$C., but depends upon the salinity). This contrasts with the situation in the case of fresh water, described above. This means that any cold water formed at the surface by its contact with the overlying cold atmosphere will sink to a level commensurate with its density, and it will be replaced by warmer water—and this will occur below 4°C. as well as above it. Arctic surface water, however, may at times have a temperature slightly lower than that of intermediate levels because of the presence of very brackish water (from melting ice) floating upon the surface. The fresher water has the characteristics of all fresh water in that it expands as it cools below a certain temperature (4° C. for absolutely fresh water). Thus it will float over the warmer salt water below.

The open ocean is sufficiently vast that there seldom develops such a thermal stratification as occurs in fresh-water lakes, and thus there is no stagnation of the lower waters, nor is there a spring or fall overturn. Certain enclosed arms of the sea, where the depth at the entrance is not as great as that of the portion away from the sea (e.g., the Black Sea, many fjords in Norway, etc.), become very stagnant in the deeper layers of the water. Frequently there comes about a complete depletion of the oxygen content, and the development of toxic quantities of hydrogen sulfide. These conditions, however, are relatively insignificant in marine waters.

Light is absorbed as it passes through water, and it penetrates only for a limited distance into the sea. There is, of course, no definite line of demarcation between the lighted portion of the sea and the unlighted portion, but usually 200 m. is arbitrarily selected as the limit of penetration (suitably sensitive instruments can detect light much lower than this, but for all practical purposes light can be said to absent below 200m.). Of course light penetration is greater at low latitudes than at high latitudes because of the greater angle of incidence of the rays of the sun nearer the equatorial regions.

Wave action is usually confined to the upper few meters, and the effects of wind upon the lateral movement of the water (wind-driven currents) become obliterated for all practical purposes at a very limited depth. All water in the sea, however, is in dynamic motion at all times—some of it in rapid motion and some of it moving very slowly. The details of the motion of the deeper waters of the sea are not as well known as in the case of surface waters, for obvious reasons, but the relatively high concentration of oxygen in bottom waters, salinity differences, etc., serve as current indicators, and show in general a sinking of water from the surface to the great depths in subarctic and subantarctic regions and a slow upwelling in regions of diverging currents and to some degree in equatorial regions, as well as along many of the west coasts of continents. The current systems are of great importance, not only in their effects upon climates, but also in that they carry enormous quantities of plankton organisms from their point of origin, frequently to their doom in regions whose conditions they cannot withstand. Upwelling water also serves to bring accumulations of plant nutrients from the depths to the photosynthetic zone where they can be utilized by plants.

Ecological subdivisions of the sea. —The two great environmental subdivisions of the sea are the benthic and the pelagic regions, the former being the bottom region (inhabited by the benthos, as described above). The pelagic region, however, comprises the vast bulk of the environments of the sea, and it is here that the nekton and the plankton live.

The pelagic region is divided roughly into the neritic province and the oceanic province. The neritic province is that portion relatively close to shore, and lying above the continental shelf (the edge of the continental shelf lies at about the 200-m. contour), while the oceanic province is that overlying the deeper regions (see fig. 8). Fresh-water drainage from the land, carrying nutrients with it, pours into the neritic province. In addition, in the neritic province, plant nutrients are not so readily lost by sinking deeper than the zone in which plant life can exist. Thus production in general, and plankton production in particular, is much higher near shore than in the oceanic pelagic region. Furthermore, physical and chemical conditions are much less stable in the neritic than in the oceanic pelagic province, because of the combination of land drainage and shallow water in the former.

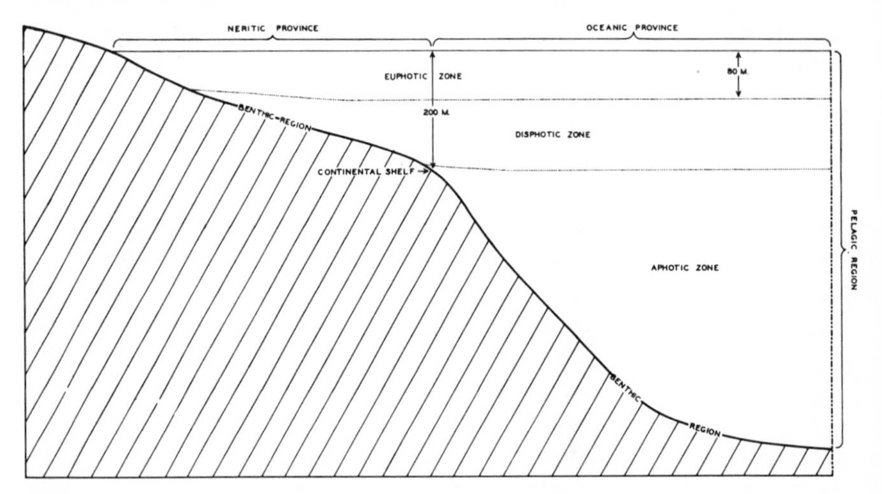

Fig. 8. Diagram to show the major ecological subdivisions of the sea.

Vertically, the water of the oceanic pelagic region can be divided into zones according to the penetration of sunlight. The upper 80 m. (the depth varying with latitude and with transparency of the water) is well lighted, and here the synthetic activities of the phytoplankton are in excess of their destructive or catabolic metabolism. This well-lighted portion of the sea is the euphotic zone. Below 80 m., in the disphotic zone, the water is poorly lighted. Although photosynthesis may occur, and plants may live at least temporarily in this zone, the catabolic activities of the plants exceed their photosynthetic activities. Below about 200 m., in the aphotic zone (which constitutes the bulk of the ocean waters), eternal darkness is the rule. Living green plants are never found in the aphotic zone, even temporarily.

The euphotic and disphotic zones extend into the neritic province as well. Here, however, the water is more turbid due to increased plankton production, the pouring of fluvial detritus into the sea, and other factors. Therefore, light does not penetrate as far into neritic water as into oceanic waters, and the boundary between the euphotic and disphotic zones is displaced upward.

Historical Aspects of Plankton Study

It was not until the invention of the microscope, under the stimulation of the Renaissance, that any of the smaller plankton organisms came to be known with any intimacy. However, seafaring men had long observed some of the more marked effects of the presence of plankton organisms. The luminescence of the sea was known ever since man first observed the sea. Sailors knew when they were approaching shore because of the change of the water from the ultramarine of the plankton-poor oceanic water to the greenish of the plankton-rich neritic province. Whalers could detect good whaling grounds by the color and consistency of the water, caused by myriads of planktonic schizopods and copepods. The Red Sea was named long before its plankton was known, yet its red color is due to a bloom of plankton organisms. Similarly, lakes were known to "slime" and "bloom" for mysterious reasons. Now we know that the "sliming" and "blooming" are caused by the plankton.

A few animals and plants of the plankton were described very early. They were captured simply by dipping them out of the water. However, it was not until around 1845 that Johannes Müller began a consistent study of the small organisms he captured by dragging a fine net through the waters of the North Sea. It was not until 1887 that Hensen coined the term "plankton" to apply to these minute floating organisms. These early investigators, and many others whose names are not so well known, but to whose efforts we owe a large part of our basic knowledge of this new cosmos, found that the plankton organisms formed a more or less balanced community. There were the plants, the herbivores, the primary carnivores, the secondary and tertiary carnivores, the parasites, commensals and symbionts, and the detritus feeders.

Research on the plankton flourished during the latter part of the nineteenth century. But in a new biological world such as this, the first research was necessarily taxonomic. Nearly every new type obtained from the plankton sample was a species new to science, and needed to be described. Concentration of research was on the marine plankton, for lakes were thought to be uniform and uninteresting. Later, however, marine techniques were adapted for the study of lake plankton as well, and again a new biological world was discovered.

After the preliminary taxonomic splurge, interest began to develop in the quantitative relationships existing among the various planktonic species. This led directly to the modern emphasis in plankton research, which is on the ecological aspects of the field, including the place of the plankton in the general economy of aquatic environments—and, indeed, indirectly, in that of terrestrial environments as well (see Riley, *et al.*, 1949).

Terminology

DURING THE period since Hensen first coined the word "plankton," there has evolved a rather complicated terminology referring to the many different aspects of plankton study. A single organism in the plankton is known as a *plankter* (the word "planktont" means the same thing, but is incorrect etymologically, and has become obsolete). All plankters are members of the *euplankton* (true plankton), as distinguished from the *pseudoplankton*. The latter often appears to the naked eye to be no different from the living animals and plants, but it consists of dead plankters and of nonliving debris—sometimes very important in the economy of the sea and of lakes, at least in local situations. The water of the oceanic pelagic province has less pseudoplankton than that nearer shore, because of land drainage, the stirring up of bottom deposits by storms, the decay of dead benthic plants, etc., in the neritic province.

The euplankton can be classified artifically and arbitrarily according to size. The largest planktonic forms, such as large jellyfish, euphausids, mysids, salps, *Corethra* larvae, and many others, are called the *macroplankton*. The macroplankton is defined arbitrarily as those forms which are larger than microscopic, and hence greater than about 3 mm. in size. Some of the scyphozoan jellyfish may reach a diameter of more than seven feet.

The *microplankton* consists of plankters which have dimensions less than 3mm., but which will be captured by a net made of the finest silk bolting cloth of #25 mesh, with openings between the meshes 0.06 mm. square. Thus the microplankton consists of organisms whose dimensions are between 0.06 mm. (=60 microns) and 3mm. Most of the animals of the plankton, and a large number of the diatoms and dinoflagellates among the plant species, belong to this category.

However, many plankters constitute the *nannoplankton*, these being the forms not captured by a #25 mesh plankton net, and hence having a diameter of less than 60μ. Most of the species involved are plant species, and especially important here are the coccolithophores, the silicoflagellates, many diatoms, the bacteria, and some of the unicellular algae. On the other hand, many Protozoa, mostly Mastigophora, among the animals are also involved. The nannoplankton is less well known than the plankton of the other categories because of minute size and the special techniques required for its study, but it appears to be of very great importance in the economy of fresh waters and the sea.*

Plankters also may be classified as to their basic nutritional requirements. Planktonic plants are called *phytoplankton.* The *phytoplankton proper* consists of chlorophyll-bearing plants, which are therefore capable of performing photosynthesis, while the *saproplankton* consists of nonphotosynthetic plants, including the bacteria and fungi. The *zooplankton* consists of those plankters with a holozoic nutrition, and thus it includes all of the planktonic animals. It is of course true that the line of demarcation between one-celled animals and one-celled plants is a very tenuous one, and this is also the case in the distinction between those Protista that comprise the phytoplankton and those that comprise the zooplankton. The dinoflagellates are correctly claimed both by botanists and zoologists. Some are typical plants in their nutrition. Others are typical animals. Still others are both animals and plants at the same time. The single dinoflagellate genus *Gymnodinium,* for example, among others, contains species showing all three conditions. However, in spite of some confusion, the terms are valid because most plankters fall readily into one of the two categories. The phytoplankters are the producers, and the zooplankters are the consumers.

Again, plankters may be classed with regard to their environments. The *limnoplankton* is the plankton of lakes, the *rheoplankton* is the plankton of rivers, the *haliplankton* is the plankton of salt water, and the *hypalmyroplankton* is the plankton of brackish water. The *hypoplankton* consists of plankters living near the bottom (e.g., certain mysids, etc.). The *epiplankton* is that plankton living in the euphotic zone, and thus includes most of the phytoplankters, as well as many animals. The *bathyplankton* lives in the aphotic zone, and therefore contains no living phytoplankters proper. The *mesoplankton* (see note below) lives in the disphotic zone, and may include a few green plants as temporary inhabitants.

* The terminology used in the paragraphs immediately above is widely used, as in Sverdrup, Johnson, and Fleming (1946), p. 819. Sometimes the term *mesoplankton* is used, however, to refer to the plankton of intermediate size and *microplankton* to refer to the smallest forms.

Furthermore, the plankton can be classified according to the life history of the organisms involved. Most of the plankters, both in the ocean and in fresh water, have no existence other than as members of the plankton. This is especially true in the oceanic pelagic region near the surface. For example, the calanoid copepod *Eurytemora hirundoides* (see Davis, 1943) is typical of many of the planktonic copepods, both fresh-water and marine. The female carries the eggs with her (many marine species do not carry the eggs, in which case the eggs themselves take up an existence in the plankton) until they hatch. The eggs hatch as nauplii (see figs. 9 A-B-C-D), which are planktonic. After five ecdyses, the nauplius metamorphoses into a copepodid stage, which resembles the adult but does not yet have all the adult structures. The copepodid is also planktonic. The copepodid also undergoes five ecdyses. The sixth copepodid stage is the adult stage. Thus the life history is completed without the animals ever leaving their planktonic life. Animals of this type are classified as the *holoplankton.*

On the other hand, many important and characteristic plankters are only temporary members of the plankton community. These constitute the *meroplankton.* Such, for example, are those benthic forms that have a planktonic larval stage, as for example *Balanus,* the acorn barnacle (Groom, 1894, 1895). *Balanus* is strictly benthic during most of its life, being immovably attached to the bottom. When reproduction takes place, the eggs develop in a brood sac until eventually they hatch as typical nauplii (very similar to the nauplii of the copepod described above). After a longer or shorter planktonic existence as a nauplius (the time depending upon the species, the temperature, and other factors), metamorphosis takes place and a second planktonic stage, the cypris larva (so-called because of its great resemblance to the ostracod genus *Cypris*) appears (see fig. 10). After a brief existence, the cypris larva settles on the bottom, or on any suitable surface, attaches itself, and a second metamorphosis produces a minute replica of the adult barnacle. Thus the barnacle has only a brief planktonic existence relative to its long benthic existence. In fresh water, the larval stage of flies of the genus *Corethra* are typical plankters. The adult, however, is not even aquatic.

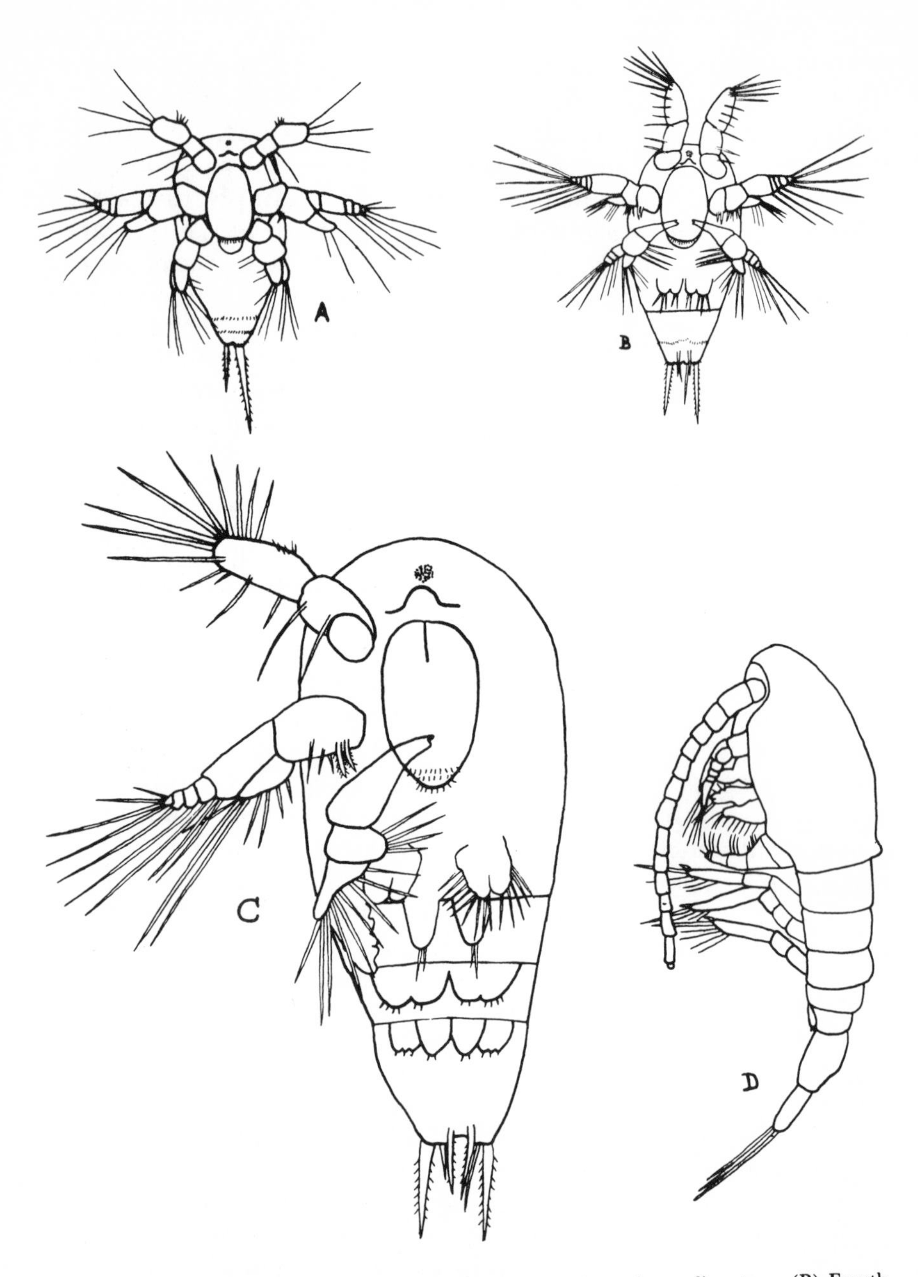

Fig. 9. Immature stages of *Eurytemora hirundoides*. (A) Second naupliar stage. (B) Fourth naupliar stage (early metanauplius). (C) Sixth naupliar stage (late metanauplius). (D) Second copepodid stage. From Davis, 1943.

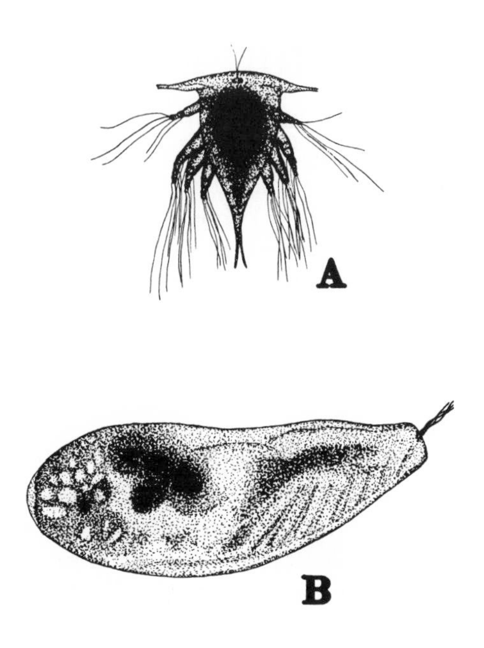

Fig. 10. Planktonic larval stages of the barnacle. (A) Nauplius stage. Note the characteristic antero-lateral horns of the carapace. (B) Cypris stage. Note the conspicuous cement gland at the anterior end. This clearly distinguishes barnacle cypris larvae from ostracods.

Finally, reference must be made to the *tychoplankton.* The tychoplankton is the very temporary and accidental plankton, and consists of animals and plants which have temporarily migrated or have been carried into the plankton from their normal benthic habitat. Usually they are so carried by strong tidal or other currents, but sometimes other means serve to displace them. Many genera of diatoms, for example *Isthmia* and *Licmophora,* as well as benthic Foraminifera, etc., are frequent members of the tychoplankton. In Biscayne Bay and Florida Bay the common sea horse, *Hippocampus,* has occasionally been captured in routine plankton hauls. *Hippocampus* probably enters the tychoplankton through the decay and detachment from the bottom of pieces of turtle grass to which it has attached itself, rather than simply through the agency of strong currents.

11. Special Adaptations to Planktonic Existence

EACH special type of environment in which organisms live has, in an evolutionary sense, called forth the development of special adaptations by means of which these organisms are better able to cope with the problems of their habitat. Plankters, suspended in the water, at the mercy of every water current, and finding no places in which to conceal themselves, have encountered several difficult problems during their evolutionary development. They have solved these problems in a number of ways:

Flotation.—The most important problem plankters have found it necessary to solve is the simple problem of flotation. Protoplasm and animal and plant skeletons are heavier than water, and will sink to the bottom unless there is some special means of preventing it. And plankters must prevent, or at least retard, sinking if they are to remain plankters.

Typical plankters maintain their planktonic existence through their small size. Except for most of the planktonic Protozoa (Protozoa are small anyway), Rotifera, and a few other groups, plankters average considerably smaller than corresponding benthic, nektonic, and terrestrial species. The nannoplankton and the microplankton constitute by far the main bulk of plankters, the macroplankton being conspicuous but comparatively minor as far as total bulk is concerned. Small size increases the ratio between surface area and volume (or weight). Thus a hypothetical cubical organism with dimensions of 1 cm. on all sides would have a volume of 1cc. and a weight equal to (say) 1.2 gm. Its surface area will be 6 sq. cm. The ratio between its surface area and its volme is 6 to 1, and that between its surface area and its weight is 6 to 1.2. If we imagine eight other hypothetical cubical organisms, with protoplasm of exactly the same density, but with dimensions just half those of the first case (i.e., whose sides are only 0.5 cm.—as though we had cut the original organism into eight equal parts), then the volume and weight of all eight organisms together will be identical to the volume and weight of the single organism, but the surface area will have increased to 12 sq. cm. (0.5 x 0.5 x 6 x 8 = 12). Thus in the smaller organisms the ratio between surface area and volume will be 12 to 1 rather than only 6 to 1, while the ratio between surface area and weight will have increased to 12 to 1.2. In other words, if the organisms in question were planktonic forms, the relative increase of surface area of the smaller forms would present twice as much frictional resistance to sinking per unit of weight as in the case of the larger one. The importance of small size will strike us if we remember that the very important nannoplankton consists of organisms whose dimensions are all less than 0.06 mm. The destructive "red tide" of the west coast of Florida in 1946-47 (see Gunter, *et al.,* 1948) was caused by a species of *Gymnodinium* (an unarmored dinoflagellate) that measured only about 0.03 x 0.03 x 0.012 mm., but it occurred in enormous numbers, samples being encountered containing as many as 60,000,000 organisms per liter. This species of *Gymnodinium* maintained its position in the plankton, among other methods, by its minute size.

A second characteristic of many planktonic organisms that helps to keep them in proper position in their environment is the presence of special spines and other processes. The genus *Chaetoceros* is among the most important of the diatom genera in the marine plankton, and each *Chaetoceros* cell is provided with four long spines at the four corners (see fig. 11-A). Such spines, of course, serve to increase the surface area without an undue increase of volume or weight, and in this way help to retard sinking. Other common planktonic diatom genera with devices similar to those of *Chaetoceros* are *Bacteriastrum, Biddulphia,* and *Ditylum* (see fig. 11-B). However, processes for the retardation of sinking are not confined to diatoms alone. Among the dinoflagellates, some species of *Ceratium* (especially tropical forms) are notable for the length of their "horns" (see fig. 11-E), and many planktonic animals are similarly provided. Many species of the foraminiferan genus *Globigerina* have numerous supporting spines. Planktonic larval stages of the polychaete worms often have enormously developed setae which aid in flotation (see fig. 11-D). Pelagic copepods such as *Calocalanus pavo* (see fig. 11-G) often have feathery furcal setae. Many pelagic copepod eggs (see fig. 11-C) and fish eggs are provided with numerous spines. Among the decapod Crustacea the larval stages of

Sergestes, of *Solenocera* (see fig. 11-H), and of the Porcellanidae (see fig. 11-I) are especially noteworthy for the extensive development of spines for flotation. Finally, the fleshy arms of the planktonic pluteus larvae of certain echinoderms should be mentioned as serving the same purpose (see fig. 11-F.).

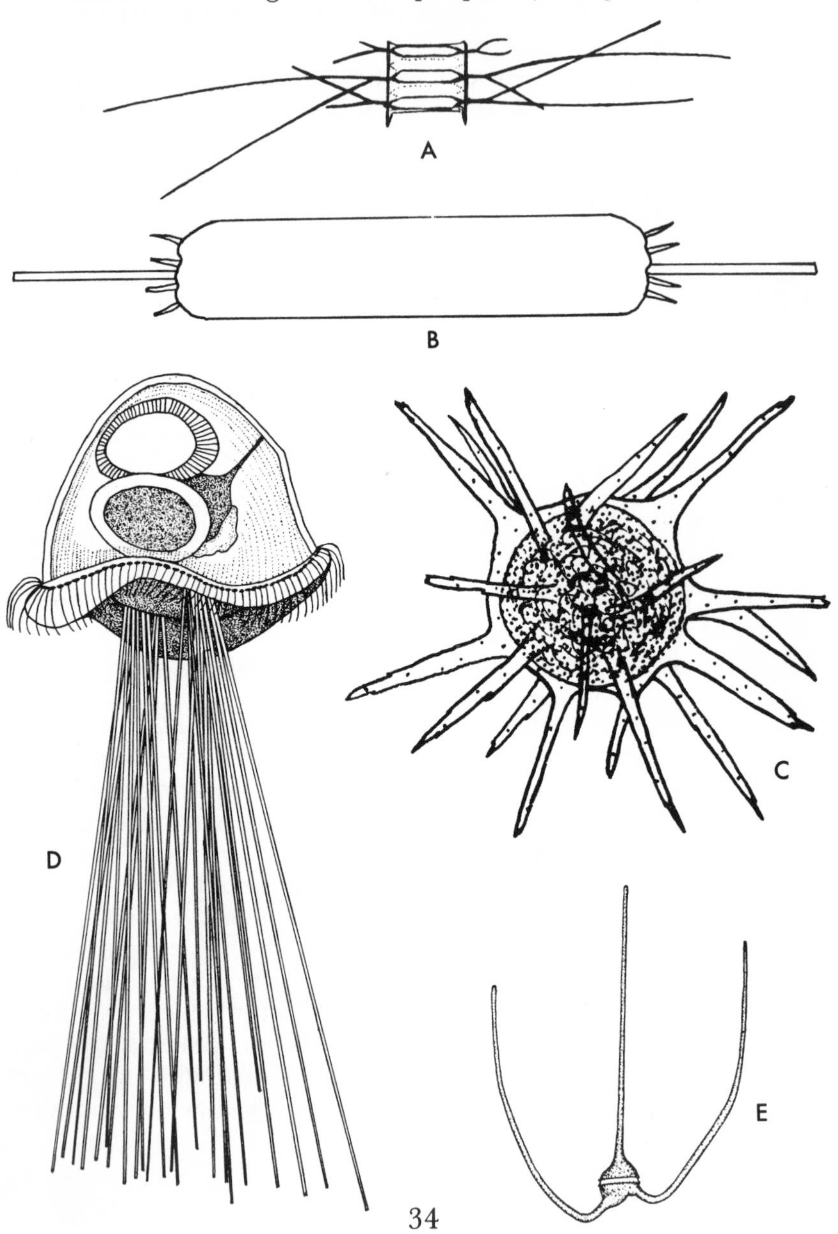

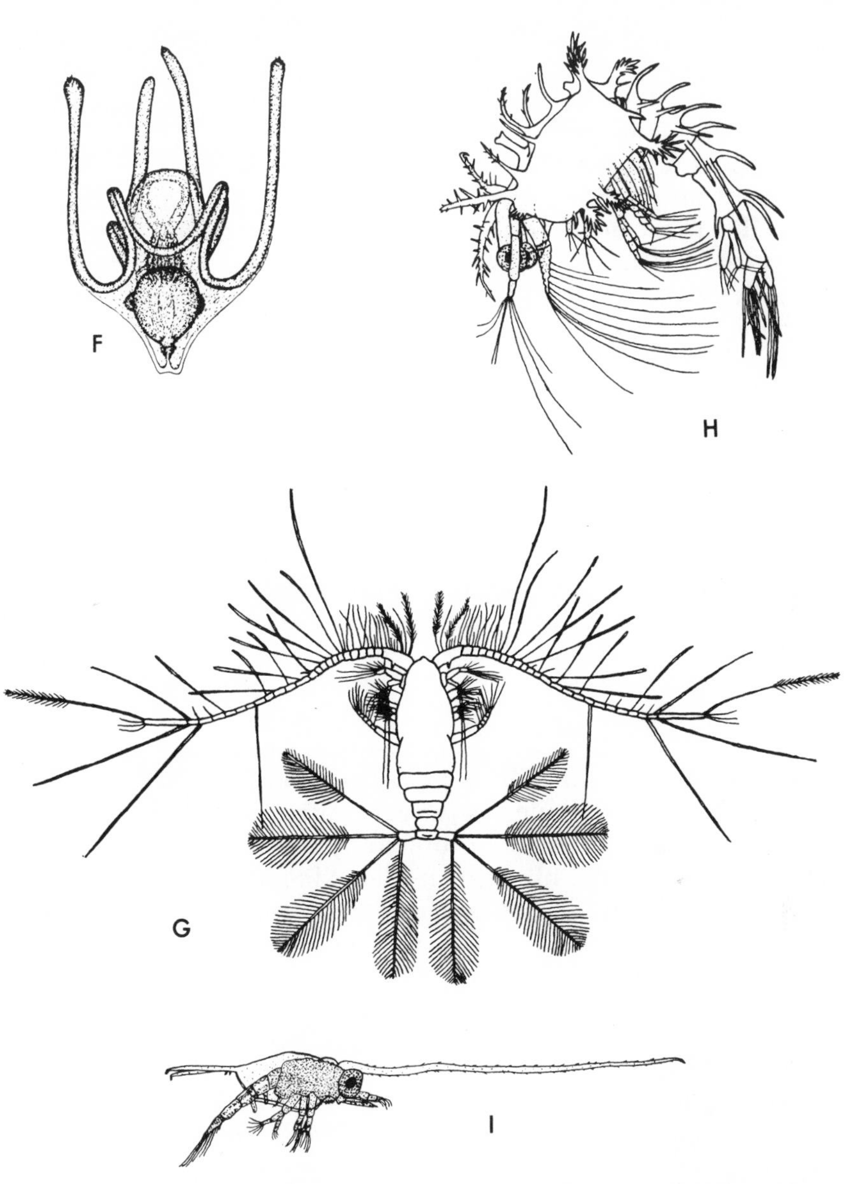

Fig. 11. Means of flotation in planktonic organisms. (A) *Chaetoceros decipiens;* (B) *Ditylum brightwellii*; (C) Pelagic egg of a copepod (*Centropages furcatus*?); (D) Larval stage of a polychaete; (E) *Ceratium* sp.; (F) Pluteus larva of an ophiuroid (i.e., ophiopluteus); (G) *Calocalanus pavo* (from Pratt, after Wheeler; reproduced by permission from *A Manual of the Common Invertebrate Animals* by H. S. Pratt. Copyright, 1935. The Blakiston Company. (H) *Solenocera* larva; (I) zoea of the Porcellanidae.

In a somewhat similar manner, some planktonic species develop a needle shape, and such organisms will sink through the water relatively slowly if they are resting horizontally. They will sink much more rapidly if they come to lie vertically, but most such forms are so balanced that they constantly lie horizontally. Among the animals, good examples of such a body form are to be encountered in the copepod *Aegisthus mucronatus* (see fig. 12-A), with its enormously developed furcal setae. Plants may also take on such a shape, and for the same reason. It is thus in the diatom *Rhizosolenia* sp. (see fig. 12-D), and in *Ceratium fusus* among the dinoflagellates.

Still other forms have evolved a pancake-like or ribbon-like shape. In such forms the rate of sinking is greatly retarded in that the organism sinks sidewise with a fluttering motion, much as a flat sheet of paper flutters groundwards when it is released from a high building. Such adaptations are shown in the diatom genera *Coscinodiscus, Actinoptychus* (see fig. 12-B), and *Rhabdonema* (see fig. 12-F), certain planktonic examples of the Platyheminthes, many of the jellyfish *(Obelia, Aequorea)*, the bryozoan cyphonautes larva (see fig. 12-C), larvae of the brachiopod *Lingula* (see fig. 12-E) and the copepod genera *Sapphirina* and *Copilia.*

Some of those algae that are essentially unicellular plants (in the sense that all the cells are exactly alike), but which have evolved a "colonial" life to give greater protection from small herbivores, have solved the problem by the formation of chains of cells. Thus in the blue-green alga *Skujaella (Trichodesmium)* (see fig. 13-A) and in the diatoms *Skeletonema, Chaetoceros* and *Bacteriastrum,* as well as in many other algal genera, most species form chains of cells. Likewise among the dinoflagellates, where most species are solitary, a few planktonic forms occur in more or less permanent and characteristic chains. Sometimes the chains of cells are contorted, as in *Anabaena* (see fig. 13-B), or twisted spirally as in *Chaetoceros debilis, Rhizosolenia stolterfothii, Eucampia* sp., etc., and this serves to increase resistance to sinking still more.

Not only the outer shape of the organism but also other structures may be evolved to aid in flotation. Thus bottom-living diatoms are usually characterized by the presence of heavy silicious walls, whereas most planktonic species have very delicate shells. Similarly, planktonic Crustacea (e.g., the bizarre decapod *Lucifer*) have much more delicate exoskeletons than do their near relatives among benthic forms. In other words, in the evolutionary development of many species of plankton, protection from enemies has been sacrificed in order that their planktonic existence might be maintained.

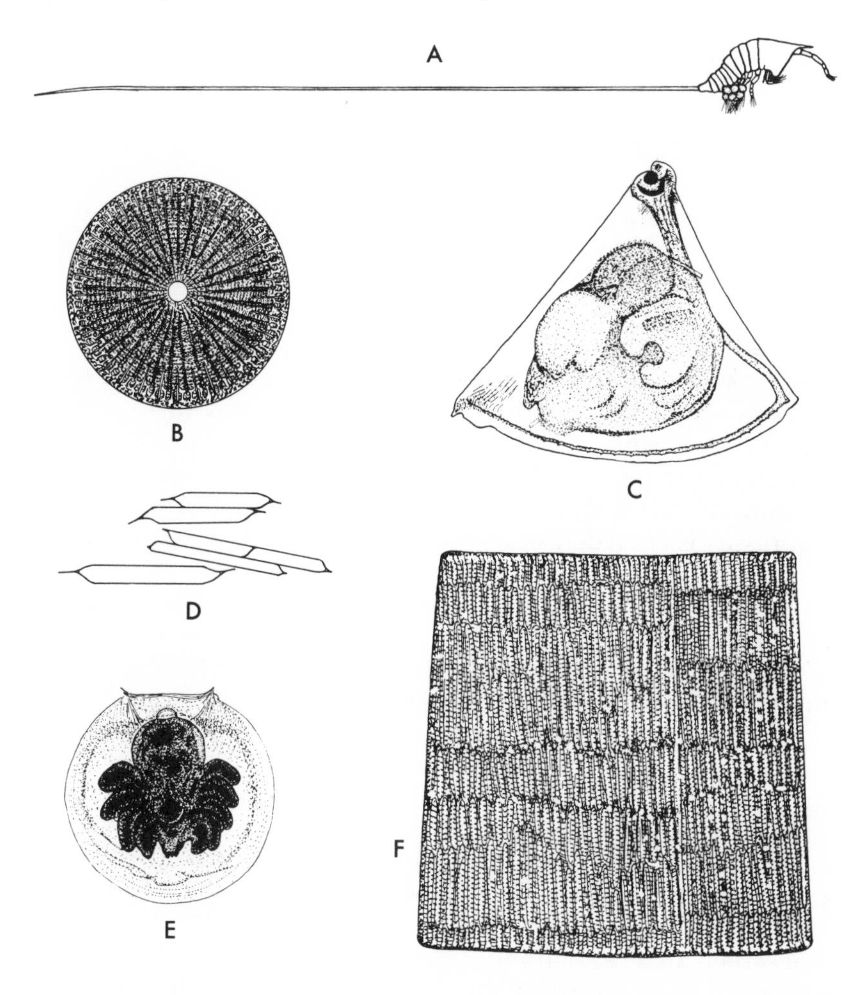

Fig. 12. Means of flotation in plankton organisms. (A) The copepod *Aegisthus mucronatus*; (B) The diatom *Actinoptychus* sp.; (C) A cyphonautes larva of a bryozoan; (D) The diatom *Rhizosolenia* sp.; (E) Planktonic larval stage of the brachiopod *Lingula* sp.; (F) The diatom *Rhabdonema* sp.

In many plankters, the specific gravity of the organism approaches more nearly to that of the environmental sea or fresh water through a high degree of vacuolization of the protoplasm, or simply through an unusually high water content. Thus the water content of jellyfish may be as much as 96.6 per cent (Hyman, 1940) of the weight of the animal. The cell of the dinoflagellate *Noctiluca,* as well as that of *Pyrocystis* (see fig. 13-C), is highly vacuolated. In these cases, the actual protoplasm of the animal is not increased in bulk or weight, but it is either spread out through a larger volume by many vacuoles containing water, or it is diluted by the presence of extra water in it. In either case sinking is retarded.

All of the flotation devices mentioned so far serve only to retard sinking. None of them is able to *prevent* sinking, and the question could legitimately arise as to the value of such structures and devices, for ultimately the organism would sink anyway. It has been also pointed out that motile animals (see below) with numerous spines would find not only that their sinking would be retarded by the spines, but also that their natatory movements upwards would be hindered just as much. Of course the point is that all of these organisms live in an environment that is dynamic and not static. If the sinking of the organisms is retarded many of them will be transplanted upwards by the turbulence of the water, and thus regain approximately their original position, whereas without the spines sinking would be too rapid to be compensated for by the turbulence. However, many plankters have evolved means of flotation involving still other mechanisms.

For instance, another common method of decreasing the specific gravity of the whole organism is by the storage of oil droplets within the protoplasm. The oil may in most cases serve either a primary or secondary function by providing food in times of need. Many blue-green algae contain such oil droplets, as do certain dinoflagellates, copepods, schizopods, and planktonic fish eggs (see fig. 13-D). In many cases (e.g., certain blue-green algae and dinoflagellates) the oil reduces the specific gravity to such an extent that the organisms are lighter than water, and will float at or near the surface. *Noctiluca* is somewhat lighter than water and floats to the surface, where it may be carried by the wind and accumulate in windrows, or be cast up against the shore. In these cases the innumerable individuals may often discolor the water to such an extent that one's boat appears to be sailing through a sea of dilute tomato soup. Although it was at one time thought that the low specific gravity of *Noctiluca* was caused by its fat content, it has subsequently been shown (Pratje, 1921) that individuals with little or no fat float as well as those with numerous fat vacuoles. Thus, though fat may aid in flotation, the main cause of the low specific gravity must reside in the cell sap vacuoles, which in *Noctiluca* are numerous and large. Krogh (1939) supports the hypothesis of Goetart and Heinsius, who considered that the only salts which would be lighter than sea water in isotonic solutions would be ammonium salts, most probably ammonium chloride, NH_4Cl. On the other hand, Gross (1934) found the cell sap to be strongly acid, with a pH of around 3, and he thought the acid was the cause of the low specific gravity. The simple chemical anlayses necessary for the solution of this problem appear not to have been attempted to date.

A few plankters, such as the fresh-water larval stage of *Corethra* (see fig. 13-E), and the marine *Physalia* (the Portuguese Man-o'-War), etc., have evolved air bladders as a means of support in the planktonic environment. *Corethra* has developed two specialized sacs in the tracheal system at each end of the body, changes in the volume of which allow the animals to adapt themselves to changes in the specific gravity of the environmental water (Wigglesworth, 1931). In *Physalia* (see fig. 14) one of the persons (pneumatophore) of this siphonophore is specialized to form the large air bladder. In *Corethra* the air apparently is originally obtained from the atmosphere, while in *Physalia* the tissue of the pneumatophore actively secretes the gas into the central cavity. Among benthic plants there are numerous brown algae that are provided with air-filled floats, but the only one of these that can be considered to come under the definition of the term "plankton" (or rather, "neuston") is *Sargassum* which may grow either attached to the bottom or floating at the surface of the water. *Sargassum* is famous for the fact that it grows in considerable profusion floating over the very deep water of the Sargasso Sea in mid-Atlantic. However, the tall tales told about ships becoming enmeshed in the weeds are highly exaggerated.

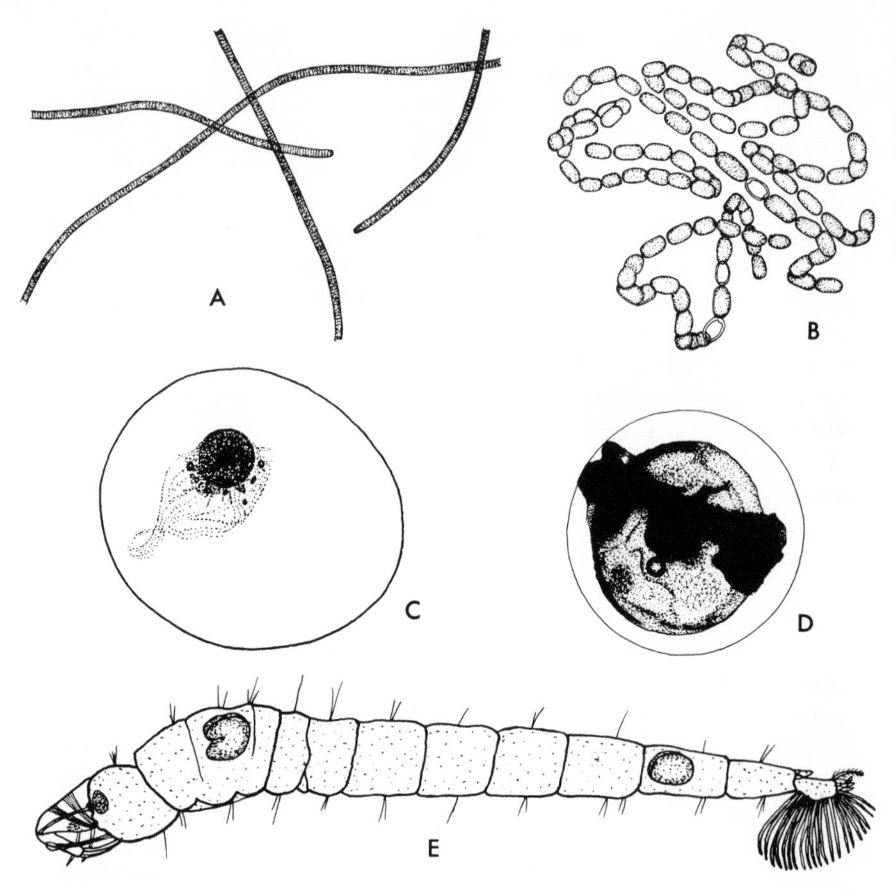

Fig. 13. Means of flotation in planktonic organisms. (A) The blue-green alga *Skujaella* [*Trichodesmium*] *thiebauti*; (B) The blue-green alga *Anabaena* sp.; (C) A pyrocystis stage of a dinoflagellate (*Pyrocystis noctiluca*); (D) A pelagic fish egg to show a conspicuous oil droplet in the yolk; (E) A planktonic dipteran larva, *Corethra* sp.

Finally, we should consider one other mechanism whereby zooplankters (and even some phytoplankters, such as *Volvox, Eudorina,* etc.) are able to maintain their planktonic position. The plankton by definition is at the mercy of the water currents, but nearly all zooplankters are capable of some sort of feeble movements, whether by general body movements or by means of cilia, flagella, or specialized appendages. The movements (as will be discussed below) may be strong enough not only to prevent sinking through the water, but even to allow for more or less extensive vertical migrations.

Fig. 14. *Physalia,* showing flotation by means of a large air sac (pneumatozooid). From Parker, Haswell, and Lowenstein. Reproduced by permission from *A Textbook of Zoology,* Vol. 1, 6th ed., by H. J. Parker, W. A. Haswell, and O. Lowenstein. Copyright, 1940, Macmillan & Co., Ltd. (London).

Protection from enemies.—In order to help maintain their floating position in the water, most plankters have, during their evolutionary development, sacrificed their heavy protective shells, as noted above, and thus have become much more vulnerable to herbivores and predators. Yet there is no hiding place in the planktonic environment. Under these circumstances certain species have evolved their protection along one of two lines. The spines that serve so admirably for flotation in the diatoms may serve also as protection against copepods and other animals that might wish to devour them. Copepods, for example, seem to devour non-spiniferous diatoms in preference to spiniferous ones. Some zooplankters, likewise, such as many of the polychaete larvae, develop special protective spines. However, a more characteristic protection has developed among many of the zooplankters whereby they have become extremely transparent. This type of "hiding" is found not only among such prey as the copepod *Copilia* (see fig. 15-A) and the decapod *Lucifer*, but also among the predators themselves, such as the Chaetognatha (which, because of their extreme transparency, are often known as the glass worms; see fig. 15-B).

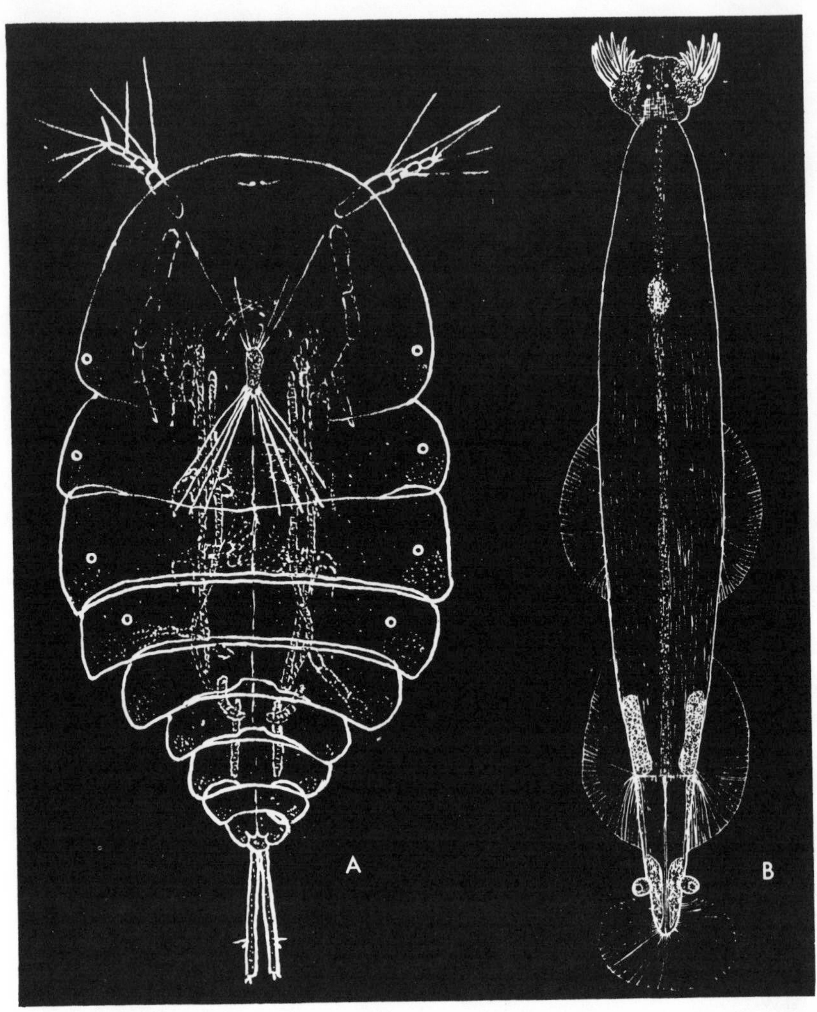

Fig. 15. Transparency in planktonic animals. (A) *Copilia mirabilis,* a copepod. (B) *Sagitta enflata,* a chaetognath.

Other animals living near the surface have developed a light blue coloration, and this likewise is a good protection from predators in the blue surface waters. An example of such an animal is the common neritic copepod of Florida, *Acartia spinata.* However, transparency and protective coloration, as a substitute for the lack of suitable hiding places in the planktonic environment, is needed only in the euphotic and disphotic zones. In the aphotic zone there is no sunlight by means of which the animals could be seen. In the deeper layers, many of the planktonic animals (as is true likewise of many of the benthic and nektonic forms) are deep red or black.

It must not be thought, however, that all zooplankters in the euphotic and disphotic zones are either transparent or bluish. Many are very brilliantly colored, and therefore very conspicuous. The copepod *Metis,* for example, is a deep red, although it is a surface form. The male of *Sapphirina* bears a most brilliant display of iridescent crystalline areas on the dorsal surface of the body. Such animals are protected from their enemies by other means, or they make up for their higher mortality by a high fecundity.

Adaptations to water movements.—In addition to finding it necessary to evolve special means of flotation to maintain their planktonic existence, and special means of mechanical protection or of transparency in an environment that offers no hiding places, plankters are more or less at the mercy of whatever water currents may be in their vicinity.

It is true that in lakes (even the largest ones) the currents are not nearly as important to the plankters as they are in the sea, except insofar as wind-driven currents may at times crowd them together towards the leeward end of the lake. Such crowding may be beneficial in some cases by facilitating mating, or by conveniently concentrating the prey to be utilized by certain planktonic predators, etc. On the other hand, it may be very deleterious and lead to such evils of overcrowding as suffocation, concentration of waste products, depletion of carbon dioxide content (in the case of phytoplankters), etc. However, a given lake has such a short (geological) life span, and the currents are so inconstant during the year because of the comparatively small area even of the largest lakes, that limnoplankters have not evolved any striking special adaptations to the currents. On the other hand, the oceans have an enormous area, cutting across the tropical, the temperate, and the frigid zones. In the oceans, a single body of water is subjected to widely varying climatic conditions with regard to temperature, rate of evaporation, amount of precipitation, direction of prevailing winds, etc. In addition, such a body of water is sufficiently large that the lag of the liquid material behind the solid land becomes noticeable in the rotation of the earth. All these factors together affect the water in the oceans in such a manner as to establish the great systems of circulation that are so characteristic and constant in the oceans. Thus in the Northern Hemisphere in both the Atlantic and the Pacific Oceans there is a clockwise circulation of the water near the surface, formed by a westward drift near the Equator, a northward deflection along the eastern shore of North America and Asia respectively, an eastward deflection across the northern portion of each ocean, and a southern deflection along the western shore of Europe or North America, as the case may be. In addition, in each case there is a northern branch of the current, the one in the Pacific being directed up into the Gulf of Alaska and the one in the Atlantic passing along the coast of Norway, eventually entering the Arctic Ocean.

In the Southern Hemisphere there is a similar counterclockwise circulation of the water in each ocean. Figure 16 is a simplified presentation of our present knowledge of these great current systems. Such currents will, of course, because of their constancy, have a great influence upon the drifting life of the sea. A tropical plankter in the Atlantic Ocean, for example, might be carried by the Gulf Stream from its normal tropical habitat into the North Atlantic where climatic conditions are very different. If it or its descendants were to live sufficiently long, it might conceivably be carried even into the Arctic Ocean. Of course, strictly tropical plankters would quickly perish under such varying conditions, but other plankters, such as the calanoid copepod *Paracalanus parvus,* have a world-wide distribution, and could undoubtedly withstand, to a considerable extent, the changed conditions.

As an adaptation to the great current systems, certain plankters have developed very specialized life histories. The most famous of these is in the case of the larval stage of the fresh-water eel, *Anguilla* (see Schmidt, 1925 and Russell and Yonge, 1928). The myriads of *Anguilla* live for the most of their life in fresh water, but at the time of sexual maturity they migrate to the sea and disappear from sight. The life histories of two closely related eels of this genus are known from the Atlantic Ocean. These are *Anguilla rostrata* of the Atlantic coast of North America and *A. vulgaris* of the Atlantic coast of Europe. It has been found that the adults of the American species migrate to an area off the West Indies, where they liberate their planktonic eggs. The eggs hatch into peculiar leptocephalus larvae, very different from the adult eel in that they are broad and flat and extremely transparent, as adaptations to their planktonic habitat (see fig. 17).

The leptocephalus larvae maintain their planktonic existence for a period of about two years. Meanwhile they have been transported by the currents from their hatching point in the Sargasso Sea, and finally they are caught up in the great Gulf Stream and carried along the coast of the United States. Since the time that elapses between the act of hatching and the transport of the larvae to the coasts of North America is also two years, they reach the coast just at the termination of their planktonic existence. The leptocephalus larvae then metamorphose into eel-like elvers, which ascend the fresh-water streams in the vicinity of their place of metamorphosis.

The European eel undertakes an even more extensive migration after it reaches salt water. As a matter of fact, it swims nearly the width of the Atlantic Ocean, to a place in the Sargasso Sea somewhat to the east of that chosen by the American form. Actually, the breeding areas of the two species overlap somewhat. Here the eggs are laid, and they hatch into leptocephalus larvae, very similar in appearance to those of the American form. They also are carried by the currents, and eventually they are caught up by the Gulf Stream. However, at the time they find themselves off the coast of North America they are not yet ready to metamorphose into elvers, for their planktonic leptocephalus stage lasts three years instead of two. Thus the Gulf Stream carries them back across the Atlantic Ocean to the shores of Europe, where they arrive at the end of their three-year planktonic life. Thereupon they undergo metamorphosis, become elvers, and migrate up the rivers and streams of Europe. The remarkable adaptations of the two species of *Anguilla* lie not in bodily structures, but in the physiological conditions that regulate the time the organisms maintain their planktonic existence, in this way preventing a mix-up of the adults of the two species.

Not quite so spectacular, perhaps, is the life-story adaptation of the California sardine (*Sardina caerulea*). The life history

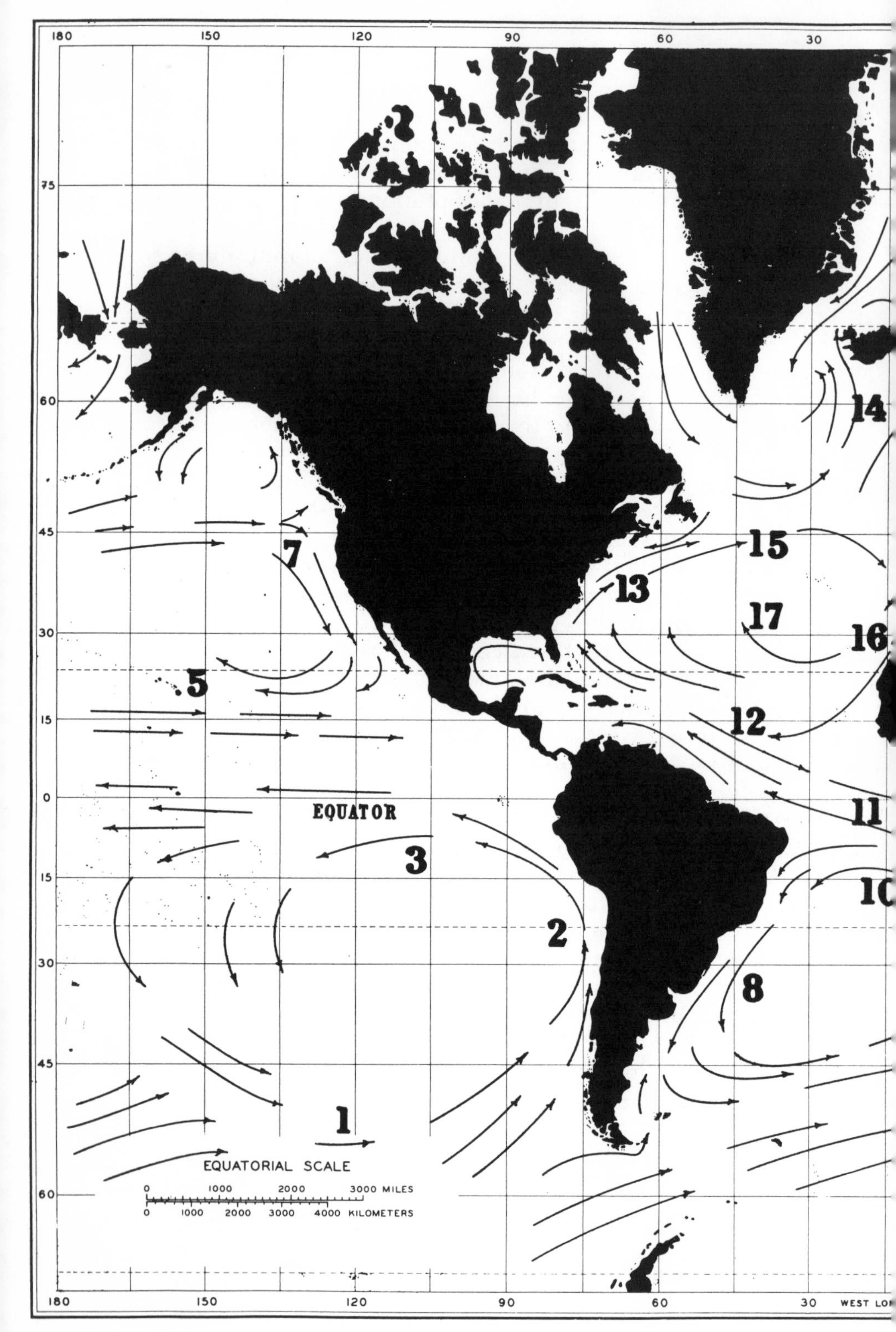

180
150
120
90
60
30
75
60
45
30
15
0
15
30
45
60
EQUATOR
EQUATORIAL SCALE
0 1000 2000 3000 MILES
0 1000 2000 3000 4000 KILOMETERS
WEST LO
1
2
3
5
7
8
10
11
12
13
14
15
16
17

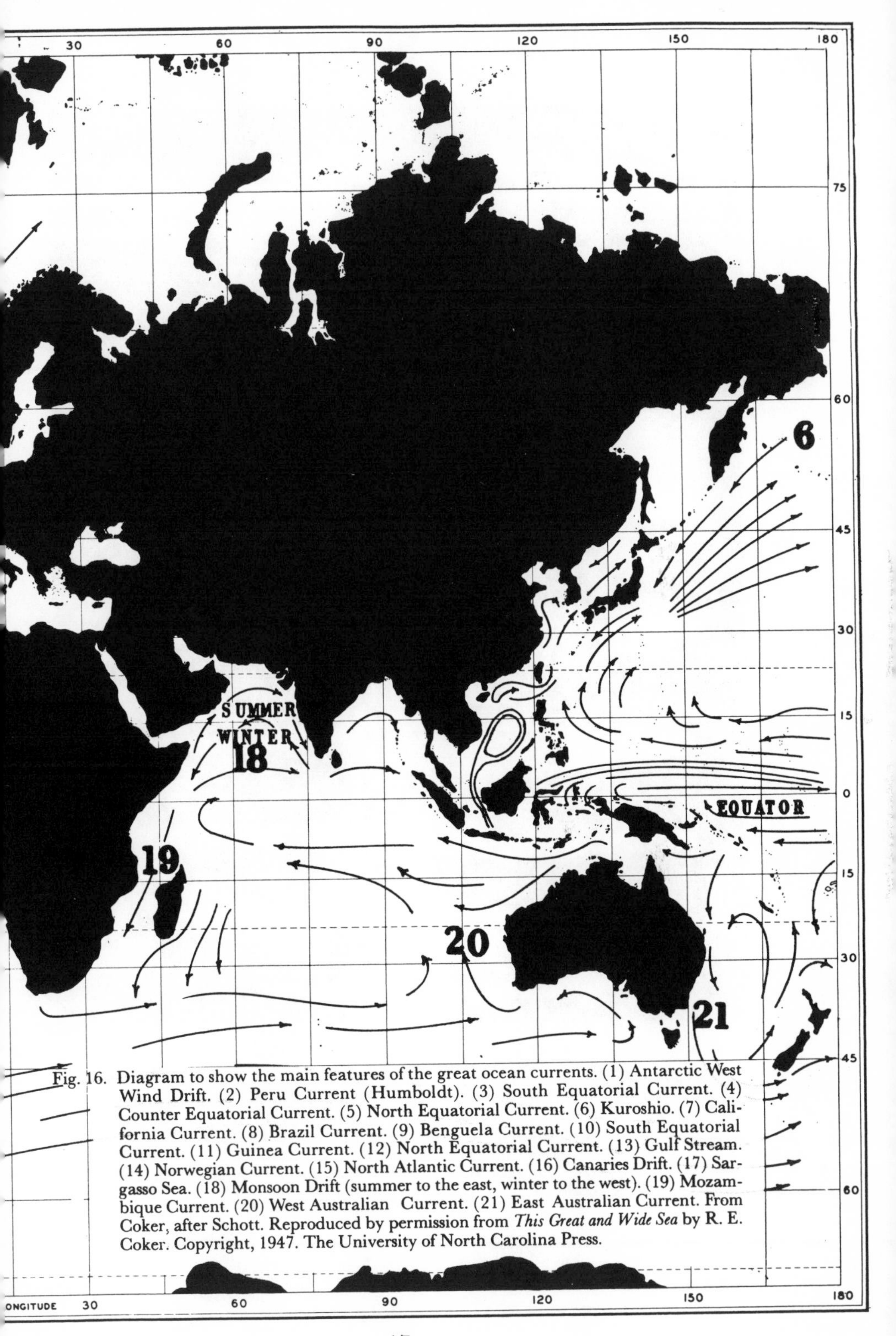

Fig. 16. Diagram to show the main features of the great ocean currents. (1) Antarctic West Wind Drift. (2) Peru Current (Humboldt). (3) South Equatorial Current. (4) Counter Equatorial Current. (5) North Equatorial Current. (6) Kuroshio. (7) California Current. (8) Brazil Current. (9) Benguela Current. (10) South Equatorial Current. (11) Guinea Current. (12) North Equatorial Current. (13) Gulf Stream. (14) Norwegian Current. (15) North Atlantic Current. (16) Canaries Drift. (17) Sargasso Sea. (18) Monsoon Drift (summer to the east, winter to the west). (19) Mozambique Current. (20) West Australian Current. (21) East Australian Current. From Coker, after Schott. Reproduced by permission from *This Great and Wide Sea* by R. E. Coker. Copyright, 1947. The University of North Carolina Press.

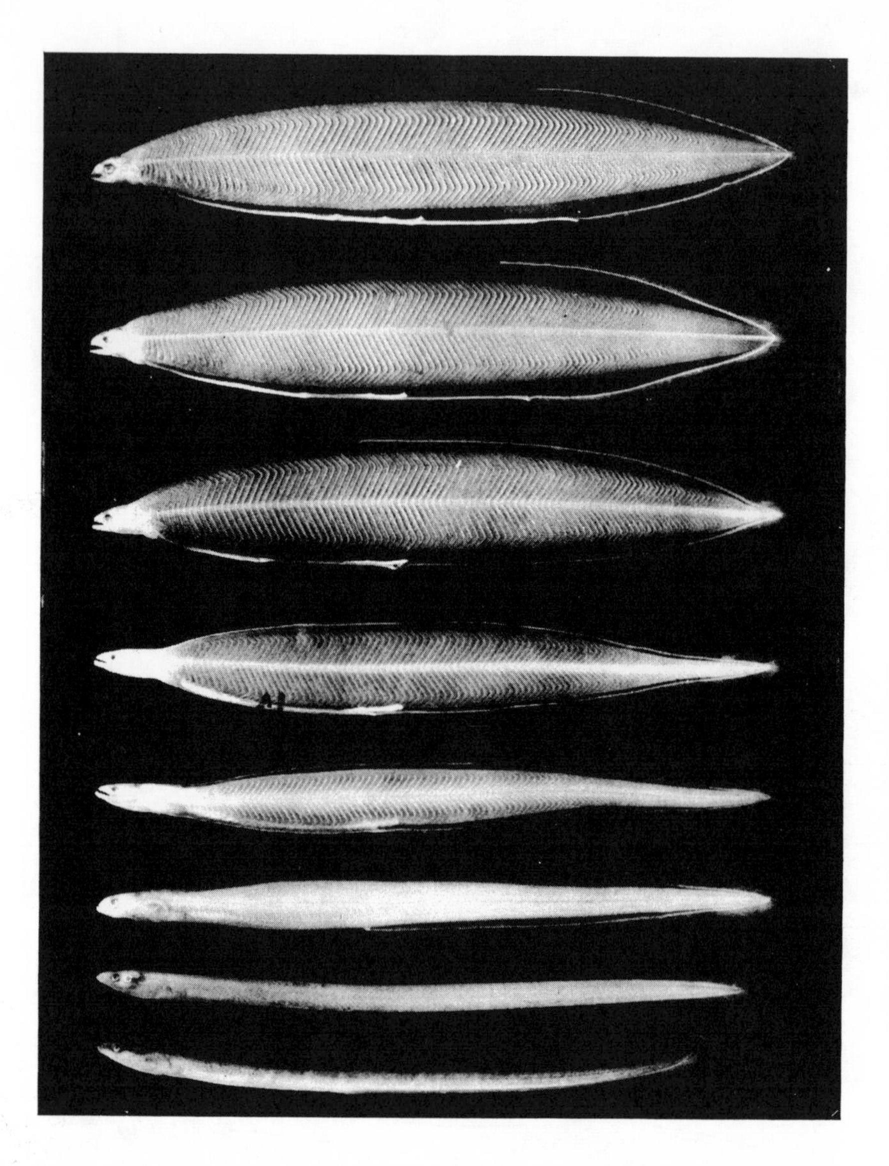

Fig. 17. The metamorphosis of the planktonic leptocephalus larva of the American eel (*Anguilla rostrata*) into its nektonic elver stage. The leptocephalus larva hatches from the egg in the Sargasso Sea and is carried to the coast of North America by currents. Here the transformation to the elver is completed, and migration into fresh-water streams occurs. From the *National Geographic Magazine*, 24 (October, 1913): 1145. International Copyright by National Geographic Society, 1940. Republished by special permission.

of this species has been studied carefully by Scofield (1934), and the current systems involved have been described by Sverdrup and Fleming (1941). The adults of the California sardine live up and down the coasts of California and Baja California, but at the breeding season, they concentrate in the area between Point Conception and San Diego, where the eggs are liberated. The concentration around this position seems to be regulated largely by temperature conditions of the water (eggs may be liberated at other places than that described if temperature conditions are favorable). The eggs and the larvae that result when they hatch take up a planktonic existence, and begin to be transported by the currents. At the time Scofield did his work on the sardine life history the currents off the California coast were known only from the reports of ships' set, and it was thought that they were such that the larvae hatched near Point Conception would be carried far to the south, and at the same time far to sea, and hence lost. Scofield was surprised, therefore, to find that the older larvae, just before metamorphosis, were to be found relatively close to shore and not too far to the south. He theorized that the larvae might have at least some powers of swimming against the currents or that there might be subsurface currents differing from the surface ones. The results of the hydrographic investigations of the Scripps Institution of Oceanography, however, offer a very simple explanation of the phenomenon. The currents from the Point Conception region do not pass uniformly southward and seaward, but first they bend landward, then northward, and here finally they make a sharp turn and pass southward again, this time much closer to shore (see fig. 18). The time consumed between the hatching of the larvae and their metamorphosis to an existence as nekters is closely synchronized with the time required for the water to circulate around the system, so that the larvae are not carried far from their normal adult habitats by southward and seaward currents during their planktonic existence.

The rheoplankton must also have adaptations to currents. Lotic systems inevitably carry their plankton content seaward, and the faster the current the more rapidly the lotic system is denuded of its plankters. This is, of course, the major reason for the sparsity of plankton in swiftly flowing streams.

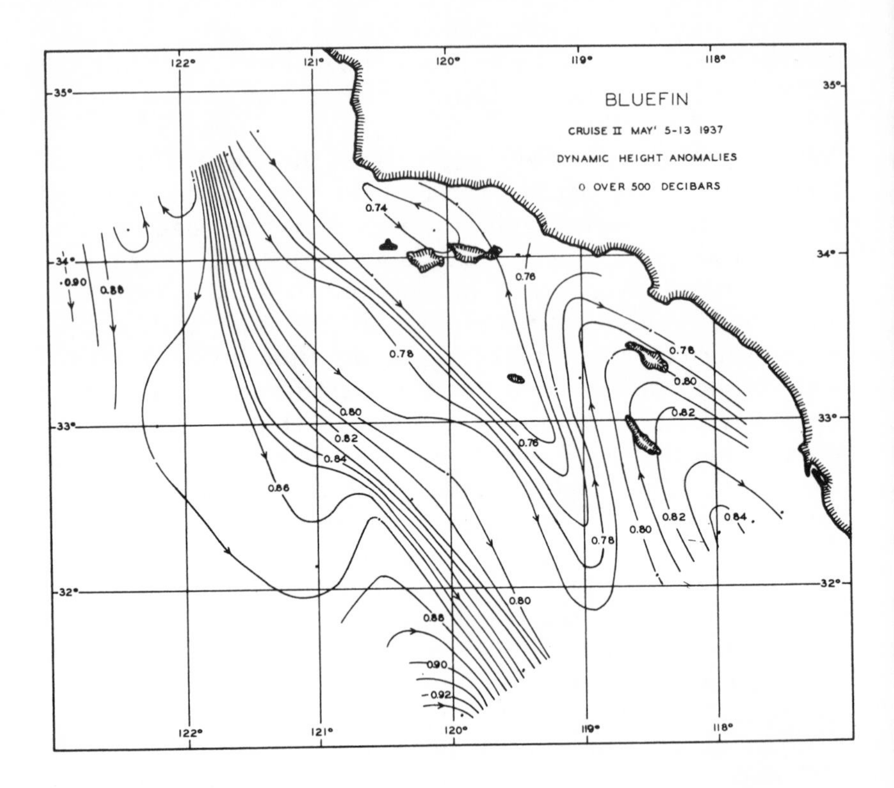

Fig. 18. The current system off the coast of Southern California, May 5-13, 1937. Dots signify the stations occupied. Arrows signify the direction of water flow. Such currents would tend to prevent planktonic larval sardines hatching west of the Santa Barbara Islands (34° N. Lat.) from being carried unduly southward before metamorphosis into nektonic young. After Sverdrup and Fleming, 1941.

However, slowly flowing rivers often have a very great plankton content, which must be constantly replenished if the river is not eventually to become barren of plankters. Replenishment may be accomplished in one of two ways. Either (1) resting spores (diatoms, etc.) or resting eggs (rotifers, copepods, etc.) are formed, which become lodged in the bottom, and which are subsequently carried upstream on the feet of wading birds, and by other means; or (2) there is a residual population in the backwaters, bordering swamps, or in the lake or pond which is the source of the river, and this serves as a constant source of repopulation. In either case the plankton will multiply rapidly in its lotic environment, and may develop into a very heavy population if conditions are favorable. Of course vast quantities of plankton are destroyed when a fresh-water river empties into the salt water of the sea, though in general the dead bodies will not be wasted, in that they may be utilized as food by detritus feeders near the mouth of the river.

Adaptations to unfavorable conditions.—Many zooplankters and phytoplankters in lakes and in the neritic pelagic area of the sea have evolved special adaptations to unfavorable chemical or physical conditions. Many lentic rotifers, for instance, produce resting eggs, which are resistant to drying or to cold. It is by means of these that life is maintained during any period during the summer when the pond in which they are living may dry up. Or the resting eggs may serve to maintain existence through the cold of winter, and through other unfavorable conditions.

Similarly, neritic pelagic diatoms in the sea are characterized by the formation of resting spores (see fig. 19). Oceanic pelagic diatoms do not form such resting spores, inasmuch as the chemical and physical conditions of the open sea are more uniform than those of the neritic province. When the neritic diatoms encounter unfavorable conditions, as they often will do in temperate or frigid climates, they form resting spores to tide them over. In shallower parts of the neritic province these resting spores (at least for certain species) may even sink to the bottom and be removed from the plankton (so that many neritic diatoms must be considered to be part of the meroplankton). Upon the resumption of favorable growing conditions, the resting spores germinate, and a new vegetative crop results.

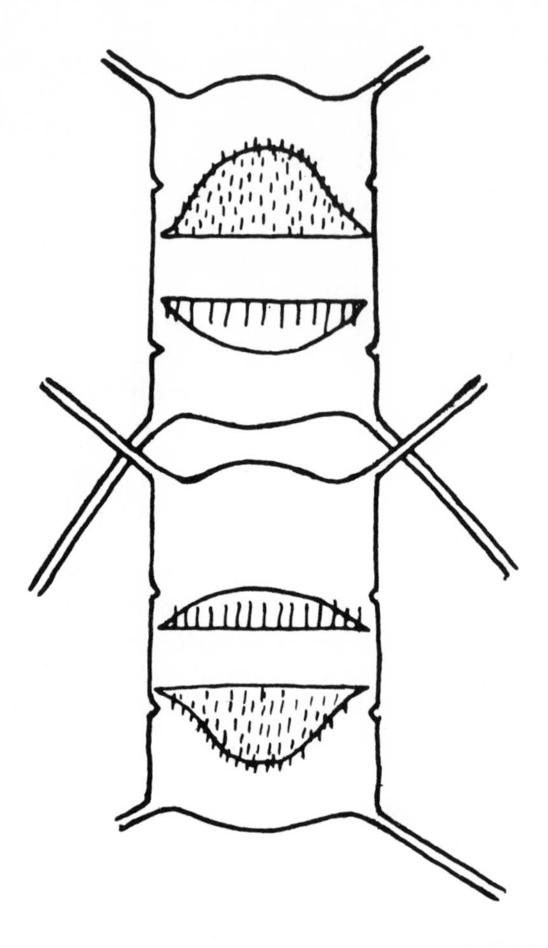

Fig. 19. Resting spores of the diatom *Chaetoceros vanheurckii.* Redrawn from Sverdrup, Johnson, and Fleming, 1946. Reproduced by permission from *The Oceans* by H. U. Sverdrup, Martin W. Johnson, and Richard H. Fleming. Copyright, 1942, by Prentice-Hall, Inc.

III. General Cycle of Production in Aquatic Environments

THE CYCLE of production in aquatic environments does not differ basically from the cycle of production in other environments. Following Pennak (1946) and others, we can distinguish the producers (the plants), the consumers (the animals), and the transformers (the saprophytes) as broad general categories. Through photosynthesis and other metabolic activities characteristic of holophytic nutrition, plants are able to build up all the complex organic compounds characteristic of living organisms (i.e., carbohydrates, proteins, fats, vitamins, etc.) from simple inorganic materials such as carbon dioxide, water, nitrates, etc. Certain species of animals, the herbivores, then graze upon these plants, digest them, and utilize the products of digestion to build their own tissues and as a source of energy for their life activities (these processes are an integral part of holozoic nutrition). Other species of animals, through holozoic nutrition, devour the herbivores. These are the primary carnivores, and they in turn may be consumed by secondary or tertiary carnivores, or carnivores of still higher level. The plants, the herbivores, and the carnivores of all levels are subject to parasitic infestation by animal or plant parasites, and all of them including the parasites are subject to death. Upon death, decay sets in, and partial decay will result in the production of humus or detritus according to circumstances. Many animals are neither herbivores nor carnivores, but subsist upon detritus or humus (e.g., earthworms, *Chironomus* larvae among the insects, certain copepods, rotifers, etc.). Finally, by the action of bacteria and other saprophytes, decay may go to completion, so that the organic material of the dead organisms is completely transformed into the materials of the bodies of the saprophytes, and into carbon dioxide, water, nitrates, phosphates, and the like—that is, into inorganic materials. These inorganic materials may be utilized in the holophytic nutrition of green plants, and in this manner the cycle is completed. Figure 20 shows a diagrammatic presentation of the phenomena described above.

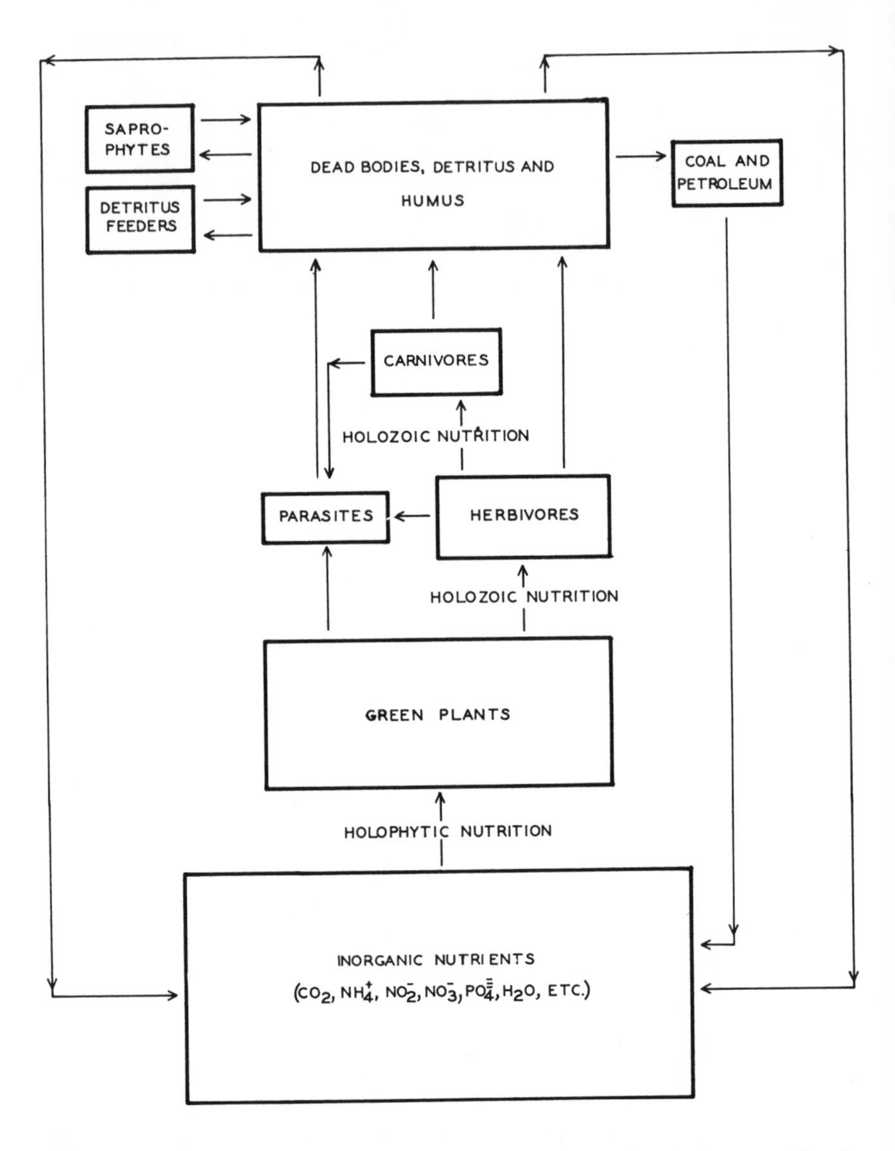

Fig. 20. Diagram to show the general cycle of organic production, both terrestrial and aquatic. See text for discussion.

As mentioned previously (p. 26), organisms we classify as plankters include all the categories described. The phytoplankton proper includes the green plants of the plankton, the saproplankton includes the saprophytes, and the zooplankton includes herbivores, carnivores of various levels, and detritus feeders.

If we consider the aquatic environment more intimately, simplifying our consideration of certain aspects of the general cycle of production, we shall see that the most important aquatic organisms among the producers are the phytoplankters, though of course their productive activities are supplemented by the activities of benthic plants. The benthic plants are relatively more important in fresh-water environments than they are in the sea, because of the proportionally smaller bottom area suitable for the growth of attached plants in the sea.

Some of the phytoplankton is consumed by zooplankters, directly or indirectly, and a considerable portion of the zooplankton is devoured by nektonic and benthic animals. To a certain degree, nektonic and benthic animals, as well as benthic plants, are harvested by the activities of some of the terrestrial animals (including man), and thus material is lost from the water more or less permanently. Terrestrial animals, by death and decay, or by defecation followed by decay, or by other processes, liberate nitrates and other inorganic materials into the soil and air, whereupon they may be utilized by terrestrial plants. Conversely, terrestrial animals consume, directly or indirectly, many terrestrial plants. Some of the products of terrestrial organisms, both animals and plants, will be washed into the hydrosphere by rains, while all of the types of aquatic organisms may die and decay and thus be transformed into the material of the saproplankton, and into inorganic nutrients, etc. The inorganic materials thus formed may be utilized by phytoplankters proper or by benthic plants in their holophytic nutrition. Other inorganic nutrients, in deeper waters, and especially in the sea, may be lost from the process of productivity for long periods by sinking into the aphotic zone. However, these may be recovered by upwelling of the water, and, given sufficient time, all or nearly all will be so recovered.

Other materials, both organic and inorganic, may be lost

permanently for all practical purposes in the formation of bottom deposits and of petroleum deposits. Many zooplankters are heavy contributors to bottom deposits, but benthic plants, benthic animals, and even phytoplankters, may also form important deposits. The phytoplankton and the nektonic and benthic animals are thought to be the most important contributors to petroleum deposits.

Figure 21 is a diagrammatic presentation of the cycle of production in aquatic environments, as discussed in the above paragraphs.

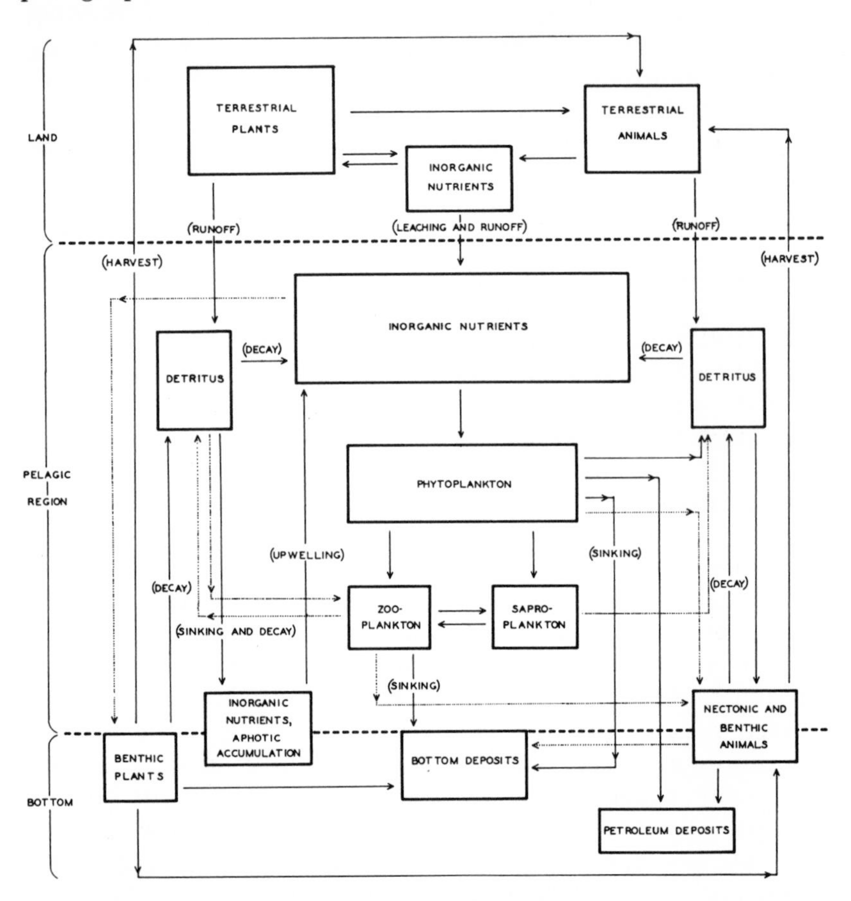

Fig. 21. Diagram to show the cycle of organic production in aquatic situations, and the interrelationships of the plankton with terrestrial and other aquatic organisms. See text for discussion.

IV. The Phytoplankton

THE Phytoplankton proper is almost universally distributed in all adequately lighted bodies of water, whether fresh or salt. There are a few local exceptions where water is poisoned by industrial pollutants or by natural poisons (e.g., arsenic), but these form an infinitesimal proportion of the whole. The wide distribution of phytoplankters, and their frequent abundance, accounts for the great importance of the phytoplankton as a major basic food material in the food cycle of aquatic situations. They are of very great importance in all sizeable bodies of water, but they are *relatively* especially important in the open sea (the oceanic pelagic province). Here the deep waters prevent the growth of attached vegetation, and land drainage cannot wash in organic debris for ultimate animal consumption to the degree that it does in the littoral pelagic province.

Latitudinal variations of distribution.—However, the phytoplankton is by no means uniformly distributed from place to place, even within short distances. It is very difficult to make accurate generalizations with regard to latitudinal variations in the distribution of the phytoplankton, both because of our lack of really detailed knowledge of conditions at very high and very low latitudes and because of wide local variations at all latitudes. This is especially true of lakes, but also of the oceans. It is generally stated that the production of phytoplankton is higher in high latitudes in the ocean (Russell and Yonge, 1928, p. 126) than in temperate regions, and that the production in tropical seas is lower. Using centrifuge methods, Lohmann (1920), as summarized by Riley *et al.* (1949), found that in the eastern Atlantic Ocean the number of phytoplankton cells in a column of water beneath 0.1 sq. m. of ocean surface varied considerably with latitude, being considerably higher in temperate than in tropical waters. Thus the average number of cells was 240 million between 50° and 40° N. Lat., 80 million cells between 40° and 30° N. Lat., 24 million cells between 30° and 20° N. Lat., 20 million cells between 20° and 10° N. Lat., and 24 million cells between 10° N. Lat. and the Equator. All of Lohmann's samples were obtained in May and June and there is no indication that samples were obtained in comparable parts of the annual production cycle at the several latitudes. Riley *et al.* (1949) compared the phytoplankton of Georges Bank with that of the Sargasso Sea. They found that in May, 1939, the phytoplankton of the Sargasso Sea beneath 1 sq. m. of water surface was equivalent to 5.44 gm. of carbon, while at Georges Bank the comparable figure was 9.96 gm. of carbon. Here, however, we are comparing nutrient-rich water lying over the continental shelf with nutrient-poor water over the abysses, and latitude is by no means the only factor involved. Both in the work of Lohmann and that of Riley *et al.*, seasonal factors are ignored, and the total annual productivity is the important factor.

In this connection, the results reported by Graham (1941) on plankton of the open Pacific are of considerable interest. He found that production in the tropics was greater than that in temperate regions because of the renewal of surface plant nutrients by upwelling of nutrient-rich water in the tropics.

It is a little dangerous to generalize about relative production in frigid, temperate, and tropical ocean waters because measurements in highest latitudes have been confined largely to the summer months, when the Arctic and the Antarctic are more available. Also, observations in the tropics have been made largely by expeditions on the high seas, and not on a year-round basis by shore stations. In the tropics, Allen (1939) shows as much as 500,000 cells per liter of diatoms in the waters of the Pacific coast of Panama, and Gunter *et al.* (1948) indicate an extremely large production of phytoplankton off the west coast of Florida in 1946-47. The latter is, however, only on the edge of the tropics.

Similarly in lakes, those of high latitudes are thought to have a high productivity of phytoplankton (Welch, 1935, p. 208) compared to temperate regions, though some (e.g., Pennak, 1946, p. 348) have stated that such lakes have low productivity. There evidently have been no year-round studies of high latitude lakes, at least none published in journals readily accessible in this country, and there have been very few observations of any kind. Tropical lakes are only slightly better known than those of high latitudes. They seem to have a somewhat lower production than temperate lakes, but here again our lack of definite knowledge is prodigious. Lakes of high altitudes are notorious for their low productivity, but this is possibly due to factors other than the simple one of altitude (e.g., low content of nutrient salts, etc.). Lakes of low altitudes are extremely variable in their productivity, depending upon their type (oligotrophic or eutrophic), locality, physical and chemical characteristics, etc. (see fig. 22).

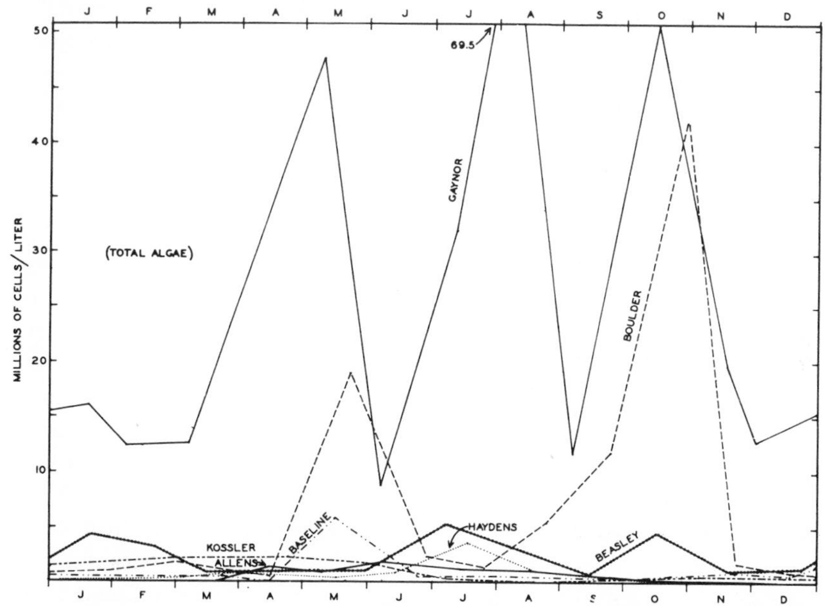

Fig. 22. Variability of production in lakes. A study of seasonal phytoplankton production in seven lakes in Colorado. Note the great variation of total production from lake to lake, and also the varying seasonal distribution in the different lakes. After Pennak, 1946.

It is obvious that much more research needs to be done both in high and low latitudes, and on both fresh-water and salt-water environments. The region of Canada recently opened up by the Alaska Highway offers a wonderful opportunity for the more intimate study of lakes of many types in high latitudes, while the numerous lakes in Florida have hardly been touched limnologically, and are very accessible for study at all times of the year. The Florida lakes are not strictly tropical in every sense of the word, but they certainly approach the tropical type, and are much more accessible than other lake areas in the tropics. Similarly, a more thorough study of production in marine environments is needed in the Arctic and Antarctic seas, and it is hoped that the current interest in these areas in Europe, Asia, and North America will bring us greatly increased knowledge of these important regions. Tropical marine shore stations need to be established or developed for the intimate study of tropical seas.

The Bermuda Biological Station for Research, the Lerner Marine Laboratory at Bimini in the Bahamas, the Marine Biological Station in Honolulu, the University of Miami Marine Laboratory, and others are in localities where much basic work can be done on productivity in the tropics. The laboratory in Miami is especially favorably located in that it is established on the shores of a continent and therefore is especially accessible.

Several hypotheses have been attempted to explain the seemingly greater productivity of the waters of high latitudes. Probably none of these is sufficient in itself as an explanation. Further detailed and year-round observations will help to clear up the questions that are raised.

First, it has been pointed out that at high latitudes the winters are long and without light for photosynthesis, so that nutrient salts would have an opportunity to accumulate. Thus when sufficient light for photosynthesis comes in the summer months, there would be a very high production of phytoplankton. Of course, most plankton investigations of waters in high latitudes have been made during summer months, and it is probable that the total production for the period of a whole year would average not as high as the apparent productivity indicated by summer analyses alone.

Second, the long daylight hours of the summer season in high latitudes have been thought to influence phytoplankton productivity by allowing 24 hours a day of photosynthesis. In the lower latitudes of the temperate and tropical regions, on the other hand, there is an alternation of day and night, with of course no photosynthesis taking place in the latter. Considered on a year-round basis, according to this explanation (assuming radiation as the only controlling factor—which it certainly is not), production at all latitudes would be much more nearly equal than is usually thought to be the case.

Third, it has been suggested that the metabolism of organisms is low in the cold waters of the high latitudes, with the result that the death of organisms is delayed. In such a case, it is possible that several generations of the organisms would accumulate, making for a large standing crop, though the actual productivity per unit time might be relatively low. Of course, in plankton observations—especially the "hit-and-run" observations heretofore usual for high latitudes—it is the standing crop that is measured.

Fourth, some have spoken of the possibility of a reduction in the quantity of deleterious radiations from the sun striking the water surface in high latitudes. Such a reduction would be due to absorption of the harmful rays by the thicker layers of atmosphere through which they must pass at high latitudes before reaching the surface of the earth. At the same time, radiation suitable for photosynthesis would be plentiful. It is well known that in lower latitudes, in the upper layers of water, the phytoplankton often is retarded by strong radiation from the sun (see the discussion of the vertical distribution of phytoplankton, p. 65 below).

Fifth, the fact that the colder waters of high latitudes have a greater capacity for the solution of gases has been cited as possibly influencing productivity. However, the soluble gas that would influence the metabolism of the phytoplankton is carbon dioxide (CO_2), because of its use in photosynthesis. Carbon dioxide is never a limiting factor in the productivity of phytoplankton in the sea, and probably seldom is in freshwater environments. Therefore, it does not seem likely that the greater solubility of CO_2 in the colder waters could be a major cause of greater productivity in waters of high latitudes.

Finally it has been suggested that the apparently large phytoplankton production in seas of high latitudes, especially the Antarctic, is not due to the latitude as such, but is caused by an upwelling of water in those regions, bringing plant nutrients to the surface. Among others, Jespersen (1935) has shown that even in tropical regions the areas near the west coasts of continents have very heavy concentrations of plankton. Jespersen showed that macroplankton volumes in the upper layers of the water (50-300 m. of wire) were greater along the west coast of Africa than at any other location investigated, and in deeper layers (1,000-5,000 m. of wire) a greater volume was found off the west coast of Panama than anywhere else, though nearly comparable volumes were encountered in the North Atlantic far from shore.

Probably none of the hypotheses described above offers an adequate explanation of higher productivity in Arctic and Antarctic waters. In fact until we have further observations, we cannot state unequivocally that such higher productivity is characteristic, at least on a year-round basis. It is possible that the first three hypotheses combined, or the last hypothesis alone, might provide an adequate explanation.

Local variations of distribution.—Heretofore we have spoken only of very broad, latitudinal variations of phytoplankton production. In addition, at all latitudes, there may be great local horizontal variation in production (or at least in the standing crop). Such variations may be caused by a number of factors.

Previously we have discussed the great variability of lakes with regard to their chemical, physical, and biological characteristics. One of the major ways in which lakes may vary biologically is in their phytoplankton production. Thus two lakes, otherwise similar in location and in all or most of their chemical and physical characteristics, may produce very different quantities of phytoplankton. This, however, is due to the isolation of the lakes from each other by more or less insurmountable obstacles. The different history of each body of water leads to differing production.

Even within one body of water, either in a marine or a freshwater situation, there may be a great horizontal variation of phytoplankton production. Certain areas may have a profuse phytoplankton population, while other areas, near by, may have a scanty population. Such local variations may have a number of different causes, some of which are as follows:

Winds, especially prevailing winds, may concentrate certain species along the shore on the leeward side of a body of water, so that locally along the shore there are enormous numbers, while elsewhere the concentration of phytoplankters is much lower. Winds, of course, affect those forms such as the blue-green algae and certain dinoflagellates that float near or at the surface more than they affect those forms floating deeper under the surface.

Converging water curents, or eddies, may serve to concentrate the phytoplankton (and of course the zooplankton as well) in certain places. Thus, in the Puget Sound region in summer such currents may serve to cause the concentration of the dinoflagellate *Noctiluca* in enormous quantities in certain localities, so that some areas of the water are decidedly colored.

Local and widespread blooms of phytoplankters will be discussed later. They may cause local horizontal differences in phytoplankton abundance.

An originally uniformly distributed population of phytoplankters may lose its uniformity due to the grazing activities of zooplankters or of other animal life. In this way, the vertical migration of a swarm of copepods into a mass of phytoplankton will result in a decimation of the phytoplankters, while a lack of copepods in a nearby spot will allow the phytoplankters there to remain unmolested.

Local variations of nutrient salts may be the cause of variations of phytoplankton production. Such local variations may be caused by a number of things. A city, for example, may pass its sewage into a body of water, locally increasing the concentration of nutrient salts. Biscayne Bay, Florida, is favorable for the study of the effects of sewage disposal. A trip by boat up and down Biscayne Bay at almost any season of the year will show great variations of plankton content. The water of the lower portion of the Bay is very clear and blue, with the bottom clearly visible, while the waters near the city of Miami are greenish and much more nearly opaque. Plankton volumes, which can be used directly or indirectly as an indication of phytoplankton volumes, are uniformly higher in the upper portion of the Bay than in the lower portion. Data published in tabular form by Smith *et al.* (1950) are shown graphically in figure 23. The higher volumes characteristic of the upper portion of the Bay are indicated by the samples obtained from the Miami Beach Boat Slips station, while the lower volumes characteristic of the lower Bay are shown by the samples from the Featherbed Bank station.

In addition, local variations of nutrient salts may be produced by land drainage, as indicated by the uniformly higher phytoplankton production of the littoral pelagic province as contrasted with the production of the oceanic pelagic province, and by the upwelling of deeper waters. The importance of land

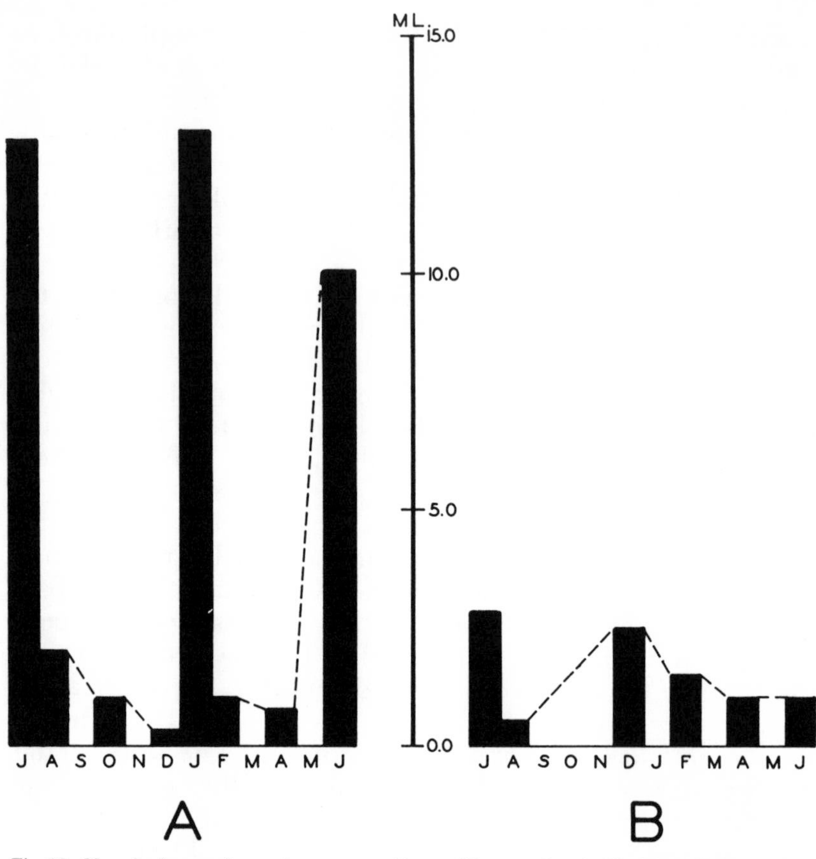

Fig. 23. Net plankton volumes in upper and lower Biscayne Bay in Florida to illustrate the effect of sewage effluents upon the richness of the plankton. (A) Miami Beach Boat Slips (upper Biscayne Bay), near the source of pollution. (B) Featherbed Bank (lower Biscayne Bay), far from the source of pollution. From tabular data in Smith *et al.* (1950).

drainage has been shown very clearly along the coast of Norway by Gran (1931). He found that the very first appearance of spring bloom conditions among the diatoms occurred each year very close to shore, and spread from there out over the banks. Similar conditions were encountered in the Barents Sea by Kreps and Verjbinskaya (1930). Nutrient salts have a tendency to accumulate in deeper waters because of the lack of photosynthetic activity there and because of the sinking of organic material through the water, followed by subsequent decay. Upwelling brings a rich supply of these salts to the surface and allows for a high phytoplankton production wherever it occurs. The west coasts of South Africa, South America and North America are characterized by upwelling and by an especially rich phytoplankton production. Thus Phifer (1934) found a standing crop of 3,000,000 diatom cells per liter in East Sound, Orcas Island, in the Puget Sound area on the West Coast of the United States, the second highest number of diatom cells ever reported from a marine area.

Finally, variations of local distribution of phytoplankton may be caused by the flow of plankton-poor water into a body of plankton-rich water, or *vice versa.* For instance, the flow of water into a lake from streams and other inlets often results in the formation of an area either rich or poor in phytoplankton near the mouth of the inlet. Early in May of 1948 (unpublished data in the author's possesion) the situation in two brackish-water lakes in Southern Florida offered a situation of this nature. The water of Middle Lake, Dade County, was green and very opaque due to the enormous abundance of a dinoflagellate of the genus *Gonyaulax.* Nearby Seven Palm Lake was very clear, and the bottom was easily visible, though the lake was deeper than Middle Lake. However, the water of certain small areas in Seven Palm Lake was similar in appearance to that of Middle Lake. The two lakes are connected to each other by a very short, rather narrow passage, through which the tide runs back and forth. Evidently the incoming tide had carried masses of *Gonyaulax*-rich water from Middle Lake to Seven Palm Lake, where it had been isolated into patches.

Vertical distribution.—The vertical distribution of the phytoplankton is also of interest. Obviously, because of the lack of light in the aphotic zone of the ocean and of the deeper lakes, there can be no green plant life there. The phytoplankton proper is confined to the euphotic and disphotic zones. It might be expected that the greatest concentration of phytoplankton would be found in the water at the very surface, and that there would then be a gradual, consistent decrease in quantity with depth, commensurate with the gradual decrease of illumination. However, this is seldom the case.

For the most part, the vertical distribution of the phytoplankton is as indicated in figure 24. Most of the production lies

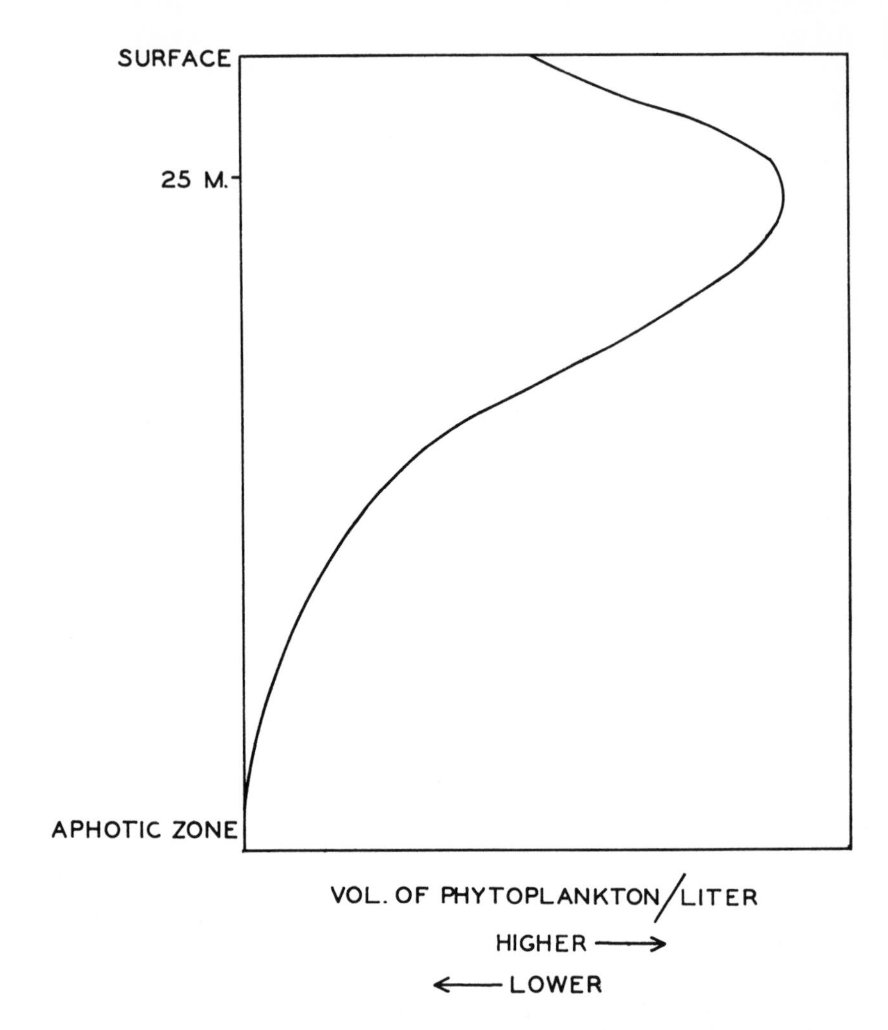

Fig. 24. Idealized diagram to show the usual type of vertical distribution of phytoplankton in the sea. The depth of the maximum varies considerably from place to place and from time to time. The curve would be similar in fresh-water lakes, but with the maximum more shallow; and, of course, few lakes are sufficiently deep to have an aphotic zone. Stagnation of the hypolimnion in lakes would also modify the curve.

within the euphotic zone, as would be expected, and there is a gradual decrease until the extinction of the phytoplankton at the beginning of the aphotic zone. The greatest production, however, lies several meters below the surface, because of the deleterious effects of strong sunlight.* Many phytoplankton organisms are either killed outright or at least retarded by excessive illumination (see Jenkin, 1937), and thus their optimum environment as far as radiation is concerned lies some distance below the surface, where much of the radiation already has been absorbed by the water. For example, Sleggs (1927), working off the coast of Southern California, found the greatest concentration of phytoplankton organisms to be between 25 and 55 m. below the surface. The depth of greatest productivity, however, varies with latitude, being greatest in the tropics and less at high latitudes where the radiation falling upon the surface of the water is less to begin with. Riley *et al.* (1949) report the greatest concentration of plant pigments in the Sargasso Sea and in Florida Strait at a depth of 100 m., where available light amounts only to 1/200 that of the surface illumination. On the other hand, the same authors showed the maximum concentration of plant pigments at only 10 m. over Georges Bank. An attempt is made to explain such vertical distribution of the phytoplankton as follows: phytoplankton production is greatest in the well-lighted upper layers of the water, but the individual plankters sink slowly through the water in such a manner that their bulk is to be found below the most productive layers. In fact, the authors point out that in the Sargasso Sea the bulk of the phytoplankton occurred at a depth where light deficiency would lower photosynthesis below the rate of respiration (hence the bulk occurred in the disphotic zone!). The continued presence of phytoplankters in the surface layers is explained on the assumption that they constantly are restocked by the turbulence of the water. The greater the turbulence of the water and the more rapid the phytoplankton production, the shallower will be the depth at which maximum occurrences of phytoplankters will be found.

* In highly turbid lakes the zone of photosynthesis does not extend to very great depths, and optimum photosynthesis occurs only slightly below the surface. As Verduin (1951) has shown, turbidity may affect the crop of diatoms very considerably. Furthermore, floating types of blue-green algae, such as *Anabaena* and *Microcystis,* appear to have their maximum photosynthetic rate near the surface, so that where these forms predominate in lake plankton the above discussion does not apply. The facts in this note were supplied to the author by J. Verduin of the Franz Theodore Stone Institute of Hydrobiology in a personal communication.

With all its merits this explanation is certainly an oversimplification in that it does not include all factors (availability of nutrients, temperature, etc.). The waters of the Gulf Stream for instance have been shown to be slightly richer in plant nutrients than those of the Sargasso Sea, and especially in Florida Strait the waters would be much more turbulent, yet the depth of maximum phytoplankton abundance is the same in both—100 m. Gillbricht (1952) shows that the maximum of diatoms occurs in the bay at Kiel, Germany, in the vicinity of the salinity discontinuity layer, and he thinks this maximum is caused by the slower rate of sinking in the more saline deeper water, with a consequent accumulation of the individuals that are constantly sinking from the less saline water above.

It is true, however, that there are some important exceptions to the vertical distribution described above. Many blue-green algae float near the surface of the water, and may be present in such vast numbers as to form a surface mat on the water of certain lakes (Prescott, 1939). The author has seen a vast swarm of *Chlorella* in the Baltimore Harbor area (Davis, 1948), where the organisms were so numerous as to make the water a murky green. However, in this example, *Chlorella* was present in large numbers only at the surface, and even a few feet beneath the surface the numbers had greatly decreased.

Seasonal variations of distribution.—In addition to the general horizontal and vertical distribution of the phytoplankton discussed above, there is a marked seasonal variation of distribution.It is conventional, both in fresh-water and marine studies, to speak of two annual phytoplankton maxima and two annual minima. The maxima occur in the spring and fall months whereas the minima occur in the winter and summer. The spring maximum is greater than the fall maximum and the winter minimum is much less that the summer minimum. Such a seasonal distribution is schematically pictured in figure 25.

1. *Marine seasonal distribution.*—

The conventional picture certainly holds for the North Sea, and for many other similar localities in north temperate marine environments, though even here there are considerable varia-

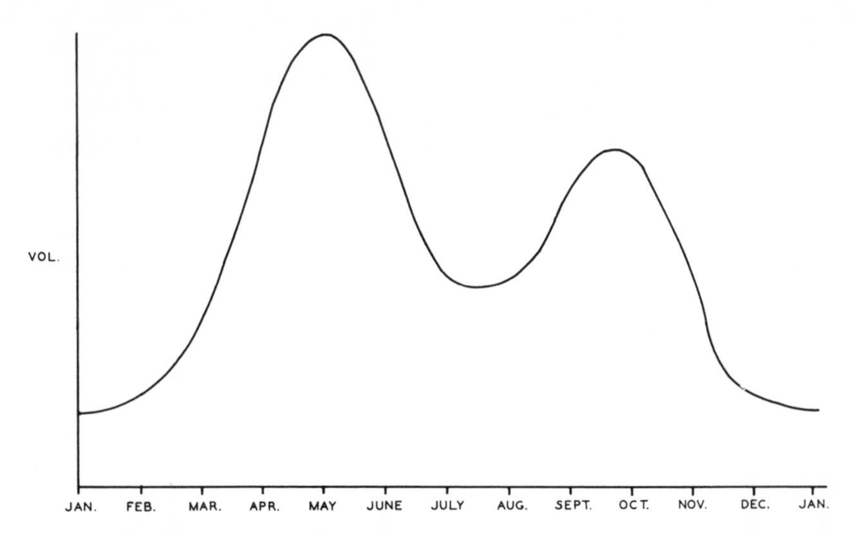

Fig. 25. Idealized distribution of the phytoplankton in temperate waters. A winter minimum is followed in turn by a spring maximum, a midsummer minimum, and a fall maximum.

tions from year to year because of more or less obscure differences in environmental conditions. However, such a distribution apparently is not found in all parts of the temperate oceans, and certainly is not the case in the ocean at very high and very low latitudes. For example, in the Arctic and the Antarctic Oceans there is no solar radiation available during the winter months because of the long winter night, and thus there can be no photosynthesis, and no phytoplankton production whatsoever. In the tropics, there are no well-marked seasons comparable to those in temperate regions.

However, practically no year-round studies have been made either at high or low latitudes. Marshall (1933) has given a rather detailed account of the seasonal distribution of phytoplankton from the Great Barrier Reef region. According to her results, the weekly station in the lagoon of the Great Barrier Reef showed one phytoplankton maximum in late August (this was a very brief but extremely prolific maximum), a second in November (a smaller but slightly more protracted maximum), and a third in March, April, and May. The third maximum did not reach as high a peak as did the first, but it was far more massive. Thus these results show smaller maxima in midwinter and spring, and a massive maximum in the fall.

Smith *et al.* (1950) show that in the Miami area of southern Florida the total net plankton volumes do not follow the conventional pattern established for temperate waters (see fig. 26).

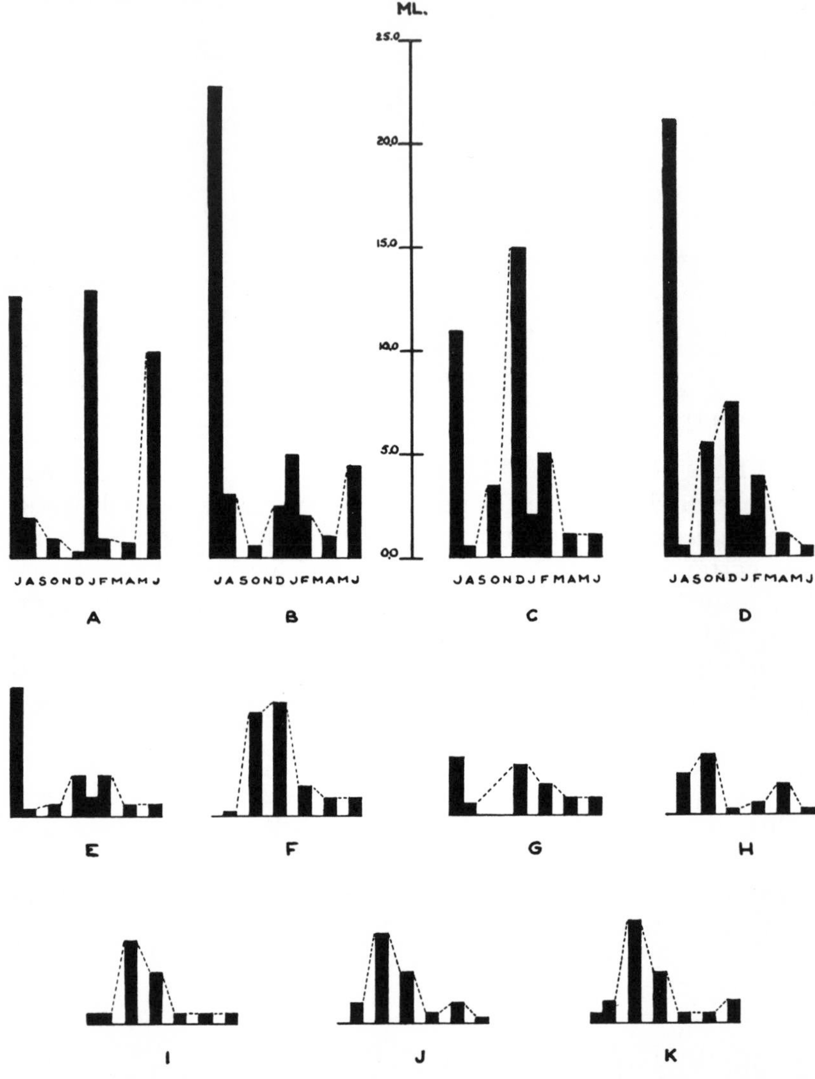

Fig. 26. Seasonal distribution of net plankton volumes in waters in the vicinity of Miami, Florida. A, B, C, and D are from upper Biscayne Bay nearest the city of Miami; E, F, G, H, and K are from lowerBiscayne Bay; I an J are from open waters some distance from the city. Note the occurrence of winter maxima, supplemented by summer maxima in stations near Miami. From tabular data in Smith *et al.* (1950).

In general, most of the stations examined in Biscayne Bay and in the nearby ocean showed a midwinter plankton maximum, though it is true that stations in northern Biscayne Bay (i.e., near the city of Miami, where pollution by sewage may have affected the results) showed a second maximum in summer.

Many attempts have been made to explain the seasonal distribution of phytoplankton in north temperate seas on the basis of simple physical, chemical or biological factors. These attempts have been surveyed, in part, by Gran (1931). None of these explanations seems to be adequate in itself, though it is probable that each explanation describes one of the several contributing factors.

It has been thought that one of the factors influencing the spring phytoplankton bloom lies in the increased solar radiation that falls upon the surface of the sea at that season. This follows a winter season in which photosynthesis has been at a minimum, thus allowing an accumulation of nutrient salts. The two factors together are thought to result in a rapid increase of the phytoplankters. But it is possible to show that on most days in the winter in the temperate region there is sufficient solar radiation for photosynthesis. In fact, there is even sufficient radiation on many winter days to be detrimental to phytoplankters that might be living in the surface layers of the water.

Others have thought the winter minimum was due to the low temperatures of the water at that season, which presumably would reduce metabolic activities of the phytoplankters to the extent that phytoplankton production would be very low. However, the cold waters of the Arctic and the Antarctic produce heavy standing crops of phytoplankton when other conditions are suitable.

The midsummer minimum has been thought to be due to the depletion of nutrient salts in the surface layers through the action of the great bloom of phytoplankters in the spring. On the other hand, it has been shown that the surface waters in temperate regions have sufficient nutrient salts at all times of the year for adequate phytoplankton growth. Furthermore, repletion of the nutrient salts in surface waters previous to the development of the fall maximum is difficult to picture, inasmuch as there is no special overturn of the water in the fall, such as occurs in many lakes.

An attempt has been made to explain the summer minimum on the basis of grazing by the zooplankton and by other animals. According to this theory the spring phytoplankton bloom produces a great deal of basic food, which results in a subsequent great increase in the zooplankton. When the zooplankton population has increased sufficiently it consumes the phytoplankters more rapidly than they can reproduce themselves. Thus the peak of phytoplankton production is passed and the summer reduction sets in. According to this hypothesis, the fall phytoplankton maximum sets in when the reduced primary food supply during the summer minimum results in a reduction of the number of zooplankters, thus allowing a resurgence of the phytoplankters. However, no such close correlation of phytoplankters and zooplankters has been proved in natural waters. Furthermore, such an explanation ignores the fact (discussed more adequatelybelow) that the spring phytoplankton maximum is due to the bloom of certain species only, while the fall maximum occurs because of a bloom of entirely different species.

It has also been suggested, but without sufficient evidence to support the hypothesis fully, that fluctuations of the phytoplankton can occur without any relationship to an advantageous change in the environment. This would, if it could be proved, signify an inherent, regular, metabolic cycle. It remains unsafe, however, to make any definite statements to this effect until all the minute environmental changes, especially the interactions of such changes with each other, have been thoroughly investigated with tools and procedures more delicate than any at present available.

It seems probable that no simple physico-chemical or biological explanation for the regular seasonal progression of phytoplankton production is possible. The basic causes determining the maximum and minimum probably include the interaction of most of the factors suggested above with other more minor influences contributing to the picture. Riley *et al.* (1949), on theoretical grounds, analyzed the seasonal distribution of the phytoplankton as a result of the interaction of the original phytoplankton (P), the photosynthetic rate (thus growth) (P_h), the respiratory rate (R_p), and the grazing rate (G). In its simplest form this interaction is expressed in the mathematical formulation:

$$\frac{dP}{dt}=P(P_h-R_p-G).$$

They then analyzed the photosynthetic rate as being a function of incident solar radiation (I_o), a photosynthetic constant (K), the extinction coefficient for the penetration of visible light into the water (k), the depth of the euphotic zone (L), nutrient depletion due to vertical turbulence (A''), and reduction of the coefficient of the photosynthetic rate due to the nutrient depletion (v_p). Hence,

$$P_h=\frac{KI_o}{kL}(1-e^{-kL})A''v_p.$$

They analyzed respiratory rate as a function of the coefficient of respiration at O° C. (R'), the coefficient of the increase of respiratory rate with temperature (s), and the temperature (T). Hence,

$$R_p=R'e^{sT}$$

They analyzed the grazing rate as a function of the grazing constant (W) and the numbers of herbivores (H). Hence,

$$G=WH$$

Thus the simple preliminary equation becomes:

$$\frac{dP}{dt}=P\left[\frac{KI_o}{kL}(1-e^{-kL})A''v_p-R'e^{sT}-WH\right]$$

They found that on Georges Bank the theoretical curve "agreed with the observed population within reasonable limits of error," namely 25 per cent. They point out, however, that the Georges Bank equations "are too simplified to be suitable for general application. Errors caused by simplification of the assumptions may be insignificantly small in one area and very large in another."

2. *Seasonal distribution in lakes.—*
Many authors have attempted to describe seasonal distribution of phytoplankton in temperate lakes by means of a curve similar to that for temperate seas. As pointed out by Pennak (1946), there have been only a limited number of year-round studies of lakes, and the curves have been constructed on the basis of results obtained in only a few lakes, all of which are relatively large lakes in the temperate zone. Pennak reported seasonal studies of five Colorado lakes (therefore temperate lakes), most of which did not follow the conventional curve at all closely, (see fig. 22, p. 59). Sawyer *et al.* (1943, 1944) reported seasonal studies of polluted Wisconsin lakes with similar results. Tropical and Arctic lakes undoubtedly do not follow such a pattern, though it is not possible to make any definite statements on this score, inasmuch as there have been, to my knowledge, no published accounts of year-round studies on regions outside the temperate zone.

Attempts have been made to explain the seasonal development of phytoplankton production in those lakes that follow the conventional curve of production. For example it has been thought that, during the winter stagnation period, when there is little photosynthesis, nutrient salts accumulate. At the spring overturn, any nutrients that have accumulated in the bottom waters are mixed with the entire water mass, and at the same time the solar radiation again becomes effective because of the removal of ice interference at the surface. As a result of these events there is a sharp upturn in the development of the phytoplankters, resulting in the spring bloom. Thereupon, it is said, the nutrient salts in the surface waters are soon depleted by the great phytoplankton production, while simultaneously there develops a stagnation of the waters of the hypolimnion, as described above (p. 13). In this way the summer minimum is produced. At the time of the fall overturn, however, the nutrients that have accumulated in the hypolimnion are again mixed throughout the water, and the stagnation of the hypolimnion is removed, resulting in a fall phytoplankton bloom, which, however, is not as vigorous as that of the spring, and which ends with the approach of the winter stagnation.

On the other hand, Birge and Juday (as reported by Welch, 1952, p. 260) have shown that in those Wisconsin lakes which follow the conventional pattern of seasonal phytoplankton production there is sufficient nitrate and other nutrient salts in the epilimnion during the whole summer, and, furthermore, that there is no increase of these salts just preceding the maxima. Chandler (1940) has described an example in western Lake Erie in which the entire vernal phytoplankton bloom took place beneath an ice cover. Thus it would seem that the conventional explanation is not adequate, and other or additional factors must be responsible. J. W. G. Lund (1949, 1950) thinks that silicon and light are the limiting factors in diatom production. He thinks that the low production of diatoms in Lake Windermere in the winter was due to inadequate light, and that the termination of the spring bloom occurred when the silicon content of the water became low. However, he does not believe that the supply of silicon provides a complete explanation of the fluctuations in diatom numbers in all lakes at all times of the year. Similarly, Wawrik (1952) has shown the importance of silicon in diatom production. An Austrian fish pond, fed by ground water, exhibited diatom blooms after each rise in water level. The incoming ground water was rich in silicon, but the available Si was used up very quickly by the diatom growth, and the bloom would disappear.

Nonseasonal phytoplankton blooms.—In addition to the regular seasonal blooms of phytoplankton that are characteristic of most bodies of water, there are occasional blooms of a different nature. With or without obvious cause the phytoplankton increases greatly in abundance until there is an enormous quantity present. Frequently the water is greatly discolored, and it may sometimes become slimy or even matted on the surface with phytoplankters (the last-mentioned phenomenon occurs, of course, only in those cases where the phytoplankters are of the floating type, such as many of the blue-green algae). Such blooms may occur at any season of the year, and either in fresh-water or in marine environments. They may build up slowly to enormous numbers and decline equally slowly back to ordinary quantities, or they may increase in numbers extremely rapidly, and as suddenly die out.

Marshall (1933, pp. 113, 115) shows an example of the latter type of increase for the phytoplankton of the lagoon of the Great Barrier Reef. In August at her weekly station in the lagoon, there were between 2,000 and 10,000 diatom cells per liter at the surface, while "on 29th August there was a sudden rise near the island to 28,000, and a further rise the following day to 167,400 diatom cells per litre—the highest figure reached during the whole year. . . . This rich diatom flora disappeared as quickly as it had appeared. On 4th September at the weekly station numbers had fallen to their usual value. . . . "

Allen (1946, pp. 150-151) describes the slower type of increase towards bloom conditions in the ocean off La Jolla, California: "The rise and fall of dinoflagellates in this occurrence followed fairly well the binomial curve so commonly observed for such phenomena, including the gradual increase from small numbers to extremely large with a following decline gradually to numbers relatively small or vanishing. . . . Throughout the four weeks that 'red water' was observed in 1945 the preponderance of *Gonyaulax polyedra* Stein was so excessive (six million cells per liter at the peak of abundance . . .) as to leave no doubt of its being the causative organism. However, there were several different days in which pier catches indicated numbers of several other species which would have been considered significant if appearing alone (not overshadowed by *Gonyaulax*). For example, *Prorocentrum micans* Ehr. was represented by thousands of cells. . . . *Ceratium furca* Clap. et Lach. showed numbers almost as large. . . "

Gunter *et al.* (1948, p. 315), in describing the progress of the phytoplankton bloom on the west coast of Florida in 1946-47, state: " . . . there was a definite sequence. First the water became turbid or slightly changed in appearance from a clear green or bluish-green to a more opaque green. . . . Then it became yellowish green and later greenish yellow. At the next stage, and possibly only in spots a few yards or a few hundred yards across, the water became a bright saffron yellow in color. . . . Gunter and Smith observed a patch of this water. . . . The water color was very striking and peculiar in contrast to the green waters of the bay. The water had other peculiar characteristics in addition to color. It was viscid like a thin oil and a bucket of it taken up as a sample had an oily appearance. It felt very slimy . . . and ran off the hand and fingers in thin gelatinous-like streams. . . . After the yellow color stage the water turned brown slowly and then 'red', the color which in the popular mind was associated with the mortality of fishes. This red water . . . appeared to be an opaque, dull, dark amber with a greenish-yellow cast. At other times it seemed to have a faint lavender tint. After the dark brown or red stage the water reverted to an opaque green or greenish blue color, characteristically the color of waters with heavy plankton growth."

In fresh waters, the most common phytoplankters forming such enormous blooms are the blue-green algae, blooms of which commonly appear in midsummer. Such a bloom has been described by Prescott (1939, p. 74) as follows: "Phytoplankton species which enter into 'waterbloom' associations thrive very successfully in warm water (25° to 30° C), particularly if there is an abundance of carbon dioxide in reserve and an adequate supply of nitrogen compounds. . . . The greatest pests are Myxophycean species, especially those species which multiply rapidly, float high in the water . . . , and which possess sticky mucilaginous sheaths or colonial envelopes. . . . In Lake East Okoboji, in Iowa, these plants periodically form dense, 'soupy' masses of vegetation and thick, floating scums which decay rapidly in the intense light and heat of mid-day. Spirit Lake, nearby, also has dense growths of blue-green algae during the summer period, but here the flora is made up of species which do not float high in the water . . . and which do not possess sticky sheaths. . . . Consequently this lake is spared the climax conditions which are very disturbing in Lake East Okoboji."

Sawyer *et al.* (1943, 1944), whose results have in part been summarized by Lackey (1949), studied certain lakes in the Madison area of Wisconsin. These lakes were subject to greater or lesser fertilization and pollution from sewage and industrial effluents and from agricultural drainage. The lakes with a higher degree of fertilization and pollution were subject to unusually frequent, nonseasonal blooms of blue-green algae and other plankters, and the blooms often were very obnoxious to the nearby residents. These authors found that the blooms were most frequent in the lakes receiving high concentrations of inorganic nitrogen and phosphorus. Organic forms of these elements were not found to be of great significance in causing blooms. The ratio between inorganic nitrogen and phosphorus was found to be of importance. In culture experiments, for instance, blooms would occur if both nitrogen and phosphorus were high, but blooms of many blue-green algae occurred even when nitrogen was low, provided phosphorus was high. This the authors explained as being due to the supposed ability of many blue-green algae to "fix" atmospheric nitrogen.

Sawyer *et al.* (*op. cit.*) found that the inorganic nitrogen and phosphorus decreased in surface waters during the course of a bloom, and that simultaneously the organic forms of these elements increased—a point which is not without interest in this connection.

Excessive phytoplankton blooms may have one of a number of different biological effects, the most conspicuous of which is extensive animal mortality. Such mortality may be due to suffocation or to the direct or indirect production of poisons of one sort or another. Suffocation may be caused, for example, during the hours of darkness when photosynthesis is not taking place, and when the metabolism of the large number of phytoplankters is using up the available oxygen, or it may be caused by the death and decay of large quantities of the phytoplankters. The process of decay, of course will use up the oxygen. Prescott (1939, fig. 4) publishes a convincing photograph of fish mortality in a lake, caused by the decay of masses of dead planktonic blue-green algae. The same author (1948) gives an extensive review of the biological and other effects of objectionable algae.

Animal mortality may also occur through poisoning caused by the phytoplankton bloom. Decay of enormous numbers of phytoplankters after their death in the crowded conditions may first produce a depletion of the oxygen supply; continuation of decay under anaerobic conditions would thereupon result in the production of toxic hydrogen sulphide gas, with death resulting among those animals which otherwise might have been able to withstand the temporary lack of oxygen.

However, in other cases there may be an actual direct production of poisons by the phytoplankters themselves, as often appears to be the case during blooms of dinoflagellates. From various parts of the world have come reports of "red tides," and similar phenomena, produced by blooms of dinoflagellates. Often these "red tides" have been accompanied by the mortality of large numbers of fish and other marine animals. An extensive "red sea" affecting Japanese waters, described by Miyajima (1934), resulted in the death of enormous numbers of pearl oysters. In 1946-47 the "red tide" (actually yellow-green) along the west coast of Florida resulted in very heavy mortality. Gunter *et al.* (1947) estimated that 50,000,000 fish had been killed, and their estimate is conservative (it was made in January, and mortality continued until June). In addition to the fish mortality, an unestimated number of bottom-living invertebrates were killed, resulting in great biological and economic loss. The nature of the poison that produced such serious results is unknown, as is the manner in which it was produced (i.e., whether as an active secretion by the cells, or as a result of the autolysis or decay of many of the billions of cells that were crowded into small areas).

In many cases where excessive blooms of phytoplankters cause no animal mortality, they may produce an obnoxious sliming of the water, or unpleasant odors (blue-green algae are especially likely to produce unpleasant odors). These odors are especially objectionable when blooms occur in reservoirs, or in other bodies of water used by cities or towns as a source of drinking water. Usually, however, such conditions are only obnoxious and not toxic to such terrestrial animals as might drink the water. There have been authenticated cases, however, of toxicity in such waters, as reported, for example by Fitch *et al.* (1934), Prescott (1948) and T. Olson (1951, 1952). Olson surveyed all known authenticated cases of (fresh-water) algal poisoning of domesticated and wild mammals and birds. He showed that not only ingestion of the algae themselves, but also the drinking of filtered water in which they had been growing is very toxic to such animals as mice. Furthermore, the liver of animals which had died of algal poisoning was very toxic to dogs, etc. He was unable to isolate a specific toxin, but was able to show that it was a dializable, water-soluble substance (or group of substances), which is heat-stable and able to resist ordinary domestic water purification methods.

On the other hand, extensive phytoplankton blooms are not always deleterious in their effects. The increased quantity of phytoplankters present may act as basic food for an increased quantity of animal life, and ultimately may result in an increase of the fish population, or of the population of commercially important invertebrates, thus benefiting man himself. In some sections of China, for example, it is the habit of the populace to fertilize fresh-water ponds with human excrement. The increase of nutrient salts from this source results in an increase of the growth of the phytoplankton, which produces an increase in the production of small animals suitable as food for fish. Because of the increased quantity of basic fish food, the fish population increases and thus benefits the diet of the humans.

Even where a phytoplankton bloom has marked deleterious effects, it may indirectly result in an increase of productivity of the water, though the harmful effects may outweigh the beneficial. For instance Gunter *et al.* (1948) describe a marked increase in the zooplankton following the disastrous bloom of dinoflagellates during the "red tide" off the west coast of Florida. The increased zooplankton ultimately would affect favorably the oncoming generation of such fish and other animals as survived the disastrous mortality that accompanied the "red tide." As a matter of fact, it is probable that the majority of phytoplankton blooms, even those that are excessive, are beneficial rather than harmful. Harmful blooms are produced only under certain special circumstances or in cases where the bloom is caused by certain specially harmful organisms, as, for example, some species of dinoflagellates and of blue-green algae.

Causes of phytoplankton blooms.—It is very difficult to determine precisely all the causes of phytoplankton blooms, as has been indicated above in the discussion of seasonal variations of phytoplankton abundance. However, in cases of general blooms of phytoplankters it seems obvious that the most frequent cause lies in an increase of available plant nutrients. Such an increase may come about in one of a number of ways. Marshall (1933, p. 122) states that in several cases the phytoplankton maxima she observed in the lagoon of the Great Barrier Reef appeared after periods of strong winds. The most extensive maximum appeared during the season when the trade winds were blowing. Her thought was that the winds stirred up the bottom deposits, liberating accumulated nutrients into the water, and in this way paving the way for increased phytoplankton production. Similar factors influencing a bloom of *Microcystis aeruginosa* in Lake Mendota have been suggested by Wohlschlag and Hasler (1951). In contrast to Marshall's observations, Chandler (1940) and Verduin (1951) have shown that turbidity in western Lake Erie, caused largely by the disturbance of bottom deposits through the agency of wave action, has a deleterious effect on phytoplankton production. Increased turbidity, they think, cuts down on the light available for photosynthesis, and thus decreases phytoplankton productivity. Verduin estimated that the light intensity would be reduced to one per cent of the surface values at a depth of 2.7 m. when the turbidity is 25 p. p. m., whereas the value would be 4.8 m. if the turbidity was 12 p. p. m. He pointed out that, because of the great turbulence of the water in western Lake Erie, diatoms are carried up and down throughout the entire water column, and that in the clearer water they would spend 6 hours out of the 24 in water illuminated by more than 1 per cent of the surface illumination, whereas in the more turbid water they would spend only 3 hours out of the 24 under those conditions. In western Lake Erie the diatom crop was 12 times as great in 1949 as it was in 1950. Verduin attributes the crop failure in 1950 to the much greater extent of turbid water in 1950.

An increase of nutrients sometimes also comes about by an upwelling of water from the aphotic zone, where lack of photosynthesis has allowed an accumulation over the course of time. Such upwellings may occur in conjunction with high offshore winds, which drive surface water away from the shore and thus force deeper water up to the surface. Or the upwelling may occur as a result of diverging currents. Sverdrup and Allen (1939) have shown that variations in the upwelling of deep water off the coast of California bring about great variations in the phytoplankton population. They say (pp. 143–144): "Without exception, few diatoms were found within water masses which had existed as surface water for a long time and large populations of diatoms were found only in water masses which had spent relatively a short time as surface waters." That the high nutrient salt content of the recently upwelled water is not the only factor favoring diatom growth was shown, however, by the fact that "water which had recently been drawn to the surface or was of mixed type did not always contain many diatoms. Other factors must . . . exist . . . in areas where the character of the waters appears favorable for large numbers of diatoms. These factors are probably of combined biological and chemical nature . . . diatoms may not have had time to develop or . . . [they may have been] removed by grazers after or prior to a depletion of necessary nutrient salts . . . "

It has been suggested (Gunter *et al.*, 1948) that seismic disturbances occasionally may lead to the enrichment of the water, either by causing upwelling of deep water or by stirring up the bottom muds. There have been no published data, however, which support this hypothesis.

Heavy rains and floods frequently carry rich supplies of nutrients from the land into the water, and thus pave the way for a blooming of the phytoplankton. Many of the more notable blooms of phytoplankton described in the literature have occurred after a period of heavy rains. The phytoplankton bloom off the coast of Texas in 1935 (E. J. Lund, 1935) was preceded by heavy rains.

A number of investigations have ascertained that the artificial use of fertilizers in ponds can increase the quantity of phytoplankton, and that this can favorably influence the productivity of fish in the pond (Swingle and Smith, 1939; Meehean, 1934, among others). That the simple addition of fertilizers without adequate precautions does not lead inevitably to satisfactory results is indicated by the findings of Ball (1948) and Ball and Tanner (1951). In these studies two lakes in Michigan were treated with an inorganic fertilizer containing 10 per cent nitrogen, 6 per cent available phosphoric acid and 4 per cent potassium. The plankton content of the water increased appreciably after each addition of fertilizer, but during the second summer heavy mats of filamentous algae caused unpleasant odors and probably destroyed the normal breeding areas of some of the fish. During the second winter both of the fertilized lakes underwent an almost complete winter kill, whereas the two nearby control lakes had no winter kill.

Brunel *et al.* (1950) have given an extensive bibliography on the subject of the culturing of algae, and have summarized much of the work that has been done in this field. They show not only the importance of the usual inorganic fertilizers for the growth of algae, but also the importance of such more or less obscure growth stimulants as occur in soil extracts, etc.

The problem of single-species blooms.—Often a phytoplankton bloom consists of a great growth of a single species, rather than a heterogeneous bloom of a number of different species together. In addition to the general problem discussed above as to the ultimate stimulating cause of the bloom, the problem of the exclusion of other species enters in. Examples of such blooms have been described above (see quotation from Allen, 1946, and Gunter *et al.*, 1948, pp. 76 and 77, above). Unpublished data in the possession of the author show that during the "red tide" on the west coast of Florida, described by Gunter *et al.* *(op. cit.)*, *Gymnodinium brevis* amounted to 99.28 per cent and 98.99 per cent of the total organisms in the water in two unconcentrated samples of sea water. The results are shown in Table 3.

TABLE 3

Composition of Organisms in Two Samples of Sea Water, Taken on the West Coast of Florida

Organism	Dock at Ft. Myers Beach, Jan. 18, 1947		Off Venice, July 23, 1947	
	Number per Liter	Per Cent of Total	Number per Liter	Per Cent of Total
Gymnodinium brevis	13,900,000	99.28	60,324,000	98.99
Other Phytoplankters	102,000	0.72	616,000	1.01

There seems to have been very little attempt to investigate the causes of this phenomenon of single-species blooms. Superficially, it would appear that physical and chemical environmental conditions were favorable for several to many species simultaneously at the time of the bloom, yet often only one species blooms. Sometimes more than one species will bloom in this manner, but often in such an instance the main species involved are closely related, or else one species blooms to such an extent that the minor blooms of other species are masked (see quotation from Allen, 1946, p. 76, above) Two explanations, which are not mutually exclusive, have been hinted at. Allen (*op. cit.*) suggests the possibility that where diatoms are abundant the dinoflagellates are scarce and vice versa, through what he calls diatom-dinoflagellate mutual exclusion. Presumably there would be a chemical deterrent. The deterrent produced by the diatoms would hinder the growth of the dinoflagellates and/or favor the growths of diatoms, and vice versa.

Harvey (1933) has pointed out that in culturing diatoms, the growth of *Nitzschia* was more rapid in those cultures where the medium was made up with sea water in which the same species previously had been growing than in those where ordinary sea water was used to make up the medium. This would indicate a chemical substance produced by *Nitzschia* which favored the growth of the same species, but which perhaps would be detrimental to the growth of other species. Rice (1954) reported that *Nitzschia* and *Chlorella*, growing in cultures, produced antagonistic substances that inhibited their own further growth. However, the antagonistic substances produced by each inhibited the growth of the other more than it inhibited its own species. For instance *Chlorella*, when grown on agar, created an area on the agar unsuitable for the growth of *Nitzschia* cells. Rice further showed that both *Nitzschia* and *Chlorella* were inhibited in culture media prepared by using water from a pond in which *Pandorina* had been abundant.

Lefevre *et al.* (1949) found that certain algae, such as *Cosmarium impressulum* and *Selenastrum minutum* are compatible, and can be grown together in the same culture, whereas many other forms, such as *Chlorella pyrenoidosa* and *Pediastrum clathratum* cannot be grown together, because the *Pediastrum* species is blocked and destroyed by the *Chlorella*. This they attributed to the presence of thermolabile antibiotic substances which inhibit growth and multiplication in algae.

Thus, single-species blooms may be caused by the self-stimulation of the individual species involved. An initial advantage obtained in one way or another by a certain species would be the stimulus to further production, which would eventually result in a peak where there would be practically 100 per cent of a single species (in extreme cases). The existence of such chemical stimulators and inhibitors should be easy to demonstrate by the use of pure cultures of phytoplankters, but to date little seems to have been published on this.

V. Phytoplankton-Zooplankton Interrelationships

WE HAVE assumed, and rightly, that the zooplankton is dependent upon the phytoplankton (see fig. 21, p. 56), for it is a general fact that holozoic organisms depend ultimately upon holophytic organisms, with only very minor exceptions. In aquatic biology the relations between zooplankton and phytoplankton, and their relation to other aquatic organisms, is often designated graphically by a triangle (see fig. 27) in which the phytoplankton occupies the base of the triangle, the zooplankton occupies a somewhat lesser area immediately above the phytoplankton, and the benthic and nektonic animals occupy the apex (as indicated previously, however, littoral and lake zooplankters may subsist at least in part upon detritus that has been derived from decaying benthic plants, or which has been washed into the water from the land).

Some investigators, notably Pütter (1909), have claimed that many aquatic animals do not depend upon particulate food to the extent that is usually believed, but that they are able to absorb dissolved organic materials from the water and use these as their primary source of food. Pütter's methods and results were subjected to severe criticism, resulting in a thorough re-examination of the problem by Krogh (1931). Krogh pointed out that dissolved organic materials occur in fairly large quantities in fresh waters (10 mg./L. or more), but that they are in the main waste products and other unassimilable substances. Experimental results obtained by Křížnecký (1925) for tadpoles, mussels, etc., and by Peters (1921) for Protozoa show that under laboratory conditions certain animals are *capable* of utilizing dissolved nutrients, but the described experimental conditions are seldom or never met with in nature, and Krogh (*op. cit.*) concluded that "there is no convincing evidence that any animal takes up dissolved organic material from natural water in any significant amount."

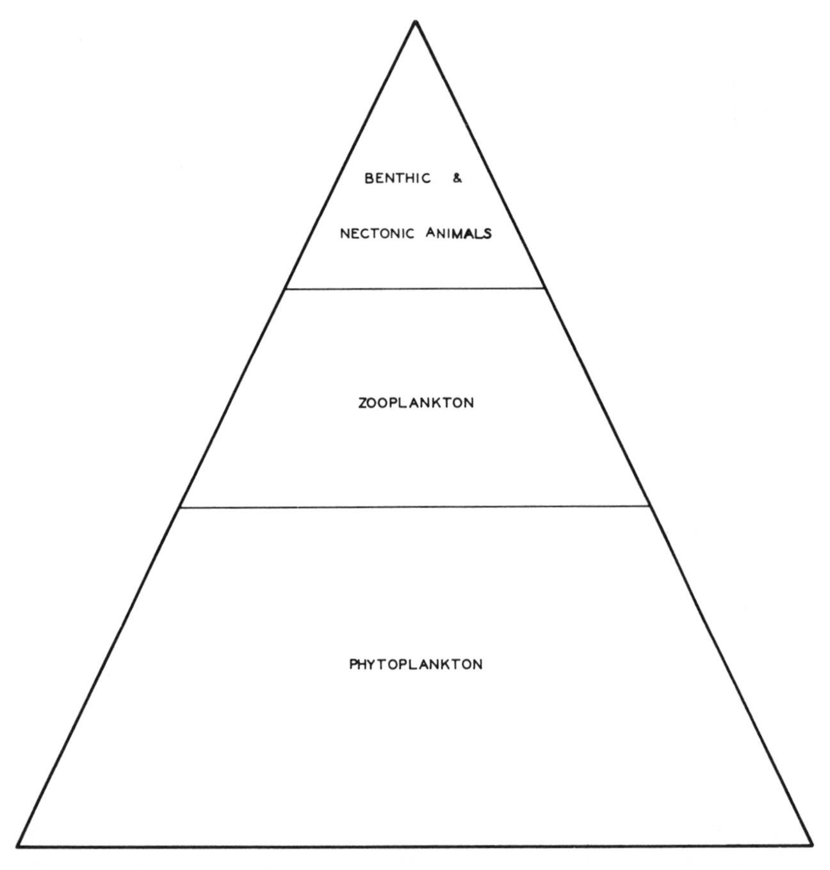

Fig. 27. Triangle to show aquatic production relationships. Adapted from Birge and Juday.

In general, it is usually stated that the abundance of the zooplankton is dependent upon the abundance of the phytoplankton, but that zooplankton production is slower, hence the peak of zooplankton production lags behind the peak of phytoplankton production (see fig. 28), as shown by Riley *et al.* (1949) for Georges Bank. However, when we come to make a close examination of the relative abundance of phytoplankters and zooplankters in the ocean, we do not always see a clear case. Bigelow (1926, p. 39), for instance, states of the plankton of the Gulf of Maine: "The most striking event in the seasonal cycle of the zooplankton of the Gulf of Maine (if a negative one) is that a very decided decrease, amounting on occasion almost to complete disappearance of the pelagic fauna, takes place early in spring over the whole area of the gulf, coincident with the tremendous vernal flowering of diatoms . . . , an event the precise date of which varies locally and from year to year . . . with all its members sharing in the impoverishment, the rare as well as the common, the less abundant forms practically disappear and the scanty catches become extremely monotonous."

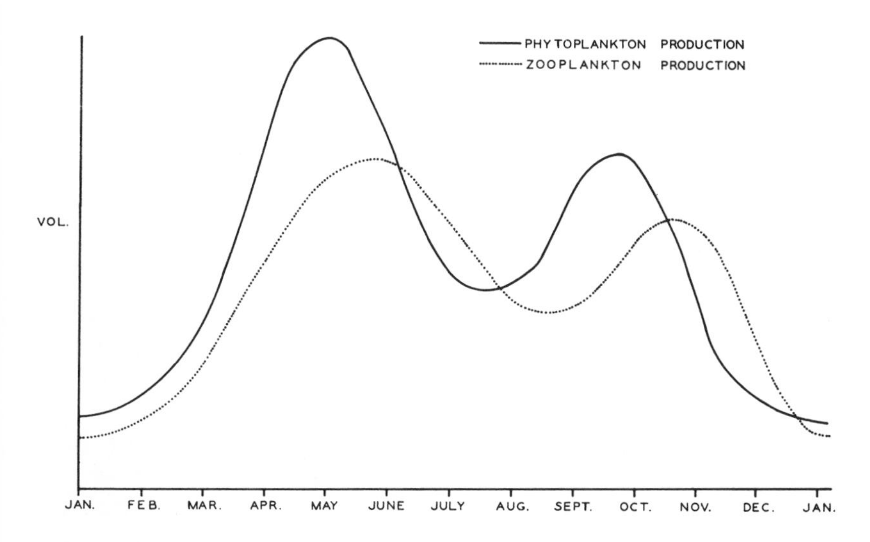

Fig. 28. Idealized diagram to show the lag of zooplankton production behind phytoplankton.

Hardy and Gunther (1935) found a similar phenomenon in their studies of Antarctic plankton. The zooplankton was most abundant in areas in which the diatom population was low. In fact, they found that the zooplankton, and the whales that fed upon it, were most abundant in water of relatively high phosphate content, showing that the water had been poor in plant growth for some time. Harvey *et al.* (1935) found in the English Channel that the phytoplankton was consistently abundant where the zooplankton was poor, and vice versa. Other observers have reported similar cases from various parts of the world. On the other hand, Steeman Nielsen (1937) has shown that not infrequently both the zooplankton and the phytoplankton occur in large quantities together. Nevertheless, general results seem to indicate a frequent inverse relationship.

It must be remembered that most of the discussions of phytoplankton-zooplankton interrelationships are based upon facts discovered by the study of collections obtained by the use of plankton nets. Thus the nannoplankters have been neglected, although the nannoplankton production may be of great importance to a solution of the problem. There is great need of a thorough study of the seasonal and local distribution of zooplankton in relation to the distribution of the *total* phytoplankton.

There have been three main attempts to explain the apparent mutual exclusion of the phytoplankton and the zooplankton. The three theories are not necessarily in opposition to each other, and perhaps all three are partly valid. These are the theory of grazing, the theory of phytoplankton interference, and the theory of differential growth rate. They have been discussed by Sverdrup, Johnson and Fleming (1946, p. 898-904) and by Clarke (1939).

The theory of grazing was proposed by Harvey (Harvey *et al.,* 1935). According to this theory, in those places where the zooplankton population is greater to begin with (perhaps because of swarming in the breeding season, etc.), the consumption of the phytoplankters takes place at such a rate that they are not able to build up large numbers. In those places where the initial population of the zooplankton is lower, the phytoplankters have a chance to multiply rapidly, and this results in a great local phytoplankton production. From the measurements of the rate of filtering by copepods and schizopods, it is possible to calculate that these zooplankters conceivably could make great inroads on a mass of phytoplankton and literally "eat holes in it" if they were present in sufficiently large numbers.

The theory of animal exclusion was proposed by Hardy and Gunther (1935). These authors thought that those zooplankters which, during the course of their diurnal vertical migrations, found themselves below a dense mass of phytoplankters would find it impossible to migrate up to the surface again. It was presumed that the factor preventing the animals from migrating into the area of dense plant growth is a chemical one. However, the possibility also exists that it is a physical one, where the animals simply find it difficult to make headway in the desired direction because of the close proximity of the phytoplankters to each other. Regardless of how the animals are excluded, however, according to Hardy's theory the currents in the lower levels where the animals are living will be different from those at the surface where the plants abound. Therefore the zooplankters eventually will be found far from the mass of plants, and then they can migrate to the surface without hindrance. In this way a discontinuous local distribution would be established. Bainbridge (1949) has obtained laboratory evidence tending to contradict Hardy and Gunther's theory. Bainbridge found that *Neomysis vulgaris* vigorously moved from diatom-poor into diatom-rich regions. The movement was most marked when using starved experimental animals.

The theory of differential growth rate was proposed by Steeman Nielsen (1937). The reproductive cycle of the phytoplankters is relatively very short compared with that of the zooplankters, so that the plants are able to increase their numbers rapidly and produce a bloom. The animals browse upon the phytoplankters, but their reproductive cycle takes so long that it is a long time after the phytoplankton bloom before the results will show among the zooplankters. By this time the currents may have carried the animals far from their point of origin, and perhaps far from the mass of phytoplankters at whose expense they grew, or else the bloom of phytoplankters may have passed. Thus there would develop a discontinuous distribution between the animals and the plants. At times, as Steeman Nielsen pointed out, the currents would not separate the two groups, nor would the phytoplankton bloom pass, and both animals and plants would then be very abundant in a single sample of the plankton.

The above discussion is meant to apply only to conditions in the sea. In fresh waters, conditions are somewhat more simple because such bodies of water, being of relatively small size, do not have extensive current systems. The abundance of zooplankters in fresh-water lakes depends much more directly upon the abundance of the phytoplankters, though their slower rate of reproduction results in a lag of the zooplankton behind the phytoplankton. Krogh and Berg (1931) showed that in Lake Frederiksborg, Denmark, there was a close correlation in the spring and autumn between the maximum production of carbohydrates by the phytoplankton and the maximum production of Cladocera. Hanuška (1949), in Czechoslovakian lakes, has shown that the quantity of the zooplankton depends on the concentration of the nannoplankton, including the bacteria. On the other hand, the zooplankton of fresh-water lakes probably depends to a greater extent upon detritus from decaying aquatic plants than is the case for the marine zooplankton.

VI. Distribution Problems

Cosmopolitanism.—The wide distribution of many plankton species is one of the most striking features encountered in the study of planktology. Fresh-water species are especially noteworthy here, though many marine forms also may exist over a very wide range. In contrast with conditions in terrestrial environments, where with few exceptions (of which man is the most noteworthy) particular animal species have a relatively local distribution, many plankters have literally a world-wide distribution. This condition, known as cosmopolitanism, applies to many species of blue-green algae, green algae, mastigophorans, rotifers, cladocerans, and, to a lesser extent, copepods. Often a single species in a genus is cosmopolitan, while in others the cosmopolitanism applies to two or three very closely-related species, which together are world-wide in distribution.

A good example of cosmopolitanism may be seen in the common fresh-water dinoflagellate *Ceratium hirundinella,* which occurs from the Arctic (where its active existence in the water is limited to a few short summer months) through the temperate region to the tropics (where it is active for the entire year). It is found on all the continents in suitable lakes and ponds.

The rotifers are notably cosmopolitan also. H. S. Jennings, writing of the rotifers in Ward and Whipple (1918, pp. 578-579), states:

> The studies thus far made of the rotifers of different regions seem to indicate that in general these animals may be said to be potentially cosmopolitan, any given species occurring wherever the conditions necessary for its existence occur. Whether any given rotifer shall be found in a given body of water depends mainly, not upon the locality of this body of water, but upon the precise conditions there found. Studies on the rotifers of Europe, Asia, Africa, America and Australia show, not different faunas in these regions, but the same common rotifers found everywhere, with merely a new form here and there, and it is an extraordinary fact that when a new rotifer is described from Africa or Australia, its next occurrence is often recorded from Europe or America. In stagnant swamps all over the world appear to be found the characteristic rotifers of stagnant water; in clear lake water are to be found the characteristic limnetic rotifers. . . . Variation in the rotifer fauna of different countries is probably due mainly to differences in the conditions of existence in the waters of these countries. . . . Two bodies of water half a mile apart, presenting entirely different conditions, are likely to vary more in their rotifer fauna than two bodies of water 5000 miles apart that present similar conditions. . . . The problem of the distribution of the Rotifera is then mainly a problem of the conditions of existence rather than of the means of distribution.*

It must not be thought that cosmopolitanism where it occurs among fresh-water organisms is always a perfect cosmopolitanism. Sometimes an animal or plant species that is found in all climates and on all continents is completely lacking over a considerable area of one of the continents. In some cases the apparent lack of perfect cosmopolitanism may be due to imperfect knowledge on our part, but in other cases it seems to be that the organisms actually are missing from certain regions, even though suitable water seems to be available.

* Reprinted by permission from *Freshwater Biology*, by H. B. Ward and G. C. Whipple, published by John Wiley and Sons, Inc., 1918.

Also, of course, many species are not at all cosmopolitan. That is, the species have only local distribution. This is the case, for example, with the fresh-water species (and the other species as well) of the interesting copepod genus *Pseudodiaptomus.* The genus is completely lacking from European waters, but has been reported from most other parts of the world. Among the fresh-water species, *P. forbesi* is limited to China, *P. poppei* to Celebes, and *P. lobipes* to India.

Cosmopolitanism is more frequent in the open ocean (see Ekman, 1935), where water currents are freer, than it is in coastal waters, but it is in general less marked in the oceans than in fresh water. A few marine species are probably truly cosmopolitan, such as, for example, the bathypelagic mysid *Gnathophausia gigas,* which is found everywhere that the ocean is of sufficient depth. Larger numbers are world-wide in their distribution except for their absence in a certain region. The common copepod, *Paracalanus parvus,* for instance, is very widely distributed in tropical and temperate waters, but it is apparently absent from the Arctic and Antarctic. Throughout its range it is often very common, as it is, for example, in Biscayne Bay, Florida, during the winter months, in Japanese waters (Mori, 1937), off Martha's Vineyard, Massachusetts (Bigelow, 1926), etc.

As in fresh water, however, large numbers of haliplankters are more limited in their distribution. Many are restricted to tropical waters, while others are temperate, boreal, arctic, etc. Currents often carry species far away from their usual environments. For example, the Gulf Stream carries tropical species as far north as Massachusetts in the summer months. Of course such forms are being carried by the currents to their doom.

The number of tropical species.—It has been stated very frequently, along with statements concerning greater productivity in waters of high latitudes (see above, pp. 57-62), that there is a sparsity of species in high latitudes and a plethora of species in low latitudes. Typical of such statements is that of Russell and Yonge (1928, p. 125): "But, while the plankton is present in greatest quantities in these northern waters, it is remarkable that compared with the plankton of the warm and tropical regions the numbers of different kinds of animals are extremely few. The catches made in the northern waters can almost be described as monotonous in composition, that is, although they are so large, they will be made up of only comparatively few species of animals. To the collector then the catches made in warmer regions prove vastly more interesting on account of the wealth of different species to be found there, even though they be present only in small numbers."

Inasmuch as there has been little thoroughgoing work (see Russell, 1934) on the inshore tropical waters, such statements are made largely on the basis of the results obtained by the various oceanic expeditions. Undoubtedly the theory is accurate for the open-sea plankton, but it is not necessarily completely correct for inshore waters. Unpublished data in the author's possession, obtained from off the coast of the state of Washington and off the coast of Florida, respectively, show nothing that would indicate a larger number of species in single tows from tropical waters than in the cold temperate waters. Table 4 shows typical results from analysis of three samples taken from the lightship "Swiftsure" off the mouth of the Strait of Juan de Fuca in 1935, compared with three typical samples from the southern Florida region. Suitable available literature on the

TABLE 4

COMPARISON OF THE NUMBER OF SPECIES OF PLANKTERS FOUND IN CERTAIN SELECTED SAMPLES FROM OFF THE COAST OF THE STATE OF WASHINGTON AND OFF THE COAST OF FLORIDA

Locality	Date	Number of Phyto-plankton Species	Number of Zoo-plankton Species	Total Number of Species
"Swiftsure" Lightship	5-14-35	50	24	74
"Swiftsure" Lightship	9-1-35	47	16	63
"Swiftsure" Lightship	11-5-35	15	11	26
Biscayne Bay, Florida	12-13-47	15	40	55
Gulf of Mexico	12-16-47	20	21	41
Florida Current	7-19-47	13	33	46

phytoplankton of the waters of the state of Washington made it possible to identify correctly more species in Washington than in Florida. Even taking this into consideration, however, it is evident that the northern waters by no means have smaller numbers of phytoplankton species than do the southern waters. On the other hand, the limited results here presented indicate a somewhat larger number of zooplankton species.

Despite all this, however, it has been shown with certainty by Russell (1935) that in most taxonomic groups in the plankton far more species are characteristic of warm waters than of cooler waters. Russell's results have been summarized by Sverdrup, Johnson, and Fleming (1946, Table 99, p. 868). Such a prevalence of warm-water species in the various taxonomic groups, however, does not mean that any one plankton tow from tropical waters will necessarily contain large numbers of species.

Bipolar distribution.—A number of plankters exhibit what is known as bipolar distribution. A species that has bipolar distribution has a discontinuous distribution, part of its population living in the northern hemisphere and part in the southern hemisphere, with an intervening region free of the species. Originally the term was intended to apply only to this kind of discontinuous distribution when the the organisms actually lived in the polar seas, but more recently it has come into use to describe also those cases where the habitats are separated by a narrower band of intervening water—i.e., where the habitats are either subpolar or temperate. Bipolar distribution may apply to an individual species, but at other times the term is used to describe the distribution of closely related species. The planktonic gastropod *Limacina,* for instance is very common and important in the northern North Atlantic, but is completely absent from the tropics. Closely related species of the same genus, however, are abundant in the southern part of the South Atlantic, and in the Antarctic Ocean.

Bipolarity may be a true bipolarity, or only apparent. In the case of apparent bipolarity, the organisms involved actually exist also at lower latitudes, but are confined to deeper waters, where the temperature conditions are similar to those found towards the poles. Only in the latter situation do the organisms live near the surface of the sea, where they are, of course, more conspicuous, and therefore better known. Apparent bipolarity is often known as tropical submergence. A second type of bipolarity that is only apparent is exhibited by animals found in cooler waters to the north and south, and also in the cool waters off the west coast of Central America, even in the tropics (see Ekman, 1935).

Vertical distribution.—Plankters, as might be expected, differ greatly in their vertical distribution. Obviously, the phytoplankters are limited to the upper lighted euphotic and disphotic zones (their distribution within these zones is described above, pp. 65-68). Like the plants, many of the zooplankters avoid the very surface waters, except at night, because of the deleterious effect of strong sunlight, which repels (or even is detrimental to) many of the animal forms. On the other hand, other zooplankters seem not to be harmed by the light, and they remain at the very surface, or near it, regardless of the time of day. Neuston animals such as *Physalia, Velella,* certain Cladocera in fresh water, etc., being directly associated with the surface film, are of this category. Many strictly planktonic animals, however, such as the immature stages of many copepods, the adult stages of other copepods, jellyfish of various species, etc., may remain near the surface. Wilson (1942) speaks of the adult copepod *Sapphirina auronitens* as follows: "The species thus shows a preference for the surface . . . but the attraction cannot be very strong since so many are left in the deeper tows" (pp. 206-207).

Inasmuch as all the phytoplankton exists in the upper zones of the water, all herbivores among the zooplankters must likewise live in these zones, i.e., within the upper 200 meters or so. However, the distinction between herbivores and detritus feeders is not marked, and in many cases the same species will ingest whichever food happens to be available. The presence of richer food supplies in the waters near the surface is clearly shown by the results of Bogorov (1946*b*), who studied the zooplankton of high latitudes in the Arctic Ocean. He found that 83 per cent of the zooplankton organisms lived between the surface and 10 m., while 12 per cent lived between 10 m. and 200 m. and only 5 per cent lived between 200 m. and 750 m.

Most of the specimens living between 200 m. and the surface were younger forms. There was a greater diversity of species living in the lowest level.

Such forms as the copepod species *Microsetella rosea* are confined to the upper zones, while other (usually larger on the average) species such as *Haloptilus plumosus* seem to be nearly or quite confined to deep water. One of the important results of the great Challenger Expedition was the establishment (Haeckel, 1887) of the fact that the Radiolaria have a very distinct and peculiar vertical distribution. Those Radiolaria living near the surface of the sea (*Acanthrometron,* etc.) are characterized by strict spherical symmetry and long, loosely arranged radiating spicules. Haeckel called these forms the "pelagic" Radiolaria. A second group of Radiolaria were found to live as a part of the hypoplankton, just above the bottom. These Haeckel called the "abyssal" Radiolaria. They are characterized by having massive solid skeletons of relatively small size, with coarse trabeculae and small pores. Spherical symmetry is rare or wanting. Between the "pelagic" and the "abyssal" radiolaria live the intermediate forms, which Haeckel called the "zonarial" Radiolaria. In structure as well as habitat, the "zonarial" forms are intermediate between the other two groups described. In his discussion of the "zonarial" fauna, Haeckel, after distinguishing five zones from the surface to the depths, (the pelagic to about 25 fathoms, the pellucid from 25 to 150 fathoms, the obscure from 150 to 2,000 fathoms, the siliceous from about 2,000 to about 3,000 fathoms and the abyssal close to the bottom in deep water) says: "So far as our isolated and incomplete observations of the zonarial Radiolarian fauna extend, it appears that the subclass Porulosa . . . predominates in two upper zones, and as the depth increases is gradually replaced by the subclass Osculosa . . . , so that the latter predominates in the two lowest zones. The obscure zone which lies in the middle is probably the poorest in species. In general, the morphological characters of the zonarial fauna appear to change gradually upwards into the delicate form of the pelagic and downwards into the robust constitution of the abyssal; so also the average size of the individuals (within the limits of the same family) appears to increase upwards and decrease downwards" (p. cliv).

Influence of the currents on distribution.—Especially in the ocean the distribution of plankters is very greatly influenced by the currents. In the first place, as indicated elsewhere, if there were no vertical or horizontal currents in the sea, the plant nutrients in the euphotic and disphotic zones very quickly would be depleted through being utilized by the plants, with subsequent sinking into the aphotic zone. Likewise, new supplies of nutrients brought in by land drainage would be very much localized because of lack of horizontal currents. In such a situation the remaining phytoplankton would be very sparse, hence the zooplankton would likewise be very rare. However, in actuality there is extensive mixing of the water, with deep water being brought constantly to the surface at certain places, surface water sinking to the depths at other places, and with vast systems of horizontal currents carrying the water to localities far distant from its point of origin.

Certain of the currents carry vast quantities of warm water, with its contained zooplankton and phytoplankton, into higher latitudes and into cooler climates, while other currents carry masses of cold water down into lower latitudes where climatic conditions are less rigorous. As the currents move, of course, the temperature of the water does not remain static, but the water cools down or warms up as the case may be. In addition, tropical waters, which have become highly saline because of a high rate of evaporation in the tropics, may become considerably diluted by land drainage, or by heavy rains; or in other ways the salinity may be changed considerably. In the great current systems, therefore, unless a particular species of animal or plant plankton is especially tolerant (as some of the cosmopolitan species are), conditions will finally be obtained that are unsuitable for the continuation of the life of the organisms involved. The currents will have carried them to their destruction. This deleterious effect of the currents in scattering plankters into unfavorable environments was first pointed out by Damas (1905). Before the final destruction of the plankters involved, however, a preliminary effect of the unfavorable conditions is often the inability of the plankter to breed. For example, the large copepod *Rhincalanus nasutus* is a tropical form, but adults have been found in the Gulf of Maine and even in the Norwegian Sea, carried there by the agency of the Gulf Stream.

However, this species is completely unable to breed in such localities, even though the adult may live there for a time. Immature stages are never found—only the adult. In such situations it is obvious that the organism itself, which is relatively short-lived, will soon die without being able to reproduce itself, and hence without being able to restock the water with its species.

There are also a number of cases known where plankters that have been carried somewhat out of their normal environment by the currents find it possible to spawn, but where the larval stages find living conditions intolerable. In such a case, one frequently finds mature and submature adults, along with eggs and sometimes certain of the youngest larval stages, but without any of the intermediate larval stages. In still other cases, certain plankters annually are able to extend their successful breeding range through being carried by the current into higher or lower latitudes in the summer or winter respectively. But when the season changes to colder or hotter, as the case may be, all individuals, both mature and immature, are killed off, and must be replaced the following year by a new current-carried invasion. Such plankters exist and breed in these temporarily suitable habitats. They may become abundant, and even extremely important in the general food cycle (e.g., as food for small fish, etc.), yet they are not at all permanent residents, and they or their offspring eventually are destined to be destroyed by unfavorable environmental conditions. An example of such a plankter is the cold-water copepod *Calanus hyperboreus*, which enters the Gulf of Maine (Bigelow, 1926) in winter and appears to breed there, but which disappears from the gulf during the warm months.

The dispersal of planktonic larvae.—We have seen that many marine benthic species depend upon current systems for the wide distribution of their numbers during their planktonic larval stages. During their planktonic life, because of their exposed and unprotected existence as plankters, enormous numbers of the larvae are eaten by predators. Thorson (1946, 1950), studying the Øresund, near København, Denmark, has shown that benthic invertebrates with a long planktonic larval stage, which depend in their larval life upon other plankters for food (i.e., planktotrophic meroplankters with a long pelagic life), fluctuate widely in the abundance of benthic adults from year to year. On the other hand, he found that similar forms with no planktonic larval stage, or those with a short planktonic larval stage, were much more stable in numbers from one year to the next. This is to be interpreted as the result of varying amounts of plankton food, and varying numbers of predators. Myriads of meroplankters are carried by currents to unfavorable situations and perish. Johnson (1939) has studied the dispersal of a form that is especially suitable in that the larvae are readily recognizable, and that they are set free from the parents in a very limited area: the larvae of *Emerita* (undoubtedly *E. analoga*) off the coast of southern California. The adults of this sand crab live between the tides on sandy beaches, where they move up and down with the flow and ebb of the tide, maintaining themselves in such a position that they are constantly washed by the food-bearing waves. In this location the eggs are produced and carried by the females until they hatch as zoeae. After passing through five zoeal stages, which Johnson estimates takes about four months, the larvae metamorphose and settle down to grow to adulthood. If at the time of metamorphosis a larva finds itself over deep water far from shore, or over a rocky bottom near shore, the young sand crab will be destroyed.

In his studies of samples of net plankton taken by the "E. W. Scripps" off the coast of California in 1938, Johnson (*op. cit.*) found that the Stage I larvae were, as would be expected, confined to the stations nearest shore, but he found a number of older stages farther from shore (with the exception that in his studies no examples of Stage V were encountered). Fig. 29 shows the results given by Johnson for Cruise IV, taken in August, 1938. In all the cruises studied by Johnson the *Emerita* zoeae (with only a single exception) offshore were to be found in water of inshore origin, often in eddies of such water. It is not conceivable that any appreciable number of the larvae in water so far offshore could ever have been caught in shoreward currents and carried back to the regions suitable for the life of the adult. Thus they were being carried to their doom by the current, which must be the fate of innumerable larvae of all sorts of littoral benthic species. Such is the price that must be paid for the benefits of wide dispersal. Nevertheless, sufficient larvae are able to overcome their hazards and grow to maturity, as is evidenced by the great abundance of littoral benthic organisms. Johnson very appropriately speaks of the *Emerita* zoeae as "many tiny drift bottles" which only have to be recovered to tell the oceanographer about the currents the "drift bottles" have encountered.

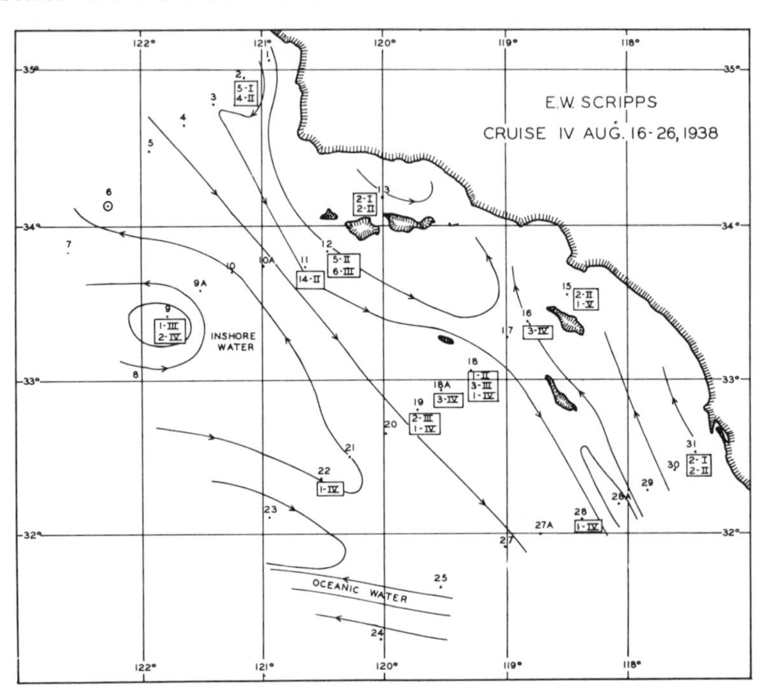

Fig. 29. Distribution of *Emerita* larvae off the coast of Southern California, August 16-26, 1938. The curved directional lines indicate calculated flow. The numbered dots indicate the stations occupied by Cruise IV of the "E. W. Scripps." Station 6 (circled) was not occupied for plankton. Arabic and Roman numerals within the boxes indicate respectively the number of *Emerita* larvae encountered and their zoeal stages. For further discussion see the text. From Johnson, 1939.

In many cases, the prevailing currents are so constituted that they will carry all the pelagic larvae of benthic animals living in particular areas away, never to return. In such cases all restocking of the area in question must come about through the agency of pelagic larvae arriving from elsewhere. In other cases, prevailing currents will determine that certain areas, otherwise perfectly suitable for the existence of certain benthic forms, will lack such species entirely. For instance, if a bank is prevailingly washed by currents from the open sea, there would be little possibility of suitable planktonic larvae reaching it, and most benthic species characterized by having planktonic larvae would be excluded.

However, Fish (1935) has shown that local rotary drifts of the water are formed as vortices. These tend to cause the planktonic larval stages of bottom-living and nektonic animals to be retained over the banks until they are ready to settle to the bottom, or until they are able to swim against the currents. Thus endemic populations of benthic and nektonic animals are maintained on banks that might otherwise in the course of time be swept bare by the currents.

Not only benthic species depend upon the ocean currents, but also many species of the nekton do likewise. We have already studied in another connection the life histories of the eel and of the California sardine (see pp. 43-49 above), which are cases in point. In addition to his work on the life history of the eel, Johannes Schmidt (1909) studied the life history of gadoid fishes along the coasts of Iceland. His results have been described by Sverdrup, Johnson, and Fleming (1946, p. 865). Ocean currents form a clockwise eddy around Iceland. At the breeding season the gadoid fishes concentrate along the west and south coasts of the island, where they produce large numbers of pelagic eggs. The eggs and the resulting larvae are carried by the currents, and metamorphosis takes place when the larvae reach the north and east coasts. Here they take up their existence in the nekton. Thus, the endemic stock on the north and east coasts is maintained from year to year. Steeman Nielsen (1935) and Jespersen (1940) have shown that in the waters of Iceland the greatest production of phytoplankton and zooplankton, respectively, occurs in the very waters where the major portion of the gadoid fishes breed. The planktonic fish larvae necessarily feed upon these plankton organisms.

The maintenance of endemic populations of holoplankters.—We have seen how currents are able to carry plankters away from their normal environments and thus away to their destruction. However, currents are also important in the maintenance of endemic populations of holoplankters. This is true primarily of tidal currents and of the great semipermanent eddies that are to be found in many localities. When tidal currents surge in and out of a semienclosed bay or sound, there will not be a complete flushing of the body of water, and a characteristic endemic fauna and flora often develops. The Scottish lochs are of this nature.

One of the most important of the eddies is in the vast area known as the Sargasso Sea. Here, although the water is by no means static, there are few consistent, permanent currents. The Sargasso Sea does gain water from the great system of currents that surrounds it, and at the same time it gives up water to these currents. Within the confines of the area there is an endemic population of plankters (although this population is noted for its sparsity), and the various currents are, so to speak, seeded with these plankters from the great eddy. Within the currents the plankters may multiply greatly, if conditions therein are suitable. Unless they escape again from the clutches of the currents, however, they may well be carried to conditions that are entirely unsuitable, where they will die. The Gulf of Mexico, although its plankton has been very little studied, probably contains another such eddy, in which an endemic population of holoplankters is maintained and which helps in the seeding of the Florida Current and the Gulf Stream.

In the Gulf of Maine there is a well-known eddy which serves to maintain an endemic plankton population. It has been studied by Redfield (1941), who found that the eddy is a counterclockwise one. During the winter and spring months a considerable amount of cold and barren water enters the Gulf of Maine from the north. Later in the season this source of water is cut off, but while it is entering there is necessarily a loss of water also, which passes out over Georges Bank towards the south (see fig. 30, taken from Redfield). The biomass of the plankton passes during the year from the northern part of the gulf, swings about towards the southern portion in the western part of the gulf, then swings around north again. Not only the biomass, however, but also individual species, such as *Limacina retroversa,* were shown by Redfield to follow the same general path. During the time the water remains in the gulf the biomass increases (except during winter months, when unfavorable conditions decimate it), and all the water that leaves over

Georges Bank is relatively rich in plankton; it is undoubtedly this water that makes Georges Bank such a favorable fishing ground.

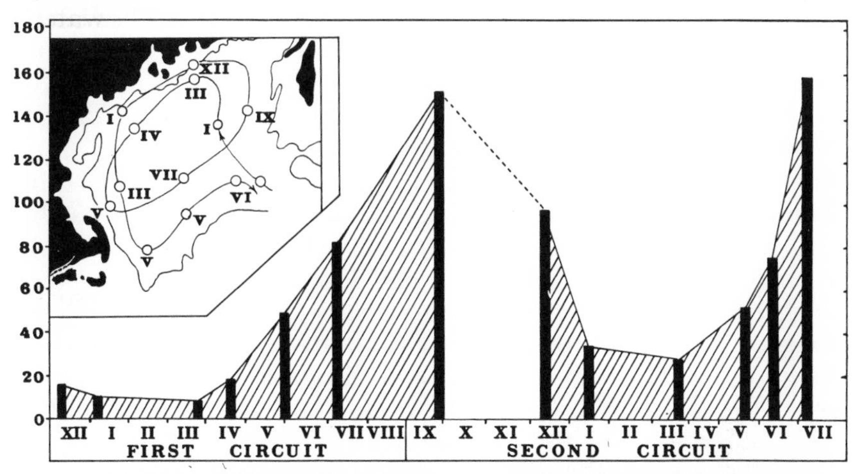

Fig. 30. Changes in the biomass of the plankton in the Gulf of Maine. As shown in the inset, northern water enters the gulf during the winter, then swings in a double counter-clockwise eddy during the following two years, finally leaving over Georges Bank (not shown). The black bars indicate the biomass (volumes) of the plankton taken at the stations and in the months indicated. Note the great increase in plankton volume during the first circuit, its considerable decimation during the winter, and its increase again during the second circuit. From Sverdrup, Johnson, and Fleming, after Redfield. Reproduced by permission of Alfred C. Redfield. Reproduced by permission from *The Oceans* by H. U. Sverdrup, Martin W. Johnson, and Richard H. Fleming. Copyright, 1942, by Prentice-Hall, Inc.

Plankters as indicators of currents.—We have already examined the effect of the currents in carrying plankters far from their normal habitat into regions where they cannot continue life indefinitely because they cannot breed, and where they often are unfavorably affected directly as well. The other side of the same coin is that the origin of a mass of water can be determined in many cases by the nature of the plankton it contains. If, for example, *Rhincalanus nasutus* is found in the water off the coast of Greenland, this means not only that the copepod is being carried to its death in an unfavorable environment but also it is direct evidence that water of tropical origin has reached the shores of this island (see Jespersen, 1939). Likewise the presence of salps or of the copepod *Microsetella rosea* in Biscayne Bay, Florida, would indicate that oceanic water has entered the bay, for these animals are oceanic, and not found as a regular thing in such inshore localities. A great deal more work needs to be done on the normal habitats of plankters in order that we may be better able to detect currents through the examination of the plankton content of the water. It has been said that the plankton can tell us more about the currents than the most refined methods of hydrography, and this probably would be true if we knew enough about the plankters themselves. For example, in the work reported by Johnson (1939) referred to above (p. 99), a single zoea of *Emerita* was found in water which hydrographic analysis showed to be of oceanic origin. However, it seems very likely that this larva was indicative of at least a minimum of admixture of coastal water with the oceanic water at the station where the larva was found. The only other possible explanation would be that the larva migrated from a mass of coastal water into the oceanic when the two masses were contiguous. Neither of these possibilities was detected by the hydrographic methods used, although the latter were of the highest caliber.

Other means of distribution.—The distribution of plankters may take place by other means than by the currents, especially in the case of fresh-water plankters, where currents are not of such great importance. Many plankters, for example, form spores (many phytoplankters and Protozoa) or resting eggs (rotifers, Cladocera, etc.), which are able to withstand extreme conditions of dessication, cold, etc. Often these spores or eggs settle into the bottom muds when their habitat dries up, and they may then be scattered by the wind as part of the dust. If they settle in a suitable locality they will germinate or develop and hatch to produce the adult forms again. Other plankters, either as spores or eggs, or even as adults, may be carried from one place to another by various animals, of which the wading birds are probably the most important. The bottom muds containing the organisms in question, or their eggs or spores, may be carried on the feet and feathers of the birds to new environments. Many fresh-water species are notoriously cosmopolitan in their distribution.

Seasonal distribution of plankton species. — We have already seen that the general picture of the plankton varies from season to season, as well as from place to place. Now let us examine the changes somewhat more closely. Each plankton species has its own seasonal history, which is more or less independent of the seasonal history of the other species. The general story is similar for fresh-water and marine species, and for phytoplankton and zooplankton species.

In any locality, there is a seasonal progression in the plankton, such that first one species is dominant and then another, at rather frequent intervals during the year. The common species in one month are often the rare species of the following month. This is clearly shown by any consistent investigation covering a period of several months, as for example the study by Gran and Braarud (1935) of the phytoplankton in the Bay of Fundy. Something of this is shown strikingly by figure 31, which is constructed from data gathered at Lake Mendota, Wisconsin. Here it is clearly indicated that the spring plankton maximum is due largely to nannoplankters, whereas the autumnal maximum is due to the entirely different species that belong to the microplankton. Table 5 shows results from La Jolla, California, as published by Graham (1943). Here there may be complicating effects due to water currents and patchiness but in large part the apparent succession is an actual succession.

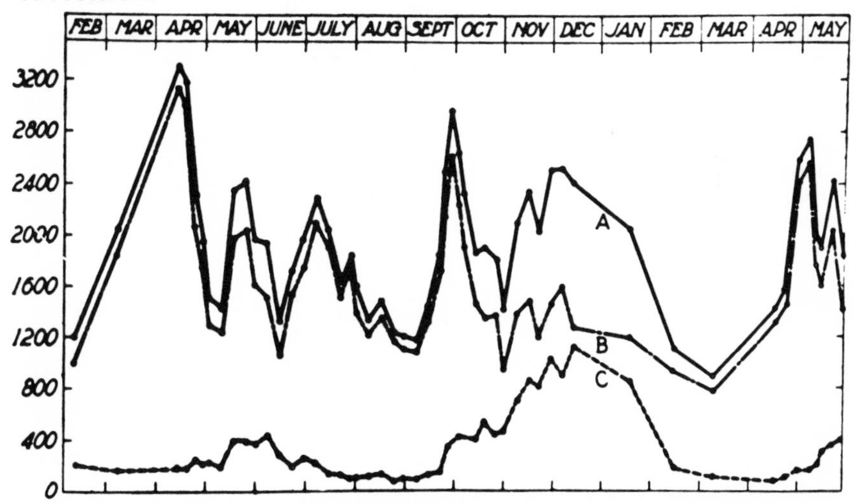

Fig. 31. Seasonal distribution of plankton in Lake Mendota, Wisconsin, in 1916 and 1917. (A) The total plankton. (B) The nannoplankton. (C) The net plankton (microplankton). Quantities are expressed in milligrams of plankton per cubic meter of water. It is obvious that most of the maxima are primarily due to the nannoplankton, but that one is primarily due to the microplankton. From Welch, after Birge and Juday. By permission from *Limnology* by Paul S. Welch. Copyright, 1935. McGraw-Hill Book Company, Inc.

TABLE 5*

COMPARISON OF TOTAL CELL COUNT AND COMPOSITION OF PHYTOPLANKTON AT LAJOLLA, CALIFORNIA

Date	Total Cell Count	Composition of Phytoplankton
1941		
August 14	11,600	50 per cent *Prorocentrum* spp.; 28 per cent *diatoms*
August 15	30,450	45 per cent *Prorocentrum* spp.; 30 per cent *Chaetoceros* spp.
August 21	219,600	95 per cent small *Chaetoceros* spp.
August 22	343,160	96 per cent small *Chaetoceros* spp.
August 26	132,000	91 per cent small *Chaetoceros* spp.
1942		
July 9	573,000	96 per cent *Nitzschia seriata*
July 10	168,000	83 per cent *N. seriata*
July 14	13,000	46 per cent *N. seriata;* 23 per cent *Peridinium* spp.
July 15	25,000	56 per cent *N. seriata;* 16 per cent large diatoms
July 16	18,000	44 per cent *N. seriata;* 44 per cent large diatoms
July 17	11,500	70 per cent *N. seriata;* 13 per cent large diatoms
July 20	2,000	50 per cent *Goniaulax* spp.
July 21	2,500	60 per cent small *Peridinium* spp.
July 22	3,500	43 per cent *N. seriata;* 29 per cent small *Peridinium* spp.
July 23	6,500	46 per cent *Goniaulax* spp.; 21 per cent *Peridinium* spp.
July 24	2,000	100 per cent dinoflagellates (25 per cent *Prorocentrum* spp.)
July 27	7,000	50 per cent *N. seriata;* 25 per cent large diatoms
July 29	18,500	40 per cent large diatoms; 22 per cênt *Chaetoceros* spp.
July 30	10,000	65 per cent large diatoms
August 3	51,000	50 per cent *Coscinodiscus* sp.; 13 per cent other diatoms
August 4	9,500	63 per cent *N. seriata;* 32 per cent large diatoms
August 7	15,000	Mixed
August 11	14,000	Mixed
August 14	25,000	36 per cent large diatoms; 28 per cent *N. seriata;* 24 per cent *Prorocentrum* spp.

A peculiarity of the seasonal variation of the plankters is that in different years the progression of species is not always the same, and sometimes it is even radically different. Table 6, taken from Johnstone, Scott, and Chadwick (1924), indicates the differences that may occur from year to year, using four species of the common diatom genus *Chaetoceros* as examples.

*From Graham (1943), p. 156, Table 1.

TABLE 6*

CHAETOCEROS SPP.—NUMBERS PER AVERAGE MONTHLY CATCH DURING THE PERIOD OF MAXIMAL ABUNDANCE. PORT ERIN, ISLE OF MAN

Year	Decipiens	Debile	Teres	Sociale
1907	104,000	6,844	1,667	667
1908	32,913	316,276	544,500	1,576,500
1909	18,987	77,262	1,457,700	118,857
1910	1,308,900	3,790,200	871,600	2,152,850
1911	1,580,356	11,697,058	2,210,016	6,042,756
1912	2,299,222	5,949,444	2,717,545	8,227,100
1913	575,500	927,167	1,765,925	806,222
1914	821,311	18,972,800	577,867	1,229,250
1915	1,248,889	1,181,778	1,084,222	1,845,611
1916	320,250	323,175	1,008,000	6,548,044
1917	453,750	22,158,075	3,043,750	5,352,889
1918	1,400,000	18,635,733	456,444	98,444
1919	1,321,750	3,023,102	31,807	266,817
1920	1,281,333	923,122	64,833	....

Verduin (1951) described an extreme variation in the species composition of the vernal phytoplankton pulse in western Lake Erie in 1949 and 1950. In 1949 the diatom community was nearly a pure stand of *Tabellaria fenestrata,* whereas in 1950 the much smaller spring bloom consisted of a mixture of *Asterionella* and *Cyclotella,* in about equal quantities, with only a scattering of specimens of *Tabellaria.* On the other hand, often there is a

*From Johnstone, Scott, and Chadwick (1924), p. 87. Reprinted with the kind permission of the University Press of Liverpool.

general repetition of the seasonal progression from year to year. Pearsall (1932), for example, states as a generality for English lakes that the diatoms are abundant in winter and spring, when the concentration of plant nutrients in the water is relatively high; that green algae and desmids occur in largest numbers in summer when nutrients are low; and that the blue-green algae are most abundant when dissolved organic material is high and nutrient salts are low.

Each plankton species has its own seasonal history, and usually there is an annual appearance of one to several maxima (pulses). Such pulses by no means affect all species at the same time (not even all species of phytoplankton at the same time, nor all species of zooplankton). Instead a pulse affects only one or a few species, and not others, which have their own pulse (if any) at some other more or less distant time.

The pulse of a particular species may or may not coincide with one of the general seasonal plankton maxima (see fig. 32).

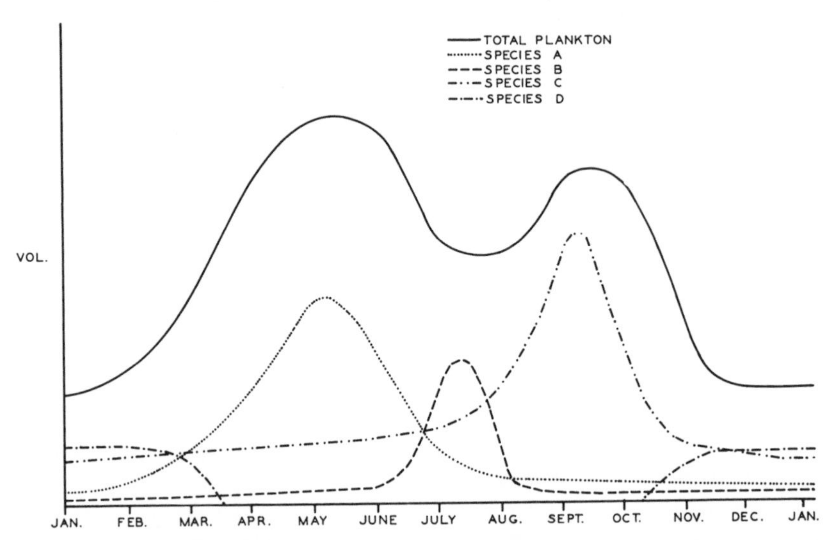

Fig. 32. Diagram of hypothetical cases of pulses of individual species in relation to general plankton maxima.

It is true that the pulses of certain species occur in the spring maximum, but often the same species do not have a pulse during the autumnal maximum. Others show a pulse in the fall and none in the spring. Still others have a pulse in one or another of the general plankton minima, while many have more than one pulse per year, and these may be distributed in such a way that one falls during a general plankton maximum and others during the minima. A few plankters show a uniform seasonal distribution, without any real pulses, but in the case of others the pulses are extreme. Figure 33 shows the seasonal pulses of several plankters from Lake Mendota, Wisconsin. It will be noted, that *Daphnia*, among other species, has practically no pulses; that *Aphanizomenon* has a marked pulse during the winter, a smaller one in spring, and a marked minimum in summer; that *Melosira* has a marked pulse in fall and a smaller one in spring; and that *Stephanodiscus astraea* is apparently absent from the plankton except in the spring.

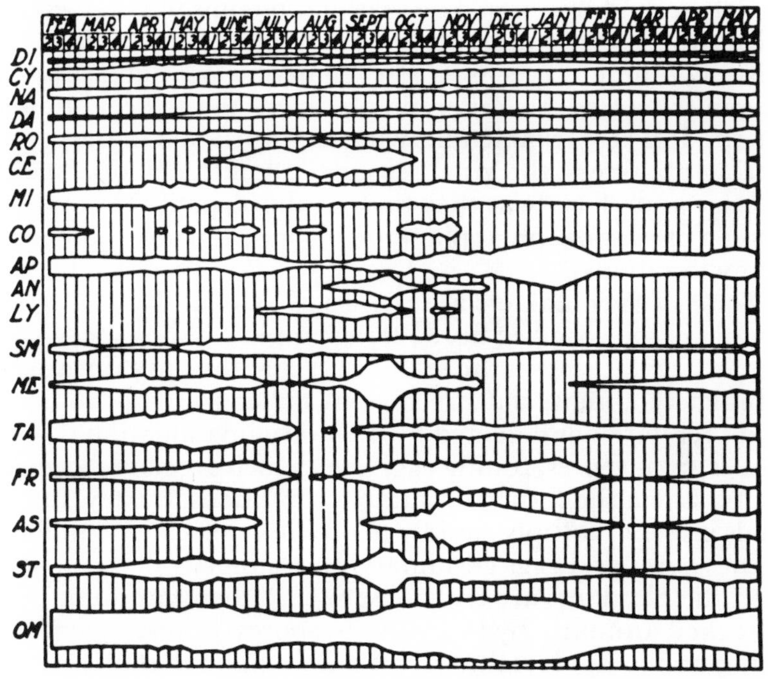

Fig. 33. Quantitative seasonal changes of the abundance of certain plankters in Lake Mendota in 1916 and 1917. Microplankters. Each species has its own seasonal distribution largely independent of other plankters and not necessarily correlated with the general plankton maxima depicted in Fig. 31. (DI) *Diaptomus;* (RO) *Rotifera;* (CA) *Ceratium;* (MI) *Microcystis*; (CO) *Coelosphaerium;* (AP) *Aphanizomenon*; (AN) *Anabaena*; (LY) *Lyngbya*; (SM) *Staurastrum*; (ME) *Melosira*; (TA) *Tabellaria*; (FR) *Fragillaria*; (AS) *Asterionella;* (ST) *Stephanodiscus;* (OM) organic matter. From Welch, after Birge and Juday. By permission from *Limnology* by Paul S. Welch. Copyright, 1935. McGraw-Hill Book Company, Inc.

The pulses may vary greatly from place to place and from year to year. Sometimes the pulse of a particular species will coincide with the pulse of another or of several other species, but in other years the pulses of the species will not coincide. Certain plankters will not show pulses in one location, but in another they will develop one or more typical pulses during the year. Sometimes a pulse develops slowly and dies away slowly (see above, p. 76, for a discussion on phytoplankton blooms—which are nothing more than excessive pulses), whereas at other times, or for different plankters, the pulse arises suddenly and ceases just as suddenly.

The great seasonal general plankton maxima are not due to the bloom of all species, but often of only a few, and sometimes of only a single species. The seasonal blooms in two localities with similar environments are often caused by the bloom of entirely different species, and even in one location the spring bloom is caused by different plankters than the autumnal bloom. It is remarkable that with such variation in the contributing plankters, the great seasonal plankton blooms should appear as uniformly as they do (see above, p. 68-75, for a discussion of seasonal phytoplankton blooms, and their absence in many situations).

Some of the plankters are apparently absent during certain seasons of the year (see fig. 33 for examples). When species are apparently lacking in the seasonal cycles, they usually are not completely absent, but are present either in small numbers, and therefore not detected by ordinary investigational techniques, or as spores, resting eggs, etc., which are not recognized easily. Upon the return of suitable conditions, the few individuals present multiply rapidly to produce detectable numbers, or the spores and eggs germinate. In a minority of cases, however, the plankters are truly absent, and the location must be restocked annually by individuals, spores, or resting eggs arriving from other localities. This is a situation that arises more frequently in the sea than in fresh-water environments, for, as indicated above (p.98), currents may carry plankters annually to environments that are only temporarily suitable for their life and reproduction.

Diurnal vertical migrations.—Many plankters, in both fresh-water and marine situations, make more or less extensive diurnal vertical migrations. Such migrations may be towards the surface in the daytime, and down into the depths (or scattered) at night. More often, however, the migration is the reverse of this, with the plankters deserting the surface in the daytime and returning surfacewards at night. Superficially it would appear that we have here a simple case of phototropism, with the animals attracted or repelled by the light of the sun. Actually, diurnal migrations are not this simple to explain (in fact they are very highly complicated phenomena), though it is universally agreed that light must have some influence on the action, inasmuch as the migrations are in one way or another synchronized with the coming and going of daylight.

Animals of which many species are known to take part in diurnal vertical migrations are the Chaetognatha, to a lesser extent the Rotifera, the Crustacea, and the planktonic larvae of *Corethra* among the Insecta. Of these, undoubtedly the most important are the Crustacea, where numbers of species of Copepoda, Cladocera, Amphipoda, Mysidacea, Euphausiacea, and pelagic Decapoda undertake the migrations (in no taxonomic group of plankters, however, do all the species exhibit diurnal migrations).

Examples of diurnal migration.—

1. The fresh-water larvae of *Corethra* (Diptera) are to be found distributed throughout the water, from the very surface to the bottom, during the night, but by the time dawn has come they have migrated to the very bottom of the water, where the older individuals bury themselves in the bottom muds, while the younger ones remain in the plankton, but close to the bottom (hence in the hypoplankton). In their diurnal vertical migrations the *Corethra* larvae pass through the thermocline with its rapid and sometimes great changes of temperature, and they migrate into the hypolimnion, which is frequently completely devoid of oxygen. Thus they are able to take up a temporary anaerobic existence. Figure 13-E shows an example of a larva of *Corethra,* taken from the surface waters of Lake Apopka, Florida, at night.

2. *Calanus finmarchicus* is the most important marine copepod in the North Atlantic because of its great abundance and large size. It undergoes a considerable diurnal vertical migration, as shown by Nicholls (1933), appearing at the surface in large numbers during the night, but migrating down until it has a midday maximum concentration at around 70m. (see fig. 34). Another common copepod, *Metridia lucens*, has been shown by Clarke (1934*b*) (see fig. 35) to have a similar vertical migration. In Clarke's studies in the Gulf of Maine, however, *Metridia* did not come all the way to the surface at night. A thermocline had developed in the

water and the copepods did not pass through this into the warmer water above.

Fig. 34. Vertical diurnal migrations of adult females of *Calanus finmarchicus.* From Sverdrup, Johnson, and Fleming, after Nicholls. Reproduced by permission of the Council of the Marine Biological Association of the United Kingdom. Reproduced by permission from *The Oceans* by H. U. Sverdrup, Martin W. Johnson, and Richard H. Fleming. Copyright, 1942, by Prentice-Hall, Inc.

Fig. 35. Vertical diurnal migrations of adult females of *Metridia lucens.* The lines indicate light intensity in microwatts. From Sverdrup, Johnson, and Fleming, after Clarke. Reproduced by permission of George L. Clarke. Reproduced by permission from *The Oceans* by H. U. Sverdrup, Martin W. Johnson, and Richard H. Fleming. Copyright, 1942, by Prentice-Hall, Inc.

3. Waterman *et al.* (1939) studied the vertical migrations of some of the larger pelagic Crustacea off the continental slope in the Atlantic Ocean. The results they obtained for the pelagic prawn *Gennadas elegans* were rather typical (see fig. 36). At midnight there was a maximum of the prawns at a depth of about 400 m. whereas in daylight there was a shift so that the concentration was at 800 m. Thus there was a vertical migration of 0.4 km., and the animals would encounter a pressure change amounting to almost 39.5 atm. in such a distance. Furthermore, there would be considerable temperature change.

Speed of migration.—The speed at which the zooplankters migrate up or down is difficult to determine with accuracy, because it is not known whether the animals perform their movements in a direct line or whether they move in a zigzag manner. Assuming that they move directly up and down, Waterman *et al.* (*op. cit.*) estimate that for the larger forms they were considering, the speed of migration varied from 29 to 125 m. per hour. In migrating downward the animals do not simply sink through the water, because this would not result in the extensiveness of the movement. Instead the animals must swim actively downwards. Of course in the

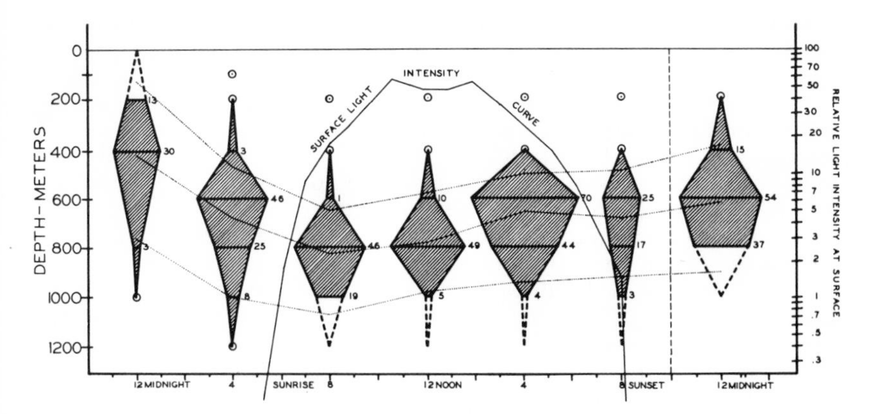

Fig. 36. Diurnal vertical migration of adults of the deep-sea prawn *Gennadas elegans.* The numbers to the right of the horizontal bars represent the actual numbers of animals of this species obtained at the depth indicated. The vertical dashed line towards the right of the figure indicates the omission of 24 hours because of stormy weather. The uppermost and lowermost dotted lines depict the upper and lower boundaries of 80 per cent of the population, while the middle dotted line depicts the center of population. After Waterman, Nunnemacher, Chase, and Clarke.

downward migrations they are aided by gravity, just as in their upward migration they must not only swim the required distances, but must also overcome the effect of gravity, and thus expend more energy and, in effect, swim farther than the observed distance. Welsh (1933) has shown experimentally in the laboratory that the copepod *Centropages* is capable of swimming towards a light at the rate of 82 m. per hour. In nature, *Daphnia longispina* has been shown to have a migration speed amounting to 10 m. per hour. Similarly, Hardy and Bainbridge (1951) have shown that the copepod *Calanus* can swim upward 50 ft. (15.2 m.) per hour, and that the euphausiid *Meganyctiphanes* can climb as much as 305 ft. (93 m.) in an hour.

Effect of age, sex, and environment upon migration.—The age and sex of the plankter has been shown to have an effect upon diurnal migration habits. Males and females may exhibit differing migratory patterns, while both may differ from the pattern shown by juveniles. For example, the female of *Calanus finmarchicus* undergoes much more extensive diurnal vertical migrations than either the males or the immature forms. Such a habit can be favorable to the species in the long run by helping assure more widespread interbreeding. The females migrate downward into water layers where the currents differ from the more superficial layers of the water in which the males and immature forms are staying. Thus the two sets will be separated, and when the females again rise towards the surface they may find themselves in among a different set of males. This type of behavior has helped greatly in the very widespread distribution (even world-wide) of certain plankters, differentiated into not more than environmental "races."

In many other cases, the nauplii and immature forms are to be found in surface waters in the daytime, while the adults of both sexes are in deeper layers. As shown, for example, in figure 37, the young of *Daphnia* in Lake Lucerne, Switzerland, undertake more extensive vertical migrations than do the adults, while in the case of *Bosmina* the adults do not migrate at all, though the immature stages migrate to a considerable extent. In the case of *Diaptomus laciniatus* the young migrate less than the adults. Other cases are also shown in the figure.

Diurnal vertical migrations may vary within one species according to external environmental circumstances. A particular species may show a certain pattern in one locality (for example, in a particular lake), and the same species may show an entirely different pattern of migration in another locality where the physical or chemical, or even biological, conditions are different. In other cases the pattern will differ within one species with the differing seasons. Water color, turbidity, breeding activities, etc., may radically alter the pattern of migration.

Uferflucht.—In passing, the phenomenon known to European investigators as "*Uferflucht*" should be mentioned. This is the condition whereby, in lakes, the limnetic zooplankters have a tendency to avoid regions close to shore, and congregate over the deeper waters. "*Uferflucht*" is thought to be determined, not by any direct lateral migration of the plankters, but by their vertical migrations. That is, a plankter that finds itself in the shoreward region of a lake will migrate to the bottom when daylight arrives. It will then continue to migrate downward along the bottom (thus towards the deeper center of the lake), and when it moves vertically again the following night it will not be so close to shore. Continued night after night and day after day this process will result in the concentration of the plankters over the deeper parts of the water.

Causes of diurnal migrations.—A number of attempts have been made to explain the stimulus that regulates vertical diurnal migrations. The most important of these are as follows:

1. Phototropism. It is obvious that an attempt would be made to explain such migrations on the basis of a simple phototropic response of the zooplankters. The animals would be repelled by the strong light near the surface in the daytime. In fact, no one has ever attempted to deny that light has something to do with the migration—otherwise the migrations, if they took place, would not long remain synchronized with the appearance and disappearance of daylight. The Soviet investigator Bogorov (1946*a*) has shown that in the Barents and White Seas during the summer months, when there is 24 hours of sunlight each day, there is no evidence of a vertical migration of the plankters. Each species maintains a uniform distribution in depth, and this in spite of the fact that in the Barents Sea there is a great deal of tidal turbulence, which requires that the animals exert considerable energy just to maintain their position. On the other hand, in the Barents and White Seas in the autumn there is a regular alternation of day and night, just as in the temperate zones

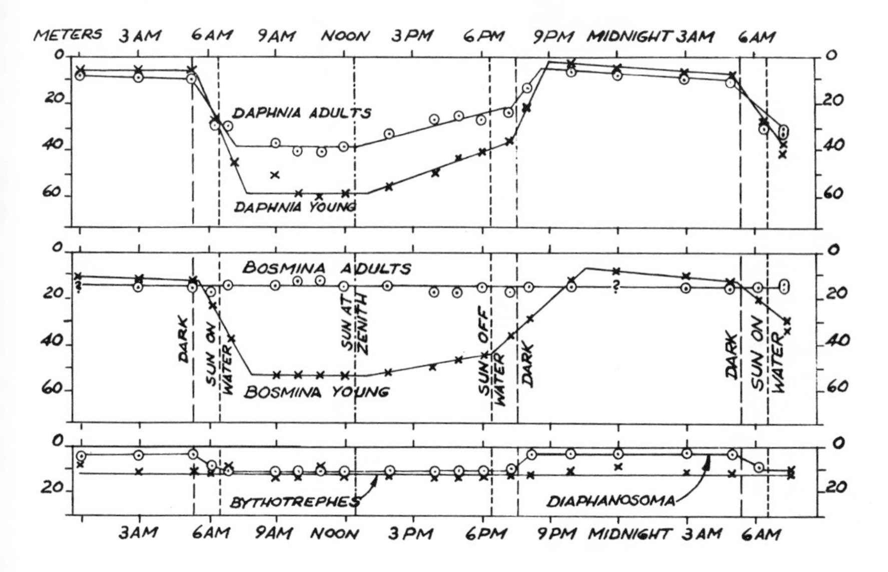

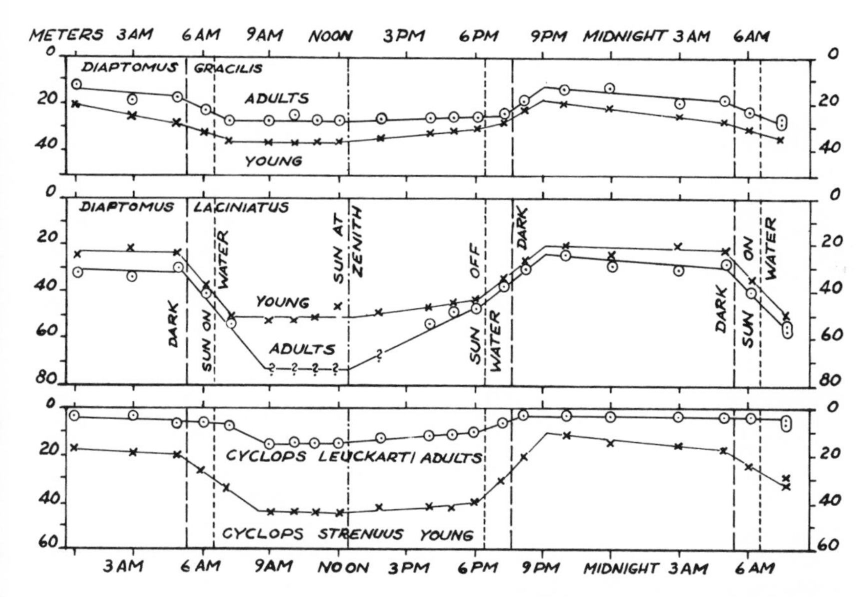

Fig. 37. Variations of the diurnal vertical migration of different age groups of Copepoda and Cladocera in Lake Lucerne, Switzerland. From Welch, after Worthington. By permission redrawn from *Limnology* by Paul S. Welch. Copyright, 1935. McGraw-Hill Book Company, Inc.

where most of the investigations of diurnal migrations of plankters have been undertaken. Bogorov found that in the autumn regular diurnal migrations of the usual type were the rule. He was unable to investigate these areas in the winter because, obviously, of weather difficulties, though such observations would have been of considerable interest.

A simple negative phototropism, however, is not adequate to explain the migrations. Otherwise the animals would migrate farther from the surface in the daytime than many of them do; furthermore, at night when there is a lack of light, they do not in most cases scatter out into all depths as would be expected—*Corethra* larvae, described above (p. 111), are an exception —but instead the bulk of the plankters remain together at essentially the same level.

An attempt has been made to modify the phototropism theory by stating that each species has its own particular optimum conditions of illumination, and it follows these conditions of illumination up and down in the water as the night and day alternate. Unfortunately, it has been impossible to find any support for this theory in nature. As mentioned above, the plankters do not usually scatter out at night, as would be demanded by this theory; more important, many cases are known where the downward migration begins to take place long before the dawn begins to break. The histogram showing the vertical migration of *Calanus finmarchicus,* shown in figure 34, for example, indicates that the downward migration began at 1 A.M.! Waterman *et al.* (1939, p. 271), in the discussion of the results of their study of the migration of certain larger planktonic Crustacea, state: "The most striking point which appears in a comparison of the biological and the light intensity data is that much of the migration apparently occurred when there was no light more intense than starlight . . . .Before sunrise most of the animals had already descended about half of their total distance downward, and several, such as *Boreomysis, Eucopia,* and *Parapasiphaë,* had reached the deepest point in their migration by sunrise. In most cases, furthermore, the greater part of the upward migration must have been accomplished between the time of sunset and midnight."

2. Phototropism-geotropism interaction. An attempt has been made to explain the migrations on the basis of an interaction between a negative phototropism and a negative geotropism, such that the animals react to swim away from the bottom, but are repelled by the light coming from above. In addition to the arguments against theories depending too much upon light as a direct stimulus, it is obvious that many plankters make such extensive migrations each day that they must actively swim downward, rather than merely sink passively. Such active downward swimming activities would not be likely against the gradient of a negative geotropism such as is thought to be acting here. In general the old concepts of rigid tropistic reactions among animals are too greatly oversimplified to be a reflection of reality.

3. Temperature changes. The least satisfactory attempt to explain diurnal vertical migrations is based upon the well-known fact that cyclic

changes of the temperature of the surface waters occur co-ordinated with the appearance and disappearance of the sunlight. The hypothesis states that the plankters seek the cooler deeper waters during daylight hours, and then at night they are free to come to the surface because the water cools when the sun goes down. However, such cyclic temperature changes are relatively slight, especially in deeper waters, and it is largely only the water very near the surface that is affected. Furthermore, many plankters migrate daily through waters that have widely differing temperatures, going from cooler to warmer layers in the upward migration and from warmer to cooler when they go downward. This is especially true in the ocean. *Corethra* larvae, and others both in fresh-water and marine situations, migrate up and down through the thermocline when this is developed, and the temperature differences encountered by one migrating plankter may amount to a number of degrees centigrade daily. In fact, the diurnal migrants in general must be classed as eurythermal animals. Any evidence that has been mustered to support the theory of thermal regulation of diurnal migration must be based upon limited observations in shallow fresh-water environments, or upon observations in the spring and fall when diurnal surface temperature changes are relatively more noticeable because of the general uniform temperature of the water throughout.

4. Hunger. Another attempted explanation is based upon the hunger drive of the zooplankters, coupled with their repulsion from the surface layers by the daylight. According to the theory, the plankters would presumably remain in deeper water, protected from the penetration of light during the daytime, but upon the cessation of daylight, hunger would cause them to swim up to more superficial layers where the food organisms (phytoplankton) are more abundant. Upon becoming replete they would then sink down again. For at least three reasons this cannot be a completely adequate explanation. First, in many parts of the ocean the greatest concentration of the phytoplankton is a considerable distance below the actual surface, yet many of the plankters migrate to the very surface or nearly so during the hours of darkness. Second, the animals do not simply sink downward of their own weight, as would be expected if their migration downwards were taking place before the appearance of daylight, and were due to repletion. Instead they actively swim downwards. Third, in the case of certain copepods, it has been shown that they continue to filter the water for food at a constant rate (see below, p. 128) regardless of the concentration of phytoplankters; and where the phytoplankton is excessively abundant, the excess passes rapidly through the intestine without being utilized. In other words, it appears that the animals do not become replete and stop their seeking after food, even when the food is overabundant.

5. Physiological rhythm. It has been claimed that the zooplankters in question have an innate periodic physiological rhythm, driving them to migrate up and down. It is claimed that the only effect of light is its influence in synchronizing the already-existing rhythm. But as shown

above, Bogorov (1946*a*) found that in northern seas where the summer light falls upon the surface 24 hours a day there is no diurnal rhythm. Furthermore, as Kikuchi shows (after Russell) in his review of diurnal migrations in plankton Crustacea, (1930), *Calanus finmarchicus* will migrate towards or to the surface on dull cloudy days, as well as at night. This does not indicate an innate diurnal rhythm (see fig. 38).

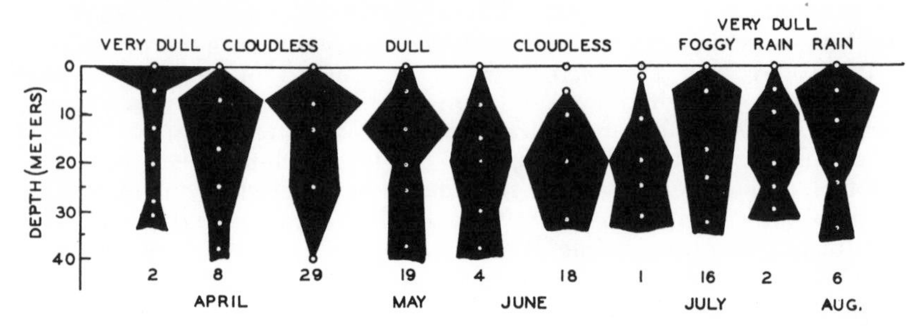

Fig. 38. Vertical position of *Calanus finmarchicus* in relation to the intensity of daylight, in the area of Plymouth, England. After Russell.

In other words, the cause of diurnal vertical migrations is not definitely known, and this remains a very important field for extensive investigations, in both marine and fresh-water environments. From the evidence it would seem probable that several factors interact to cause the migrations. Possibly in one set of environmental circumstances the immediate causative stimulus will be found to be one factor or combination of factors, while in another set of circumstances the cause is something else again.

VII. Geographic and Seasonal Variations of Structure

MANY individual plankton species show distinct variations in size or in structure from one locality to another, or from one season to another. In many cases, with our present knowledge, it is difficult to say with any certainty whether certain variations are hereditary or environmental in character. Many organisms which were originally described as separate species are now known to be local or seasonal variants of one and the same species, and it may well be that many species still considered to be distinct will be found in the future to be synonymous. The only answer to such problems lies in further field observations, supplemented by laboratory culture and experimentation.

Geographic variations.—There have been very few detailed measurements of diatoms occurring at various latitudes, but one of the striking features of the phytoplankton off the coast of Florida is the small size of the diatoms present, compared with examples of the same species found elsewhere. For example, *Skeletonema costatum,* which seems to be cosmopolitan, is much larger in the cold waters of the Puget Sound area of the state of Washington than it is in Florida waters. The same is true of *Chaetoceros debilis.* If this smaller size is universally true of diatoms living in tropical waters, then the smaller size may be considered either to be an adaptation to the lesser density (or lesser viscosity) of the water, thus reducing the tendency of the diatom cells to sink, or it can be considered to be simply the result of the more rapid cell divisions that take place at the higher temperatures. In the latter case the result will be the same with respect to the sinking rate, for smaller forms present a greater surface area relative to their volume, as noted above (p. 32).

On the other hand, Wimpenny (1936) made detailed measurements of diatoms (especially *Rhizosolenia styliformis* and *R. alata*) in English waters and found some evidence that the warmer waters contained larger individuals. He was surprised by these results, and upon investigating the literature he found that in many genera of diatoms the larger species inhabited warm waters, while smaller species inhabited the colder. It is evident that more detailed work needs to be done on this problem in other parts of the world, following Wimpenny's lead.

Similarly, many phytoplankton species that live both in warmer waters and in cooler waters show longer spines in the case of the former (see Böhm, 1931). Even in one locality the spines will be shorter during the cooler season than in the warmer season. Such adaptations are obviously suited to better flotation in the warm water, whereas this increase of flotational ability is not needed in the colder waters.

In fresh-water environments, variations of plankters from one locality to another may be very marked. This is especially true among the rotifers and the Cladocera. Certain species of Cladocera show such extremes of local variation that almost every lake may have its own peculiar race, differing from all others in the size of the crest, length of spines, etc.

These differences are partly hereditary but mostly environmental. Dürken (1932) cites the interesting case of *Daphnia longispina,* which bears only a low crest when it is reared in the cold, but which develops a high crest when it is reared in warm water (as in a hothouse) with plenty of food. The high-crested form, after it is placed in a cold-water situation, will produce one more generation with the high crest, but all subsequent generations revert to the low-crested condition. *Daphnia cucullata* shows similar variation, with the production of low-crested forms in cold water and high-crested forms in warm water (see fig. 39).

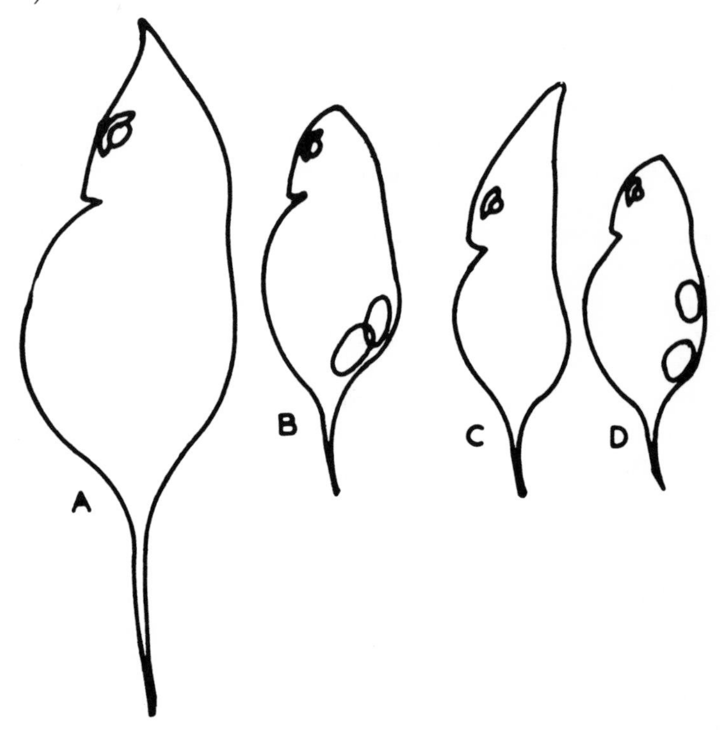

Fig. 39. High-crested Cladocera from warm-water environments and low-crested forms from cold-water environments. (A and B) *Daphnia longispina.* (C and D) *D.cucullata.* After Woltereck.

Seasonal variations.—Many plankters vary in structure with the season. For instance *Keratella cochlearis,* a rotifer, shows a large number of forms from time to time. It varies from larger and smoother in the winter, with a well-developed posterior spine, to smaller and rougher in the summer with a very short spine or none at all. The anterior spines also vary from season to season (see fig. 40). Field studies show that *Ceratium hirundinella* often develops a fourth spine in the summer, whereas individuals have but three in the winter. Huber-Pestalozzi (1923) has shown that cysts of this species grown at 9° C. give rise to a large number of individuals with three spines, whereas those grown at 15-23° C. develop for the most part into individuals with four spines. Species of the Cladocera also vary from season to season. Figure 41 shows the variation of the Cladocera *Daphnia cucullata* and *Bosmina coregoni,* as well as of the rotifer *Asplanchna priodonta,* throughout the seasons.

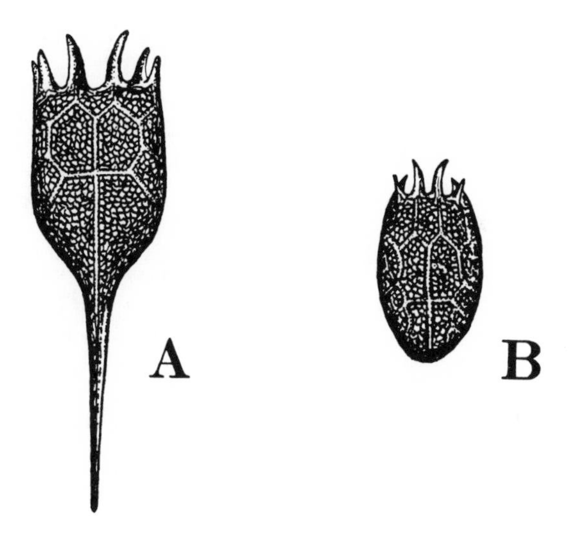

Fig. 40. Seasonal variations of *Keratella cochlearis.* (A) Winter form; (B) Summer form. From Ward and Whipple, after Lauterborn. Reproduced by permission from *Freshwater Biology* by H. B. Ward and G. C. Whipple. John Wiley & Sons, Inc., 1918.

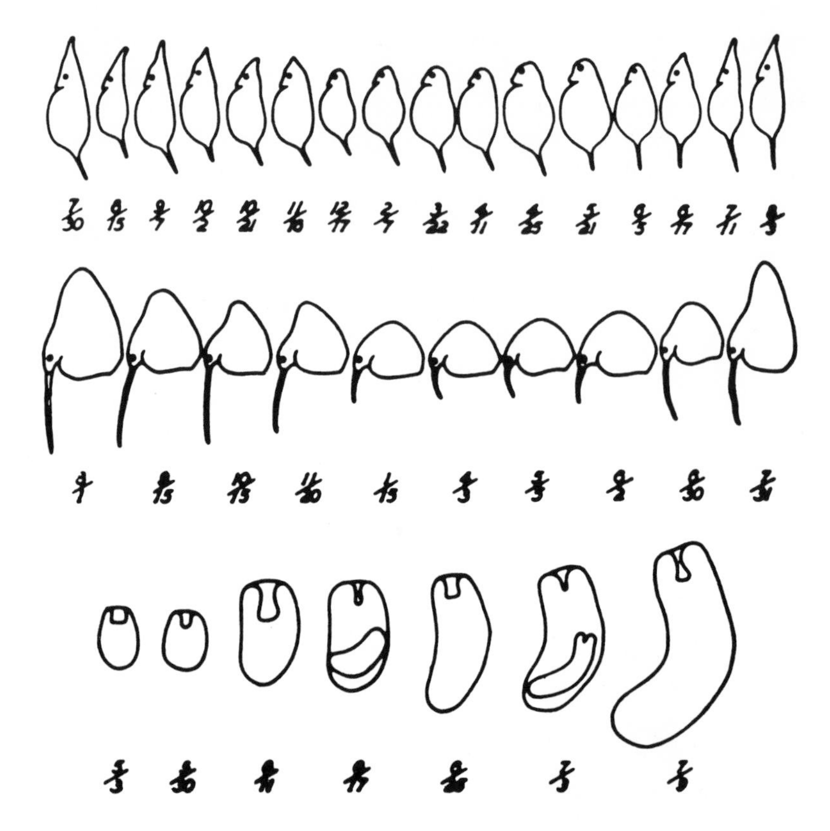

Fig. 41. Seasonal variations of form in *Daphnia cucullata* (uppermost series), *Bosmina coregoni* middle series), and *Asplanchna priodonta* (lowermost series). The numerator of each fraction indicates the month of the year and the denominator indicates the day of the month at which each body form occurred. From Welch, after Wesenberg-Lund. By permission from *Limnology* by Paul S. Welch. Copyright, 1935. McGraw-Hill Book Company, Inc.

Wesenberg-Lund (1926), as summarized by Welch (1935, pp. 244-245), has made the following pertinent observations on seasonal variations of form in fresh-water plankters:

1. Seasonal change of form is best developed among the perennial species; least, if at all developed among those summer forms which disappear during the winter.

2. Seasonal changes of form may take a different aspect in different lakes due to local variations.

3. Transition from the winter to the summer form . . . occurs almost abruptly, often requiring only 2 or 3 weeks; in the autumn, however, the transition from the summer to the winter forms is a gradual one.

4. Regardless of the extent to which local variation may diversify the character of the summer forms of the perennial species in different lakes, winter brings a convergence in all of the species to the same type of body form, which in each species is invariably the same. . . .

5. The change of form appears in the passage of one generation into one or more succeeding generations and not by growth changes within a single generation.

6. Once the summer body shape has been attained, the form is not appreciably changed during the summer season.

7. Seasonal change of body form among the Cladocera and the Rotifera seems to be restricted to the female sex.

8. A few plankters exhibit a seasonal change of body form which appears to be a reversal of the usual sequence, e.g., in the rotifer *Keratella* (*Anuraea*) *cochlearis*. . . .

9. Coincident with appearance of the summer forms of the perennial species, the periodic summer species (not present during winter) begin to appear. . . .

10. It has been claimed that these seasonal forms are not developed in arctic, alpine, and such northern lakes as do not have wide, annual variations of temperature but are restricted to the low-lying lakes of the temperate zone.*

Causes of seasonal variations.—There are at the present time two main theories which attempt to explain the phenomena of seasonal variations of body form. The first of these is that of Wesenberg-Lund (1900) and Ostwald (1902), who attempt to show that the variations are adaptations for flotation. The water in the summer is less dense and less viscous than in the winter, hence such devices as the crest of *Daphnia,* the fourth spine of *Ceratium hirundinella*, etc., are advantageous through increasing the general surface area. Likewise, the smaller size of many summer forms would be a flotational adaptation. The theory is generally known as the buoyancy theory. Thus the rotifer *Keratella cochlearis* is smaller in the summer than in the winter, in agreement with the buoyancy theory. On the other hand, this species has smaller spines, in direct contradiction to

* Reprinted by permission from *Limnology* by Paul S. Welch. Copyright 1935. McGraw-Hill Book Company, Inc.

the theory. In many other cases there is likewise a reversal of the theoretical changes that should take place. An even more serious objection is based upon the fact that the changes of body form often do not coincide with the changes of the temperature of the water. Sometimes the changes come about before the temperature change and sometimes the old body structure remains considerably after the temperature has changed. Furthermore, in many cases in nature, the cold-water races of certain Cladocera have higher crests than warm-water races of the same species.

Other objections that have been raised to the buoyancy theory include the fact that Cladocera living in water that is of greater density because of high salt content have higher crests than those living elsewhere, and the fact that the crest does not necessarily affect the sinking rate of the Cladocera anyway because Cladocera hold themselves in a vertical position in the water rather than horizontal.

A second theory was set forth by Woltereck (1928), who thought it was not the temperature that caused the changes, but the changes in the food of the animals during the seasons. It is true that the kind of food present might be influenced by the temperature, among other factors, but the structural changes were not due to the direct effect of temperature.

More recently Brooks (1946) has re-examined the problem of cyclomorphosis, using species of *Daphnia* occurring in Bantam Lake, Connecticut. He combined field studies with laboratory culture experiments and was able to show that in *D. retrocurva* the relative length of the head to the carapace in newly-hatched individuals has a close correlation with the temperature. This was shown both in his field observations and in his culture experiments. However, in the subsequent growth into adulthood his field observations and his culture experiments showed contradictory results. In the field the head and the carapace grew at the same relative rates as long at the temperature was below 19° C., but the head grew faster than the carapace at temperatures higher than this. At such higher temperatures, on the other hand, he found that the rate at which the head growth exceeded the carapace growth was independent of both the temperature and the food supply. In the laboratory, contrary to the above results, the head always grew more slowly relative to the growth of the carapace, regardless of the temperature. Brooks concluded that the embryonic and the post-embryonic stages can vary independently of each other.

It is possible that factors as yet unkown, other than temperature and food, are also of importance in the seasonal variations of size and shape of plankters.

VIII. The Food of Plankters; Plankters as Food

The food of the plankton organisms.—There is a very great deal that is not known concerning the food of the various plankton organisms, and here lies another great field for future investigations. Of course the phytoplankters proper have a holophytic nutrition, and the saproplankton is saprophytic. The former were discussed above (pp. 57-85), and the latter need not particularly concern us here, beyond the fact that saproplankters are present at all depths, and that they subsist largely on organic detritus, such as the dead bodies of other plankters. Many, however, exist by utilizing the dissolved organic material in the water. This is of less importance in the sea than in many fresh-water environments, for in the latter we find by far the higher content of dissolved organic material. The saproplankters have not been studied adequately either in freshwater or in the sea.

Animal meroplankters that hatch from relatively large eggs containing much yolky material often depend upon the yolk supply for the whole of their planktonic existence, hence they consume no food at all from any outside source of supply. Thorson (1946) speaks of such planktonic larvae as "lecithotrophic pelagic larvae," and he has shown that they may exist as meroplankters for as much as 18 days—"probably often just as long as the larvae of the planktotrophic type with a long pelagic life" (p. 438). Many sponges, lamellibranchs, echinoderms, polychaetes, etc., have such lecithotrophic planktonic larvae.

The food of zooplankters other than the lecithotrophic meroplankters can be determined in one of two ways: either by the direct observation of the act of ingestion by the animals, or by an examination of the stomach contents. Superficially the latter method would seem to be more satisfactory, because interference by rapid motions of the mouth parts is not involved; but often, unfortunately, much of the material ingested is so very delicate that it becomes unrecognizable almost as soon as the animal ingests it, so that often the enteric contents are nothing more than an indeterminable mass of granular material.

Filter feeders.—Filter feeding is rather common among zooplankters. Many of them ingest other plankters, organic debris, and even inorganic debris indiscriminately. Others, however, are able to reject, with greater or lesser efficiency, those materials that are unsuitable as food. It will be observed that it is not feasible to draw a sharp line of demarcation between those filter feeders that live upon other plankters and those that feed upon detritus. Often the actual food that is used depends upon what food is available at the particular moment.

Important plankters that utilize a filter-feeding mechanism of one sort or another include certain of the Cladocera, many Copepoda, euphausiids, appendicularians, the salp, and *Doliolum*. Two examples are given in some detail below.

1. Copepods. By no means are all of the copepods filter feeders. On the other hand, many of them, including the important genera *Diaptomus* and *Calanus,* which inhabit respectively fresh water and salt water, feed by this means. In such filter-feeding copepods (see fig. 42) the second pair of maxillae is provided with many long, often feathery setae, and the two appendages are held in such a position as to form a basket-like filter. The second antennae, mandibular palps, and first maxillae, aided by the maxillipeds (posterior to the second maxillae), move at rates as high as 600 times per minute in *Calanus finmarchicus* (Cannon, 1928). The movements set up in the water in this manner, aided by the general swimming movements set up by the use of the first antennae and the swimming legs of the thorax, carry food particles into the trap formed by the second maxillae, where they are filtered out, conveyed to the mouth, and swallowed.

Observations of the stomach contents of filter-feeding copepods show little that is recognizable except fragments of diatom and dinoflagellate shells, but there is usually a mass of greenish or yellowish granular or amorphous material, as described, for example, by Naumann (1923). The indeterminate material might be the remains of small delicate phytoplankters or other objects. Naumann showed that nauplii filter everything larger than 1μ in diameter out of the water, and that they are indiscriminate, as are even the adults of *Diaptomus.* However, he found that other copepods were not indiscriminate.

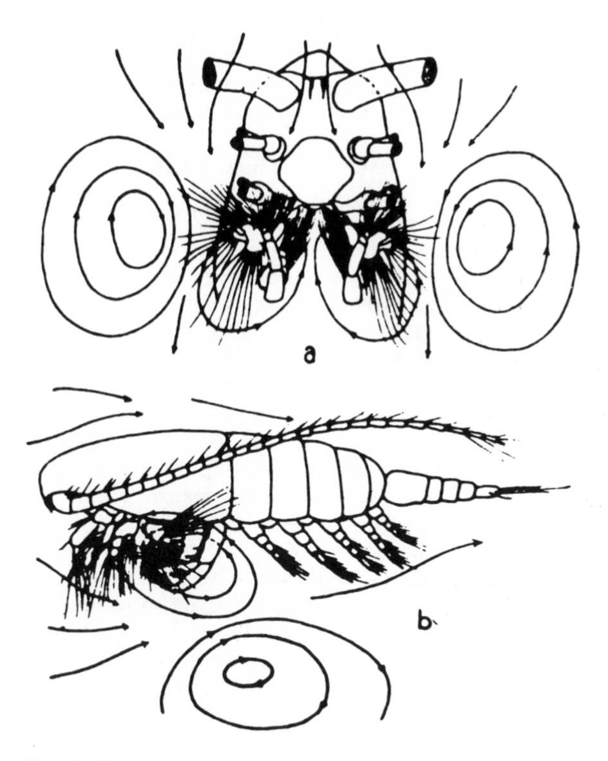

Fig. 42. "The filter-feeding mechanism of *Calanus finmarchicus*: (a) ventral view, anterior portion of the animal, distal parts of 1st and 2nd antennae and mandibles removed; (b) lateral view of entire animal. The lines with arrows indicate the vortices set up." From Sverdrup, Johnson, and Fleming, after Cannon. Reproduced by permission of the Company of Biologists, Limited (Cambridge, England). Reproduced by permission from *The Oceans* by H. U. Sverdrup, Martin W. Johnson, and Richard H. Fleming. Copyright, 1942, by Prentice-Hall, Inc.

Fuller and Clarke (1936) have shown that *Calanus* cannot live on bacteria alone in nature, though preliminary experiments by Clarke and Gellis (1935) had indicated that these might be of some importance. From the experiments reported in these two papers it would appear that the main food of *Calanus* is those myriads of nannoplankters that swarm in the waters at all times, supplemented by microplanktonic diatoms, but apparently not much by the larger diatoms and dinoflagellates. Yet it is precisely these larger forms that are the only ones detectable in the examination of stomach contents. The filtering rate for *Calanus* was calculated by Fuller and Clarke (*op. cit.*), on the basis of the speed of removal of carmine particles from a suspension of known concentration, and was found to be approximately 5 cc. per day. At the same time, they calculated that each *Calanus* at Vineyard Sound (where *Calanus* was very common) would find it necessary to filter 72 cc. per day in order to obtain enough food to maintain itself, judging by the total measured standing crop of plankton present in this locality. Obviously, either the filtration rate or the standing crop was measured by means that were insufficiently accurate, as is admitted by Fuller and Clarke, who speak of their results as "very preliminary."

On the other hand, Harvey (1937) has shown that *Calanus* removes phytoplankters more efficiently the larger they are, and this would explain the discrepancies obtained by Fuller and Clarke. Thus, according to Harvey, a single *Calanus* could filter 4.6 cc. per day when feeding upon *Nitzschia closterium* forma *minutissima,* 48 to 96 cc. per day when feeding on *Lauderia borealis,* and 168 to 240 cc. per day with *Ditylum brightwellii.* Riley *et al.* (1949) found that each milligram of the general zooplankton removed the phytoplankton from between 34 and 52 cc. of seawater per day, a figure somewhat comparable to Harvey's results with *Calanus.* Obviously, more work is needed on such problems, both in the sea and in fresh water.

There is general agreement that the copepods are of extreme importance in the food cycle of the sea. Their role as the chief intermediaries between the phytoplankton and larger animals has been emphasized by many investigators of the plankton and of the fisheries (see Clarke, 1934*a*).

2. Appendicularians. *Oikopleura* among the Appendicularia has a very unusual mechanism for its filter-feeding activities (see fig. 43). The animal lies within a delicate gelatinous "house" that it has secreted. On the upper surface of the "house" there is a grilled sieve, through which sea water enters, propelled by the action of the "tail" of the animal inside.

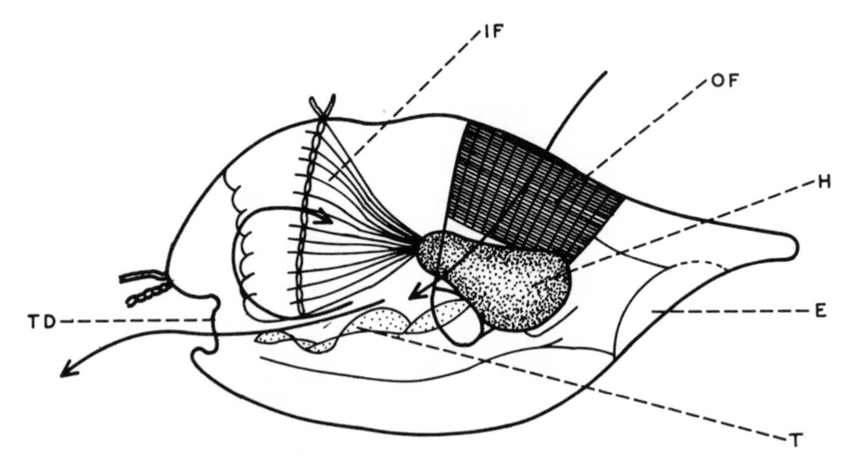

Fig. 43. The filter-feeding apparatus ("house") of the pelagic urochord *Oikopleura.* See text for discussion. (E) exit; (H) "head" or body of the animal; (IF) inner filter; (OF) outer filter; (T) "tail" of animal; (TD) trap door. Redrawn from Sverdrup, Johnson, and Fleming. Reproduced by permission from *The Oceans* by H. U. Sverdrup, Martin W. Johnson, and Richard H. Fleming. Copyright, 1942, by Prentice-Hall, Inc.

The grill serves to prevent any larger particles from entering the "house," though its meshes are sufficiently coarse to allow the nannoplankters to enter. The water is then carried around to a much finer sieve inside the "house," where the smaller plankters are strained out, subsequently to be ingested. When the process of filtration first begins, the water cannot leave the "house," but by the time the pressure has built up somewhat, a trap door at the posterior end suddenly opens and the water rushes out, propelling the animal, "house" and all, forward by jet propulsion. In this way the animal constantly reaches new feeding grounds. Within a few hours the filters of the "house" become clogged, or sometimes damage occurs in one way or another, and the animal then escapes by an "escape hatch" at the anterior end, and secretes a new "house." Living Coccolithophores were first discovered (previously the coccoliths had been known only from geological deposits) in the material collected by the Challenger Expedition, and they were found only in the filtering apparatus of appendicularians, having as nannoplankters escaped capture by the plankton nets used in the expedition.

3. Other filter-feeding mechanisms need not be detailed here. Suffice it to say that the methods vary greatly in the different groups of animals. Some of the euphausiids for example, create a water current by the movement of their abdominal appendages (swimmerets), and the long, biramous thoracic appendages co-operate to form a filtering net or basket. In *Doliolum,* on the other hand, a current of water is set up by means of strong cilia on the cells lining the pharynx. The water enters through the oral opening, passes out the digestive tract through the numerous gill slits in the pharynx, and finally leaves the animal through the atrial aperture. As the water passes through the gill slits, food materials are strained out of it, and they eventually are carried to the esophagus and thus to the remainder of the digestive tract.

Sedimentation feeders.—Sedimentation feeders also are common among plankters. In this case the food itself does not differ from that of the filter feeders—that is, it consists of other plankters or of detritus, as the case may be. In sedimentation feeding, a current is set up in the water, by one means or another, in such a manner that suspended food particles are carried to the animal and then tend to settle out of the region of the mouth or cytostome. Sedimentation feeding is to be found among plankters in many Protozoa (*Paramecium, Stentor,* etc.), in most rotifers, and some Mollusca (pteropods), and is especially prevalent in the sea among the pelagic larval stages of many of the phyla.

The rotifers present a rather typical case. The ciliary crown around the mouth creates currents (see fig. 44) or eddies, which cause the food to settle within the ring of cilia, hence near the mouth. Ingestion ensues. Many rotifers have the power to reject materials that they find unsuitable as food, though others do not have this power.

Predators.—Many plankters are predators, spreading a "net" of tentacles or otherwise seeking actively after their prey, and subduing it after a struggle. The prey may then be chewed up or swallowed whole. Many pelagic Protozoa, the Chaetognatha in general, some rotifers, and coelenterates, the annelid *Tomopteris,* and some copepods are included among the active predators.

1. The chaetognaths are aptly named for their powerful bristle jaws, which they can use to very good effect in the capture of other zooplankters. The prey is swallowed whole, and the frequency with which the chaetognaths are encountered with whole copepods or whole decapod larvae inside the intestine is a good indication of the vast destruction these creatures can inflict upon their fellow plankters.

2. Many copepods, such as *Candacia,* are predators rather than filter feeders. An examination of the intestinal contents of such copepods will show remains of the exoskeletons of other copepods and similar prey. The prey is grasped by the mouth parts (especially the maxillipeds), which are very strong and provided with heavy setae, instead of being more delicate with finer setae as in the filter feeders. The prey is chewed to pieces before ingestion.

Plankters as food of other organisms.—Many benthic and nektonic animals feed directly upon the plankton or upon detritus. As in the plankton itself, there are filter feeders, sedimentation feeders, and predators.

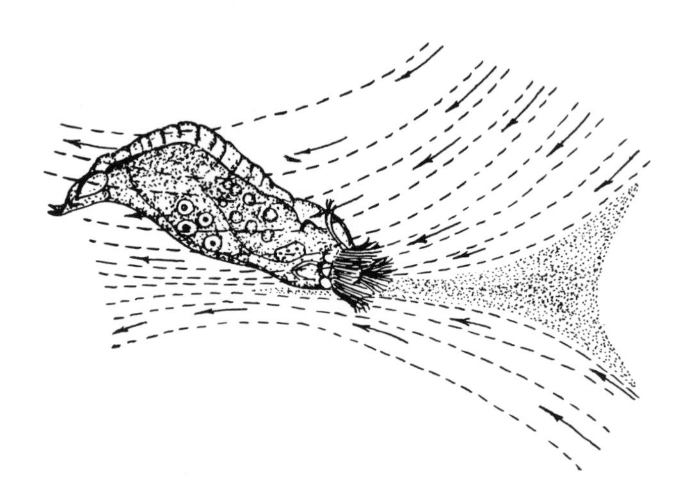

Fig. 44. Sedimentation-feeding activity in a rotifer, *Proales sordida.* The dotted area depicts how food particles from a considerable volume of water are concentrated in the vicinity of the mouth by water currents set up by the activity of the cilia. From Ward and Whipple, after Dixon-Nuttall. Reproduced by permission from *Freshwater Biology* by H. B. Ward and G. C. Whipple. John Wiley & Sons, Inc., 1918.

Filter feeders.—Filter feeders constitute the bulk of the plankton feeders. Here we must include some of the largest known animals, including the whales and the basking sharks, as well as many smaller forms.

Among benthic plankton feeders obtaining their food by this method, the clam (and similar bivalves), the barnacle, and the tunicates will be described. The clam (see fig. 45) brings a current of seawater in through the incurrent siphon by means of the action of the myriads of cilia covering the general body surface and the inner surface of the mantle. This water spreads out over the gills on each side, but the gills are very porous and the water is drawn through the pores into inner passages, whence it is carried to the suprabranchial canal above the gill and finally to the outside through the dorsal siphon. Both oxygen and food are brought to the gills by the water current. The food particles are filtered out, become enmeshed in the mucus covering the gills, and then cilia carry the food and mucus together anteriorly. Finally the materials are carried to the palps, where sorting of the suitable material from the unsuitable takes place, again by means of cilia. The suitable food is then carried into the mouth and ingested.

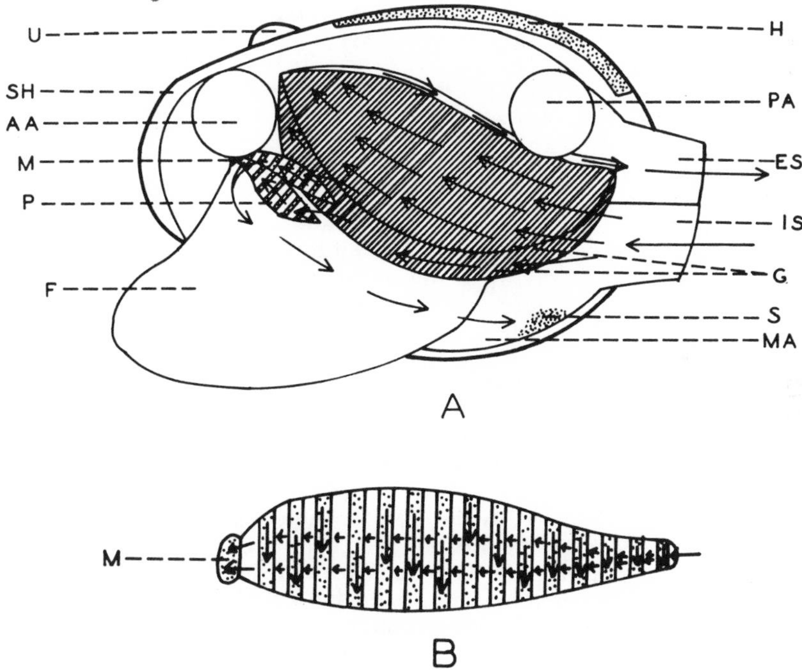

Fig. 45. Diagram to show the feeding of the clam. (A) Diagram of the soft parts of the clam. lying in the right valve. (B) Diagram of a single palpus. (AA) anterior adductor muscle; (ES) excurrent siphon; (F) foot; (G) gills; (H) hinge ligament; (IS) incurrent siphon; (M) mouth; (MA) mantle; (P) palp; (PA) posterior adductor muscle; (S) accumulation of sand; (SH) shell; (U) umbo.

Obviously, plankton-feeding bivalves will find their most suitable livelihood where proper plankters are present in large numbers. Kincaid (1942) has described a case where man has made commercial use of this fact: "The oyster growers of France discovered that the plankton of the open coastal waters did not contain sufficient foods for the massed volume of oyster life developed through an artificial system of culture, and as a result they developed highly specialized methods involving construction of claires or shallow ponds in which they learned to control the production of certain plankton organisms and were thus able to multiply the food supply of the oysters far in excess of the normal limits. It thus became possible to produce a marketable product from that which, without artificial control, would have been of little value . . . " (p. 212) The results of controlled experiments on the fertilization of ponds have been mentioned above (p. 82).

The barnacle is a strictly sessile feeder upon plankton and detritus, and it depends entirely upon whatever food is brought to it by water currents, for it is unable to move about and seek better feeding grounds. While it is immersed in the water, a barnacle will perform continuous and characteristic feeding movements. The carapace is opened and the rami of the several thoracic appendages are extended far out into the water and then quickly pulled back in. The rami are filiform and provided with numerous hairs, and together they form a sort of basket which acts as a filter to remove suitable food particles from the water.

The filter-feeding apparatus of the tunicates is very similar to that of the pelagic urochord *Doliolum*, as might be expected in animals so closely allied. Water is taken in through the mouth and filtered through the large number of gill slits. When the food is filtered out of the water, it is entangled in mucus secreted by the endostyle. Then it is carried by ciliary currents dorsally by means of the circumpharyngeal bands and cilia on the general inner surface of the branchial basket. Finally it is carried posteriorly by the cilia of the hyperpharyngeal band to the esophagus.

Usually it is thought that the only function of gill slits in the Chordata is respiration, but it seems likely that their fundamental and primitive function was filter feeding, and that they adopted a respiratory function only secondarily. Later in evolutionary development, of course, in larger and more active chordates, the respiratory function became relatively more important, and usually the chief function of the gill slits in the fishes is respiratory. Many of the fish, however, continue to obtain their food by a filter-feeding apparatus. The California sardine (see fig. 46), for instance, has a number of fine gill rakers that act as a filtering apparatus. The animal feeds by swimming through the water with its mouth open, and its food consists of whatever plankters are present in the water at the time. Thus it may feed upon either phytoplankton or zooplankton indiscriminately, as shown by Lewis (1929). Larger sardines may become predators to a limited degree.

The whalebone whales are perhaps the most noteworthy of the plankton feeders because they are the largest animals that have ever been known to exist on the earth, yet they feed upon tiny plankters. It is true that the whales do not feed upon the smallest plankton organisms, but seem to like especially the copepods, the euphausids and mysids, and the pteropod *Limacina.* All of these they eat in enormous quantities, obtaining them by passing the water over the baleen or whalebone (see fig. 47), where the frayed edges capture the organisms in large quantities. They seek out the water richest in these plankters, and the whalers, long before anything was known scientifically or otherwise about the plankton, had learned to recognize water suitable for whales by its very color and consistency.

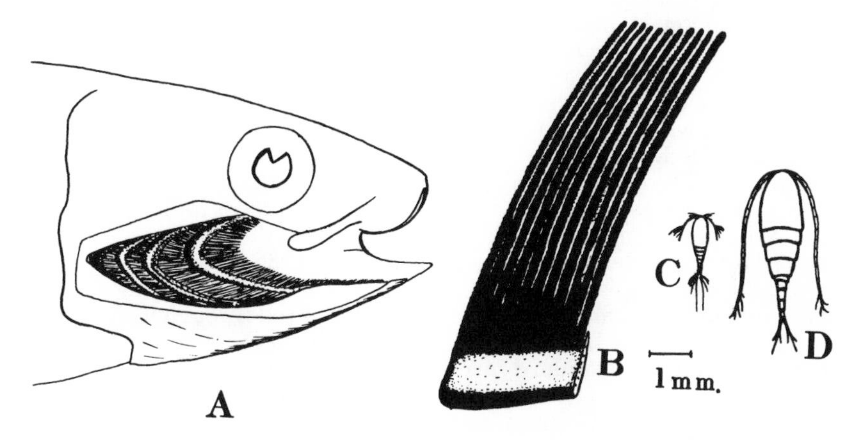

Fig. 46. "The filter-feeding apparatus of the California sardine; (A) gill cover and gills removed to show one side of branchial sieve formed by gill rakers; (B) enlarged drawing of section of branchial sieve; (C) *Oithona plumifera,* a small copepod, drawn to the same scale as (B); (D) *Calanus finmarchicus,* a medium-sized copepod, drawn to the same scale as (B)." From Sverdrup, Johnson and Fleming. Reproduced by permission from *The Oceans* by H. U. Sverdrup, Martin W. Johnson, and Richard U. Fleming. Copyright, 1942, by Prentice-Hall, Inc.

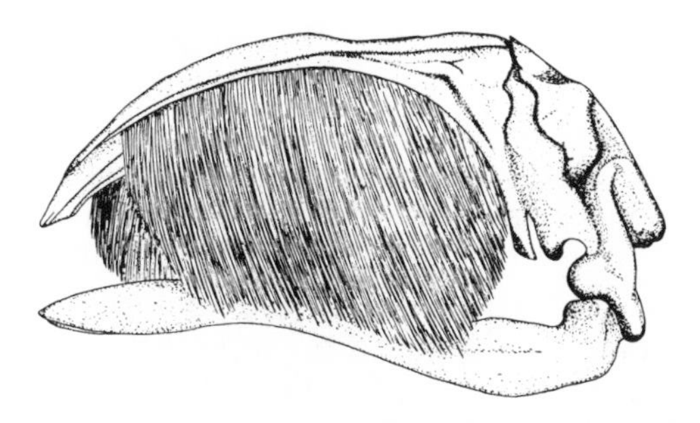

Fig. 47. The filter-feeding apparatus of a whalebone whale. *Balaena cisarctica.* The skull, showing the baleen plates (whalebone) in place. Plankton organisms are enmeshed in the frayed portion of the baleen plates as water is forced past. From the *Standard Natural History,* Vol. 5. Copyright, 1884. D. E. Cassino and Company.

Sedimentation feeders.—Sedimentation feeders are much less important among benthic forms, and nonexistent among nektonic forms. Sedimentation feeding is largely confined to small forms, such as attached rotifers and Protozoa.

Predators.—Predators among nektonic plankton feeders are mostly confined to small or middle-sized fish. The larger specimens of the California sardine, for example, feed upon schizopods by seizing each one individually, though they only supplement their diets in this way, feeding mainly upon the general plankton. Young fish of many commercial and non-commercial varieties feed upon plankters as predators.

The herring is an extremely important commercial fish that is somewhere between a filter feeder and a predator. The herring certainly has a tendency to capture individual copepods and other plankters, but cases are known where as many as 60,000 copepods have been found in the stomach of a single herring, all of which were certainly not captured individually. The herring has coarse gill rakers, which serve as supplementary aids in the capture of food. The herring never feeds indiscriminately upon the plankton, but seeks after areas in which the copepods are most abundant. With this fact in mind, Hardy *et al.* (1936) developed a plankton indicator, consisting of a cylinder in which a disk of bolting cloth is mounted in such a manner that sea water is filtered through it. After the plankton indicator has been dragged behind the boat for a standard length of time at a standard speed, the disk is examined. If it is green, a rich phytoplankton is indicated, and such areas are very poor herring fishing grounds. If the disk is red it indicates copepods, and with only slight magnification the numbers of copepods is easily estimated in a rough fashion. The potential herring catch is larger, the larger the number of copepods captured. Hardy *et al.* (*op. cit.*) report that those fishermen using the indicator consistently brought larger catches to market than those who did not do so, simply because they knew more definitely which localities the herring would be likely to avoid. Hence, they did not waste time by casting their nets in those places where there were no fish.

The use of the plankton by man.—Indirectly the plankton is of very great value to man because of its basic place in the general food cycle of aquatic environments. On the other hand, man has very seldom in the past thought of using the plankton directly for his own purposes. In the Orient, the Japanese and Chinese have long eaten the scyphozoan jellyfish *Rhopilema esculenta* and *R. verrucosa.* These large jellyfish (diameter 45 cm.) are preserved in alum and brine, or between the steamed leaves of one of the native oaks. Then before serving they are soaked for a period, and flavored with spices. Jellyfish, however, are notably deficient in solid material, being largely water, so that their food value is not great. They are eaten more as a pickle or appetizer than as a basic food. On the other hand, the very nutritious mysids are often eaten in the Orient (see Zimmer, 1926–28): *Mesopodopsis orientalis* is widely used in India, and *Neomysis awatschensis* is the basis of an important fishery in Japan. In northern China a "shrimp paste" is made from the plankton (see Clarke, 1949), which is used all over China as a "main accessory food."

Whale-bone whales, which subsist entirely upon plankters, are born about 16 ft. long, and in a matter of only two years they grow to adulthood, with an average length of 75 ft. As soon as these phenomenal facts were ascertained the question inevitably arose whether food that was obviously so nutritious for the whale could not be used directly by man, either for his own food or at least to feed his stock. A number of chemical analyses have been made of the plankton, to determine its actual nutritive value. Johnstone (1908) found that marine copepods had the following proportions of certain important organic materials (expressed here in percentages of the dry weight of the copepods):

Protein	59	per cent
Fat	7	per cent
Carbohydrate	20	per cent
Chitin	4.7	per cent
Ash	9.3	per cent

Birge and Juday (1922) found that certain fresh-water copepods varied considerably in their chemistry. Their figures for various copepods show:

Protein	44 to 61	per cent
Fat	16 to 40	per cent
Ash	4.1 to 5.8	per cent

It is certain that the fat content of copepods (and therefore the proportional amounts of the other substances) varies considerably on a seasonal basis, depending upon the amount and kind of food available, as well as upon the physiological condition of the animals and the period of their life history.

Diatoms have been found to have a much lower percentage of protein than the copepods, and, of course, a much higher percentage of ash, because of their siliceous skeletons. Table 7, from Birge and Juday, shows the chemical constitution of various fresh-water phytoplankters and zooplankters.

Table 7

Chemical Analyses of Certain Fresh-Water Plankters*

(Figures are Given as Percentage of Dry Weight.)

Lake	Date	Organism	N	Crude Protein (N×6.25)	Ether Extract	Pento-sans	Crude Fiber	Nitrogen-Free Extract	Ash	SiO_2
Mendota	9-30-1911	*Microcystis*	8.60	53.75	4.55	....	2.11	32.05	7.54	0.38
Monona	7-11-1914	*Microcystis*	9.27	57.94	2.67	4.97	0.26	34.82	4.31	0.13
Monona	10-17-1917	*Microcystis*	6.32	39.50	2.75	....	0.65	52.09	5.01	....
Waubesa	7-7-1917	*Chiefly Microcystis*	8.35	52.19	5.02	7.80	....	....	7.81	1.62
Mendota	9-19-1914	*Anabaena*	8.27	51.69	1.11	4.81	0.63	39.40	7.17	0.95
Devils	10-8-1913	*Anabaena* and *Coelosphaerium*	8.35	52.19	2.05	6.15	1.17	39.93	4.66	0.27
Mendota	7-11-1915	*Aphanizomenon*	9.30	58.12	3.72	2.04	0.53	30.12	7.51	1.16
Mendota	12-3-1913	*Aphanizomenon* and *Anabaena*	9.94	62.12	4.34	3.42	1.30	25.72	6.52	0.17
Monona	7-20-1915	*Lyngbya*	9.17	57.31	2.36	5.25	3.42	31.24	5.67	0.15
Monona	7-24-1915	*Lyngbya*	8.21	51.31	1.38	3.76	7.39	34.74	5.18	0.20
Culture	5-13-1915	*Ankistrodesmus*†	8.32	52.00	8.78	2.41	9.67	29.55	....	....
Mendota	8-28-1915	Diatoms	3.66	22.87	13.60	2.87	1.43	22.60	39.50	30.78
Monona	7-6-1916	*Volvox*	7.61	47.56	5.54	1.00	6.32	34.30	6.28	0.24
Fowler	9-28-1918	*Diaptomus*	10.38	64.87	8.01	....	8.58	12.60	5.94	....

Lake	Date	Organism	N	Crude Protein (N×6.25)	Ether Extract	Pento-sans	Crude Fiber	Nitrogen-Free Extract	Ash	SiO_2
Mendota......	12-10-1918	*Cyclops*	9.57	59.81	19.80	....	10.07	4.58	5.74	....
Mendota......	5-1-1912	*Diaptomus* and *Cyclops*	9.27	57.93	16.74	....	5.92	13.59	5.82	0.36
Mendota....	5-8-1918	Same as above	9.87	61.69	17.68	....	5.58	9.47	5,58	0.01
Green........	8-22-1913	*Limnocalanus*	7.18	44.88	39.90	0.58	3.96	7.16	4.10	0.10
Devils........	7-2-1913	*Daphnia pulex*	7.45	46.56	3.90	1.32	9.02	14.67	25.85	0.73
Devils........	10-1-1915	*D. pulex*	8.27	51.69	2.82	1.92	8.51	12.25	24.73	1.16
Devils........	5-15-1916	*D. pulex*	6.55	40.94	4.60	....	7.25	24.04	23.17	2.84
Devils........	5-25-1917	*D. pulex*	5.82	36.38	12.07	....	6.96	25.19	19.40	1.46
Devils........	8-30-1917	*D. pulex*	7.58	47.37	3.10	....	10.89	21.77	16.87	0.14
Monona......	4-4-1914	*D. pulex*	8.63	53.94	21.25	0.80	3.34	13.85	7.62	0.07
Monona......	6-17-1916	*D. pulex*	7.91	49.44	3.45	....	5.58	23.32	18.21	1.95
Monona......	8-5-1913	*D. Pulex* and *D. hyalina*	8.51	53.19	8.42	1.58	8.31	15.44	14.64	0.08
Waubesa......	5-10-1917	*D. pulex*	7.55	47.19	27.90	....	4.53	8.25	12.13	1.02
Mendota......	8-28-1917	*D. hyalina* and *D. retrocurva*	8.35	52.19	4.55	....	8.35	19.02	15.89	0.67
Kawaguesaga..	8-14-1913	*Holopedium*	8.35	52.19	10.65	6.71	5.80	23.72	7.64	0.28
Monona.......	8-8-1912	*Leptodora*	7.69	48.06	25.93	1.03	9.55	8.46	8.00	0.20
Monona.......	8-5-1913	*Leptodora*	9.28	58.00	8.70	0.72	4.60	13.20	15.50	0.22
Mendota......	9-3-1917	*Leptodora*	9.84	61.50	5.88	....	8.80	12.51	11.31	0.23
Mendota......	6-9-1916	*Corethra punctipennis*	10.74	67.12	9.45	....	6.15	9.32	7.96	0.16

*From Birge and Juday (1922), Table 49.
†ash-free.

It will be noted that such zooplankters as the copepods have a chemical constitution not inferior to that of beefsteak and other meats. The phytoplankters have a chemistry that is not inferior to that of many common vegetables. Furthermore, the flavor of the zooplankters is excellent. That of some of the phytoplankters, however, can be very obnoxious, and phytoplankters can even be dangerous (see Clarke, 1949, and Prescott, 1948). Scientists have often eaten samples of the plankton, and occasionally man has eaten plankton of necessity. Kincaid (1942) reports that the Soviet explorers who landed on the North Pole in 1937, and who did so much to advance our knowledge of the far north during their drift southward to the east of Greenland, ate plankton during one of the periods in which their food supplies ran low. During their trip by raft across the Pacific Ocean from Peru to the South Sea Islands, the travelers on the raft Kon-Tiki experimented with eating the plankton, and found it good (Heyerdahl, 1950). On the other hand, Clarke (1949) found that white rats died after being fed for a time on a diet of marine plankton, apparently through not being able to utilize more than a small fraction of the energy contained in the food.

It is a matter of practical interest to inquire whether there is sufficient plankton in the water to provide an economical yield in any proposed commercial operations. Clarke (1939) thinks there is not sufficient to make collecting it by ordinary means worth while. He estimates that there is 0.1 gm. of dry obtainable plankton per cu. m. of sea water in the coastal waters between Cape Cod and Chesapeake Bay, and that 7,500 cu. m. of sea water would have to be filtered to obtain sufficient plankton to provide food for one person for one day. This would mean, he says, drawing a 2-m. net through the water at a speed of two knots for 2½ hours—obviously impractical.

Birge and Juday (1922), on the other hand, find that the Wisconsin lakes they investigated have a much higher yield than this. According to these authors, Lake Waubesa (the richest they examined) would give an annual yield of 2,592 tons of dry plankton per acre. The same lake now yields about 295 pounds per year per acre in the form of commercial and sports fish. They estimate that half of the plankton crop could be removed without injury to the fish population.

The investigations of Juday (1943) also show a high content of protein and fat in the plankton of the lakes he studied. Considering the nannoplankton and the net plankton together, and expressing the results in grams per cubic meter, he found that Lake Mendota contained 0.878 gm. of protein and 0.149 gm. of fat (i.e., 1.027gm of nutritive substance) per cubic meter of water, that Lake Monona contained 1.821 gm. of protein and 0.170 gm. of fat (i. e., 1.991 gm. of nutritive material), and that Lake Waubesa contained 2.401 gm. of protein and 0.228 gm. of fat (i.e., 2.629 gm. of nutritive material). These figures are much higher than Clarke's, and show a feasible amount of material. Undoubtedly, similar quantities of plankton exist at least in certain parts of the sea, perhaps especially in such regions as the Scottish Lochs, Puget Sound, and in the great whaling grounds.

There is no doubt concerning the nutritive value of the plankton, and in many places there seems to be an adequate supply for commercial purposes, but there remains a great practical difficulty in obtaining general plankton in commercial quantities. Drawing a standard plankton net through the water does not seem to be practical for three reasons. In the first place, the largest practical net for such purposes is only about 2 m. in diameter, and therefore strains only a limited quantity of water. In the second place, a power boat is needed to draw the nets through the water, running up the operating expenses to a degree not commensurate with the yield. In the third place, it has been found that the meshes of an ordinary plankton net very quickly become clogged with the plankton, reducing the filtering efficiency to an estimated 1 per cent at the end of 3 minutes.

Hardy (1941) has discussed this matter and proposes to overcome the first two objections by the construction of huge nets with an opening 18 ft. square and a body 100 ft. long. He proposes to set these nets by securing them to an anchored boat or to buoys, in such a manner that they can swing with the tide. Tidal currents are utilized to bring the water through the nets. Obviously, such nets could be set only in particular locations, such as in the tidal currents in the entrance ways of the Scottish Lochs. Hardy estimated that it would take 2 men to tend a fleet of 10 such nets, and that 200 men operating 100 fleets of nets would be able to secure a yield of 26.25 tons dry weight of plankton per day. Hardy, however, would not by his method be able to overcome the third objection—the clogging of the net—and this might mean that his calculations are somewhat optimistic.

Juday (1943) has suggested the obtaining of plankton in commercial quantities by setting up batteries of large centrifuges. However, he cites no experimental evidence that would substantiate the practicability of such a method.

Shropshire (1944) has devised a much more efficient apparatus, consisting of a rotating cylinder (see fig. 48) through which water is borne, either by the tidal currents or by force.

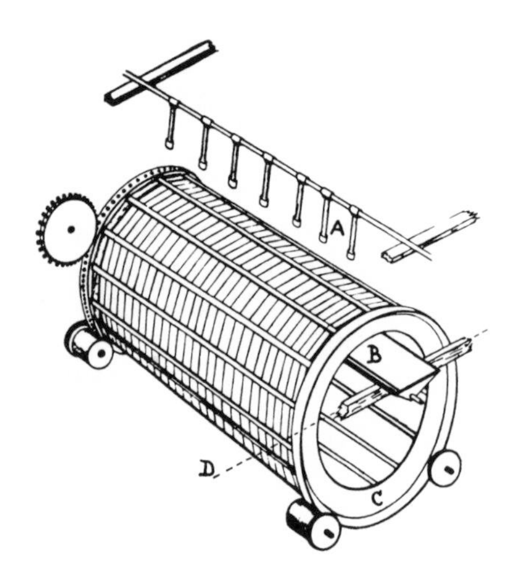

Fig. 48. Shropshire's plankton harvesting machine. (A) spray nozzle; (B) collecting trough; (C) flange to prevent backwash; (D) water line. From Shropshire, 1944.

The walls of the cylinder are covered with bolting cloth, and the water that enters the cylinder at the open end is forced to come out through the bolting cloth, thus filtering out the net plankton. In order to prevent clogging of the net, the upper portion of the cylinder is not immersed in the water, and a series of spray nozzles is directed against it in order to form a backwash and loosen the plankton from the net. A trough within the cylinder serves to collect the plankton. Shropshire used two such cylinders in tandem, the first being 8 ft. in diameter and 16 ft. long, while the second was 4 ft. in diameter and only 15 in. long. The plankton material coming out of the trough of the second cylinder was almost gelatinous, and was taken to a filter press and formed into cake. Using this apparatus with a pump forcing 1,000 gal. per minute through the first cylinder (the estimated capacity of the cylinder was 25,000 gal. per minute), he succeeded in obtaining cakes as indicated in Table 8.

TABLE 8

PLANKTON COLLECTED BY THE SHROPSHIRE PLANKTON HARVESTER*

Flow in Gallons	Weight of wet cake in grams
32,384	1,021
32,384	1,214
48,576	1,279
23,074	2,546
12,144	1,529
16,192	1,474

It does not seem impossible that the plankton eventually will become a useful supplement to the diet of man or to that of his domestic animals. However, much further study must be made of the problems involved. It is possible that a practical commercial plankton fishery could develop in conjunction with an industry such as the one removing magnesium and bromine from sea water. A Shropshire harvester mounted in the incoming stream of water conceivably could filter out large quantities of plankton, which could be pressed into cakes and fed to cattle or hogs, or even used by man himself.

However, the main value of the plankton to man will undoubtedly remain an indirect value for a long time to come, in its role as the primary food for larger animals.

*From Shropshire (1944).

IX. The Kinds of Plankton Organisms

No attempt has been made in the foregoing pages to deal taxonomically with the organisms that occur in the plankton, yet inevitably the plankton student is confronted with the necessity of identifying the many different kinds of plankters that he encounters during his examination of samples. Inasmuch as members of almost every major group of animals and plants occur in the plankton, at least during developmental stages, the task of identification becomes a rather difficult and complicated one.

In some aspects of planktological study, complete identifications of the organisms being dealt with are not necessary, yet for many purposes the careful planktologist will need to identify the forms with which he is dealing with the maximum of detail and accuracy. To do so, he must become an expert in the taxonomy of several major groups as different from each other as are blue-green algae and the Copepoda. In order to determine even common forms to species the investigator would need not only extensive training and years of experience, but also the immediate availability of a larger volume of taxonomic publications than it is commonly possible to have. These publications are often widely scattered in obscure journals, and in several foreign languages. Obviously, such determinations will be impossible for most investigators, even if they devote their entire research time to the taxonomic side of their problems.

The general investigator who is interested in problems beyond those of taxonomy finds it necessary often to make only partial identifications of his material in certain groups of organisms, making specific determinations only for those forms that are of the greatest importance to him. Often, even for these identifications, he must rely upon experts who are specialists in the taxonomy of a particular group.

On the other hand, in regions where the taxonomy of particular groups of plankters has been studied and published in some detail, the general investigator will find it possible with relative ease to determine the specific names of the more important species with which he is dealing. In the Woods Hole, Massachusetts, region, for example, because of the existence of Wilson's (1932) publication on the Copepoda, the general investigator, after a minimum period of self-training or group training in copepod anatomy and taxonomy, could identify rather easily the more common and significant copepod species that he might encounter. The situation is similar in marine waters of the San Diego region in California, where the diatoms can be determined relatively easily because of the existence of Cupp's (1943) diatom study. In studies conducted in a particular locality, any such local lists and local taxonomic works must be consulted for accuracy and completeness.

However, even the general investigator or the plankton student with a limited taxonomic background can succeed in making reasonably accurate determinations of plankters, in some cases to genus, in others to larger taxonomic categories, with the aid of a little perseverance and the expenditure of a little time. It is proposed to discuss in the following pages each of the major taxonomic groups existing as plankters, giving for each, in addition to ecological notes, general hints as to structure and aids to the recognition of the various taxonomic subdivisions, insofar as this is possible. It will be realized that in a work of this sort, no attempt can be made to attain completeness. However, those taxonomic categories down to the rank of genus which are commonly encountered in plankton samples will be considered. It is to be understood that those planktonic genera common in one locality need not necessarily also be common in other localities. Representatives of certain genera may thrive, and even dominate, in certain waters, yet be rare or entirely missing in other places which superficially are similar in environmental characteristics. Even a genus which, considered from a global point of view, is too rare to be of any importance may when conditions are suitable form a strictly localized bloom of considerable magnitude. It is manifestly impossible in the keys that follow to consider all such genera. An attempt is made to include all true planktonic forms that are regularly and commonly encountered in plankton samples, regardless of where they may have been obtained. It is inevitable that there shall be unfortunate omissions, and wherever possible the reader should check his identifications by consulting the literature in the field. References will be made from time to time to other publications which will be useful to the student in the identification of plankton forms. Some of these will be taxonomic works dealing primarily with particular localities, but they also will be useful to investigators all over North America and the world.

Among the more useful general references dealing with the plankton as a whole (or with plankton forms incidentally while dealing with broader subjects) are Johnstone, Scott and Chadwick (1924), Ward and Whipple (1918), Needham and Needham (1930), Pennak (1953), Jespersen and Russell (1939-1952), Pratt (1935), Dawydoff (1928), MacBride (1914), and Thorson (1946), as well as the important series of volumes included in *Nordisches Plankton, Die Tierwelt der Nord- und Ostsee, Süsswasser-Flora Deutschlands* and *Süsswasser-Fauna Deutschlands* (see entries by author in the bibliography).

Johnstone, Scott, and Chadwick deal only with the marine plankton of English waters. In addition to an interesting and valuable discussion of the plankton of the area they are considering, they give a useful set of plates illustrating the more important plankton types. The plates have also been reprinted separately. Ward and Whipple deal with fresh-water organisms of all the major taxonomic groups occurring in that environment. In doing so, they incidentally deal at considerable length with plankton organisms. The book, based primarily upon North American forms, contains elaborate keys for the determination of genera, and even to a considerable extent for the determination of species, though it cannot be relied upon for the latter because the keys are admittedly by no means complete. Similarly, Pennak (1953) deals with fresh-water invertebrate animals many of which are planktonic. Pennak includes keys to species for those groups whose taxonomy in the United States is well known, and to genera where the taxonomy is less well known.

Needham and Needham similarly deal with North American fresh-water organisms in general, and only incidentally with plankton forms. The work is a small paper-covered pamphlet, containing many useful keys and plates. Its small size makes it convenient to use, but also limits the number of forms that can be considered. Pratt in his useful book considers common invertebrate forms in general, including many that are planktonic in habitat. Jespersen and Russell edit a series of leaflets on the identification of marine plankton organisms.

Dawydoff and MacBride write about invertebrate embryology, hence do not deal specifically with planktonic forms. Their books however, consider invertebrate larval forms, and are therefore extremely valuable for the study of the marine plankton, where large numbers of invertebrate larvae are planktonic. Thorson deals with the larvae of marine bottom invertebrates of Denmark, many of which have a world-wide distribution as well.

Because the plants are the basic producers, in the plankton as elsewhere, it was felt that it would be desirable to consider first the phytoplankton, and subsequently the zooplankton. However, many plankton organisms are unicellular, and among such forms are many organisms that are transitional between the plant kingdom and the animal kingdom. Botanists and zoologists, from their different points of view, classify these one-celled forms somewhat differently, and from an ecological point of view neither system is satisfactory. In the pages below, an attempt is made to discuss the plants, from the simplest to the most complex, then to discuss forms that must be considered to be transitional, or closely related to transitional forms (which means that some organisms, definitely producers in their nutritional behavior, are considered at this point, though they are simpler in their organization than other plants mentioned previously). Finally, organisms that are strictly animals are considered, from the simplest forms to the most complex.

The organisms of the plankton will, therefore, be considered in the following order:

1. Cyanophyta
2. Chrysophyta: Bacillariaceae
3. Chlorophyta
4. Protozoa
5. Porifera
6. Coelenterata
7. Ctenophora
8. Platyhelminthes
9. Nemertea
10. Nemathelminthes
11. Rotifera
12. Bryozoa
13. Brachiopoda
14. Phoronidea
15. Chaetognatha
16. Annelida
17. Arthropoda
18. Mollusca
19. Echinodermata
20. Chordata

It will be observed that there is no consideration of the bacteria, nor of the fungi. These forms, insofar as they occur in the plankton, are not easily recognized except by the use of more or less complicated culture methods, and it is for this reason that it was not felt desirable to discuss them herein. The bacteria have been studied in some detail by ZoBell (1946). The groups of organisms discussed below can be determined by the following key.

Artificial key to the phyla and other major groups of planktonic animals and plants

1. One-celled, or if many-celled, with little or no differentiation of cells 2

   Many-celled, with a great deal of specialization and differentiation of cells. Usually with well-developed tissues, organs, and organ systems 11

2. Cells provided with pseudopodia, cilia, or flagella during the major portion of the life cycle, by means of which motion can take place 3

   If cells or organisms are in motion, movement is not by means of pseudopodia, cilia, or flagella during the major portion of the life cycle 8

3. Cells with flagella during the major portion of the life cycle 4
   Cells without flagella during the major portion of the life cycle ... 6
4. Cells with one transverse flagellum in a girdle, and a second trailing flagellum. Phylum Protozoa. Order Dinoflagellata .... (p. 175)
   Cells with flagellum or flagella arranged otherwise 5
5. If multicellular, all cells alike or nearly so. Phylum Protozoa. Class Mastigophora .... (p. 170)
   Multicellular, with small flagellated cells at one pole and large nonflagellated cells at the other pole. Phylum Porifera (amphiblastulae) .... (p. 199)
6. Cells with pseudopodia during the major portion of the life cycle. Shelled forms. Phylum Protozoa. Class Sarcodina 7
   Cells with cilia during entire life cycle. Class Infusoria. Family Tintinnidiidae .... (p. 196)
7. With a central capsule, and with a silicious or strontium sulfate shell composed of radiating spicules and/or spicules forming one or more concentric lattice frameworks. Order Radiolaria .... (p. 184)
   Without a central capsule. Shell usually forming chambers of various shapes, often forming looser or tighter spirals. Order Foraminifera .... (p. 181)
8. Cells with photosynthetic pigments 9
   Cells without photosynthetic pigments. Phylum Protozoa. Order Dinoflagellata .... (p. 175)
9. Cells without choroplasts, and without true nuclei. Phylum Cyanophyta .... (p. 150)
   Cells with chloroplasts and true nuclei 10
10. Cells with rigid silicious cell walls arranged to form two halves (valves) which fit into one another like the two parts of a pill box. Phylum Chrysophyta. Class Bacillariaceae ...... (p. 155)
    Cells without silicious cell walls, and cell walls of the usual arrangement. Phylum Chlorophyta .... (p. 163)
11. Without organ systems; cell differentiation slight 12
    With some organ systems at least moderately well developed 13
12. Cells forming a hollow ball; cells at one pole flagellated and small, while those at the other pole are large and not flagellated. Phylum Porifera (amphiblastulae) .... (p. 199)
    An external layer of small ciliated cells covers a solid or hollow central core of large endodermal cells. Phylum Coelenterata (planulae) .... (p. 199)
13. Radially or biradially symmetrical forms 14
    Bilaterally symmetrical forms, or asymmetrical 15

14. Radially or biradially symmetrical forms, sometimes forming complicated colonies of individuals. Diploblastic. Phylum Coelenterata . . . . (p. 199)
    Biradially symmetrical forms with eight rows of ciliary combs arranged the length of the body. Triploblastic. Phylum Ctenophora . . . . (p. 210)
15. With a vertebral column or a notochord, or both. Phylum Chordata . . . . (p. 270) . . . 16
    With neither a vertebral column nor a notochord 18
16. With a vertebral column. Subphylum Vertebrata . . . . (p. 277)
    With a notochord only 17
17. Notochord confined to the tail region. Subphylum Urochorda . . . . (p. 272)
    Notochord extends the length of the body. Subphylum Cephalochorda . . . . (p. 277)
18. With a well-developed external skeleton and with segmented appendages. Phylum Arthropoda . . . . (p. 224) . . . 19
    Without segmented appendages 32
19. Carapace forming a bivalved shell covering the entire body 20
    Carapace not thus 21
20. With a well-developed cement gland. Subclass Cirripedia (cypris larvae) . . . . (p. 241)
    Without a well-developed cement gland. Subclass Ostracoda . . . . (p. 230)
21. Three pairs of segmented appendages. Class Crustacea (nauplius larvae) . . . . (pp. 231, 241, 249, and 252)
    With more than three pairs of appendages 22
22. Carapace forming a bivalved shell which does not cover the head. Class Crustacea. Order Cladocera . . . . (p. 225)
    Carapace not thus 23
23. Head, thorax, and abdomen fused together. With four pairs of legs. Class Arachnoidea. Order Acarina . . . . (p. 259)
    Head, thorax, and abdomen not fused 24
24. With a carapace covering (though not necessarily attached to) part or all of the thoracic segments 25
    Without a carapace 30
25. Some of the thoracic appendages are uniramous 26
    All thoracic appendages are biramous 27
26. With two pairs of well-developed thoracic appendages, the second developed to form a large raptorial claw. Class Crustacea. Order Stomatopoda (pseudozoeae) . . . . (p. 257)
    With eight pairs of thoracic appendages, of which the last five pairs are uniramous. Class Crustacea. Order Decapoda (p. 252)

27. Adults 28
Larval stages. Class Crustacea. Orders Mysidacea, Euphausiacea, Decapoda, and Stomatopoda . . . . (pp. 244, 249, 252, and 257)
28. Carapace small, not covering the last five thoracic segments. Order Cumacea . . . . (p. 242)
Carapace large, covering entire thorax 29
29. Carapace fused with all the thoracic segments; thoracic gills present, not covered by the carapace. Order Euphausiacea . . . . (p. 249)
Carapace fused only with the first three thoracic segments; thoracic gills, if present, covered by the carapace. Order Mysidacea . . . . (p. 244)
30. Most of the thoracic appendages are uniramous and leg-like 31
Most of the thoracic appendages are biramous and swimmeret-like. Subclass Copepoda . . . . (p. 231)
31. Thoracic legs all of one kind; body usually flattened dorso-ventrally. Order Isopoda . . . . (p. 242)
Thoracic legs of more than one kind; body usually flattened laterally. Order Amphipoda . . . . (p. 243)
32. Body barrel-shaped and transparent; conspicuous rings of muscle in body wall; internally with gill slits in a pharynx. Phylum Chordata. Subphylum Urochorda (salps etc.).(p. 272)
Not thus 33
33. Body with paired lateral setae. Phylum Annelida. Subclass Polychaeta . . . . (p. 221)
If present, setae are not paired and lateral 34
34. With exoskeleton or lorica 35
Soft-bodied, without exoskeleton 40
35. Body clearly segmented, worm-like. Phylum Arthropoda. Class Insecta. Order Diptera (larval forms) . . . . (p. 258)
Body not segmented 36
36. Body very thin; flat and triangular. Phylum Bryozoa (cyphonautes) . . . . (p. 217)
Body not thus 37
37. With a bivalved shell covering all or most of the body 38
Not thus 39
38. With a lophophore. Phylum Brachiopoda . . . . (p. 218)
Without a lophophore. Phylum Mollusca. Class Pelecypoda. . . . . (p. 260)
39. Body with a cap-shaped univalve shell. Phylum Mollusca. Class Gastropoda . . . . (p. 260)
Body with a lorica, which does not form a cap as above. Phylum Rotifera . . . . (p. 214)

40. Larval forms, with one or more bands of cilia encircling the body 41
    Larval forms and adults; not as above 43
41. Ciliated bands (prototroch) simple. Phyla Annelida and Mollusca (trochophores) . . . . (pp. 221, 260)
    Ciliated band not thus 42
42. Ciliated band forming two lateral lappets. Phylum Nemertea (pilidium larvae) . . . . (p. 212)
    Ciliated band forming eight median lobes; digestive tract incomplete. Phylum Platyhelminthes (Müller's larvae) (p. 212)
    Ciliated band forming several to many median tentacles; digestive tract complete. Phylum Phoronidea (actinotroch larvae) . . . . (p. 218)
43. Larval forms, with a more or less complicated ciliated band around the mouth, and with a second band anterior to the mouth. Phyla Echinodermata and Chordata (pluteus-type larvae) . . . . (pp. 266, 270)
    Adults, not with the above characteristics 44
44. With a corona of cilia surrounding the mouth. Phylum Rotifera . . . . (p. 214)
    Not thus 45
45. Flat, leaf-like forms. Phylum Platyhelminthes . . . . (p. 212)
    Not thus 46
46. Body cylindrical and worm-like, with no trace of segmentation, and with no well-developed head. Phylum Nemathelminthes . . . . (p. 213)
    Body not thus 47
47. Body elongate, with paired lateral fins 48
    Body not thus, Phylum Mollusca. Class Gastropoda . . (p. 260)
48. With strongly developed chitinous jaws laterally on the head. Phylum Chaetognatha . . . . (p. 218)
    Without jaws. Phylum Nemertea . . . . (p. 212)

1. Cyanophyta (Blue-Green Algae)

The Cyanophyta are the simplest of the plants we group together for convenience as the algae. They appear to be most closely related to the bacteria, differing from these forms mainly in the possession of photosynthetic pigments (i.e., chlorophylls, etc.). The pigments and the associated carotinoids are not, however, confined within special cytoplasmic bodies, the chromatophores, as in other autotrophic plants. Instead the whole peripheral region of the protoplasm contains the pigmentation.

In addition to chlorophyll and carotinoids (which are found also in other green plants), the Cyanophyta usually contain a bluish pigment, phycocyanin, and often also a reddish pigment similar in appearance to the phycoerythrin of the red algae. The group is commonly known as the blue-green algae because the combination of green chlorophyll and blue phycocyanin produces this coloration. Many forms—indeed, most of them—are actually blue-green in color, but variations of the relative quantities of chorophyll, carotinoids, phycocyanin, and phycoerythrin, combined sometimes with pigmented sheaths around the cells, and with light refracted by pseudovacuoles, cause many of them to be other colors, such as red, bright green, yellow, brown, purple, and even black. At times, and in some localities, a great abundance of such colored forms in the plankton may cause the water in which they are growing to take on a colored cast. This has happened in the Red Sea, which obtained its name from the reddish color the water assumes from time to time because of the existence of the "blue-green" alga *Trichodesmium erythraeum.*

The Cyanophyta are similar to the bacteria in that they have no nucleus in the usual sense of the term. There is chromatic material within the cell, and usually this is congregated towards the center of the cell, but there is no nuclear membrane, nor any nucleolus. In many cases the pigments described above do not invade the center of the cell where this chromatin material is located. The colorless "central capsule" thus delimited has a superficial resemblance to a true nucleus, but by no means does it correspond to the nucleus in the cells of higher forms, both plants and animals.

Also as in the bacteria, none of the Cyanophyta exhibit sexual reproduction in any form. Asexual reproduction occurs, especially in one-celled species, by means of binary fission. In many-celled species it may take place by means of the breaking up of the colony or filament of cells; by means of the formation of one-celled to several-celled filaments, the hormogones, which break free and produce new plants; or by means of the formation of resting spores consisting of one to several cells. These spores are produced from the usual vegetation cells by a process of growth and transformation. None of the blue-green algae ever reproduce by means of flagellated swarm spores, and this serves to differentiate them from the other algal forms.

In some groups of blue-green algae there is a difference in structure among the cells of the filament. Certain cells, after being produced in the usual manner by the division of ordinary vegetative cells, become somewhat enlarged, then die and lose their content of protoplasm. These cells are known as heterocysts. Their function is not clear, though in some species (but by no means all) they are a weak point at which the filament breaks up readily into shorter filaments in asexual reproduction.

The cell wall in the Cyanophyta consists of two main layers, the inner of which is thin and firm. Previously this portion of the cell wall was thought to be chitinous, but more recent investigations have indicated that it is not chitinous, but is composed of cellulose (or perhaps a pectin-like hemicellulose). The outer portion of the cell wall is usually relatively very wide and gelatinous. Sometimes it is firm, whereas at other times it is soft and "watery." Commonly it is stratified or laminated. It may enclose a single cell, or may come to enclose a colony of individuals. This outer portion, which is often colored, is called the sheath, and its consistency and thickness are useful in the identification of the various genera. In planktonic forms the sheath serves as an aid in flotation. It also may cause individual filaments or colonies to adhere together to form loose aggregates.

The cells of many species of blue-green algae contain structures known as pseudovacuoles, or gas vacuoles. The precise nature of these structures is not clear, nor is their function definitely known, though several hypotheses are extant. Some think they are tiny gas vacuoles, whose function is to give buoyancy to the forms that possess them, while others think they are merely storage products. Such pseudovacuoles may occur in planktonic species of the genera *Anabaena, Coelosphaerium, Lyngbya,* and *Microcystis.* In some forms they may refract the light in such a manner as to give color to the organisms (e. g., *Skujaella erythraeum* in the Red Sea).

Many "water blooms" of serious and damaging proportions appear to be associated with blue-green algal species which have especially well developed sheaths and which contain gas vacuoles. Such forms float at or near the surface, and the innumerable colonies or filaments adhere together to form dense mats at the surface. The death and decay of some of this dense growth results in obnoxious odors, oxygen depletion, and the production of toxic products, including hydrogen sulfide.

Many Cyanophyta (e.g., *Oscillatoria*) are capable of active movements over a substratum. Others are unable to move. The nature of the movement is disputed, but the most widely accepted theory is that it takes place by means of secretion of mucoid materials, which swell rapidly upon contact with water. In planktonic species such movements can occur only in connection with the surface film.

The Cyanophyta occur in both marine and fresh-water situations, but in fresh water they are relatively much more important. In both salt and fresh water they may be benthic, epiphytic, or in rare instances symbiotic on other organisms, or they may be planktonic. *Nodularia spumigena* often occurs in brackish water, and may cause sliming in such places as the Gulf of Bothnia in the summer.

In fresh water there are numerous planktonic species. Even in relatively small numbers they may at times produce strong odors in the water, so that if the water is used for domestic purposes, it becomes distasteful. The control of blue-green algae in reservoirs and other sources of water supply has become one of the major problems of communities, and has been given a great deal of study. In those cases where a pond or lake has developed a "waterbloom" of blue-green algae, the water often is avoided by cattle and other beasts because of the obnoxious taste. As mentioned above (p. 79), occasionally the water actually becomes poisonous to those terrestrial animals that may drink it. The nature of the poison is unknown.

"Waterblooms" of blue-green algae usually occur in hard-water lakes or ponds, and they appear almost entirely during the late summer and early fall months. Different species are to be found in soft-water lakes and ponds, but in such locations they seldom or never predominate in the plankton, hence they seldom form blooms.

The Cyanophyta are often divided (as in Smith, 1950) into the orders Chroococcales, Chamaesiphonales, and Hormogonales. Only the first and the last, however, contain planktonic species (other than the tychoplankters). The Chroococcales contain one-celled or many-celled species, where the cells never form filaments. Reproduction is by binary fission or by fragmentation of the colony of cells. In the Hormogonales the several cells are arranged in filaments, and reproduction is by hormogones. In addition, in many cases there are heterocysts and/or resting spores.

Artificial key to genera of Cyanophyta commonly occurring in the plankton

1. One-celled or colonial, never forming filaments. Chroococcales . . 2
   Many-celled, forming definite filaments. Hormogonales 11
2. Several to many cells forming a flat, rectangular colony one cell deep. All cells imbedded in a common colonial envelope of gelatinous material 3
   One-celled to many-celled. In many-celled forms, the cells never all in one plane in the colony 4
3. Cells very regularly arranged, in a geometrical pattern . . . . *Merismopedia* (fig. 49)
   Cells not regularly arranged . . . . *Holopedium* (fig. 50)
4. Colony spherical, with cells arranged in a layer somewhat below the surface 5
   If colonial, the cells not arranged thus 6
5. Cells tend to remain in two's and four's, and they are rather irregularly arranged . . . . *Gomphosphaeria* (fig. 51)
   Cells single and regularly arranged . . . . *Coelosphaerium* (fig. 52)
6. Cells spherical or oval, never elongate 7
   Cells elongate 9
7. Many-celled forms, with masses of cells close together . . . . *Microcystis* (fig. 53)
   Many-celled forms, or single-celled. Cells farther apart 8
8. Many-celled forms. Practically all cells are spherical. No trace of individual sheaths around cells . . . . *Aphanocapsa* (fig. 54)
   One-celled, or relatively small colonies. Cells mostly hemispherical. Usually the sheaths surrounding the individual cells are apparent . . . . *Chroococcus* (fig. 55)
9. Cell is very long and closely spiral . . . . *Spirulina* (fig. 58)
   Cell not so long, and not spiral 10
10. Individual cells with distinct sheath enclosed in the sheath of the mother cell . . . . *Gloeothece* (fig. 56)
    Sheaths of the individual cells are completely fused together and indistinguishable . . . . *Aphanothece* (fig. 57)
11. Heterocysts present 12
    Heterocysts absent 17

12. Cells not as long as they are wide . . . . *Nodularia* (fig. 59)
    Cells at least as long as they are wide 13
13. Filaments gradually tapering towards the tip . . . .
    . . . . *Gloeotrichia* (fig. 61)
    Filaments uniform in diameter, not tapering 14
14. Filaments single, not forming associations . . . *Anabaena* (fig. 60)
    Filaments joined to form associations 15
15. Contorted filaments in a firm matrix . . . . *Nostoc* (fig. 62)
    Filaments not in a matrix 16
16. Filaments usually not contorted. Laterally confluent to form small fascicular colonies . . . . *Aphanizomenon* (fig. 63)
    Filaments not thus . . . . *Anabaena* (fig. 60)
17. Sheath conspicuous or absent, end of the filament with extended cell wall . . . . *Lyngbya* (fig. 64)
    Sheath inconspicuous 18
18. Filaments often joined laterally by the inconspicuous sheaths to form colonies . . . . *Skujaella* (*Trichodesmium*) (fig. 65)
    Filaments not joined thus . . . . *Oscillatoria* (fig. 66)

Interested persons should consult Geitler (1925, 1930–31), Huber-Pestalozzi (1938), Oltmanns (1922–23), Prescott (1951), Smith (1950), Tilden (1910), Ward and Whipple (Olive, 1918), and Wolle (1887) for further information on the Cyanophyta.

2. Chrysophyta: Bacillariaceae (Diatoms)

With the possible exception of the insufficiently known nannoplankters, the microplanktonic diatoms are the most important and basic plants in the sea, and in fresh water also they are among the most significant. Their importance lies not in their size as individuals, for they are all microscopic, the largest being barely visible to the naked eye. Their importance lies rather in their enormous abundance and their widespread occurrence. They may be found in nearly any aquatic environment where there is sufficient light to maintain their photosynthetic activities. Many are benthic, others are even aerial or terrestrial, some being found in soils as much as four feet below the surface (see Smith, 1950). Still others are epiphytic or epizoic. Of these, a few attain some indirect importance in the plankton by the fact that they are epizoic on zooplankters (e.g., *Licmophora* growing epizoically upon various copepods in the plankton).

However, considering the aquatic environments of the world as a whole, those diatoms that in their own right are planktonic are obviously of greatest significance in the total food chain. In the ocean the water is too deep, except for an extremely narrow coastal fringe, to support a benthic flora, so that in the vast reaches of the sea, the only plant population must of necessity be a planktonic population, and the diatoms are the most conspicuous part of this floating plant life. Even in the parts of the oceans near shore the total productivity of the planktonic diatoms usually exceeds that of all other plant forms together, and the total productivity here is especially high, considering the relatively small area involved.

On the other hand, on the high seas in tropical oceans, the dinoflagellates replace the diatoms in considerable part. It appears, however, that this does not hold for many inshore tropical waters—see Marshall (1933) and Smith *et al.* (1950). In bodies of fresh water, also, the diatoms often, though by no means always, take a role subordinate to that of other plants, especially the Cyanophyta and the Chlorophyta.

The diatoms are essentially one-celled plants, though some may form simple colonies (usually as chains of identical or nearly identical cells). The cell differs radically in structure from that of other algae. The cell contents are enclosed in a cell wall that consists of two parts, called valves. The valves are usually aptly described as fitting together like the parts of a pill box. The upper valve is known as the epitheca, and the lower as the hypotheca. The lateral region where the two valves overlap is known as the cingulum or girdle. The valves are normally more or less silicious, though it has been found possible to obtain growth of diatoms in the laboratory in a medium devoid of silicon. The silicification shows considerable variation, depending in part upon the richness of the natural waters in silicon, and in part upon the nature of the diatom species involved. Benthic forms, for example, tend to have very heavy cell walls, whereas planktonic forms, as an adaptation to planktonic existence, have extremely thin valves. The valves usually are more or less ornamented with radially or bilaterally (or sometimes asymmetrically) arranged pores and slits. It appears that the cell contents derive their only contact with materials in the surrounding environment through these structures, for where the silicious cell wall is thick and imperforate, it is impervious to the exchange of respiratory and photosynthetic gases, as well as to nutrient salts. The several pores on a valve are usually arranged symmetrically. In one of the orders, the Centrales, the pores are arranged radially around a central point, while in the only other order, the Pennales, they tend to be arranged in transverse rows, bilaterally symmetrical with reference to a central straight or sigmoid slit or canal, the raphe. In some of the Pennales there is a clear area in the center of the valve, called the pseudoraphe, which superficially resembles the true raphe. The raphe clearly is associated with the peculiar type of jerky gliding movements where these occur in the diatoms. Only those pennate species which have a true raphe are capable of movement, and none of the centric forms can move. The movement, according to the most widely accepted theory, is caused by a streaming motion of the cytoplasm along the raphe, supplemented by the extrusion of mucilage.

The arrangement of the pores, the presence or absence of a raphe or pseudoraphe, and the configuration of the raphe when present are characters that are used in the determination of many of the species of diatoms. In fact, though the number and form of the chromatophores is used in some cases, practically the entire system of taxonomy used for the Bacillariaceae is based upon the structure of the cell wall (or frustule, as it is usually called), rather than upon the soft parts of the cell. This not only is convenient, but was necessitated by the existence of large numbers of fossil species of diatoms, for which, of course, no soft parts are available for study. Diatom specialists feel that the study of the hard parts is adequate anyway in the classification of the group, for the intricate and often beautiful ornamentation of the frustule is directly due to the nature of the protoplasm, including its physiology.

In addition to the structures of the frustule mentioned above, other features of taxonomic importance are the number and arrangement of the intercalary bands (rings or partial rings of silicious material between the valves on the girdle side), the presence or absence and arrangement of spines and protrusions on the valves, the presence of a central nodule dividing the raphe into two halves, the presence of internal septae, etc.

Often specimens of diatoms are boiled in dilute nitric acid, then washed free of acid and mounted in a medium with a high index of refraction (hyrax, etc.) in order to bring out these structures clearly. Such a technique is often unsatisfactory for planktonic forms because in general they have such delicate cell walls that the boiling process destroys not only the protoplasmic parts, but the frustule as well. Frequently the valve structure of planktonic forms can be studied satisfactorily by strewing them on a microscope slide and incinerating the organic matter. However, this also is too rough treatment for many of the forms, and often it is suitable simply to dry the frustules on a microscope slide, preliminary to mounting them.

The valvular markings are often almost or quite invisible in simple water mounts. However, in any particular part of the world, once the forms that occur in the plankton have been definitely determined, they can usually be recognized with fair certainty by their general shape and growth forms, without the necessity of minute examination of their ornamentation.

In the northern portion of the North Pacific Ocean, and to a much greater extent in the Antarctic, the bottom of the sea is covered with deposits so rich in diatom frustules that they must be classified as diatom oozes. The total bottom area covered with such diatom oozes amounts to over 30,000,000 sq. kil. The oozes have been formed, as have the other types of pelagic oozes, by the rain of empty shells from the pelagic surface regions above. Diatomaceous earth, which is so widely used commercially, is fossilized diatom ooze.

The internal parts of the diatom cell are used to a lesser degree in the taxonomy of the group. As is universally true in the algae other than the Cyanophyta, there is a well-defined nucleus. This is usually imbedded in the cytoplasm beneath the cell wall, instead of being central in the cell. Extending through the mid-region of the cell is a large vacuole, and the main body of the cytoplasm is distributed in a layer beneath the cell wall. One or more chromatophores are imbedded in the cytoplasm. In the order Pennales there are usually only one or two chromatophores, which may be lobed or perforate, while in the Centrales there are usually several to many. The chromatophores are most often brownish or yellow-brown in color. They contain the usual chlorophyll pigments, including the carotinoids. In addition, it has frequently been claimed that there is a special pigment, diatomin, which gives the diatoms their characteristic coloration. This has been questioned, however, by investigators who think that the coloration is due to the usual chromatophore pigments, present in different proportions than usual (see Cook, 1945).

Reproduction in the diatoms usually takes place by simple cell division, whereby one half of the protoplast remains in the epitheca and the other half in the hypotheca. Each half then secretes a new half-frustule, but the new half is always a new hypotheca, so that one of the two cells produced by cell division is normally somewhat smaller than the other. Through repetition of cell divisions over and over, it is obvious that the size of most of the individuals would gradually decrease until the size would be incompatible with continued life. However, before this point is reached a peculiar growth process, by means of auxospores, takes place, with the result that the original size is reconstituted once more. In this auxospore formation the protoplast within individual frustules grows to become several times its initial size, secreting around itself a wall of distensible pectin-like material. The auxospore then pushes its way out of the old frustule and secretes two new valves according to the nature of the species involved, whereupon ordinary cell divisions may recommence.

Such an increase of size by auxospore formation has been shown to occur in a number of diatom species. However, it appears not to be universal. In some forms although cell division occurs as described above, there is no evidence of a gradual decrease in size of the subsequent generations. This is due to the ability of the frustules of many forms to spread or stretch to a certain degree during cell division and during the resulting formation of a new hypotheca. Consequently the new cells formed by cell division remain the same size as the parent cells. Many of the planktonic diatom species, with their thin cell walls, would appear to fall into this category, and auxospores are comparatively seldom found in diatoms taken from natural plankton.

Reproduction also may take place in other ways among diatoms. Especially in marine neritic species, but also in certain fresh-water species, resting spores may be formed. These are usually considerably smaller than the original diatom cells, and become surrounded with very heavy walls, which are characteristic in shape and ornamentation for each species. Such resting spores may be liberated from the old frustule by disintegration of the latter, or they may remain within the intact frustule. In either case they may continue planktonic life or they may slowly settle to the bottom and become a part of the bottom muds. In this condition they appear to be able to withstand unfavorable growing conditions, for example during the winter months, but upon the resumption of favorable conditions they will germinate.

Finally, reproduction may take place sexually, either by cells similar in appearance to the auxospores described above, or by means of flagellated microspores (usually isogametes). There still remains a great deal to be learned concerning the life cycle of the diatoms, both planktonic and benthic.

Artificial key to common planktonic genera of the Bacillariaceae

1. Frustule usually cylindrical or drum-shaped, sometimes angular; ornamentation on each valve arranged in a radial pattern around a central point. No raphe or pseudoraphe present. Order Centrales 2
   Valves usually elongate, and frustule not cylindrical; ornamentation arranged in a bilateral pattern on each valve around a longitudinally placed raphe or pseudoraphe. Order Pennales 20
2. Single-celled and drum-shaped, without horns or long spines 3
   Frustule not thus 6
3. Valves with wide hyaline expansion or wing all around, with radial rays . . . . *Planktoniella* (fig. 67)
   Valves without such a hyaline expansion 4
4. Valves with a striated margin . . . . *Cyclotella* (fig. 68)
   Valves without a striated margin 5
5. Cells with inconspicuous submarginal spines, or none. Valve margin from girdle view simple . . . . *Coscinodiscus* (fig. 69)
   Cells with conspicuous marginal spines. Valve margin from girdle view undulating . . . . *Stephanodiscus* (fig. 70)
6. Cells disc-shaped, without horns. Usually held together in loose chains by one or more strands of mucus 7
   Cells not thus 8

7. Cells held together by several mucilaginous strands. Cells without spines at margin of valves........ *Coscinosira* (fig. 71)
Cells held together by one mucilaginous or cytoplasmic strand, or imbedded in an amorphous gelatinous mass. Cells usually with marginal spines and marginal mucilaginous strands........................... *Thalassiosira* (fig. 72)

8. Cells cushion-shaped from girdle view; with heavy horns at the corners; usually with spines in center of valves; valve view with 2, 3, 4, or 5 poles.................................... 9
Cells not thus..10

9. Cell walls very delicate. Horns short. Sculpturing delicate .... *Cerataulina* (fig. 73)
Cell walls heavier. Horns heavy. Sculpturing usually pronounced.......................... *Biddulphia* (fig. 74)

10. Cells single or forming chains. Margin of valves with long processes. Usually no spine in middle of valve end.............11
Cells without long processes on margin......................13

11. Margin of each valve with two long processes. The processes from adjacent cells of colonial species fuse firmly near the base .... *Chaetoceros* (fig. 75)
Margin of each valve with several long processes..............12

12. Cells single. Processes are delicate and unbranched......... .... *Corethron* (fig. 76)
Cells usually in chains. Processes strong. Processes from adjacent cells are fused together for a considerable distance, and thus they often appear bifurcate........ *Bacteriastrum* (fig. 77)

13. Cells cylindrical and usually very long in valve view. Each end of the cell with a single valvular spine, centrally or eccentrically located..14
Cells not thus..15

14. Spine eccentric. Many intercalary bands. No marginal spines on valves......................... *Rhizosolenia* (fig. 78)
Spine central. Intercalary bands inconspicuous. A circle of delicate spines on the valvular margins...... *Ditylum* (fig. 79)

15. Margin of valves with small spines or with well-developed horns...16
Margin without spines or horns.............................18

16. Margin of valves with small spines......... *Lauderia* (fig. 80)
Margin with two prominent horns.........................17

17. Horns relatively long, each with one or more hyaline claws at the end............................. *Hemiaulus* (fig. 81)
Horns usually relatively short. No hyaline claws........... .... *Eucampia* (fig. 82)

18. Cells in chains, and held together by long hollow spines........19
Cells in chains, not held together thus...... *Melosira* (fig. 83)

19. Cells with hexagonal areolations....... *Stephanopyxis* (fig. 84)
    Cells without hexagonal areolations...... *Skeletonema* (fig. 85)
20. Valves without a raphe. A pseudoraphe usually present.........21
    Both valves with a true raphe.....................................28
21. Cell with one end enlarged, forming star-shaped clusters....
    .... *Asterionella* (fig. 86)
    Cell not thus. May at times form star-shaped clusters..........22
22. Cell in girdle view rectangular. Many intercalary bands.....
    .... *Striatella* (fig. 87)
    Cell, if rectangular, without numerous intercalary bands.......23
23. Cells circular and large. Bent in saddle shape.............
    .... *Campylodiscus* (fig. 88)
    Cells not thus...24
24. Cells united into bands by the whole valve side...........
    .... *Fragillaria* (fig. 89)
    Cells seldom or never united thus..........................25
25. Cells in girdle view with four distinct longitudinal lines. Valve with an inflation in the center........... *Tabellaria* (fig. 91)
    Cell in girdle view without such lines. Valve without inflation....26
26. Girdle view rectangular. Fresh-water.......................27
    Girdle view narrow and linear. Marine..................
    .... *Thalassiothrix* (fig. 90)
27. Cells very long and narrow................ *Synedra* (fig. 92)
    Cells shorter, usually attached to one another to form zigzag chains............................. *Diatoma* (fig. 93)
28. Raphe and valves sigmoid............. *Pleurosigma* (fig. 94)
    Raphe and valves not thus.................................29
29. Valves with a central nodule.............. *Navicula* (fig. 95)
    Valves without definite central nodule.......................30
30. Valves usually with a wing on one or both sides.... *Tropidoneis*
    Valves without wing. With keel...............................31
31. Rapheal fissure with a conspicuous row of circular pores....
    .... *Nitzschia* (fig. 96)
    Rapheal fissure not thus....................................32
32. Valve view symmetrical.................. *Surirella* (fig. 97)
    Valve view asymmetrical and lunate....... *Amphora* (fig. 98)

Interested persons should consult Boyer (1927), Cupp (1943), Lebour (1930), Elmore (1922), Hustedt (1930), and Ward and Whipple (Snow, 1918) for further information on the diatoms.

3. Chlorophyta (Green Algae)

The Chlorophyta are algae in which the proportions of the chloroplast pigments are much as they are in the higher plants, with the consequence that, except in rare cases, the plants have a bright green color. They are distinguished from the Chrysophyta by the lack of a silicious cell wall. Instead, the cell wall of the true green algae is cellulose and pectinoid in nature. They are distinguished from the Cyanophyta by the presence of chloroplasts and a true nucleus.

There is considerable confusion in the definitions given for the group by various authors in biology. Most botanists include in the Chlorophyta a number of green flagellated genera that are considered by most zoologists to be flagellated Protozoa, and which the zoologists believe to be much more closely related to the Dinoflagellata or the Polymastigina (for example) than to any of the nonmotile plants. For the purposes of the present work all flagellated Protista will be considered together as a unit, which will be designated with the zoological term "Mastigophora." The Chlorophyta, as considered here, will therefore include only those forms without flagella. Even with this arrangement some difficulty is met with in drawing a line of demarcation between flagellated and nonflagellated forms, inasmuch as many of the algae have a flagellated stage in their life history, and in a few cases the flagellated stage is quite lengthy. Conversely, many flagellated species have an alga-like, nonflagellated, colonial palmella stage in their life history. Only those forms, however, in which the major vegetative stage is nonflagellated are considered here as Chlorophyta.

Also, there has been confusion among botanists in the past concerning the position of the desmids. Frequently they are considered to be a group entirely separate from the Chlorophyta, at other times they are classified as a separate order in the class Chlorophyta, and still other authors consider them to be closely related to such green algae as *Spirogyra.* The author is not in a position to make any independent judgment on the merits of the arguments that have led to these varying judgments by phycologists, and will therefore follow Smith (1950) in assigning the desmids to the Chlorophyta.

The Chlorophyta, as delimited above, occur very widely in damp terrestrial environments, in fresh water, and in the sea. They are, however, more characteristic of fresh water. Large numbers of species of green algae are either sessile, epiphytic, commensal, or symbiotic, but most species in lakes and ponds are either typical or facultative plankters. In the sea there are no pelagic Chlorophyta, and consequently the marine green algae need no longer concern us.

Planktonic Chlorophyta vary considerably in their structure, though none are of great complexity. Many are single-celled, while others consist of two to several cells. Often the cells are imbedded in a thick pectinaceous matrix, while in other cases the cell walls are much thinner and firmer. Many-celled species, both with and without a matrix, may be filamentous, or else they may form spherical colonies. The cell wall usually consists of a cellulose inner portion, surrounded by a pectinaceous sheath. The latter, at least on its outer border, is somewhat soluble in water, and therefore must be constantly renewed. In the desmids the pectin appears to be renewed through definite pores in the cell wall, but in other forms such pores are lacking and the method of renewal is more problematical. The solubility of the outer layers of the cell wall in filamentous forms (i.e., in those forms large enough to manipulate grossly in the hands) gives many of them a typical slippery feeling.

Reproduction of the Chlorophyta may be either asexual or sexual. For the most part asexual reproduction is of a very simple kind, consisting of accidental or natural fragmentation of colonies, or, in some one-celled species, of simple binary fission. However, asexual reproduction may also take place by means of flagellated zoospores. Sexual reproduction may be isogamous or heterogamous. Frequently, especially in the more primitive species, sexual reproduction may take place by means of flagellated zoogametes, which are indistinguishable in the main from the several kinds of holophytic flagellated organisms that will be discussed below (pp. 170-172) under Mastigophora. There are one to several flagella, an eye spot, contractile vacuoles, etc. (see discussion of palmella stages of flagellated species on p. 171). These zoogametes, which are usually, but not always, isogamous, fuse together in pairs, and the zygote thus formed, either immediately or after a resting stage, germinates to form new individuals. In a few of the Chlorophyta there is a true alternation of generations, comparable to that of higher plants, but in most the zygote develops four zoopores, which in turn grow into individuals similar to the parents that produced the gametes.

In other cases sexual reproduction takes place by means of conjugation (as in the filamentous genus *Spirogyra,* etc.) where cells of two individual plants, in a typical case, produce copulation tubes, which fuse together when they touch each other. In many cases one of the two cells then acts as a male cell and migrates by amoeboid movements through the copulation tube into the other cell, whereupon the two cells fuse to form a zygote. In other cases both cells migrate into the copulation tube, and fuse half way between the two original cells.

Planktonic green algae may become extremely abundant, especially in small lakes and ponds where the water is relatively undisturbed by winds. When forms with a large amount of pectinaceous substance in the cell sheath become abundant they often lead to sliming of the water, while filamentous species may cause heavy mats of "water scum," the death and decay of which cause deleterious conditions. Blooms of the green algae are second only to those of the blue-greens in fouling natural fresh waters.

Artificial key to the commoner genera of planktonic Chlorophyta

1. Usually unicellular, but sometimes forming filaments or small nonfilamentous colonies. Cells bilateral, with one half a mirror image of the other. Desmidiaceae 2
   One-celled, filamentous or nonfilamentous colonies. Cells not bilateral 14
2. One-celled forms 3
   Filamentous colonies 10
3. Cell without a deep constriction in the middle 4
   Cell with a deep constriction dividing it into two parts 5
4. Ends of cell rounded or truncate . . . . *Netrium* (fig. 99)
   Ends of cell bluntly pointed . . . . *Closterium* (fig. 100)
5. Cell halves much longer than broad . . . *Pleurotaenium* (fig. 101)
   Cell halves nearly or quite as broad as long, or even broader 6

6. Cell halves smoothly rounded, lobed or with small spines only . . . . *Cosmarium* (fig. 102)
   Cell halves with well-defined sharp spines, or deeply and sharply incised 7
7. Margins of half cells deeply incised . . . . . *Micrasterias* (fig. 103)
   Margins of half cells with thicker or thinner spines, but not incised 8
8. Front of each half cell with a thickened area 9
   Front of each half cell not thus . . . . *Arthrodesmus* (fig. 104)
9. Cell usually strongly compressed, with simple spines at the angles . . . . *Xanthidium* (fig. 105)
   Cell usually with processes extending in three dimensions. Spines often very heavy . . . . *Staurastrum* (fig. 106)
10. Half cells deeply incised . . . . *Micrasterias* (fig. 103)
    Half cells not incised, relatively simple 11
11. Median constriction of cell very slight. Delicate transverse ridges just median to the cell apices . . . . . . *Hyalotheca* (fig. 107)
    Median constriction of cell is conspicuous 12
12. Filament spirally twisted . . . . *Desmidium* (fig. 108)
    Filament not spirally twisted 13
13. Cells with apical processes . . . . *Sphaerozosma* (fig. 109)
    Cells without apical processes . . . . *Spondylosium* (fig. 110)
14. Cells cylindrical and longer than broad. Forming filaments 15
    Cells single, or, if colonial, not forming filaments; or if forming short filaments, the cells are not cylindrical and are not longer than broad 19
15. Cells with rounded ends, embedded in a thick gelatinous sheath, very loosely arranged into a filament . . . . *Geminella* (fig. 111)
    Cells flattened at the end, except the terminal cells of the filament. No thick gelatinous sheath 16
16. Cell walls composed of H-shaped pieces articulated together. Reproduction by means of zoospores and zoogametes. No pyrenoids . . . . *Microspora* (fig. 112)
    Cell walls not thus. Reproduction by conjugation. Pyrenoids present 17
17. Chloroplast (or chloroplasts) narrow and ribbon-like 18
    Chloroplasts stellate . . . . *Zygnema* (fig. 113)
    Chloroplasts a broad or narrow flat, axial band . . . . *Mougeotia* and *Debarya* (fig. 114)
18. Chloroplasts definitely spiral . . . . *Spirogyra* (fig. 115)
    Chloroplasts straight or nearly so . . . . *Sirogonium* (fig. 116)
19. Cells broader than long. Forming a short filament (usually of four to eight cells) . . . . *Scenedesmus* (fig. 117)
    Cells not forming a filament 20

20. Cells neither with spiniferous processes, nor with a gelatinous sheath. Cells in groups of four (or rarely eight). Several groups of four often are held together by remains of old cell walls to form more or less consolidated plates of cells 21
If cells are in groups of four, they are either spiniferous or embedded in a gelatinous sheath 22
21. Cells of groups of fours are quadrately arranged, and all of the groups lie in the same plane . . . . *Crucigenia* (fig. 118).
Cells of the groups of four not necessarily quadrately arranged and the several groups of four lie at different planes . . . . *Westella* (fig. 119)
22. Cells imbedded in a gelatinous sheath, which may be a colonial sheath or individual sheaths around each cell 23
Cells and colonies without a conspicuous gelatinous sheath 34
23. Cells of colony held together within the sheath by threads, which may be dichotomously branched or cruciate . . . . *Dictyosphaerium* (fig. 120)
Cells not thus 24
24. Cells spherical or subspherical 25
Cells not thus 29
25. Mature cells with flat, polygonal, peripheral chloroplasts . . . . *Planktosphaera* (fig. 121)
Chloroplasts not thus 26
26. Cells arranged in twos, fours, or rings of eight within the gelatinous envelope . . . . *Tetraspora* (fig. 122)
Cells arranged individually within the gelatinous envelope 27
27. Chloroplast single, stellate . . . . *Asterococcus* (fig. 123)
Chloroplast not thus 28
28. Colonial envelope homogeneous . . . . *Sphaerocystis* (fig. 124)
Cells single or colonial. In colonial forms, each cell with its individual sheath . . . . *Gloeocystis* (fig. 125)
29. Cells lunate, with the tips of the cell nearly touching . . . . *Kirchneriella* (fig. 126)
Cells not thus 30
30. Cells arranged in groups of four, and several groups united together by branching remains of old cell walls . . . . *Dimorphococcus* (fig. 131)
If cells are in groups of four, not thus united 31
31. Cells elongate and in groups of four or eight lying parallel to each other within the colony . . . . *Quadrigula* (fig. 127)
If cells elongate, not thus 32
32. Cells elongate and they lie parallel to each other and to the longitudinal axis of the colony. Cells often in twos, end to end . . . . *Elakatothrix* (fig. 128)
If cells elongate, not arranged thus 33

33. Cells usually in colonies of eight. Cells elongate and somewhat curved or bean-shaped. . . . . *Nephrocytium* (fig. 129)
Cells single or in colonies of two, four, eight, or sixteen. Cells ellipsoid, with pointed or rounded ends. . . . *Oöcystis* (fig. 130)
Cells in large groups, ovoid or cuneate. . . *Botryococcus* (fig. 152)

34. Cells in groups of four. Several groups of four united by thread-like remains of old cell walls. . . *Dimorphococcus* (fig. 131)
Cells not thus. 35

35. Cells elongate, with pointed ends. 36
Cells, if elongate, without pointed ends. 39

36. Ends of cells without definite spines, although they may taper to fine points at the poles. 37
Ends of cells with definite spines. 38

37. Cells lunate or arcuate, with convex faces apposed. . . . . *Selenastrum* (fig. 132)
If cells lunate, not in regular colonies with convex faces apposed . . . . *Ankistrodesmus* (fig. 133)

38. Spines at ends of cell short and stout. . . . *Closteridium* (fig. 134)
Spines very long and thin. Sometimes the spine at one end of the cell is bifurcate, with the rami long and at right angles to each other. . . . . *Schroederia* (fig. 135)

39. Cells spherical, solitary or in daughter colonies of four. Nucleus held in center of cell by cytoplasmic strands. Chloroplasts small, numerous, lying immediately below cell wall. . . . . *Eremosphaera* (fig. 136)
Cells, if single, not as above. 40

40. Cells forming a flat, circular plate. . . . . . . *Pediastrum* (fig. 144)
Cells not forming a flat circular plate. 41

41. Cells elongate and in colonies of 4, 8, or 16, radiating from a common center. . . . . *Actinastrum* (fig. 140)
Cells not thus. 42

42. Cells in groups of four. Each cell with one or more long spines. . . 43
Cells not in groups of four. With or without spines. 44

43. Groups of four cells almost always solitary. Spines relatively short. . . . . *Tetrastrum* (fig. 137)
Groups of four cells almost always united with other groups to form relatively large colonies. Spines relatively long. . . . . *Micractinium* (fig. 138)

44. Cells colonial, pyramidal in arrangement. Free faces of the cells with very long, tapering spines. . . . . . . *Errerella* (fig. 139)
Cells single or colonial, but not as above. 45

45. Cells rounded, or with blunt points. 46
Cells with sharply-pointed spines or setae. 49

46. Cells single. 47
Cells forming colonies. 48

47. Cells 5μ or less in diameter, spherical. . . . . *Chlorella* (fig. 141)
Cells much larger, angular. . . . . *Tetraëdron* (fig. 145)

48. Cells forming a sac-like reticulum. . . . . *Hydrodictyon* (fig. 142)
Cells closely apposed to each other, forming a hollow, spherical colony. . . . . *Coelastrum* (fig. 143)

49. Spines short and stout. . . . . *Tetraëdron* (fig. 145)
Spines or setae long. 50

50. Cells single, each with four stout blunt processes. . . . . *Pachycladon* (fig. 146)
Cells single or colonial, not as above. 51

51. Cells smoothly rounded. 52
Cells angular. 54

52. Long setae inserted towards poles of cell and sometimes equatorially as well. . . . . *Lagerheimia* (fig. 147)
Setae inserted on all sides of the cell. 53

53. Chloroplast single and cup-shaped. . . . . . *Golenkinia* (fig. 148)
Chloroplast, if single, is laminate. . . . . *Franceia* (fig. 149)

54. Each angle of the cell with four to six delicate setae, or if there is only a single seta, it is not hyaline. . . . . *Polyedriopsis* (fig. 150)
Each angle of the cell with a single heavy hyaline spine. . . . . *Treubaria* (fig. 151)

Interested persons should consult Fritsch (1935), Oltmanns (1922-23), Pascher (1913), Prescott (1951), Smith (1950), Ward and Whipple (Snow, 1918), and Wolle (1892) for more detailed information on the Chlorophyta.

4. Protozoa

Protozoa are characteristic of all sorts of aquatic environments, and they may be planktonic, epiphytic, epizoic, benthic, or parasitic. In addition, some may live in damp mosses, soils, fecal materials, or other similar locations. The class Sporozoa, of course, has only parasitic members, and though some species are parasitic upon plankton organisms, it does not seem advisable to consider them here, inasmuch as they are not true plankters themselves. Among the Sarcodina and the Infusoria a number of soft-bodied forms occur in the plankton (*Actinophrys, Paramecium,* etc.). Infusorian nannoplankters are especially frequently encountered. In addition, some forms, such as the sarcodinian Arcella (either as living organisms or as dead shells), are at times of considerable significance as members of the tychoplankton. However, as true plankters, the Mastigophora (especially the Dinoflagellata), the sarcodinian orders Foraminifera and Radiolaria, and the infusorian family Tintinnidiidae (in the order Spirotricha) are the most characteristic. Only these will be discussed below.

a. MASTIGOPHORA (other than Dinoflagellata)

The animal and plant kingdoms merge into one another among the Mastigophora. Many of these have been shown to have a true holophytic nutrition (e.g., *Volvox*), while others have only true holozoic nutrition (e.g., *Monas*). *Menoidium* is representative of the saprophytic genera of the Mastigophora, *Trypanosoma* is one of the saprozoic genera, and Schoenborn (1946) has shown that *Astasia* has a chemotrophic nutrition. Thus all known nutrition types occur among the members of this group.

Furthermore, many of the flagellates have a mixed nutrition. *Ochromonas*, for example, appears to have holozoic, holophytic, and saprophytic nutritions simultaneously, while species of *Euglena* have saprophytic and holophytic nutritions together. *Peranema* is both saprophytic and holozoic, and *Chrysamoeba* is both holophytic and holozoic.

It has usually been customary in the past to divide the Mastigophora into two subclasses, the Phytomastigina and the Zoomastigina, on the basis of the presence or absence of chromatophores. However, such a system is highly artificial, though at times it is a convenience for certain particular purposes. Many genera classified as Zoomastigina are more closely related to genera in the Phytomastigina than they are to other Zoomastigina, while within both "subclasses" there are many groups that have only a very distant relationship to each other. Furthermore, in some genera, such as *Euglena* and *Euglenamorpha,* some species are green and others colorless.

Among those Mastigophora with chromatophores, large numbers have approximately the same pigments in the same proportions as do the higher plants, so that the chromatophores are a bright green. In fact, some botanical authors (see Smith 1950) consider many of the organisms here listed as Mastigophora to be green algae (Chlorophyta), for some of them not only have green chromatophores, but also produce starch as a storage product, and have cell walls composed of cellulose and hemicellulose. The chromatophores in other forms, on the other hand, may be any of various colors. In some cases they are yellow, in others brown, in others red, blue-green, or bluish, depending upon the proportions of the usual chlorophyll pigments, and upon the presence of other secondary pigments.

The Mastigophora under consideration here, then, form a transition between typical animals and typical plants in their nutrition. In addition, some of them are transitional in their structure between the Protista and the higher algae. Certain of the "Phytomastigina" spend part of their life cycle in the free-swimming stage, then settle down, produce a gelatinous sheath or cellulose cell wall around themselves, and by a process of cell division transform themselves into a so-called palmella stage. The palmella is in no wise basically different from some of the lower forms among the Chlorophyta. Any cell whatsoever of the palmella may at times revert to the flagellated state and swim away by itself as an independent organism.

Only those palmellae in which the major portion of the life cycle is flagellated are considered at this point, those in which the major portion of the cycle is passed as the palmella being considered under the Chlorophyta. There are, however, numerous transitional cases, and it is impossible to draw any sharp line of demarcation, as it so often is when man attempts to classify natural objects for his own convenience.

The Mastigophora abound in various aquatic situations. Many are parasitic, but most are free-living. A large proportion of the free-living species is sessile, others are epiphytic or epizoic, while still others are so regularly found floating freely in the water that they are probably second in importance in the plankton only to the green algae in fresh water and to the diatoms in the sea. Planktonic forms are perhaps of approximately equal importance in marine and fresh-water environments, though, in general, different types are involved in the two cases. Mastigophora constitute one of the most important groups of organisms involved in the nannoplankton, for many of them are extremely small. Their importance in the nannoplankton is clearly indicated by experiments in growing pure culture of phytoplankters (e.g., see Gross, 1937). In the attempt to remove a single alga cell or colony of cells from a mixed collection of the plankton, the tiny plant is picked up along with a small quantity of water in a capillary pipette, and then is washed repeatedly up to ten times in sterile water. Yet even under this careful procedure it is not at all unusual for the investigator to find that his culture degenerates after a short time because of the enormous multiplication of green or colorless flagellates. In fact, unless extreme care is used it is difficult to obtain cultures that do not contain flagellates.

Unfortunately, because the nannoplankton has remained relatively little explored, most of the groups of smaller Mastigophora are relatively poorly known. Hence, in addition to the many important and basic problems concerned with the ecological importance of the nannoplanktonic Mastigophora, there also remain on a relatively large scale even such comparatively simple and fundamental problems as the taxonomy of many of the organisms involved.

Artificial key to some of the more important planktonic Mastigophora *

1. Cells with chloroplasts or chromatophores. "Phytomastigina" 2
   Cells without chloroplasts or chromatophores. "Zoomastigina" . . . .(representative genus: *Platytheca*)

*Many important genera are omitted from this key, because, among other reasons, of our imperfect knowledge of them.

2. One-celled organisms 3
Colonial forms with two to many cells 29
3. Cells with a calcareous skeleton, composed of plates and knobs. Mostly marine forms; family Coccolithophoridae 4
Cells without such a calcareous skeleton. Marine and fresh-water forms 20
4. Coccoliths imperforate; subfamily Syracosphaerinae 5
Coccoliths perforate; subfamily Coccolithophorinae 17
5. Shell consists of a covering of coccoliths only 6
Shell with diverging coccoliths, forming appendages 8
6. Shell without an orifice . . . . *Pontosphaera* (fig. 153)
Shell with an orifice 7
7. Coccoliths disc-shaped . . . . *Syracosphaera* (fig. 154)
Coccoliths cap-shaped . . . . *Calyptrosphaera*
8. Shell composed of both covering and diverging coccoliths 9
Shell only with diverging coccoliths . . . . *Lohmannosphaera*
9. Diverging coccoliths of the shell arranged around the orifice as radiating flotational rays or as a flotational collar 10
Diverging coccoliths of the shell arranged otherwise 14
10. Covering coccoliths well developed 11
Covering coccoliths lacking or very delicate; with radiating coccoliths around the orifice . . . . *Halopappus* (fig. 155)
11. With radiating rays around the orifice 12
With a flotational collar around the orifice . . . . *Petalosphaera* (fig. 156)
12. Radiating rays short . . . . *Najadea*
Radiating rays long 13
13. Radiating rays with few joints . . . . *Michaelsarsia* (fig. 157)
Radiating rays with many joints . . . . *Ophiaster*
14. Radiating coccoliths forming an equatorial girdle around the shell 15
Radiating coccoliths arranged otherwise; each with a long calcareous needle . . . . *Acanthoica* (fig. 158)
15. Radiating coccoliths leaf-shaped . . . . . . *Deutschlandia* (fig. 159)
Radiating coccoliths otherwise 16
16. Radiating coccoliths tube-like . . . . *Thorosphaera*
Radiating coccoliths goblet-shaped . . . . *Scyphosphaera* (fig. 160)
17. Shell consists of covering coccoliths only 18
Shell consists of diverging coccoliths only 19
18. Coccoliths consisting of a basal disc, a short tubular element, and a second external disc . . . . *Coccolithophora* (fig. 161)
Coccoliths consist of a basal disc and a short tubular element, the latter thickened considerably at the distal end . . . . *Umbilicosphaera* (fig. 162)

19. Radial coccoliths stick-shaped or club-shaped . . . .
. . . . *Rhabdosphaera* (fig. 163)
Radial coccoliths trumpet-shaped . . . . . *Discosphaera* (fig. 164)
20. Cells with a reticulate silicious skeleton. Marine forms only. Silicoflagellata (representative genera: . . . .
. . . . *Distephanus, Dichtyocha*) (figs. 165, 166)
Cells without such a silicious skeleton. Marine and fresh-water forms 21
21. Chromatophores brown, or other colors than green 22
Cell with green chloroplasts 25
22. Cell with one apical flagellum 23
Cell with two apical flagellae, of different lengths . . . .
. . . . *Ochromonas* (fig. 167)
23. Cell with pectinaceous case, which in most cases bears a number of long needle-like processes . . . . *Mallomonas* (fig. 168)
Cell without such a case 24
24. Cell with a shell . . . . *Chrysococcus* (fig. 169)
Cell naked . . . . *Chromulina* (fig. 170)
25. One flagellum 26
Two flagella . . . . *Chlamydomonas* (fig.172)
Four flagella 28
26. Cell flat, with a caudal spine; spirally twisted . . . .
. . . . *Phacus* (fig. 185)
Cell not thus 27
27. Cell with a lorica . . . . *Trachelomonas* (fig. 186)
Cell without a lorica . . . . *Euglena* (fig. 171)
28. Cells subrectangular to broadly elliptical. Brackish water . . . .
. . . . *Platymonas* (fig. 173)
Cells spherical, cylindrical, or elliptical. Fresh water . . . .
. . . . *Carteria* (fig. 174)
29. Chromatophores brown 30
Green chloroplasts 33
30. Colonies forming gelatinous masses 31
Colonies not forming gelatinous masses 32
31. Marine forms . . . . *Phaeocystis*
Fresh-water forms . . . . *Uroglena* (fig. 175)
32. Colonies branching; cells with a lorica . . . . *Dinobryon* (fig. 176)
Colonies compact and not branching; cells without a lorica
. . . . *Synura* (fig. 177)
33. Colony enclosed in a gelatinous sheath 34
Colony without a gelatinous sheath . . *Spondylomorum* (fig. 178)
34. Colony flattened 35
Colony spherical or ellipsoidal 36
35. Colony small, four-sided. All flagella on one face . . . .
. . . . *Gonium* (fig. 179)

Colony somewhat larger. Flagella on all faces . . . . *Platydorina* (fig. 180)

36. Colony very large (of more than 500 cells) . . . . *Volvox* (fig. 181)
    Colony much smaller than this 37

37. Cells in adult colony of two distinct sizes . . *Pleodorina* (fig. 182)
    Cells all nearly the same size 38

38. Cells usually crowded close together . . . . . . *Pandorina* (fig. 183)
    Cells spaced far apart . . . . *Eudorina* (fig. 184)

Interested persons should consult Huber-Pestalozzi (1941), Kudo (1946), Lohmann (1920), Pennak (1953), Schiller (1926), Schultz (1928), Smith (1950), Ward and Whipple (Conn and Edmondson, 1918) for further information on the Mastigophora.

b. DINOFLAGELLATA

The dinoflagellates are second in importance only to the diatoms as basic food-producers (phytoplankters) in marine plankton. They are especially important in the open sea in tropical regions, but at times they are of great significance in all marine waters. In fresh waters, though certain species may be extremely abundant in particular localities at particular times, the group is not so significant as it is in marine situations.

Although many dinoflagellates are photosynthetic and hence producers, they are also important as consumers (saproplankters and zooplankters), for, as in many of the other Mastigophora, the group is one of those that is transitional between the animal kingdom and the plant kingdom. It is treated, therefore, both by the botanists, who often classify the dinoflagellates as members of the Pyrrhophyta—comparable to the Chlorophyta or the Chrysophyta—and by the zoologists, who usually consider them to be an order or subclass (the Dinoflagellata) of the protozoan class Mastigophora.

Various dinoflagellates possess holophytic, saprophytic, holozoic, and mixed nutritions, and in some instances different types of nutrition are known to occur among species so closely related (as indicated by their structure) that they belong to the same genus. In the genus *Gymnodinium*, for example, *G. brevis* clearly is strictly holophytic, and therefore a producer, while *G. incisum* is holozoic. On the other hand, *G. fulgens* has been observed to contain both chromatophores and food vacuoles, showing that it has a mixed nutrition. A number of species, including *G. aureum*, appear to be saprophytic. According to Kofoid and Swezy (1921) the majority of dinoflagellate species, and individuals, are consumers rather than producers in the economy of the sea. Relatively more of the thecate (armored) forms are holophytic than is the case with the nonthecate types. Because the thecate forms are more easily studied, and hence are better known, many authorities have been under the impression that the dinoflagellates are almost entirely plant-like (see, for example, Kudo, 1946, who considers the dinoflagellates to be part of the Phytomastigina).

The Dinoflagellata are flagellated Protista, characterized by having two flagella, attached near each other, usually in the center of the cell. One flagellum is thread-like and trails posteriorly, and the other is held transversely, and usually is ribbon-like. The latter flagellum, in the vast majority of cases, is located in a special transverse groove on the body, called the girdle. The girdle is usually, but not always, more or less spiral, so that its left end is more posterior than the right end, when the organism is seen from the ventral side. In a number of forms, however, there is no displacement of the girdle, while in others the left end is more anterior than the right end. The longitudinal flagellum is in a somewhat less well-defined and usually broader longitudinal furrow, the sulcus. Both girdle and sulcus are almost always provided with distinct lips, and in the thecate forms with conspicuous skeletal flanges, thus making the group on the whole very easy to recognize.

Many species of the dinoflagellates possess only a protective flexible pellicle, but others have a well-defined porous theca, composed of cellulose. In *Prorocentrum* the theca is simple and bivalved, whereas in *Ceratium* and most other thecate genera, it is composed of numerous sculptured plates, whose arrangement and structure are of considerable importance in the determination of species. Many of the thecate forms are very ornate and beautiful, and are readily recognizable in preserved material. On the other hand, naked dinoflagellates, though often very beautiful in the living state because of their coloration and their many shapes, are extremely delicate, and hardly recognizable as dinoflagellates after they have been preserved. In practically no case is it feasible to determine the species of unarmored dinoflagellates from preserved material, and only occasionally can genera be recognized under such conditions.

In the genus *Polykrikos* the cytoplasm undergoes incomplete division, resulting in conjugate individuals with more than one girdle. In other cases, (e.g., *Cochlodinium*) the whole body of the cell has undergone torsion, or twisting, in such a manner that the girdle encircles the body more than once, and the sulcus takes on a spiral character, instead of being strictly longitudinal.

Each flagellum originates in a pore on the ventral side of the cell. The pore of the transverse flagellum is relatively anterior and that of the longitudinal flagellum relatively posterior. A large number of species, especially those which are saprophytic, have a peculiar sac-like vacuole, the pusule, which is connected with the exterior through one or both of the pores. The pusule has been observed to take in water from the exterior by a gradual or sudden expansion, but its functions are hypothetical.

As in other groups of flagellates, a number of the Dinoflagellata possess stigmata, or eyespots, which are small, light-sensitive accumulations of pigment within the cell. In the genus *Pouchetia* the photoreceptor is especially highly developed, for it consists not only of an accumulation of pigment, but also of a large hyaline lens, which serves to focus the light upon the sensitive area. The photoreceptor of *Pouchetia*, because of its specialization, is known as an ocellus, and it is comparable in many respects to the ocellus found in species of the Coelenterata.

In one other respect, certain of the dinoflagellates resemble the Coelenterata: in the presence of nematocysts in the genus *Polykrikos.* The nematocysts do not appear to differ essentially in structure or in function from those of the Coelenterata, yet it is clear that they were evolved by the organism itself, and that they are not derived from ingested fragments of medusae or hydroids. It does not seem likely, however, that these two similarities between certain dinoflagellates and the coelenterates indicate a direct evolutionary development of the coelenterates from the dinoflagellates, or vice versa.

Reproduction is largely asexual, by means of binary fission, and this may in some species result in temporary chains of individuals. However, practically none of the dinoflagellates produce permanent colonies, though *Polykrikos* appears to consist regularly of a small number of individual cells permanently associated together in a small syncytium. The entire life cycle is only very incompletely known in most forms, and there are many interesting and basic problems yet to be solved. Conjugation similar to that occurring in some of the green algae has been observed in *Ceratium*, and other types of sexual reproduction have also been described, or assumed, though little is known of them.

Usually during binary fission in the unarmored dinoflagellates, a typical cyst is formed, for apparently the organisms require protection and rest during this critical period. The cyst is usually very delicate, so that the organism easily escapes from it, though in some forms the cyst is tough and resistant, as in the case of the organisms previously classified as *Gymnodinium lunula* and *Pyrocystis* spp. (see fig. 187 and 188). So little is known concerning dinoflagellate life histories, however, that it is not clear to which species these forms belong. In fact, the pyrocystis is thought to be an encysted stage in both the thecate and non-thecate species. In many cases it shows no evidence of cell division, and it may well be that at least in some instances the major portion of the life cycle is spent in the pyrocystis stage. Much clarification is needed here.

"Fiery seas" have been known to man ever since he first became associated with marine environments, but only comparatively recently was it determined that "phosphorescence" in the ocean is produced by living organisms (the phenomenon, therefore, is more accurately known as bioluminescence). Light production in the sea is found in certain bacteria, dinoflagellates, coelenterates, annelid worms, crustaceans, chordates, etc., but the major portion of the bioluminescence that strikes the eye in the open sea and in coastal waters is due to the activities of the Dinoflagellata. The very generic names of some of the forms are indicative of this fact—*Noctiluca* (night light) and *Pyrocystis* (flaming cyst), for example. Luminescence is not, however, confined to these two groups; it is produced also by such forms as *Ceratium, Peridinium, Gonyaulax,* and *Gymnodinium.* Kofoid and Swezy (1921) describe a case of bioluminescence in the sea at La Jolla, California, as follows (p. 53): "During an outbreak of yellow water near the Biological Station caused by the presence of enormous quantities of *Gymnodinium flavum* ... the breakers along the shore were brightly luminous. A few forms of both *Gonyaulax* and *Noctiluca* were found in the hauls, but not in sufficient numbers to account for this display.... It seems, therefore, certain that the phosporescence observed was due to this organism [*Gymnodinium flavum*]...."

The nature of the production of bioluminescence and its functions need not be discussed here. These problems have been discussed in considerable detail by Harvey (1940, 1952) and Dubois (1928).

The matter of the blooming of plankton organisms, including single-species blooms, has been discussed above (pp. 75-78). Such blooming of the Dinoflagellata has frequently been noticed. Many of the dinoflagellates are highly colored, and when they are present in vast numbers they cause a distinct discoloration of the water. The colors most commonly resulting during dinoflagellate blooms are red, yellow, and green, though various other colors may be produced. The production of large masses of *Noctiluca,* such as occurs frequently in Puget Sound in summer, results in water whose general appearance is that of dilute tomato soup. Blooms of *Gonyaulax polyhedra* along the California coast result in a red water, which may extend for hundreds of miles. *Gymnodinium flavum* on the California coast and *G. brevis* in Florida waters produce water of a brilliant yellow color. These blooms of dinoflagellates may or may not be accompanied by mortality of marine organisms. In those cases where mortality occurs, the destruction may be considerable, as indicated above (pp. 78–80).

Artificial key to the commoner genera of planktonic Dinoflagellata

1. Transverse flagellum terminal, or nearly so; no girdle or sulcus present. Cell without a large and conspicuous tentacle 2
   Flagella in distinct girdle and sulcus at some stage of life cycle .. 3
2. Thecate forms .... *Prorocentrum* (fig. 189)
   Nonthecate forms .... *Haplodinium* (fig. 190)
3. Cell large and spherical, girdle and sulcus suppressed in adult. There is a single large and conspicuous tentacle .... *Noctiluca* (fig. 191)
   Cell not thus 4

4. Thecate forms 5
   Nonthecate forms, or theca very delicate 9
5. Epitheca very small and surrounded by the wide lips of the girdle. Marine forms only . . . . *Dinophysis* (fig. 192)
   Epitheca usually larger (i.e., girdle nearly equatorial). If small, not surrounded by lips of girdle 6
6. Apex of epitheca rounded or truncate, never acutely symmetrically pointed 7*
   Apex of epitheca with an acute point, or definite horn 8
7. Epitheca small, cap-like. Hypotheca with strong spines . . . . *Ceratocorys* (fig. 193)
   Girdle usually equatorial. Spines, if present, very short . . . . *Gonyaulax* (fig. 194)
8. Apex with a single horn, usually very long Hypotheca with one to four long horns, which may be strongly recurved . . . . *Ceratium* (figs. 195, 196)
   Apex with one short pointed process. Hypotheca usually with two similar points . . . . *Peridinium* (fig. 198)
9. Cell provided with an ocellus, with a well-developed lens and pigment area . . . . *Pouchetia* (fig 199)
   Cell without an ocellus 10
10. Cell is a small syncytium. With more than one complete girdle, and usually with two or more nuclei. Nematocysts present . . . . *Polykrikos* (fig. 200)
    Cell not a syncytium. One nucleus and one girdle. No nematocysts 11
11. Cell body with great torsion, so that girdle has at least 1.5 turns. Girdle widely displaced. Marine forms only . . . . *Cochlodinium* (fig. 201)
    Cell body with little or no torsion. Marine and fresh-water forms 12
12. Epitheca very short, at least on dorsal side . . . . *Amphidinium* (fig. 202)
    Hypotheca very short . . . . *Hemidinium* (fig. 197)
    Girdle more nearly equatorial 13
13. Girdle displaced less than 0.2 the total length of the body 14
    Girdle displaced more than 0.2 the total length of the body . . . . *Gyrodinium* (fig. 205)
14. Cell unarmored; sulcus extends into the epitheca; mostly the marine forms . . . . *Gymnodinium* (fig. 204)
    Cell with a very delicate theca; sulcus confined to the hypotheca; mostly fresh-water forms . . . . *Glenodinium* (fig. 203)

*Some species that have been described as belonging to the genus *Peridinium* have a rounded apex. There is doubt (see Graham, 1942) that they truly belong here.

Interested persons should consult Graham (1942), Graham and Bronikovsky (1944), Kofoid and Swezy (1921), Kiselev (1950), Kudo (1946), Lebour (1925), and Schiller (1933–37) for more detailed information on the Dinoflagellata.

c. Foraminifera.

The sarcodinian order Foraminifera is almost exclusively confined to the sea, though certain primitive species occur in fresh water as well. The latter have simple, flexible, gelatinous coatings around the cell, while the marine forms have a much more highly developed skeleton (test). In the majority of the marine forms the skeleton is calcareous, though in many it is chitinous or ferruginous, and in a few it is even silicious. The test is characteristically provided with numerous pores. Often in the species with chitinous or ferruginous skeletons the animal uses these materials to cement together various foreign particles. Such shells are known as arenaceous, inasmuch as the particles are usually tiny sand grains, or mud particles. Some of the arenaceous species, however, use materials other than sand and mud. One, for example, uses only tiny plates from disintegrated brittle stars (Ophiuroidea); another selects only mica flakes; a third utilizes sand grains for the most part, but incorporates almost always a single long, sharp, unbroken sponge spicule. The power of selectivity involved in the construction of such arenaceous shells as these is inexplicable with our present knowledge, especially when we consider that the first example is a species which invariably is permanently attached to submerged plants, while the particles it uses would be nearly confined to the bottom deposits.

In some of the Foraminifera the test consists of a single chamber, within which the cell body lies. However, in the majority of species the test consists of several chambers. When several chambers are present they are of different sizes, and of them all, the smallest is the first one that was produced by the animal. As its life cycle continues, it produces more and more chambers, each one larger than the former. Often, though by no means always, the chambers are arranged in a more or less tight spiral coil, so that many of the species (e.g., *Polystomella*) have a general appearance very similar to that of the nautiloid

Cephalopoda. In fact, when these organisms were first discovered, some of them were placed in the genus *Nautilus*.

The test of most species occurs in two forms, namely, one in which the primary chamber (the proloculum) is very small, and a second in which it is relatively larger. The former tests are said to have a microspheric proloculum, and the latter a megalospheric proloculum. The two types are regular stages in the life history of the Foraminifera. This will be described below.

The soft parts of the animal consist of a single cell, which may often, however, have several to many nuclei within a single mass of protoplasm. In general the protoplasmic mass is very large for the phylum Protozoa, though only rarely is it greater than 1 mm. in diameter. Foraminifera are known with tests 190 mm. in length (*Neusina*), but hardly any species are larger than about 5 mm. in diameter, and the average is only about 0.5 mm. in existing species. In general the test is an exoskeleton, though numerous pseudopodia are extended through the apertures and pores, and they characteristically anastomose to such an extent outside the shell that they veritably form an external layer of living substance. Food (largely vegetable material in most species) is captured by the pseudopodia, but is never taken inside the shell. Instead, digestion and absorption take place exterior to the hard parts.

The microspheric form is said to have a large number of nuclei, of varying sizes proportionate to the size of the chamber in which they are located. Eventually each nucleus becomes the center of a small mass of cytoplasm, and the small cell thus formed escapes from the parent test to take up an independent existence. Upon being freed, the new cell secretes around itself a new proloculum which is much larger than that of its parent. Thus it becomes megalospheric. As it grows, new chambers are produced, but there remains only a single nucleus, which maintains a position in one of the central chambers. When full size has been attained, the nucleus fragments, each fragment becomes separated from the others within a small mass of cytoplasm, and the cells escape from the parent test. These cells are flagellated, and are known as swarmers. They are isogametes, and they fuse together in pairs to form zygotes. The zygotes then secrete a new microspheric proloculum, and the cycle begins again. Thus the microspheric form reproduces asexually, forming a sexual megalospheric generation.

The vast majority of the Foraminifera are bottom-living, or are attached to submerged plants, etc. In shallow water, or near the bottom, many of these at times may be important tychoplankters. Certain important species, however, are planktonic in their habitat. These have tests that are in general thinner than ordinary, and with larger pores to make them lighter. Furthermore, many of them have developed long spines by means of which frictional resistance to sinking is accomplished. In the deeper parts of the sea, the planktonic genera find it possible to live in regions near the surface, where food is far more abundant than on the floor. Here they may live in vast numbers, and as they pass through the various stages of their life cycle the old shells are discarded. These then sink to the bottom, where over periods of many thousands or millions of years they have built up thick deposits (see Kuenen, 1941, 1950). Such deposits are named (after the most common pelagic genus) *Globigerina ooze*, though the shells of *Globigerina* are by no means the only type to be found there. Other pelagic genera contribute, and likewise benthic species live in the ooze in abundance. *Globigerina* oozes are the characteristic bottom deposits under about one-third of the area of the oceans of the world—proportionately more in the Atlantic Ocean, and less in the Pacific. They do not occur on the floor in the deepest parts of the ocean, presumably because the pelagic tests dissolve completely in falling through deep layers of water. The fossils of Foraminifera, laid down in such deposits in past ages, have often become consolidated into soft limestones known as chalk. One of the most famous of such fossil deposits is exposed as the White Cliffs of Dover in England.

Artificial key to the planktonic genera of Foraminifera*

1. Test consisting of a number of small chambers, which form a low cone, but with the last chamber or chambers bulbous and larger than all the others together...... *Tretomphalus* (fig. 206)
   Test not thus.. 2

*For the determination of tychoplanktonic genera, consult the references given at the end of the key.

2. Only two chambers showing externally, the larger partially enclosing the smaller. Other chambers internal . . . . *Chilostomella* (fig. 207)
   Only one chamber showing externally, with numerous small pores and one larger orifice. Other chambers internal . . . . *Orbulina* (fig. 208)
   Several chambers showing externally 3
3. Numerous apertural openings along borders of chambers where they come in contact with chambers formed earlier . . . . *Candeina* (fig. 209)
   Without such apertural openings 4
4. Test nautiloid. Younger chambers visible from both sides, but older ones hidden on one side by the more recently formed chambers . . . . *Pulvinulina* (fig. 210)
   Test not thus, loosely spiral or not spiral at all 5
5. Only the last volution of chambers is visible . . . . *Pullenia* (fig. 211)
   All chambers visible 6
6. Test coarsely perforate . . . . *Globigerina* (fig. 212)
   Test not thus . . . . *Hastigerina* (fig. 213)

Interested persons should consult Cushman (1910, 1911, 1913, 1914, 1915, 1917, 1918, 1920, 1922, 1923, 1924, 1928), Galloway (1933), and Kudo (1946) for further information on the Foraminifera.

d. Radiolaria

In contrast with the Foraminifera, the Radiolaria are entirely confined to the sea. They differ from all other Sarcodina in the presence of a sharply delimited central capsule. This consists of a small portion of cytoplasm towards or at the center of the cell, and it is surrounded by a nonliving membrane which is thought to be pseudochitinous or mucinoid. The capsule membrane in the suborders Acantharia and Spumullaria is provided with numerous fine pores, uniformly scattered over its surface. In the suborder Phaeodaria, however, the pores are concentrated in three positions on the capsule known as pore fields. The suborder Nassellaria has only one such pore field.

Within the membrane of the central capsule lies the nuclear material. Ordinarily there is but a single nucleus (which sometimes is colossal, for it may have a diameter of 1-2 mm.), but preceding multiplication several nuclei may be formed. In fact, in some species the increase in the number of nuclei takes place a considerable time before multiplication, resulting in an organism which is multinuclear for a significant portion of its life. The cytoplasm surrounding the nuclear material is richly supplied with vacuoles, oil droplets, and crystals, and it is said (Haeckel) that it is the central capsule that gives off light in the luminescent species.

Just outside of the capsule membrane there is a layer of relatively undifferentiated cytoplasm, in which there lies, in the Phaeodaria, a mass of pigment granules, the phaeodium (around one of the pore fields of the capsule membrane). The significance of the mass of pigment granules is not clear, but it is thought by some to be a mass of waste products. The digestion of food in the food vacuoles also seems to take place in this portion of the cytoplasm.

Outside of the layer of cytoplasm described above is a much thicker layer, the calymma, which may be semiliquid or semisolid in consistency in the various species. Its function is probably twofold: to protect the underlying portions of the cell body, and to act as a hydrostatic organ (see below). The calymma has been variously described by different authors, and its exact nature appears to need further study. Some have described it as a heavy, usually alveolar, gelatinous layer, through which run the strands of cytoplasm that distally form pseudopodia. Others describe it as a network of cytoplasm enclosing numerous alveoli filled with a gelatinous substance. Still others describe it as gelatinous for the most part, forming the walls of large vacuoles filled with a watery fluid. The numerous alveoli usually found in the calymma give it a characteristic foamy appearance.

Outside of the calymma is a thin layer of cytoplasm and the numerous pseudopodia. The latter appear to arise from the deeper layers of the extracapsular cytoplasm and pass through the calymma before radiating into the surrounding sea water. They are of two sorts, namely, ordinary fleshy pseudopodia and the more rigid axopodia, in which there is a stiff central axis. The pseudopodia are capable of slow movements, and animals resting upon a solid surface are capable of slow locomotion by means of them. It does not seem probable, however, that the pseudopodia are of any importance as swimming organs, though they continue slow waving movements during flotation (presumably as a means of seeking for food).

Practically all of the Radiolaria have a spiculate or solid skeleton, which is imbedded in the soft parts of the cell, and hence is a true internal skeleton. In most forms the skeleton is silicious, but in the suborder Acantharia it is composed of strontium sulfate or of calcium aluminum silicate. In this same suborder the skeletal spicules are arranged in an entirely different fashion from that in other forms. The spicules radiate from the center of the central capsule, then pass outward through the extracapsular protoplasm, and extend into the surrounding water. Secondary spicules, in many cases, form one or more lattice shells concentrically outside of the central capsule. In the remaining suborders, the radiating spicules, if present at all, never penetrate the central capsule. In all the suborders there are many cases in which the individual spicules of the lattice shell or shells have fused together to form a unified, rigid framework. This may extend completely around the central capsule, or it may only partially surround it, leaving one side open. The skeletons of some of the Radiolaria are among the most exquisite of all natural structures, as indicated by such generic names as *Theocalyptra* (divine veil).

Many of the Radiolaria living near the surface of the sea have been observed to contain (mostly in the extracapsular cytoplasm, but sometimes even inside the capsule) what the early investigators described as "yellow cells." More recent study has shown that these are symbiotic plants (zooxanthellae), each containing a couple of yellowish chromatophores. Some of the zooxanthellae are now known to be modified dinoflagellates and chrysomonads, and possibly all of them belong to these two groups. Apparently some Radiolaria are able to live almost indefinitely in the absence of organic food, inasmuch as they can depend upon their symbiotic zooxanthellae as a source of nutriment. Of course deep-water species are devoid of such holophytic symbionts, and depend entirely upon their own activities to obtain food.

A number of species are colonial in their structure, and some of these may attain a considerable size (more than a centimeter in diameter). Colonial species begin life as a relatively simple cell, with a structure similar to that described above (p. 184).

However, after a time the nucleus within the central capsule divides, to be followed by a division of the central capsule itself. Repeated divisions of this nature produce a syncytial cell with many central capsules, but with only a single extracapsular mass of cytoplasm. The large cell secretes a single unified skeletal framework. In other cases, true colonies of individuals are formed, each secreting its own shell.

The central capsule has been shown to be capable of independent existence, at least for a period of time, inasmuch as both nuclear and cytoplasmic materials are present. Isolated central capsules are able to regenerate all the remaining parts of the cell. Without a nucleus, however, the extracapsular cytoplasm is unable to continue life independently for any considerable length of time, nor is it able to regenerate a new central capsule.

The life cycle of the Radiolaria is very poorly understood. Binary fission, multiple fission, and budding have been described. In binary fission first the nucleus, then the central capsule, divides. The extracapsular cytoplasm and its associated skeletal parts are likewise more or less equally divided in those cases where a rigid skeletal framework does not preclude such an event. Each of the two daughter cells then regenerates the lost half of the skeleton. In those cases where the skeleton is rigidly fused to form a single unit, one of the daughter cells escapes completely and forms an entirely new skeleton.

In many cases, and perhaps universally, reproduction also takes place by means of biflagellated swarmers. These are produced within the central capsule, first by repeated nuclear divisions, then by corresponding cytoplasmic divisions. Such swarmers have been observed to be liberated from the central capsule, but their further history is unknown. Some workers have reported these flagellated swarmers to be of two different sizes, which would indicate the possibility of heterogamous (or anisogamous) sexual reproduction. Others have reported only a single size of swarmer, and it remains unclear whether these are isogamous sexual, or simply asexual, cells. It has been suggested that the large cells among the swarmers may be nothing more than escaped zooxanthellae, but this is a matter that requires further elucidation before any definite statement can be made.

Radiolaria have an important place in the economy of the sea. Practically all of them are planktonic in their habits, and they live at all depths in the water. They feed upon the microplankton, especially upon copepods, diatoms, and upon other Protozoa. They in turn are eaten in vast numbers by various plankton feeders among the nektonic, benthic, and planktonic animals. One of the best places to search for Radiolaria is in the digestive cavities of the medusae and other plankton feeders.

The Radiolaria reach their greatest abundance in waters of medium salinity (i.e., in typical salinities of the open ocean), but they also occur in such localities as the Baltic Sea, where salinities are much lower, as well as in those tropical regions where the salinity of the upper layers becomes relatively high because of evaporation. When they become locally abundant, they often form great masses.

Many species are confined to surface waters, while others are characteristic of deeper regions, and never occur near the surface. Most of the more superficial species belong to the suborders Acantharia and Spumellaria, and they are characterized in general by relatively light skeletons with large pores and thin spicules, as well as by a relatively large size. The forms which habitually live very close to the bottom in very deep water are for the most part members of the suborders Phaeodaria and Nassellaria. These have a relatively massive skeleton with small pores and heavy skeletal bars, and in general they are smaller forms. Spines and other similar structures are less well developed. The species characteristic of intermediate layers of water are intermediate in their structure between these two extremes.

Surface-living species have a certain power of vertical movement. During storms or when the surface water is excessively warm, they move down to a considerable depth to find quieter or cooler water. The mechanism by means of which they perform this migration is not well understood, and calls for further investigation. However, many Acantharia (which are largely surface forms) bear a ring of myofibrils around each of the radial skeletal spines at the point where it leaves the cell body. One end of each myofibril is attached to the spine, and the other to the soft parts of the cell. It has been thought (e.g., by Haeckel) that by the contraction of these fibrils the volume of the calymma is increased, thus reducing the over-all specific gravity of the animal and allowing vertical migration upward. Relaxation of the myofibrils would then allow the elastic calymma to contract, reducing its volume, and the animal would sink. However, it does not appear probable that this can be the correct explanation, for in this case the maintenance of a position near the surface would require constant contraction of the myofibrils. Furthermore, expansion of the volume of the calymma would result in a reduction of specific gravity, but could not reduce it below that of sea water, and thus sinking might be retarded, but the animal would not rise.

Kudo has suggested that the myofibrils are used to contract the calymma. This would avoid the weaknesses of the hypothesis described above; but unfortunately, located as they are, it would be physically impossible for the myofibrils to contract the calymma. More probably their main, or only, function is to move the radial spines.

The constant rain of dead radiolarian shells over a period of millions upon millions of years has resulted in a vast accumulation on the floor of the sea. Radiolarian shells are found in abundance in many terrigenous deposits, in all the so-called oozes, and in the deposits of red clay in the deepest basins of the oceans. However, they are relatively most abundant on the floor at depths of from 4,300 to over 8,000 m., where they often form typical radiolarian oozes, in which over 50 per cent of the deposited material consists of radiolarian shells. Such deposits cover from 3 to 4 per cent of the bottom area of the seas. Fossil radiolarian oozes are known and form rocks such as chert. Perhaps the most extensive deposits of fossil Radiolaria are in the Barbados Islands in the Carribbean, and in the tertiary rocks of California. However, fossils of this group are very widespread, and they are thought to occur even in pre-Cambrian rocks, thus being among the oldest of all fossils.

Artificial key to certain of the commoner genera of the Radiolaria*

1. Skeleton consists of spicules radiating from within the central capsule, often also with a lattice shell of spicules outside the central capsule. Composed of strontium sulfate as a rule. Acantharia 2
   Skeleton usually present, but occasionally absent. Not arranged as above. Silicious 24
2. Without a complete lattice shell 3
   With a complete lattice shell 11
3. Radial spines from 10 to 200. With irregular arrangement 4
   Radial spines 20. Arranged according to Müller's Law † 5
4. Radial spines numerous, not fused together at the base . . . . *Actinelius* (fig. 214)
   Radial spines variable in number; opposite spines fused together at the center of the cell . . . . . . *Acanthochiasma* (fig. 215)
5. All of the radial spines similar 6
   Some radial spines differ from the remainder in size and shape . . . 9
6. Radial spines simple 7
   Radial spines provided with apophyses . . . . *Astrolonche*
7. Radial spines cylindrical . . . . *Acanthrometron* (fig. 216)
   Radial spines not thus 8
8. Radial spines compressed, elliptical in cross section . . . . *Zygacantha* (fig. 217)
   Radial spines quadrangular in cross section . . . . *Acanthonia* (fig. 218)
9. Two opposite equatorial spines are larger than all the remaining 18 spines . . . . *Amphilonche* (fig. 219)
   Four equatorial spines are larger than the remaining 16 spines . . . 10
10. Part or all of the radial spines with two opposite lateral branches or apophyses . . . . *Lithoptera* (fig. 220)
    Radial spines simple, sometimes forked, but never with lateral branches or apophyses . . . . *Acanthostaurus*
11. Two or four radial spines much larger than the remaining ones . . . 12
    All radial spines of equal size 15
12. Six radial spines larger than the remaining ones . . . . *Hexalaspis* (fig. 221)
    Two radial spines larger than the remaining ones (which are at times rudimentary) 13

*This key is based largely upon Haeckel's (1887) arrangement of the Radiolaria.

†Haeckel (1887) states Müller's Law as follows (p. 717); "Between two poles of a spineless axis are regularly disposed five parallel zones, each with four radial spines; the four spines of each zone are equidistant from one another, and also equidistant from each pole; and the four spines of each zone are so alternating with those of each neighboring zone, that all twenty spines together lie in four meridian planes, which intersect one another at an angle of 45°."

13. Shell with two opposite cone-shaped funnels, or nearly cylindrical. The funnels are the sheaths of the two enlarged spines . . . . *Diploconus* (fig. 222)
    Shell ellipsoidal, no cone-shaped funnels 14
14. Forty parmal pores on the shell . . . . *Belonaspis* (fig. 223)
    Eighty to 2,000 or more parmal pores . . . . *Phatnaspis* (fig. 224)
15. Shell single, with only one lattice shell 16
    Shell double, composed of two concentric lattice shells 23
16. Shell composed of a pavement of innumerable very small plates, pierced by 20 or 80 aspinal pores . . . . *Sphaerocapsa* (fig. 225)
    Shell composed of a latticework of branching apophyses, two or four originating from each radial spine 17
17. Each radial spine with two primary apophyses, which, however, branch 18
    Each radial spine with four primary apophyses 21
18. Apophyses of radial spines form lattice plates 19
    Apophyses of radial spines do not form lattice plates. No parmal pores 20
19. Forty parmal pores (two in each plate) . . . *Dorataspis* (fig. 226)
    Eighty to 200 or more pores (two in each plate, plus coronal pores) . . . . *Coscinaspis* (fig. 227)
20. With secondary spines on the apophyses . . . *Pleuraspis* (fig. 228)
    Without such secondary spines . . . . *Phractaspis* (fig. 229)
21. With secondary spines on the apophyses . . *Lychnaspis* (fig. 230)
    Without such secondary spines on the apophyses 22
22. Eighty parmal pores (four on each plate) . . . . *Tessaraspis* (fig. 231)
    160 to 300 or more pores . . . . *Icosaspis* (fig. 232)
23. All of the radial spines are similar . . . . . . *Phractopelta* (fig. 233)
    Some of the radial spines with and some without apophyses in the free external portion . . . . *Dorypelta* (fig. 234)
24. Fundamental shape of the shell is spherical. Spumellaria 25
    Fundamental shape is ovoid or elongate 68
25. Skeleton lacking or imperfect. No lattice shell 26
    Skeleton with well-developed lattice shell or shells 33
26. Skeleton lacking 27
    Skeleton present, consisting of individual scattered spicules 30
27. Noncolonial forms 28
    Colonial (or syncytial) forms . . . . *Collozoum* (fig. 235)
28. Neither central capsule nor calymma with alveoli . . . . . *Actissa*
    Alveoli present in the calymma 29
29. Nucleus spherical . . . . *Thalassicolla* (fig. 236)
    Nucleus branched or covered with radial sacs . . . *Thalassophysa*

30. Noncolonial forms 31
Colonial (or syncytial) forms 32

31. Spicules simple . . . . *Physematium*
Spicules branched . . . . *Thalassoxanthium* (fig. 237)

32. Spicules all of one kind . . . . *Sphaerozoum* (fig. 238)
Spicules of two or more kinds, some simple and some branched
. . . . *Rhaphidozoum*

33. Lattice shell spherical or composed of concentric spheres 34
Lattice shell not thus 48

34. Shell not armed with radial spines, though it may be thorny
or rough 35
Shell armed with two to numerous radial spines 40

35. Solitary cells 36
Colonial forms 38

36. Skeleton partly or wholly spongy in structure . . . . *Spongodictyon*
Skeleton of one, two, or three concentric lattice shells 37

37. Skeleton of a single lattice shell . . . . *Cenosphaera* (fig. 239)
Skeleton of two concentric lattice shells . . . .
. . . . *Carposphaera* (fig. 240)
Skeleton of three concentric lattice shells . . . . *Thecosphaera*

38. Outside of shell smooth, without spines or tubuli . . . .
. . . . *Collosphaera* (fig. 241)
Outside of the shell not thus 39

39. Outside of shell with solid spines . . . . *Chaenicosphaera* (fig. 242)
Outside of shell with irregular, solid tubuli . . . .
. . . . *Siphonosphaera* (fig. 243)
Outside of shell with irregular tubuli, open on both ends and
with fenestrated walls . . . . *Solenosphaera* (fig. 244)

40. Shell surface with two, four, or six main radial spines 41
Shell surface with numerous main radial spines 45

41. Two main radial spines 42
Six main radial spines 44

42. Only one lattice shell 43
Skeleton consists of two concentric lattice shells . . . .
. . . . *Stylosphaera* (fig. 245)

43. Both polar spines equal in size . . . . *Xiphosphaera* (fig. 246)
Polar spines differ in size or form . . . . *Xiphostylus* (fig. 247)

44. Only a single lattice shell . . . . *Hexastylus* (fig. 248)
Two concentric lattice shells . . . . *Hexalonche* (fig. 249)
Three concentric lattice shells . . . . *Hexacontium* (fig. 250)

45. Shell spongy in construction . . . . *Spongosphaera*
Shell not spongy in construction 46

46. Only a single lattice shell 47
Two concentric lattice shells . . . . *Haliomma* (fig. 251)
Three concentric lattice shells . . . . *Actinomma* (fig. 252)

Five or ten or more concentric lattice shells . . . . *Arachnosphaera* (fig. 253)

47. Spines all simple . . . . *Acanthosphaera* (fig. 254)
    Spines branched or forked . . . . *Elaphococcus*

48. Shell ellipsoid, or cylindrical with transverse constrictions 49
    Shell not thus 52

49. No transverse constrictions 50
    Three parallel transverse constrictions . . . . *Panartus* (fig. 255)

50. Both polar spines are equal . . . . *Lithatractus* (fig. 256)
    The polar spines are not equal 51

51. Shell is simple . . . . *Druppatractus* (fig. 257)
    Shell is double . . . . *Xiphatractus* (fig. 258)

52. Shell in the form of a disc, or of a biconvex lens 53
    Shell with three different axes at right angles to each other 62

53. Shell of ordinary structure 54
    Shell divided internally into a number of chambers or compartments, which are arranged spirally or otherwise 56

54. Equatorial margin of shell with a latticed flat girdle . . . . *Trigonactura*
    Equatorial margin without girdle, though there may be a ring of spines 55

55. Margin of disc with eight radial spines . . *Heliosestrum* (fig. 259)
    Margin with 10 to 20 or more spines . . . . *Heliodiscus* (fig. 260)

56. Shell surface spongy 57
    Shell surface with porous or nonporous sieve plates, but not spongy 59

57. No radial appendages on the spongy disc . . . . . . *Spongodiscus*
    With spongy radial arms 58

58. Three marginal arms . . . . *Rhopalodictyum*
    Four marginal arms . . . . *Spongaster*

59. Margin of disc simple, without spines or other appendages . . . . *Porodiscus* (fig. 261)
    Margin of disc with spines or other marginal appendages 60

60. Margin of disc with solid radial spines . . . . *Stylodictya* (fig. 262)
    Margin of disc with radial chambered arms 61

61. Shell radially symmetrical . . . . *Hymeniastrum* (fig. 263)
    Shell bilaterally symmetrical . . . . *Euchitonia* (fig. 264)

62. The two poles of each axis are equal. Shells regular and symmetrical 63
    The two poles of an axis are dissimilar. Shell spiral or irregular . . 67

63. Shell completely latticed . . . . *Larnacantha* (fig. 265)
    Shell with symmetrical gaps in the latticework 64

64. Medullary shell simple . . . . *Trizonium* (fig. 266)
    Medullary shell trizonal, with three girdles 65

65. Cortical shell with two girdles 66
Cortical shell with three girdles . . . . *Pylonium* (fig. 267)

66. The four gaps in the shell are simple . . . . . . *Tetrapyle* (fig. 268)
The four gaps in the shell are bisected by a septum . . . .
. . . . *Octopyle* (fig. 269)

67. Shell with a spiral arrangement . . . . *Lithelius* (fig. 270)
Shell arranged irregularly . . . . *Phorticium* (fig. 271)

68. With a mass of pigment (phaeodium) in the extracapsular cytoplasm. Central capsule with three pore fiields, one covered by the phaeodium. Phaeodaria 69
Without a phaeodium and with a single pore field. Nassellaria 80

69. Skeletal spicules do not form a lattice sphere, but are arranged radially in the extracapsular cytoplasm 70
Skeletal spicules forming one or more lattice shells 71

70. Radial spicules unbranched . . . . *Aulacantha* (fig. 272)
Spicules with terminal branches . . . . *Aulographis* (fig. 273)

71. Skeleton is bivalved, the valves usually completely separated 72
Skeleton not bivalved 73

72. The valves of the shell are firm and thick and regularly latticed
. . . . *Conchopsis* (fig. 274)
The valves are thin and fragile and hardly latticed . . . .
. . . . *Coelodendrum* (fig. 275)

73. A large shell mouth present 74
No such shell mouth present 76

74. A corona of foot-like spines around the shell mouth . . . .
. . . . *Gazelletta* (fig. 276)
No corona around the shell mouth 75

75. Shell with marginal spines . . . . *Challengeron* (fig. 277)
Shell without marginal spines . . . . *Challengeria* (fig. 278)

76. Shell with many hollow tangential cylindrical tubes 77
Shell not thus 79

77. Shell with tent-shaped elevations . . . . *Sagoscena* (fig. 279)
Shell smooth or spiny, without tent-shaped elevations 78

78. Surface smooth . . . . *Sagena*
Surface spiny . . . . *Sagosphaera* (fig. 280)

79. Meshes of the network triangular, either regular or nearly so
. . . . *Aulosphaera* (fig. 281)
Meshes of the network polygonal, and irregular . . . .
. . . . *Aulonia* (fig. 282)

80. Skeleton without a complete lattice shell 81
Skeleton with a complete lattice shell 86

81. Skeleton consists of a single ring only 82
Skeleton consists of a vertical ring and a horizontal basal ring . . . 84
Skeleton consists of two vertical rings and often a horizontal basal ring as well 85

82. Three to four basal feet on the skeletal ring . . *Cortina* (fig. 283)
No feet on the ring 83

83. Ring with branched spines . . . . *Lithocircus* (fig. 284)
Ring without branched spines, often smooth . . . . *Zygocircus* (fig. 285)

84. Basal ring with regularly placed basal feet . . . . *Cortiniscus* (fig. 286)
Basal ring not thus . . . . *Clathrocircus* (fig. 287)

85. Four gates in the skeleton . . . . *Zygostephanus* (fig. 288)
Six gates in the skeleton . . . . *Eucoronis* (fig. 289)

86. The shell is constricted into two halves sagittally 87
The shell has several lobes . . . . *Botryocampe* (fig. 290)
The shell is simple, and without lobes 89

87. Three basal feet . . . . *Tripospyris* (fig. 291)
Six basal feet . . . . *Liriospyris*
Seven to twelve or more basal feet 88
No basal feet . . . . *Dictyospyris* (fig. 292)

88. Apex with one horn . . . . *Petalospyris* (fig. 293)
Apex with numerous horns . . . . *Ceratospyris* (fig. 294)

89. Shell with no transverse constrictions 90
Shell with one transverse constriction 91
Shell with two transverse constrictions 98
Shell with three to six or more transverse constrictions 104

90. Shell with a horn . . . . *Cornutella* (fig. 295)
Shell without a horn . . . . *Cyrtocalpis* (fig. 296)

91. No radial apophyses 92
Three radial apophyses 94
Numerous radial apophyses 96

92. Shell conical, gradually dilated . . . . *Sethoconus* (fig. 297)
Shell cylindrical or ovoid. Mouth constricted or tubular 93

93. Shell without a horn . . . . *Dictyocephalus* (fig. 298)
Shell with a single horn . . . . *Sethocorys* (fig. 299)

94. Three radial ribs present, partially enclosed in the wall of the shell 95
Three radial ribs present, completely free from the shell . . . . *Eucecryphalus*
No ribs are present, but there are three radial feet around the mouth . . . . *Lychnocanium* (fig. 300)

95. Ribs prolonged into feet . . . . *Dictyophimus* (fig. 301)
Ribs prolonged into lateral wings . . . . . . *Lithomelissa* (fig. 302)

96. Three ribs in the shell . . . . *Sethophormis* (fig. 303)
No ribs in the shell 97

97. Shell with an apical horn . . . . *Anthocyrtium* (fig. 304)
Shell without an apical horn . . . . *Carpocanium* (fig. 305)

98. No radial apophyses 99
Three radial apophyses 102
Numerous radial apophyses . . . . *Calocyclas* (fig. 306)

99. Mouth of shell closed by a latticed plate . . *Theocapsa* (fig. 307)
Mouth of shell is wide open 100

100. Mouth more or less constricted . . . . *Theocorys* (fig. 308)
Shell gradually dilates towards the mouth 101

101. Abdomen of shell conical . . . . *Theoconus* (fig. 309)
Abdomen of shell flatly expanded . . . . . *Theocalyptra* (fig. 310)

102. Three wings or limbs on abdomen . . . . . . *Podocyrtis* (fig. 311)
Three wings or limbs, mainly or entirely on thorax 103

103. No free external appendages on abdomen . . . .
. . . . *Pterocorys* (fig. 312)
Free appendages on abdomen . . . . *Pterocanium* (fig. 313)

104. Mouth of the shell distinctly constricted 105
Mouth of shell not constricted 106

105. Shell with a horn . . . . *Eucyrtidium* (fig. 314)
Shell without a horn . . . . *Lithocampe* (fig. 315)

106. Shell with a horn . . . . *Lithostrobus* (fig. 316)
Shell without a horn . . . . *Lithomitra* (fig. 317)

Interested persons should consult Haeckel (1887) and Haecker (1908) for further information on the Radiolaria.

e. Tintinnidiidae

In marine waters the most prevalent of the ciliates occurring in the plankton are the tintinnids, and they also are frequently encountered in fresh water. Various authorities classify them differently within the Infusoria. Kofoid and Campbell (1929), for example, consider them to be members of a special suborder, the Tintinnoina, which belongs to the order Heterotrichida, while Kudo (1946) and Hyman (1940) consider them to be members of the family Tintinnidiidae, in the suborder Oligotricha and in the order Spirotricha. For present purposes, the latter classification will be adopted, though it is realized that any classification used will be open to certain objections.

At times the tintinnids become very abundant in marine coastal waters, but they are rather susceptible to relatively slight environmental changes, and their numbers fluctuate widely. Certain genera are characteristic of coastal waters, and others of the high seas. Only a few of the genera are found in fresh waters, but none of them is confined to this environment.

The cell of the tintinnids is usually cone-shaped, and it is attached at the aboral end to the inner portion of a cone-shaped or tubular case, the lorica. Around the broad oral end of the cell there is a left-hand spiral of well-developed membranelles, one end of which terminates in the oral depression. It is by means of this organelle that the animal propels itself and its lorica through the water, and that food particles are collected for ingestion. Cilia other than the membranelles are relatively sparse. In some genera there is a line of strong cilia on the gullet side of the body, and usually, especially in the upper portion of the body, there are scattered long body cilia.

By means of myofibrils the cell is able to extend or contract itself, and thus protrude the oral end out into the surrounding water through the mouth of the lorica, or else pull itself protectively back into its shell. Observation of living animals shows these movements, and those of the cilia and membranelles, to be complex and highly co-ordinated. Co-ordination is attained by neurofibrils emanating from a relatively large motorium.

The cell, as always among the Infusoria, contains two types of nuclei—macronuclei and micronuclei. In the tintinnids the number of nuclei of the two kinds is usually the same, but from species to species the number of each kind varies from one to a hundred. The macronuclei are usually sausage-shaped, while the micronuclei are tiny and spherical. Reproduction is by means of binary fission and conjugation.

All known tintinnids bear a lorica, which is not closely adherent to the cell, but surrounds it at some distance except at the point of basal attachment. The lorica may be gelatinous or pseudochitinous, and in many cases these substances are used to glue together numerous tiny grains of sand, portions of diatom shells, or shells of coccolithophores. In either case the shell may be simple or double, toothed around the mouth or plain, or it may be entire or alveolate. The basic material of the lorica is secreted by the cell, but the details of its arrangement seem to be determined by the activities of the animal, which shape the material into the form characteristic of the particular species involved. The shape of the lorica and its other characteristics are so very constant from one individual to another of the same species that the lorica alone is used in the determination of genera and species. This is convenient in plankton studies, inasmuch as the commonly used preservatives always result in extreme distortion of the soft parts of the animal, and usually they also result in the complete separation of the soft parts from the lorica, so that, as a general rule, in such preserved samples the only part of the animal available for study is the lorica. The presence or absence of foreign material, the nature of the lip of the shell, the presence and structure of alveoli, and the twisting and ornamentation of the shell are important characteristics in the determination of genera and species. Size, which varies from 20μ to 640μ, or more, is greatly influenced by external conditions, and therefore is useful only in a general way.

Artificial key to the commoner genera of the Tintinnidiidae

1. Lorica open at both ends, more or less cylindrical .... *Eutintinnus* (fig. 318)
   Lorica vase-shaped or sac-like. Usually closed at aboral end 2
2. Collar well developed, with large windows. Bowl swollen. No aboral horn, and no spiral structure ..... *Dictyocysta* (fig. 319)
   Lorica without windows in collar 3
3. Lorica with longitudinal ribs connected by fenestrae .... *Rhabdonella* (fig. 320)
   Lorica not thus 4
4. Lorica elongate, clarinet-shaped, usually with flaring lip around mouth. Aboral end contracted, but always open .... *Salpingella* (fig. 321)
   Lorica not thus 5
5. Oral end of lorica with closely arranged annular or spiral ornamentation 6
   If such ornamentation is present, either it is not confined to the oral end of the lorica, or it is not closely arranged 7
6. Collar sharply defined from an oval or spherical bowl .... *Codonellopsis* (fig. 322)
   Aboral region conical .... *Helicostomella* (fig. 323)
7. With hyaline collar and a patterned bowl .... *Stenosemella* (fig. 324)
   Collar, if present, not hyaline. Bowl patterned or not 8
8. Lorica with spiral lamina .... *Coxliella* (fig. 325)
   Lorica not thus 9
9. Lorica with agglomerated foreign materials 10
   Lorica without foreign materials 11

10. Collar separated from remainder of lorica by a well-defined nuchal constriction or an internal nuchal shelf. . . . . *Codonella* (fig. 326)
No nuchal groove or nuchal shelf. . . . . . *Tintinnopsis* (fig. 327)
11. Suboral region with one or two bulges. . . . *Ptychocylis* (fig. 328)
Lorica not thus. 12
12. Lorica with reticular portion aborally. . . . *Epiplocylis* (fig. 329)
Reticulation, if present, not thus. 13
13. Collar separated by a band from the bowl proper. . . . . *Cyttarocylis* (fig. 330)
Lorica not thus. 14
14. Wall with a regular polygonal structure. . *Parafavella* (fig. 331)
Wall not thus. 15
15. Lorica with an aboral horn. 16
Lorica without an aboral horn. 18
16. Horn in form of a lancet. . . . . *Xystonellopsis* (fig. 332)
Horn not thus. 17
17. Wall hyaline, or with a simple primary structure. . . . . *Parundella* (fig. 333)
Wall not thus. . . . . *Favella* (fig. 334)
18. Lorica with an inner collar. . . . . *Proplectella* (fig. 335)
Lorica without an internal collar. . . . . *Undella* (fig. 336)

Interested persons should consult Campbell (1942) and Kofoid and Campbell (1929, 1939) for further information on the Tintinnidiidae.

5. Porifera

All adult sponges are sessile, and consequently they do not appear in the plankton. Sponge spicules, on the other hand, derived from disintegrated sponges, often occur in plankton samples. Their presence, however, is accidental. In addition, in marine waters, larval stages of sponges, called amphiblastulae (fig. 337), are occasionally encountered. The asexual reproductive bodies, called gemmules (fig. 338), of fresh-water sponges may be tychopelagic at times, but they are never true members of the plankton.

6. Coelenterata

The coelenterates are rather simple animals, of a low grade of evolutionary development. In simpler cases the body wall consists of two thin layers of epithelial cells, the ectoderm and the endoderm. The two are separated by a noncellular, gelatinous stratum, the mesoglea, which in many forms among the Scyphozoa and the Anthozoa is especially thick, and often very fibrous. In these forms there has been a migration of amoeboid cells into it so that it takes on some of the characteristics of a simple connective tissue. The body wall encloses a primitive cavity, the gastrovascular cavity (or coelenteron), in which are combined two major body functions, digestion and circulation. There have been relatively few kinds of cells evolved among the coelenterates, and many of them have more than one function to perform. The most unusual cells present are the cnidoblasts, or stinging cells, each containing a complicated nematocyst. Upon suitable stimulation by an enemy or prey (or by chemical or other artificial means), the nematocysts are discharged as long, often barbed, threads. Some of them penetrate into the tissues of the enemy or prey, carrying poison along with them, so that in smaller animals paralysis results. In man a nettle-like rash is produced by many forms (*Cyanea,* etc.), while a few forms can cause severe pain (*Physalia*), or even occasional deaths (*Lobonema*).

Basically the members of the phylum are radially symmetrical, though among the Anthozoa a biradial symmetry has developed in many cases, through the elongation of the mouth and the development of such biradially arranged organs as the siphonoglyphs.

Considering the phylum as a whole, a very large number of the species of the Coelenterata are colonial, though this is true of relatively small numbers of the planktonic forms. The individuals of the colony are not completely separate from each other, but remain organically attached to each other by means of both skeletal and soft parts. Usually the gastrovascular cavity of one individual of the colony connects directly with that of the other individuals, so that if one individual fails or is physically unable to obtain food for itself, it is fed by those members of the colony which do feed. In this way each zooid (person) of the colony has many of the characteristics that we usually associate with an individual, while at the same time it has lost its individuality to the extent that it is an integral part of a whole colony. Meanwhile, the colony as a whole begins to take on some of the rudiments of individuality. The process of the subordination of the person to the colony reaches its extreme among the siphonophores (see below, p. 203), where the several persons occupy a position in the colony very comparable to that of individual organs in the body of the higher Metazoa.

Within the Coelenterata there are two main types of persons or individuals. The polyp type typically is elongate on the oral-aboral axis, and is an attached (or at least benthic) form. The medusa type is considerably or greatly shortened on the oral-aboral axis, and is a planktonic form. However, some polyps are pelagic (e.g., the hydroid *Pelagohydra*) and some medusae are benthic (*Cassiopeia, Eleutheria, etc.*).

Among the colonial coelenterates the persons of the colony are not usually all of one kind. For example, in the hydroids (Hydrozoa) there are often three different types, two of which are polyps, and one a medusa. Such a production of more than one type of individual is known as polymorphism, a condition for which the coelenterates are especially notable among animal phyla. The typical feeding individuals (hydranths or gastrozoids) are polyps, and have a ring or rings of tentacles around a mouth. On the colony, however, there are often also asexual reproductive individuals (gonangia), which have no mouth and no tentacles, and which depend upon the hydranths for their supply of food. The third type of individual is a medusa, and it is the sexual reproductive individual. Typically the medusa has a functional mouth and tentacles, because in most cases it takes up an existence in the plankton completely independent of the other persons of the colony.

The relation of the types of individual to one another is clearly seen in the life cycle of a typical and familiar example, *Obelia*. Eggs and sperm are produced by the medusae, and fertilization takes place externally in the sea water. The zygote thus formed develops into a free-swimming larva, called the planula (fig. 339), which consists of an external layer of ectodermal cells surrounding a solid central core of endodermal cells (in some planulae the endoderm is hollow rather than solid, but the cavity is not connected to the exterior through a mouth). After swimming around for a period of time the planula settles down and becomes attached to some solid surface. Meanwhile it develops a cavity in the endoderm, the cavity breaks through at the unattached end of the larva, and a circlet of tentacles develops around the newly formed mouth.

The hydranth thus formed produces asexual buds in the same manner that these are produced in asexual reproduction in hydra, but in *Obelia* the buds remain attached intimately to the parent hydranth, thus forming a colony. There is also a growth of root-like structures over the substratum, forming the hydrorhiza. After the colony is well formed, special buds, formed similarly to those that produce the new hydranths, appear at the angles between the hydranths and the stem of the colony. These buds develop into gonangia. Each gonangium produces a number of small asexual buds on its club-shaped central blastostyle. Eventually these buds drop off completely as tiny immature medusae, whereupon they take up an independent existence in the plankton. When they become mature, the medusae develop gonads and they reproduce sexually.

In *Obelia* the life cycle is divided approximately equally between the attached and the free-living stages. However, all possible variations can be found among the hydroids with respect to their life histories. The hydra, for example, has only a hydranth, which reproduces both asexually and sexually. *Pennaria* produces degenerate medusa buds, which become sexually mature and usually shed their eggs and sperm before being liberated from the colony. After liberation they live only very briefly, for they cannot partake of food. In *Obelia* the hydranth and medusa stages are approximately equal in importance. In *Gonionemus* and *Hybocodon* most of the time of the life cycle is spent in the medusa stage. In fact, *Hybocodon* is able to reproduce asexually by budding in the medusa stage, in addition to the usual sexual means. Finally, *Rathkea* exists only as a medusa.

Among the members of the class Scyphozoa the life cycle is similar to that of the Hydrozoa, except that the medusa stage is always dominant. Medusae bud off from a strobila (a hydranth) to form planktonic ephyra larvae (fig. 340), and these grow gradually to take on the adult characteristics.

Polymorphism is carried to extemes among the siphonophores (*Hydrozoa*), such as the familar Portuguese man o' war (*Physalia*) and the sail-by-the-wind (*Velella*). A siphonophore may consist of some or all of the following types of individuals, each completely different in structure and function from all the others:

polyps:

gastrozoids, which are the feeding individuals;
dactylozoids, which are the tactile individuals;
gonozoids, which are the asexually reproducing individuals.

medusoids:

swimming bells, which are the locomotory individuals;
bracts, which are protective individuals;
gonophores, which are the sexual reproductive individuals produced by the gonozoids (in a few cases these are set free, as are the medusae of many of the hydroids);
pneumatophore, which is a gas-filled float.

The phylum Coelenterata is commonly divided into three classes: the Hydrozoa, the Scyphozoa, and the Anthozoa. The Hydrozoa are characterized by having either a medusa stage or polyp stage or both, with a velum in the medusa and an undivided gastrovascular cavity in the polyp. The Hydrozoa include the hydras, the hydroids, the velate jellyfish (hydromedusae), the stinging corals (Hydrocorallina), and the siphonophores. The Scyphozoa are characterized by having a nonvelate medusa stage and a polyp stage in which the gastrovascular cavity is subdivided into four or more parts by vertical partitions, or mesenteries. The medusa stage is always dominant, and in some forms there is no polyp whatsoever. The Scyphozoa include the nonvelate giant jellyfish. None of the Anthozoa has a medusa stage, and the polyp is characterized by the presence of a few to many vertical mesenteries dividing the gastrovascular cavity into partitions. The Anthozoa include the true corals, the sea anemones, the sea pens, etc.

The majority of the coelenterates are attached and benthic, at least during a part of their life cycle. The medusa stages are characteristically pelagic and planktonic, however, and a few nonmedusoid forms likewise are planktonic. Because of its lack of medusae, the class Anthozoa is almost unrepresented in the plankton. However, many of the hydrozoan medusae occur in the plankton, and the siphonophores are either members of the plankton or of the neuston. The Scyphozoa are largely planktonic during their medusa stage, and they are among the largest of the plankters, some individuals reaching a diameter of seven feet.

Many of the pelagic coelenterates are highly luminescent, and often they are spectacularly so, especially because of their large size. The mechanism of luminescence, so far as is known, does not basically differ from that of other animal groups (see Harvey, 1940).

Planktonic coelenterates are to be found both in the sea and fresh water, but they are very rare in fresh water, where only four rare genera of medusae are known (the most common genus being *Craspedacusta*).

Artificial key to the commoner planktonic genera of the Coelenterata

1. Individual medusae 2
   Highly complex colonial forms, consisting of both polyps and medusae. Siphonophora 72
2. With a velum; no gastric tentacles. Average size relatively small. Hydromedusae 3
   Without a velum; gastric tentacles present. Average size relatively large. Scyphomedusae 60
3. Margin divided into intertentacular lappets. Tentacles not marginal 4
   Margin not thus. Tentacles, if present, usually marginal or nearly so 10
4. Stomach margin without peripheral stomach pouches 5
   Stomach with peripheral stomach pouches on the outer margin ... 6
5. Gonad forms a simple ring on the subumbrellar surface below the stomach .... *Solmaris* (fig. 341)
   Gonad in a series of subumbrellar saccules below the stomach .... *Pegantha* (fig. 342)
6. With two tentacles .... *Solmundella*
   With four to six tentacles 7
   With eight or more tentacles 8
7. With four to six peronial strands .... *Aegina* (fig. 343)
   With eight peronial strands .... *Aeginopsis* (fig. 344)
8. Stomach pouches cleft .... *Aeginura* (fig. 345)
   Stomach pouches uncleft 9
9. With otoporpae above the lithostyles ...... *Cunina* (fig. 346)
   Without otoporpae above the lithostyles ... *Solmissus* (fig. 347)

10. Statoliths in statocysts and tentaculocysts of endodermal origin 11
Statocysts either absent or of ectodermal origin 22

11. With four or six radial canals and four or six gonads 12
With eight gonads 16

12. Some or all of tentacles with adhesive discs 14
None of tentacles with adhesive discs 13

13. With four gonads . . . . *Liriope* (fig. 348)
With six gonads . . . . *Sergonia*

14. With one ring of tentacles, placed a little above the bell margin . . . . *Gonionemus* (fig. 349)
With two rings of tentacles, one at the bell margin and one slightly above 15

15. Without blind centripetal canals . . . . *Cubaia* (fig. 350)
With blind centripetal canals . . . . *Olindias* (fig. 351)

16. Stomach mounted on a peduncle 17
Stomach without a peduncle 18

17. Gonads are on the subumbrella . . . . *Aglantha* (fig. 352)
Gonads on the peduncle above the stomach . . . . *Aglaura* (fig. 353)

18. Tentacles all of one kind 19
Tentacles of two kinds 20

19. Gonads forming a continuous band around the stomach and extending along the radial canals . . . . . . *Homoeonema* (fig. 354)
Gonads developed diffusely over the radial canals . . . . *Pantachogon* (fig. 355)

20. Tentacles grouped in clusters . . . . *Botrynema* (fig. 356)
Tentacles not thus 21

21. Mouth simple, without lips; distal portion of tentacles stiff and spinelike . . . . *Halicreas* (fig. 357)
Mouth with lips; tentacles of usual form . . . . *Rhopalonema* (fig. 358)

22. Gonads lie along the radial canals 23
Gonads lie in the ectoderm of the manubrium. (Anthomedusae). 40

23. Lithocysts absent 24
Lithocysts present 28

24. Radial canals unbranched 25
Radial canals branched 27

25. Four radial canals 26
Eight radial canals . . . . *Melicertum* (fig. 359)

26. With marginal clubs or cirri . . . . *Laodicea* (fig. 360)
Without marginal clubs or cirri . . . . . . *Chromatonema* (fig. 361)

27. Branches of radial canals end blindly . . . . . *Polyorchis* (fig. 362)
Branches of radial canals connect with the circular canal . . . . *Dipleurosoma* (fig. 363)

28. With fewer than eight radial canals 29
With eight or more radial canals 38

29. Stomach without a peduncle 30
Stomach mounted on a gelatinous peduncle 35

30. Mouth a cruciform slit extending down the four radial canals . . . . *Staurophora* (fig. 364)
Mouth not thus 31

31. Short tentacles at the base of, or between, the longer tentacles . . . 32
All tentacles more or less alike in size; sometimes a few rudimentary smaller tentacles in addition 33

32. Lithocysts in open folds of the velum . . . . *Mitrocomella*
Lithocysts not thus . . . . *Eucheilota* (fig. 365)

33. With eight lithocysts 34
With sixteen or more lithocysts . . . . *Phialidium* (fig. 369)

34. Endodermal cores of tentacles project inward into gelatinous layer of the bell . . . . *Obelia* (fig. 366)
Endodermal core of tentacles not thus . . . . . *Tiaropsis* (fig. 367)

35. With eight lithocysts 36
With more than eight lithocysts 37

36. Margin with numerous cirri or warts . . . . *Eutima* (fig. 373)
Without numerous marginal cirri or warts . . . . *Eutonina*

37. Numerous marginal warts or cirri . . . . *Tima* (fig. 370)
Without marginal cirri or warts . . . . *Eirene* (fig. 371)

38. Radial canals arise from the stomach pouches in four groups . . . . *Halopsis* (fig. 372)
Radial canals do not arise in groups 39

39. With papilla-like protuberances on the subumbrella . . . . *Zygodactyla* (fig. 374)
Without such structures . . . . *Aequorea* (fig. 375)

40. Marginal tentacles branched, or with a linear series of nematocyst-bearing filaments 41
Tentacles simple 42

41. With simple oral tentacles . . . . *Cladonema* (fig. 376)
Without oral tentacles . . . . *Zanclea* (fig. 377)

42. Gonad or gonads ringlike, encircling the manubrium 43
Several gonads, radially segregated around the manubrium 48

43. With meridional lines of nematocysts over exumbrellar surface . . 44
No such lines of nematocysts 45

44. With one or more tentacles all at base of one of the radial canals; five lines of nematocysts . . . . *Hybocodon* (fig. 378)
With two or more tentacles at base of two or four radial canals; eight lines of nematocysts . . . . . . *Ectopleura* (fig. 382)

45. With four rudimentary tentacle bulbs . . . . *Pennaria* (fig. 381)
With one or four well-developed tentacles 46

46. With one tentacle and three rudimentary bulbs . . . .
. . . . *Steenstrupia* (fig. 380)
With four well-developed tentacles 47

47. Tentacles knobbed . . . . *Dipurena* (fig. 379)
Tentacles not knobbed . . . . *Sarsia* (fig. 383)

48. Marginal tentacles hollow; no oral tentacles 49
Marginal tentacles solid; knobs of nematocysts or tentacles on lips 54

49. Two well-developed and many rudimentary tentacles 50
Four or more well-developed tentacles 51

50. Stomach on a peduncle . . . . *Stomotoca* (fig. 384)
Stomach not on a peduncle . . . . *Amphinema* (fig. 385)

51. Gonads only partially separated on the four radii; with meridional lines of nematocysts . . . . *Pandea* (fig. 386)
Gonads completely separated; no lines of nematocysts 52

52. Radial canals attached to the stomach by mesenteries 53
Radial canals without mesenteries . . . . . . *Halitholus* (fig. 387)

53. Gonads basically horseshoe shaped . . . . . *Leuckartiara* (fig. 388)
Gonads not thus . . . . *Catablema* (fig. 389)

54. Tentacles in clusters 55
Tentacles not in clusters 58

55. Oral tentacles arise above the margin of the mouth 56
Oral tentacles are a prolongation of the lips of the mouth . . . .
. . . . *Rathkea* (fig. 390)

56. Each cluster of tentacles consists of two median clavate tentacles and a series of filiform tentacles . . . . . . *Nemopsis* (fig. 391)
Tentacle clusters not thus 57

57. With four radial clusters of tentacles only . . . .
. . . . *Bougainvillia* (fig. 392)
With radial clusters of tentacles and interradial solitary . . . . . tentacles . . . . *Lizzia* (fig. 393)

58. Four marginal tentacles . . . . *Cytaeis* (fig. 394)
Eight or more marginal tentacles 59

59. Lips with simple oral tentacles . . . . *Podocoryne* (fig. 395)
Lips with nematocyst knobs . . . . *Turritopsis* (fig. 396)

60. Bell cubical . . . . *Carybdea* (fig. 397)
Bell discoidal or cone-shaped; not cubical 61

61. With a single mouth 62
With numerous mouths 69

62. With an annular furrow and marginal pedalia 63
Without an annular furrow and without pedalia 65

63. With four interradial rhopalia . . . . *Periphylla* (fig. 398)
With four perradial and four interradial rhopalia 64

64. With sac-like gastric pouches on the subumbrella . . . .
. . . . *Linuche* (fig. 399)
Not thus . . . . *Nausithoe* (fig. 400)

65. Tentacles arise from the sides of the exumbrella, above the margin . . . . *Aurelia* (fig. 401)
Tentacles arise from the floor of the subumbrella 66
Tentacles arise from the bell margin 67

66. Eight clusters of tentacles . . . . *Cyanea* (fig. 402)
Sixteen clusters of tentacles . . . . *Phacellophora* (fig. 403)

67. Interradial canals branched. A ring canal present . . . . *Discomedusa* (fig. 404)
No ring canal. Radial pouches all unbranched 68

68. Eight tentacles and sixteen marginal lappets . . . . *Pelagia* (fig. 405)
Twenty-four tentacles and thirty-two lappets . . . . *Chrysaora* (fig. 406)
Forty tentacles and forty-eight lappets . . *Dactylometra* (fig. 407)

69. Each of the oral arms bears a pair of wing-like scapulets on dorsal surface near origin 70
Oral arms not thus . . . . *Cephea* (fig. 411)

70. Oral arms fused along their sides . . . . . . *Stomolophus* (fig. 408)
Oral arms free 71

71. Each oral arm terminates in a club . . . . . . *Rhizostoma* (fig. 409)
Oral arms with numerous clubs or filaments . . . . *Rhopilema* (fig. 410)

72. With an enormous gas-filled pneumatophore, acting as a float; floating on surface of the water only * . . . . *Physalia* (fig. 412)
Without such an enormous sac; on or beneath the surface 73

73. Medusa-like, with central axis (coenosarc) greatly shortened 74
Not medusa-like; central axis more or less elongate 75

74. Pneumatophore with the chitinous disc raised vertically to form a sail; surface forms . . . . *Velella* (fig. 413)
Without a sail . . . . *Porpita*

75. All nectophores alike in structure; sometimes only one nectophore . 76
The oldest nectophore differs from the others in the presence of a somatocyst 82

76. Colony with an apical float 77
Colony without an apical float 79

77. Colony nearly as broad as long . . . . *Physophora* (fig. 414)
Colony much longer than broad 78

78. Tentillae on tentacles of gastrozoids tricornuate . . . . *Agalma* (fig. 415)
Tentillae unicornuate . . . . *Stephanomia* (fig. 416)

* Preserved siphonophores such as are commonly encountered in plankton samples often are broken into pieces, so that with many genera it is usual to see only the separate persons of the colony (swimming bells, bracts, etc.). The present key is intended only for the identification of entire or reconstructed individuals, and does not apply to single parts of the colonies.

79. With one or two (rarely three or four) nectophores 80
With many nectophores 81

80. With one nectophore . . . . *Sphaeronectes*
With two or more nectophores . . . . *Rosacea*

81. Nectophores angular or, if rounded, with two dorsal prominences . . . . *Vogtia* (fig. 417)
Nectophores rounded, and with four dorsal prominences . . . . *Hippopodius* (fig. 418)

82. Nectophores rounded 83
Nectophores pyramidal or prismatic 84

83. Nectophores with basal teeth . . . . *Sulculeolaria* (fig. 419)
Nectophores without basal teeth . . . . *Galetta* (fig. 420)

84. Upper nectophore prismatic and smaller than the lower one 85
Nectophores pyramidal 87

85. Anterior nectophore with a rectangular apical facet . . . . *Abyla* (fig. 421)
Anterior nectophore with an apical ridge 86

86. Posterior nectophore with a dorsal ridge well developed . . . . . hydroecium open . . . . *Abylopsis* (fig. 422)
Posterior nectophore without dorsal ridge, hydroecium closed . . . . *Bassia* (fig. 423)

87. Prominent teeth on the oral margin 88
Oral margin without such prominent teeth 89

88. Upper nectophore with nine prominent angles . . . . *Enneagonum* (fig. 424)
Upper nectophore not thus . . . . *Diphyes* (fig. 425)

89. Mid-sector of bell quadrangular * . . . . *Chelophyes*
Mid-sector of bell pentagonal or polygonal 90

90. Divided basal plate with square-cut or rounded outer angles 91
Same with outer angles produced as wings . . . . *Eudoxoides* (fig. 426)

91. Hydroecial cavity extends well above oral level of nectosac . . . . *Muggiaea* (fig. 427)
Same does not extend appreciably above oral level of nectosac . . . . *Lensia* (fig. 428)

Interested persons should consult Bigelow (1911), Bigelow and Sears (1937), Hartlaub (1907-15), Kramp (1919, 1926, 1947), Mayer (1910), and Vanhöffen (1906) for further information on the Coelenterata.

* Sections of this key are taken from or modified from Bigelow and Sears (1937).

7. Ctenophora

The Ctenophora (sea walnuts, or comb jellies) are exclusively marine, though a few species occur in waters that are rather brackish. Most of the species are planktonic in their habitat, though a few, such as *Coeloplana,* have adopted creeping habits. All adult specimens of planktonic species are members of the macroplankton, even the smallest being a few millimeters in diameter, and some of them ranging up to 20 cm. Immature specimens, such as the common cydippid larva (fig. 429), may be members of the microplankton.

Ctenophores are very delicate jelly-like animals, whose tissues are so fragile in many species that it is next to impossible to preserve them in perfect condition, nor can they be handled at all roughly while living, else they will break into many pieces. Most are so tender that rough weather would be the cause of their demise if they stayed near the surface. Such forms invariably migrate from the surface to a sufficient depth to find still water when storm winds begin to blow.

Most of the ctenophores are biradially symmetrical, and in simpler forms they are subspherical. At the aboral end there is a prominent sense organ, and radiating orally from this are typically eight rows of strong ciliary combs (membranelles). These are the locomotory organs of the animal, and often when the transparent and colorless animal is swimming in the water, the beautifully irridescent flickering of the ciliary combs is all that can be seen. Normally the ciliary combs of each row beat in a co-ordinated manner, the effective stroke being away from the mouth and towards the aboral sense organ. This causes the animal to swim slowly through the water with the mouth forward. In some forms there is a pair of tentacles, or there are several smaller tentacles. These are never used as locomotory organs, but as food gatherers. The mouth leads into a lobed digestive tract, the ramifications of which need not be considered here. There are anal pores at the aboral end of the digestive tract, and these rarely may be used for defecation. However, indigestible remnants of the food usually pass out through the mouth.

In the order Lobata the lips are produced into two large, sometimes muscular, oral lobes. The four rows of ciliated combs on the lip sides of the animal are then longer than the remaining four. At the oral end of each of the shorter combs there is a fleshy finger-shaped protrusion, the auricle. Sometimes the auricles are large and appear like fleshy tentacles.

The body bears a superficial resemblance to that of the Coelenterata, in that it is very jelly-like in consistency. At one time the ctenophores were considered to be a class in the phylum Coelenterata. However, instead of having a simple mesoglea as in the jellyfish (where at most there were a few wandering amoeboid cells) the ctenophores have a definitely developed mesodermal layer of cells. Within this intermediate layer lie specialized mesodermal muscles.

Without known exception the ctenophores are carnivorous. At times and locally they swarm in vast numbers in the plankton, and at such times they are devastating enemies of other, smaller zooplankers. It is not a rare experience to gather a plankton tow at a place where ctenophores are numerous, and find that practically no plankters are present except the ctenophores. *Pleurobrachia*, because it is so frequently abundant and because it has long and efficient tentacles, is especially harmful to other plankters. Even small fish may be taken. On the other hand, ctenophores are also eaten by larger fish. Because of the delicate nature of the ctenophores, they are seldom or never seen in the examination of fish stomachs, but the frequent presence in *Beroë, Pleurobrachia*, etc., of the larval stages of cestodes whose adults almost certainly live in the intestines of fish is good evidence that they are eaten regularly.

As in the coelenterates, the comb jellies are among the more spectacular luminescent organisms in the plankton.

Artificial key to the commoner planktonic genera of the Ctenophora

1. With two long conspicuous tentacles in a tentacle sheath. Without oral lobes 2
   Tentacles, if present, without a tentacle sheath. Oral lobes present or absent 3
2. Body laterally compressed . . . . *Mertensia* (fig. 430)
   Body not compressed, egg-shaped . . . . *Pleurobrachia* (fig. 431)
3. With tentacles, at least in juvenile stages. These often are small . . 4
   With neither tentacles nor oral lobes . . . . *Beroë*

4. Body compressed into a band-like form ..... *Cestum* (fig. 432)
   Body not thus 5
5. Auricles long and thick and coiled in a helix ....
   .... *Leucothea* (fig. 433)
   Auricles not thus 6
6. Oral lobes large and muscular .... *Ocyropsis* (fig. 434)
   Oral lobes smaller, and not unusually muscular 7
7. With four deep lateral furrows extending upward from the level of the mouth on the margin of the oral lobes almost to the level of the apical sense organ .... *Mnemiopsis* (fig. 435)
   Without such furrows .... *Bolinopsis* (fig. 436)

Interested persons should consult Chun (1880, 1898), Krumbach (1927), Mayer (1912), and Moser (1903, 1909) for further information on the Ctenophora.

8. Platyhelminthes

The phylum Platyhelminthes is seldom encountered in the plankton. In the marine plankton the characteristic Müller's larva (fig. 437) of the polyclad worms is encountered from time to time, usually in small numbers. More advanced specimens of polyclads are also found occasionally, though usually only as tychoplankters. In some waters, adult acoels and rhabdocoels are found in small numbers, cercaria stages of many flukes are temporary members of the plankton, and various larvae of parasitic flatworms attack planktonic organisms (chaetognaths, ctenophores, etc.) with considerable regularity. Inasmuch as most of these forms are either tychopelagic or otherwise not true plankters, however, no attempt will be made here to identify their kinds.

Interested persons should consult Graff (1905, 1913) for further information on the Platyhelminthes of the plankton.

9. Nemertea

The nemertean worms, or proboscis worms, apparently are very closely related to the Platyhelminthes, although as a rule they are no longer considered to be a class of that phylum. Nemerteans are usually very long forms, often extremely so. They have a complete digestive tract, but no trace of a true coelomic cavity. There is a primitive circulatory system, which usually consists of three anastomosed longitudinal contractile vessels. All forms are characterized by a large proboscis at the anterior end, which can be extended or protracted at will. When retracted it lies within an extensive proboscis sac, which opens at the anterior end close to the mouth. The proboscis, although aiding in the capture of food, is completely independent of the digestive system.

The Nemertea are to be found in many different habitats. A few are terrestrial in moist tropical forests, while some are benthic in fresh water. Most, however, are marine, the vast majority being benthic and littoral. Many of these have a typical planktonic larval stage, the pilidium larva (fig. 438). Quite a number of others are commensal or parasitic. A few rare and aberrant forms are planktonic. These are greatly shortened and flattened, with a pair of latero-posterior fins and a single caudal fin, which aid in swimming. All known planktonic nemerteans have been found only in very deep waters, from 200 to 3,000 m. under the surface. At these depths they seem to live at considerable distances apart, and consequently relatively few specimens have ever been captured. There is a sexual dimorphism in the adults of some species, for the male has a pair of strong tentacles towards the anterior end. The function of the tentacles appears to be to clasp the female during copulation.

Because of the rarity of the pelagic nemerteans, no attempt will be made here to construct a key for their identification. The most common genus is *Nectonemertes* (fig. 439), and among the remaining genera *Pelagonemertes* and *Planktonemertes* have been encountered most frequently.

Interested persons should consult Burger (1909), Brinkman (1917*a*, 1917*b*), Coe (1926), and Schmidt (1937) for further information on the pelagic nemerteans.

10. Nemathelminthes

A great many of the nematode worms are parasitic in habit. Many, however, are free-living, being very common in bottom muds and bottom sediments both in marine and fresh waters. Others are found in damp soils, or even in more strictly terrestrial environments. Nematodes are found only occasionally in the plankton, and it seems probable that almost invariably those that are found are simply tychopelagic.

Persons who wish to determine tychopelagic nematodes are referred to Ditlevsen (1926), Goodey (1951), and Ward and Whipple (Cobb, 1918).

11. Rotifera

The rotifers are tiny microscopic animals with a relatively simple structure. Although multicellular, they are composed of a relatively small number of cells. It has been found that many of the species consist of a definite and invariable number of cells, which is an unusual condition among animals. The phylum is usually considered to be a minor one in the animal kingdom because there are less than 2,000 known existing species. Many of the species, however, are extremely common, and particular species often dominate their surroundings.

Rotifers, or "wheel animalcules," were first observed by Anton van Leeuwenhoek in 1675 (see Knobloch, 1948, pp. 14-15), during his observations of the microscopic inhabitants of water. They are characterized, as both the technical and common names suggest, by having a corona or "wheel" of cilia surrounding the mouth. The vibration of the cilia of the corona produces an illusion of rotation. Posterior to the mouth a short narrow buccal cavity leads into a wider organ, the mastax, within which are jaw-like trophi. The mastax and its enclosed trophi are of great importance in the determination of species among the rotifers, because they differ in each species, and are the only solid portion in the body of an organism that is otherwise extremely delicate. Behind the mastax a short esophagus leads into a stomach, from which an intestine extends to a cloaca and then to the anus.

The anus usually is not terminal in the Rotifera, but opens to the exterior at the base of a so-called foot. The foot typically terminates posteriorly in a pair of finger-like processes, the toes. Sometimes one or both toes, or even the entire foot, is wanting. The foot and toes are the organs of attachment in those forms that are temporarily or permanently sessile, while in some of the free-swimming species they are used as a rudder.

The internal organs of the body are relatively simple. In the more primitive forms there is a pair of reproductive organs, but in most species only a single gonad remains. The simple excretory organs open into a contractile bladder, which then opens into the posterior portion of the digestive tract, i.e., into the cloaca. There is no coelomic cavity, but a pseudocoelom is well developed. For this reason the rotifers are often considered to be more closely related to the Nemathelminthes than to any other phylum except the Gastrotricha.

Externally many rotifers are provided with a rigid lorica, which often bears long spines or is otherwise highly ornate. Other forms are entirely devoid of such structures. Some genera have movable appendages extremely reminiscent of the appendages of crustacean nauplius larvae. A few types construct protective tubes within which they are attached. There is often an extreme sexual dimorphism, with the male much smaller and more degenerate than the female. Often the male occurs only during a particular season of the year, the female reproducing parthenogenetically most of the time. In many cases the male is very rare even during the correct season, and in not a few very common species, males are entirely unknown.

Rotifers are found in nearly every body of fresh water: in tiny, temporary puddles, in rivers and swamps, and in the largest lakes. They are much less commonly encountered in the sea, and in marine situations they appear to be confined to coastal regions, reaching their greatest abundance in brackish waters. The species found in the sea often differ from fresh-water forms, but there are no genera that are confined to the sea. A few species are parasitic, but most are free-living. The free-living forms can be found, often in great abundance, living between the grains of sand on sandy beaches, or attached to or creeping over the bottom and over vegetation attached to the bottom. And, finally, many are members of the plankton. Among the fresh-water zooplankters, the rotifers are second in importance in the food chain only to the Crustacea.

As described above (p. 129), the rotifers are largely sedimentation feeders, using the corona as an organ for creating water eddies and currents. Most of the planktonic species feed upon other plankters. A few are detritus feeders, or else they are indiscriminate in their selection of food. Of the plankton feeders, most are simple sedimentation feeders, and they feed largely upon the phytoplankton. A minority, on the other hand (e.g.,

Ploesoma), feed directly as predators upon other individual zooplankters. The sedimentation feeders, having no particular need for precision of movement in gathering their food, swim in a spiral path through the water, while the predators, requiring great precision if they are to capture a swiftly moving prey, have evolved the foot into a rudder, by means of which their movements have become controlled and directional.

The Rotifera, especially in fresh waters, occupy an ecological niche in the plankton that is of extreme importance. They constitute one of the major primary consumers of plant materials, or else they feed upon other plankton animals. In turn small fish, as well as other small animals, feed voraciously upon the rotifers.

Artificial key to the commoner planktonic Rotifera

1. Foot present (may be retractile) 2
   Foot absent 10
2. Foot large, with no toes 3
   Foot various in size, with toes 5
3. Mouth opening central . . . . *Collotheca* (fig. 440)
   Mouth opening eccentric 4
4. Often solitary in plankton samples, but normally building spherically symmetrical colonies; dorsal sense organ lacking . . . . *Conochilus* (figs. 441, 442)
   Solitary or forming irregular colonies; dorsal sense organ present . . . . *Conochiloides* (fig. 443)
5. Left toe usually more than half the length of the body; right toe shorter than the left, often rudimentary . . . . *Trichocerca* (fig. 444)
   Both (or all) toes nearly equal in size 6
6. With a lorica 7
   Without a lorica . . . . *Synchaeta* (fig. 445)
7. Foot ventral 8
   Foot nearly or quite posterior 9
8. Lorica smooth, with a well-developed cephalic shield . . . . *Ploesoma* (fig. 446)
   Lorica rough, without a cephalic shield . . . *Gastropus* (fig. 447)
9. Anterior border of lorica generally with spines; toes shorter than foot . . . . *Brachionus* (fig. 448)
   Anterior border of lorica without spines; toes longer than foot . . . . *Euchlanis* (fig. 449)

10. Body nauplius-like, with six fleshy, spined appendages . . . . . .
. . . . *Pedalion* (fig. 450)
Body not thus 11
11. Body with two large lateral bristles and a third similar bristle ventrally towards the posterior end . . . . *Filinia* (fig. 451)
Body not thus, though it may have spines on a lorica 12
12. Body with lorica; intestine and anus present 13
Body sac-like and hyaline, no intestine or anus . . . .
. . . . *Asplanchna* (fig. 452)
13. Anterior border of lorica with six spines 14
Anterior border of lorica not thus 16
14. Dorsal portion of lorica with reticulate ornamentation . . . .
. . . . *Keratella* (fig. 453)
Dorsal portion of lorica with longitudinal striations, or unornamented 15
15. Posterior and anterior spines extremely attenuated . . . .
. . . . *Notholca* (fig. 455)
Posterior and anterior spines not thus . . . . . *Kellicottia* (fig. 454)
16. With one red eye . . . . *Chromogaster* (fig. 456)
With two red or blackish eyes . . . . *Pompholyx* (fig. 457)

Interested persons should consult Collin *et al.* (1912), Harring (1913), Jennings (1901), Pennak (1953), Rylov (1935), and Ward and Whipple (Jennings, 1918) for further information on the Rotifera.

12. Bryozoa

The Bryozoa are, almost without exception, sessile animals. The few exceptions are creeping forms. The adult animals, therefore, are not encountered in the plankton unless they are attached to small pieces of floating seaweeds and other similar objects. In fresh waters, the reproductive statoblasts (fig. 460) may be found temporarily as members of the tychoplankton, while in the sea the cyphonautes larva (fig. 461) is a characteristic member of the true meroplankton.

The cyphonautes larva is an extremely flat, triangular larva, encased in a rigid chitinous shell. In the main, the shell is open at the oral end of the animal. It must from its structure be considered to be a highly modified trochophore larva.

Interested persons should consult Dawydoff (1928), Lohmann (1910), and MacBride (1914) for further information on the cyphonautes larvae of the plankton.

13. Brachiopoda

As in the case of the Bryozoa, all the adult members of the phylum Brachiopoda are benthic. However, unlike the Bryozoa, there are no known fresh-water representatives of the phylum. Rarely, the larval forms of brachiopods are found in the meroplankton (see fig. 462).

14. Phoronidea

Just as in the case of the previous two phyla, the adults of the phylum Phoronidea are bottom dwellers, for the worm-like creatures live buried in the mud and sand in marine situations. As is the case with many marine benthic animals, however, there is a planktonic larva, which in this case is called the actinotroch larva (fig. 463). The actinotroch is obviously a somewhat modified trochophore larva, in which the metatroch is developed into a series of tentacles.

Further information on actinotroch larvae can be gleaned from Dawydoff (1928) and MacBride (1914).

15. Chaetognatha

The Phylum Chaetognatha is one of the smallest phyla of the animal kingdom as far as the number of species involved is concerned, for there are only around thirty known species, belonging to six genera. All of the known species, except one or two in the genus *Spadella*, are strictly planktonic, and the exceptions themselves are partly free-swimming and partly creeping forms. All chaetognaths are marine, though a few, especially juvenile stages of certain species, are found in somewhat brackish water. In spite of the small number of species involved, they often are extremely abundant, and constitute an extremely important part of the marine plankton.

Because of their shape the chaetognaths are called the arrow worms, while they are also known as glass worms because of their characteristic extreme transparency. There has been and remains considerable confusion as to the phylogenetic relationships of these forms, and from time to time they have been classified by various authors in nearly all the phyla of the

Metazoa, not even excluding the Chordata. In general shape, in the presence of only longitudinal muscles (except in *Heterokrohnia*) in the body wall, and in the simplicity of the digestive tract they show many similarities to the Nemathelminthes; and in fact they have rather consistently, by many authors, been considered as a class of that phylum. However, the chaetognaths have a true coelomic cavity rather than a pseudocoelom, and the body is definitely divided into three segments by the presence of two cross partitions. Consequently it is clear that they must be placed in a phylum by themselves, and indeed there is much doubt that they have any close affinities whatsoever to the nematodes.

The cross partitions divide the coelomic cavity and the body as a whole into three main parts, the head, the trunk, and the tail. The head coelom, however, is nearly obliterated in the adult by the development of a complicated mass of muscles. The function of these muscles is to move a series of lateral head setae, which act as seizing jaws (hence the origin of the name Chaetognatha). The seizing jaws are supplemented by one or more series of smaller teeth. The mouth, which lies subventrally between the jaws, is capable of great temporary enlargement in order to accomodate itself to the ingestion of food organisms that often are nearly as great in diameter as the arrow worm itself.

Other important structures of the head include the large and flexible fleshy cap, the dorsal brain and its nerves, the two simple eyes, and the numerous taste buds. The fleshy cap normally covers the jaws, but when the jaws are needed during the capture of the prey, it can be retracted readily.

The remainder of the body is much less complicated than the head. Frequently there is a smaller or larger collarette immediately behind the head, and there is always a flat caudal fin accompanied by either one or two pairs of lateral fins on the trunk and tail. Each fin, which is immovable, is provided with a number of parallel fin rays. The function of the fins is the stabililization of the animal in the water, and they are only incidentally used in locomotion. All movements of the animal from place to place are accomplished by the alternate contraction of the dorsal and ventral bands of longitudinal muscles in the trunk region.

The whole of the intestine lies in the trunk region, and as a rule it is completely undifferentiated. In some forms, however, there are two small intestinal diverticula at the anterior end. The anus lies on the ventral surface just anterior to the septum that separates the trunk from the tail. The large ventral ganglion of the nervous system also lies on the ventral surface of the trunk, usually somewhere in the anterior half. It is connected with the brain in the head by means of two large connectives, and it gives off a number of larger and smaller nerves to the general body.

Chaetognaths are universally hermaphroditic. The two ovaries lie, one on each side of the intestine, in the very posterior portion of the trunk. Each contains its own independent oviduct, which opens laterally just anterior to the tail septum. Mature specimens are easily detectable by the size of the ovary and by the presence of large egg cells.

The testes, on the other hand, lie in the tail segment of the animal, one on each side of a single vertical partition dividing the tail into right and left halves. Two lateral seminal vesicles are conspicuous in mature chaetognaths. Here the mature sperm are stored until needed.

The glass worms, with their extreme transparency, are very well adapted to their usual planktonic existence, for it is almost impossible to see them in their native habitat, although as adults they range from 10 to 80 mm. in length. They are all carnivorous, and are voracious feeders. With their large jaws, extreme transparency, and active movements they are veritable scourges for other zooplankters. Judging by their intestinal contents, which are always clearly visible, they feed mainly upon copepods, larval decapods, young fish, other chaetognaths, etc. They in turn, as shown by Lebour (1923) and others, are eaten by fish, medusae, pelagic worms, etc.

Artificial key to the genera of the Chaetognatha

1. The whole of the trunk is encased in an extraordinarily thick collarette . . . . *Pterosagitta* (fig. 464)
   Collarette, if present, not thus 2
2. Cap on head with two lateral tentacles . . . . . *Spadella* (fig. 465)
   Cap without such tentacles 3

3. With two pairs of lateral fins . . . . *Sagitta* (fig. 466)
   With one pair of lateral fins 4
4. Lateral fin reaching to last third of tail segment . . . .
   . . . . *Krohnitta* (fig. 467)
   Lateral fin not reaching beyond middle of tail segment 5
5. Tail segment constitutes 19–31 per cent of total body length. One row of teeth on each side of head . . . . *Eukrohnia* (fig. 468)
   Tail segment constitutes 32–40 per cent of the total body length. Two rows of teeth on each side of head . . . .
   . . . . *Heterokrohnia* (fig. 469)

Interested persons should consult Kuhl (1938), Michael (1911), and Thompson (1947) for further information on the Chaetognatha.

16. Annelida

The annelid worms, with their well-developed coelomic cavity, the clear segmental arrangement of the body, and, in many cases, the presence of fleshy, segmentally arranged appendages (the parapodia), have much closer affinities to the higher, more complex phyla than do any of the other phyla heretofore considered. It seems apparent that their closest affinity is to the great phylum of the Arthropoda, if we except the minor phylum Onychophora. There is a well-defined head, consisting of the modified first segment. The head often, but not always, bears special sense organs, including photoreceptors and tentacles. At the other end of the animal the anal segment is also somewhat different from the remaining segments of the body, primarily because it bears the anus. Usually there is a pair of anal cirri, which are apparently sensory in function. In generalized forms, such as *Nereis*, there is little differentiation of the segments lying between the head and the anal segment. In other cases, however, there may be a great deal of differentiation, as in *Chaetopterus*.

Internally the annelids show varying elaboration of the major organ systems. It does not seem necessary to consider internal anatomy here, inasmuch as planktonic species, unless they are given special treatment, are not sufficiently transparent to make these structures clearly visible.

Annelids may be terrestrial (*Lumbricus*) or aquatic (most forms). Aquatic species occur both in fresh water and in the sea, but for the most part in the latter. The vast majority of forms are benthic, or else they are parasitic, as in the case of most leeches. In fresh water there are no examples of the phylum Annelida in the plankton, but in the sea there are a few species that are planktonic as adults, and large numbers have planktonic larval stages. Pelagic adults are in general characteristic of the open reaches of the sea, while larval forms are more common in inshore waters, where the water is more shallow, and where the benthic adults occur in larger numbers than elsewhere. Larvae occur in waters of low, as well as of high, salinity.

Of the classes of the phylum, the Archianellida (as larvae), the Gephyrea (as larvae), the Myzostomaria (as larvae), and the Chaetopoda (subclass Polychaeta, both as adults and as larvae) occur in the plankton.

After a period of embryonic development within the egg, many of the marine annelids hatch in the form of typical trochophore larvae. In *Polygordius* (Archianellida), and in many others, this trochophore larva (fig. 470) is more or less oval and bears a ciliated sensory apical organ. A complete digestive tract is present, including a mouth, an esophagus, a stomach, an intestine, and an anus. The mouth and the anus usually open on the same side of the body, but some distance from each other, or else the anus is at the terminus of the antapex. Anterior to the mouth there is a ring of long cilia, the prototroch, encircling the body. Between the mouth and the anus lies a second ring of cilia, the metatroch. Finally there is a telotroch in the region of the anus.

However, the typical trochophore larva, such as described above, shows many variations as it is present in the development of different annelids. In *Eupomatus*, for example, the only ring of cilia is the prototroch, while in *Chaetopterus* only the metatroch occurs. In *Ophryotrocha* and others, on the other hand, there are several rings of cilia, and therefore the larva is called a polytroch larva (fig. 471). The mitraria larva is also a modified trochophore larva (fig. 472), where the portion of the body below the prototroch is greatly shortened, and bears a number of very long setae. The mitraria larva is a characteristic type among some of the tubeworms.

Subsequent development of the trochophore proceeds by means of the elongation of the posterior half of the body, and the development therein of segmentation. Each segment develops bilaterally arranged setae. In *Nereis*, and in many other genera, the larval stage so formed is known as nectochaete larva (fig. 473). In other forms at this stage, there appear long, nonsegmental, temporary setae on the anterior, unsegmented portion of the body (i.e., anterior, but below the prototroch). There are no existing adult annelids with setae of this nature, but they are known in certain fossil forms, and thus in all probability are to be considered as primitive holdovers. The setae have a function in the larvae in that they aid in flotation during their planktonic existence. Such larvae (fig. 474) are characteristic of *Spio, Nerine,* and other genera.

The origin of the rostraria larva (fig. 475) is unknown, but they are open sea forms, with very long setae, arranged segmentally. The cephalic lobe is developed in the form of a rostrum, and there is a single pair of long contractile tentacles.

Some of the larvae of the tube worms bear the characteristic cephalic tentacles and other structures so characteristic of tthe adults. A few such larvae remain for a considerable period in the plankton, investing themselves with a gelatinous case which aids in flotation. However, eventually all such forms settle to the bottom as members of the benthos.

Very few of the Annelida occur in the plankton as adults. Many of the Errantia among the Polychaeta may temporarily find their way into the plankton in shallow waters as members of the tychoplankton. Perhaps the most famous such case is that of the palolo worm (*Leodice*), which lives normally in crevices in the bottom. At the breeding time, on a predictable date and at a definite phase of the moon, the posterior end of these worms breaks off from the remainder and swims up to the surface in the plankton, where it remains temporarily until spawning is completed. True planktonic species, however, remaining plankters at all stages of their development, are confined to four families, namely the Phyllodocidae, the Alciopidae, the Typhloscolecidae, and the Tomopteridae. These forms are never extremely common in the plankton, though they may occur with considerable regularity in many locations.

Artificial key to the commoner genera of adult planktonic Annelida

1. Eyes very large, occupying a major portion of the head 2
   Eyes small or lacking 3
2. Parapodia with one appendage . . . . *Callizona* (fig. 476)
   Parapodia with two appendages . . . . *Greefia* (fig. 477)
3. With many setae near the acicula . . . . *Pelagobia* (fig. 478)
   With few setae or none . . . . *Tomopteris* (fig. 479)

Interested persons should consult Dawydoff (1928), Gravely (1909), Thorson (1946), and Reibisch (1905) for further information on the Annelida of the plankton.

17. Arthropoda

Members of the phylum Arthropoda, with their high degree of evolutionary development and their over-all efficiency, can and do live in almost every conceivable ecological niche of the earth and seas, wherever conditions are compatible with the existence of life. In basic structure the arthropods are extraordinarily like the Annelida except in the development of the coelomic cavity, which is rudimentary in the arthropods, and in the presence of typical arthropod jointed appendages, whereas in the annelids, if appendages are present, they are simple, fleshy structures, the parapodia. Arthropods also differ from annelids in their lack of cilia.

Of the huge classes of the phylum, the Insecta are primarily terrestrial, though there are many (mostly fresh-water) aquatic species as well. The modern Arachnoidea are also primarily terrestrial, though certain primitive forms have remained aquatic in salt water (*Limulus*), while a few others, such as the water mites (Hydrachnidae), have migrated back to water. On the other hand, the Crustacea are primarily aquatic, and only secondarily have a few of them migrated to land. All the other classes, which are comparatively minor ones, are terrestrial.

Planktonic Insecta and Arachnoidea are relatively unusual, though the larvae of the insect *Corethra* may occur in vast swarms in fresh-water lakes. Planktonic Crustacea, however, are beyond question the most important and successful of all zooplankters, both in fresh water and in the sea. Of the Crustacea, the Cladocera, Ostracoda, Copepoda, Cirripedia, Cumacea, Isopoda, Amphipoda, Mysidacea, Euphausiacea,

Decapoda, and Stomatopoda are the groups most typically found in the plankton, either as adults or larvae or both. In the Insecta, the *Coretha* larva is the only typically planktonic type. In the Arachnoidea, the Hydrachnidae often are planktonic.

a. Cladocera (Water Fleas).—The crustacean subclass Phyllopoda contains the most primitive of the existing Crustacea, namely the Brachiopoda, or fairy shrimps. However, the phyllopod order Cladocera is in many ways less primitive. Typically the animals are compressed laterally, and the dorsal body wall takes the form of folds that cover the body and limbs on each side, thus forming a sort of bivalved shell. The head, however, remains uncovered and free. The valves of the shell are firmly grown to each other dorsally, and there is no hinge. Nor are there any adductor muscles to move the valves. At the posterior end, above the body of the animal, the shell is formed so as to produce a brood sac, where the young develop. In a few aberrant forms the bivalve shell is lacking (e.g., *Polyphemus*), its vestiges being confined to the formation of a brood sac.

The head is often set off from the remainder of the body by a distinct line, the cervical sinus. On the head there is usually a pair of conspicuous and movable compound eyes, one on each side, and in addition there is a single median ocellus. In some forms the ocellus is lacking, while in at least one genus (*Monospilus*) the ocellus is the only photoreceptor present. The head is often bent downwards, and usually it has a ventroposterior pointed process, the rostrum. In addition to the inconspicuous mandibles, maxillae, and labrum, there are two conspicuous pairs of appendages on the head, namely the first and second antennae. The first antennae are uniramous and usually small. They are located ventrally near the rostrum. The second antennae are biramous and very large, constituting the main swimming organs. In many species the exoskeleton of the head is thickened to form a ridge, the fornix, which supports the muscular movements involved in the vigorous swimming movements of the second antennae. Much of the interior of the head is occupied by the powerful striped muscles needed for the swimming movements, but also contained are the most anterior portion of the digestive tract and the brain.

Posterior to the head, underneath the shell, lies the remainder of the body. A series of five or six paired appendages is to be found on the anterior portion. All of these are foliaceous and very much alike in the more primitive members of the order, but in many forms the first two pairs of legs are prehensile. The main function of the foliaceous legs is to create food-bearing water currents, while the prehensile legs appear to be used by nonplanktonic species as aides in clinging to weeds and other objects.

The legs are borne on an indistinctly segmented portion of the body. Just posterior to them, and separated from the rest of the body by a decidedly movable joint, is the unsegmented postabdomen. Ordinarily this is sharply bent forward, and may be used, especially in creeping forms, as a secondary locomotory organ. Probably it also serves as a keel in swimming species. It is usually highly spinose, with a pair of strong terminal spines, at the base of which there frequently are a number of basal spines, and in addition two to four rows of marginal and lateral teeth and spines. With these several processes, the postabdomen is admirably fitted for its additional function of keeping the trunk appendages free of debris and parasites.

The digestive tract is relatively very simple, and in a great many forms, it passes straight from the head to the end of the postabdomen without making any loops or major turns other than those necessary to follow the natural curves of the body. In other forms, however, the intestine makes at least one loop before it reaches the postabdomen. In some genera there is a distinct digestive gland associated with the anterior portion of the intestine. In others this is not clearly visible. There is also a looped band of tissue, the shell gland, whose function is thought to be excretory. Finally, dorsal to the intestine and just anterior to the brood sac, is the simple pulsatory heart, usually clearly visible through the transparent shell and body. There is no system of blood vessels, however. The heart pushes the blood anteriorly, whence it passes through a complex series of sinuses, and eventually returns to the heart, into which it enters by simple ostia.

The Cladocera occur both in fresh water and in the sea, but they are far more important in the former, where they constitute one of the major types of smaller animals. Most genera and species are to be encountered among weeds in lakes, ponds, and swamps. On the other hand, a few forms are true bottom dwellers in the mud, and some are planktonic. Although a relatively small number of species are planktonic, nevertheless the cladocerans often dominate the plankton in fresh waters, becoming in many cases the most important grazers, as well as the most important source of food for small fish.

Practically all of the Cladocera are either plant feeders or detritus feeders, and such forms obtain their food by means of the constant action of their foliaceous appendages. The feeding movements take place constantly, the food particles obtained being carried in a constant stream to the mouth, where they are ingested. On the other hand, a few planktonic species are predatory, as are *Polyphemus* and *Leptodora*. For the most part, cladocerans are less than 3 mm. in length, but *Leptodora kindtii* of the Great Lakes of North America reaches a length of 18 mm. It must be a formidable foe for other fresh water zooplankters.

As mentioned above, the eggs in the female are deposited in a brood sac under the shell. Usually a relatively large number is deposited at a time, up to 20 or 30, and a new batch of eggs is produced every few days during the warm season. In some cases the embryos develop so rapidly that they themselves begin the production of parthenogenetic eggs even before they leave their mother's brood sac. It has been estimated that a single female of *Daphnia pulex* would leave 13,000,000,000 descendants in a period of 60 days, providing all its descendants lived and reproduced at a normal rate. Of course, *D. pulex* is such a delectable morsel for small fishes and other predators that vast numbers of them succumb, and the population normally only holds its own. However, a slight upset in the balancing factors permits the production of vast swarms of individuals within a very short time, whereas conversely, a slight imbalance in other direction results in a sudden disappearance of all but scattered individuals. This aspect of the ecology of the Cladocera serves well to emphasize their extreme importance as intermediaries in the aquatic food chain.

In most Cladocera the eggs within the brood sac at times develop parthenogenetically. In a few, this type of development is the regular and almost (or quite) invariable method. In others, and this appears to include most of those occurring in temperate regions, parthenogenesis is the rule in the warm months of the year, but is relieved by sexual reproduction under certain adverse conditions. During most of the breeding season, then, parthenogenetic females produce nothing but parthenogenetic females. When deleterious conditions such as crowding occur, however, the parthenogenetic eggs give rise to a certain proportion of males. (It has become evident that male production is not associated with some internal physiological rhythm, as originally postulated by Weismann, and uncritically accepted by many subsequent workers.) Shortly after the appearance of males, though possibly from causes other than simple crowding, the females produce eggs that are considerably larger than the parthenogenetic eggs formed at other times, or by other generations. The egg attains its larger proportions at the expense of all the other eggs in the ovary, taking into its own body the yolk that is produced by the others. Finally, this special egg is fertilized by the males, and is placed in the brood sac, usually along with a second one from the other ovary. A special case, the ephippium, is formed, as a rule, around the eggs. The ephippium is shed at ecdysis and takes up an independent existence. The contained eggs develop embryologically for a short period, then go into a prolonged period of rest. In this protected resting condition the eggs in the ephippium are able to withstand extremes of cold and dessication. Only in *Leptodora* do the resting eggs hatch in a nauplius stage—all others are much more advanced when they leave the ephippium.

In alpine and far northern waters there is a smaller number of parthenogenetic generations between sexual generations, as would be expected because of a shorter warm season. Moreover,in some warmer waters, at least in the case of certain species, more than one sexual generation develops during the course of a year. This condition is known as polycycly. The condition usually appears to be associated with species living in small ponds subject to drying as well as to cold, and polycycly thus has an obvious survival value for the species. In very deep, cool lakes, where there is relatively little temperature change during the course of the year, it is not unusual for there to be a complete lack of sexual generations (acycly).

Artificial key to the commoner genera of planktonic Cladocera

1. Shell covers the body and trunk appendages 2
   Shell greatly reduced, forming only a brood sac 11
2. Shell encased in a gelatinous covering. Second antennae uniramous . . . . *Holopedium* (fig. 480)
   Shell without gelatinous case. Second antennae biramous 3
3. All six pairs of trunk appendages foliaceous 4
   Five or six pairs of appendages. First two more or less prehensile . . . 5
4. Dorsal ramus of second antenna with three segments . . . . *Sida* (fig. 481)
   This ramus with two segments . . . . *Diaphanosoma* (fig. 482)
5. Fornices and rostrum forming a beak, which partly or wholly covers the first antennae 6
   Fornices not covering the first antennae 7
6. Spherical or ovate species . . . . *Chydorus* (fig. 483)
   Form subquadrate . . . . *Acroperus* (fig. 484)
7. First antennae large, immovable . . . . *Bosmina* (fig. 485)
   First antennae small, often rudimentary, not fixed 8
8. Head and rostrum large, no cervical sinus . . *Daphnia* (fig. 486)
   Head small, with cervical sinus; rostrum present and small, or absent 9
9. With a rostrum 10
   Without a rostrum . . . . *Ceriodaphnia* (fig. 487)
10. Ventral and posterior margins straight . . . . *Scapholeberis* (fig. 488)
    Ventral and posterior margins more convex . . . . *Simocephalus* (fig. 489)
11. Brood sac very greatly reduced in size; first of six pairs of feet very long and cylindrical . . . . *Leptodora* (fig. 490)
    Brood sac larger, first pair of feet not thus 12
12. Brood sac rounded 13
    Brood sac cone-shaped and pointed, Marine . . . . *Evadne* (fig. 491)
13. Marine forms . . . . *Podon* (fig. 492)
    Fresh-water forms . . . . *Polyphemus* (fig. 493)

Interested persons should consult Herrick and Turner (1895), Keilhack (1910), Kiser (1950), Pennak (1953), and Ward and Whipple (Birge, 1918) for further information on the Cladocera.

b. Ostracoda.—The ostracods are small entomostracans equipped with a calcified (usually), clam-like bivalved shell. The body and head are completely enclosed in the shell, and usually only the ends of the appendages are visible when the shell is open. However, in many marine planktonic species the second antenna is larger and protrudes permanently from the shell, where it acts as a natatory organ. The appendages, in most cases, can be withdrawn completely within the shell, and the latter can then be closed tightly by means of special adductor muscles. Thus a considerable degree of protection is offered by the shell.

Typically there are seven pairs of appendages, five of them on the head. There is a massive pair of uniramous first antennae, and a similarly prominent pair of second antennae, between which there lies the labrum, or upper lip. Just posterior to the labrum lies the mandible, with its mandibular palp. Behind this lie two pairs of complicated maxillae, which have, in addition to other uses, a respiratory function. The two pairs of legs on the thorax are uniramous. The second one is small and has the function of cleaning the shell of debris. In addition to the paired appendages, there is at the posterior end of the body a movable furca. There is a simple eye (or eyes) towards the anterior end of the body.

The ostracods are easily recogized as such by most biologists. However, the genera and species are not commonly so well known because of the small average size of the specimens (in fresh water they average about a millimeter in length) and because of the necessity of making dissections before generic or specific determinations can be made with surety. Most of the 2,000 known species live in the sea, though individuals are extremely common in fresh water, and practically all bodies of fresh water, especially those rich in decaying organic material, contain many ostracods.

Practically all species are bottom forms, burrowing in the mud, creeping over submerged aquatic vegetation, etc. There are few or no holoplanktonic species in fresh water, though some species of *Notodromas* and *Cypris* wander regularly into this habitat. *Notodromas* (fig. 494) has two simple eyes, while *Cypris* (fig. 495) has but a single eye, and the two are most easily distinguished from each other by this structure.

In the sea, on the other hand, many species are holoplanktonic, though here too the vast majority are benthic. Many of the planktonic species are larger and rather spectacular, one reaching a length of a centimeter. The most common marine planktonic genus is *Conchoecia* (fig. 496).

Interested persons should consult Müller (1894, 1900), Pennak (1953), Skogsborg (1920, 1928), and Ward and and Whipple (Sharpe, 1918) for further information on the Ostracoda.

c. Copepoda.—The crustacean subclass Copepoda includes a large number of forms, many of which differ very widely from each other. The order Branchiura is so different from the remaining forms (order Eucopepoda) that there has been some question whether the two should be considered as being closely related at all. The Branchiura are provided with compound eyes, and there is a heart and consequently a circulation of the blood. The Eucopepoda, on the other hand, have simple eyes, and many have no heart. All of the Branchiura are parasitic, and therefore need not concern us further, though it must be mentioned that occasional specimens are captured in plankton nets. These are captured as they swim through the water from one host to another.

The Eucopepoda are usually divided into seven suborders: the Calanoida, the Harpacticoida, the Cyclopoida, the Monstrilloida, the Caligoida, the Notodelphyoida, and the Lernaeopodoida. Of these, the last three are entirely parasitic or commensal, and therefore none of them are planktonic. As in the case of the Branchiura. however, the caligoids are occasionally captured by plankton nets as they swim from one host to another. The Monstrilloida are all internal parasites of marine invertebrates during most of their immature stages, but are free-living and planktonic as adults. They are usually very rare, however. Of the remaining three suborders, most of the genera of the Harpacticoida and Cyclopoida are benthic or parasitic. A very few forms are found in damp moss far from any body of water. Certain of the harpacticoid and cyclopoid genera are planktonic, on the other hand, and often they are extremely abundant, especially the genera *Microsetella, Oithonina, Corycaeus,* and *Cyclops.*

It is in the suborder Calanoida, however, that the greatest number of planktonic genera is to be found. Among the members of this suborder, in fact, planktonic existence is the rule rather than the exception. Planktonic copepods, especially calanoids, live and often are dominant both in fresh waters and in the sea. They range in size from less than 0.5 mm. to about 15 mm., the average being around 2-3 mm.

The calanoids are probably the most generalized in their structure. The body consists of the usual crustacean head, thorax, and abdomen, and certain appendages and structures can be assigned to each subdivision. However, this fundamental division of the body is not superficially clear. It is usually clearly divided into two portions, the metasome and the urosome. The metasome is a broad, thick anterior portion of the body. It consists of the head and, in the important planktonic suborders, of the first four or five segments of the thorax. The metasome bears the important appendages of the body, namely the antennae, all the mouthparts, and the swimming legs, except, in some cases, the rudimentary posterior pair or pairs. The urosome, on the other hand, is usually a narrower and shorter posterior part of the body. It may include (Harpacticoida, Cyclopoida) the fifth thoracic segment and its pair of appendages, and it always includes the genital segment, which many authorities consider to be a part of the thorax. In addition, behind the genital segment there are from one to four true abdominal segments. The last of these is called the anal segment, because it bears the termination of the digestive tract. The anal segment, in all planktonic forms, bears a pair of furcal rami, each of which usually is provided with several well-developed furcal setae. At times (e.g., *Aegisthus* and *Heterorhabdus*) these setae are of extreme length and help the animal, as flotation organs, to maintain its planktonic existence. In other cases (e.g., *Calocalanus* and species of *Oithona*) the setae accomplish the same purpose by the development of a high degree of feathering.

Most planktonic and other free-living copepods have a total of eleven pairs of appendages. The most anterior, the first antennae, are uniramous and typically long, with up to 25 segments. Their main function is natatory in planktonic species. Often many of the segments bear special fleshy sense organs, the aesthetasks. In the females of all planktonic species, and also in the males of many, the first antennae are both alike, and simple. In the males of the majority of the species, however, one (Calanoida) or both (Cyclopoida, Harpacticoida) of these appendages are geniculate, and modified for grasping the female. In some cases among the calanoids (e.g., *Labidocera*) the geniculate antenna is very highly modified and extremely different from the first antennae of the female, as well as from the corresponding antenna on the other side of the male himself. In other cases there is relatively little differentiation.

Posterior to the first antennae lie the second antennae. These are typically smaller than the preceding pair, but in some of the parasitic forms (such as *Ergasilus*) that are occasionally found swimming in the plankton, the second antennae may be much larger than the first. Usually the second antennae are biramous, but by no means always. The third pair of appendages is a pair of mandibles. The basal segment of each is developed into a chewing organ, the masticatory portion of the one on the left side being juxtaposed to that of the right side in the region of the mouth opening. The mandible typically has a biramous palp, provided with a number of setae. Behind the mandibles lie the biramous first maxillae, and the uniramous second maxillae and maxillipeds. In filter-feeding species these three pairs of appendages act as a collection basket, whereas in predatory forms they are used in the capturing of the prey. They differ greatly from genus to genus, but are not very frequently used as determining structures in the taxonomy of the group.

All of the above appendages are located on the cephalic region of the animal. The thorax typically bears five pairs of biramous swimming legs. In primitive forms such as *Calanus* both the endopod and the exopod of all five of the pairs of legs are three-segmented, while the basipod has two segments. In the majority of genera, however, there is a reduction of the number of segments, especially of the endopod. Such a reduction occurs more frequently in the first and second pairs than in the third and fourth. The fifth pair of legs is more subject to modification than any of the others. In females the fifth legs tend to be reduced in size and in the number of the segments, both of the endopod and the exopod. Very frequently the endopod is entirely lacking, resulting in a uniramous appendage. In many other cases the whole uniramous leg of the female is rudimentary, consisting perhaps of only one, two, or three tiny segments, or even in some cases of a single seta. In still others the female fifth pair of legs is entirely wanting. In the male, on the other hand, the fifth legs are used in the transfer to the female of the complicated spermatophore that is characteristic of the group. It is true that, as in the female, the endopod is often lost, but in such cases the exopod is usually highly modified, and often complex. Furthermore, in other genera the endopod also is highly developed and specialized. Usually the male fifth legs are asymmetrically developed, sometimes with one degenerate and the other highly developed. In many, one leg is developed into an extremely complicated hand-like structure for the handling of the spermatophores.

In the Harpacticoida and the Cyclopoida, the fifth legs are usually rudimentary in both sexes, and they occur on the first segment of the urosome, whereas in the Calanoida they are located on the last segment of the metasome. In some copepods of the first two suborders mentioned, there is in addition a very tiny pair of sixth legs. In all suborders the segmentation and other characteristics of the legs are used in identifications.

During mating activities,* the male clasps the female with the modified first antennae, and tranfers to her (in the Calanoida by means of the fifth legs) a spermatophore. He fastens the spermatophore in the general vicinity of the genital opening of the female, and then usually leaves her. Fertilization takes place when the eggs are laid. In the various species, the eggs are either extruded individually into the surrounding water, where development takes place (e.g., *Tortanus, Centropages*), or else the eggs are carried until hatching time in one or two egg sacs attached to the genital segment (e. g., *Cyclops, Corycaeus, Pseudodiaptomus*). In planktonic species the eggs invariably hatch in a typical nauplius stage (figs. 497, 498). There are usually six naupliar stages followed by six copepodid stages (similar to the adult in general form), the sixth of which is the adult.

* Males are unknown in the common genus *Mormonilla,* and are rare in several other genera, the eggs of the females developing parthonogenetically.

Many of the copepods are filter feeders (*Calanus*, etc.), though the details of their food have not yet been clearly established. It does not appear that they consume many of the larger planktonic plants, but that they feed mainly upon nannoplankters, including to a limited extent the bacteria. Other species are predacious (*Euchaeta*, etc.), feeding upon other copepods and small zooplankters in general. When food is abundant, the copepods store large amounts of fat within their bodies. This fat serves not only as storage for lean times, but also aids in the flotation of the animals.

Because of their great nutritional value and their large numbers, the copepods are very important as food for larger organisms. Herring, for example, as well as chaetognaths, jellyfish, minnows, juveniles of most fishes, sardines and even the large basking sharks and the whalebone whales utilize large numbers of pelagic copepods. It is apparent that planktonic copepods occupy a most important place in the general economy of aquatic environments. In the sea they are second to no other group of zooplankters in ecological importance, while in fresh waters they share their place with the Cladocera.

Artificial key to the more important genera of planktonic Copepoda

1. Nonmarine forms 2
   Marine forms 14
2. Fifth thoracic segment and fifth legs (if present) form the posterior part of the metasome 3
   Fifth thoracic segment is the first segment of the urosome 10
3. Endopod of third and fourth legs two- or three-segmented 4
   These endopods are one-segmented . . . . . . *Epischura* (fig. 499)
4. Endopods of third and fourth legs three-segmented 5
   These endopods two-segmented . . . . *Eurytemora* (fig. 500)
5. Males, with fifth legs modified for transferal of spermatophores . . 6
   Females, without such modified fifth legs 8
6. Endopod of both fifth legs three-segmented . . . . ♂ *Limnocalanus* (fig. 501)
   Endopod of fifth legs one-segmented, or lacking on one side 7
7. Endopod of first legs two-segmented . . . ♂ *Diaptomus* (fig. 502)
   Endopod of first legs three-segmented . . . . ♂ *Pseudodiaptomus*(fig. 503)

8. Fifth legs biramous 9
   Fifth legs uniramous . . . . ♀ *Pseudodiaptomus* (fig. 503)
9. Endopod of first legs two-segmented . . . . ♀ *Diaptomus*(fig. 502)
   Endopod of first legs three-segmented . . . . ♀ *Limnocalanus* (fig. 501)
10. Fifth legs one-segmented 11
    Fifth legs two-segmented 12
11. Rami of first to fourth legs two-segmented . . . . *Microcyclops* (fig. 504)
    Rami of first to fourth legs three-segmented . . . . *Eucyclops* (fig. 505)
12. End segment of fifth legs trilobed, each lobe with either a seta or a spine . . . . *Macrocyclops* (fig. 506)
    End segment of fifth legs not thus 13
13. End segment of fifth legs with two subequal setae . . . . *Mesocyclops*(fig. 507)
    End segment of fifth legs with a long apical seta and a short inner spine . . . . *Cyclops* (fig. 508)
14. With no appendages between the first antennae and the first swimming legs . . . . *Monstrilla* (fig. 509)
    Appendages not thus 15
15. Fifth thoracic segment and fifth legs (if present) form the posterior portion of the metasome 16
    Fifth thoracic segment is the first segment of the urosome 73
16. Endopods of third and fourth legs with three segments 17
    Endopods of third and fourth legs with two segments 67
    Endopods of third and fourth legs with one segment . . . . *Mormonilla* (fig. 510)
17. Endopods of first legs with three segments 18
    Endopods of first legs with two segments 31
    Endopods of first legs with one segment 35
18. With a small brown luminescent organ on one or the other side of the anterior portion of the metasome . . . . *Pleuromamma* (fig. 511)
    Without such a luminescent organ 19
19. With a deep invagination on the first endopod segment of the second leg . . . . *Metridia* (fig. 512)
    Without such an invagination 20
20. Females, with genital segment (first segment of urosome) more or less swollen 21
    Males, with genital segment not swollen 26
21. Fifth legs uniramous . . . . ♀ *Pseudodiaptomus* (fig. 503)
    Fifth legs biramous 22
22. Fifth legs similar to the other legs 23
    Fifth legs not similar to the others 24

23. Second segment of second exopod deeply notched . . . . ♀ *Undinula* (fig. 513)
Second segment of second exopod not thus . . . . ♀ *Calanus* (fig. 514)
24. Second inner terminal seta of left furcal ramus enlarged and greatly elongate . . . . ♀ *Heterorhabdus* (fig. 515)
Second inner terminal seta not thus 25
25. Fourth and fifth segments of metasome fused . . . . ♀ *Lucicutia* (fig. 516)
Fourth and fifth segments of metasome separate . . . . ♀ *Centropages* (fig. 517)
26. Fifth legs similar to the others . . . . ♂ *Calanus* (fig. 514)
Fifth legs not similar to the others 27
27. Second segment of second exopod with a deep notch . . . . ♂ *Undinula* (fig. 513)
Second segment of second exopod not thus 28
28. Endopods of both fifth legs three-segmented 29
Endopods of both fifth legs not three-segmented 30
29. Second inner terminal seta on left caudal ramus enlarged and elongate . . . . ♂ *Heterorhabdus* (fig. 515)
Second inner terminal seta of left caudal ramus not thus . . . . ♂ *Centropages* (fig. 517)
30. Right fifth endopod with two segments, left with three segments . . . . ♂ *Lucicutia* (fig. 516)
Right fifth leg uniramous, endopod of left with one segment . . . . ♂ *Pseudodiaptomus* (fig. 503)
31. Furcal rami at least six times as long as broad . . . . *Temora*
Furcal rami at most three times as long as broad 32
32. Middle segment of third and fourth endopod with two setae, and terminal segment with seven setae 33
Middle segment with one, and terminal segment with five setae 34
33. Outer margin of second, third, and fourth exopods smooth . . . . *Calocalanus* (fig. 519)
Outer margin of these exopods toothed . . *Paracalanus* (fig. 520)
34. Exopod of first legs with two segments . . . *Rhincalanus* (fig. 521)
Exopod of first legs with three segments . . *Eucalanus* (fig. 522)
35. Endopod of second legs with three segments . . . . *Mecynocera* (fig. 523)
Endopod of second legs with one or two segments 36
36. Fifth legs lacking. Females only 37
Fifth legs present. Both sexes 47
37. Posterior surfaces of rami of third and/or fourth legs with transverse rows of spines or at least with a row of spines on the inner border of the first basipod segment of the fourth legs 38
Posterior surfaces of these rami unarmed and smooth 39

38. Head fused with first thoracic segment; no spines on the inner border of the first segment of the fourth basipod . . . . ♀ *Scolecithrix* (fig. 524)
Head separate from the first thoracic segment; with a row of spines on the inner border of the first segment of the fourth basipod . . . . ♀ *Euchirella* (fig. 525)

39. Posterior corners of metasome rounded 40
Posterior corners of metasome forming a sharp angle or forming definite spines 43

40. Inner seta on caudal rami lengthened and geniculate . . . . ♀ *Euchaeta* (fig. 526)
Without geniculate setae on caudal rami 41

41. Second endopod with two segments 42
Second endopod with one segment . . ♀ *Undeuchaeta* (fig. 528)

42. Furcal rami as wide as long . . . . ♀ *Microcalanus* (fig. 530)
Furcal rami much longer than wide . ♀ *Pseudocalanus* (fig. 527)

43. Forehead with a prominent median spine . . . . ♀ *Gaetanus* (fig. 529)
Forehead without spine. Crested or rounded 44

44. Posterior corners of metasome with spines 45
Posterior corners without spines . . . . . . ♀ *Chirundina* (fig. 531)

45. Rostrum bifurcate and well developed . . . ♀ *Aetideus* (fig. 532)
Rostrum weak, its rami usually lacking 46

46. Exopod of second antenna only silghtly longer than endopod . . . . ♀ *Gaidius* (fig. 533)
Exopod of second antenna much longer than endopod . . . . ♀ *Chiridius* (fig. 534)

47. Posterior surfaces of the rami of the third and fourth legs with spines 48
Posterior surfaces of these legs without spines 56

48. Fifth legs uniramous and symmetrical. Females only 49
If uniramous, the fifth legs are asymmetrical 52

49. Terminal segment of fifth leg broadly oval and flattened . . . . ♀ *Scolecithricella* (figs. 535, 536)
Terminal segment of fifth leg not thus 50

50. Exopod of second antenna approximately equal to endopod . . . . ♀ *Lophothrix* (fig. 537)
Exopod of second antenna from one-third to one-half longer than endopod 51

51. Fifth legs three-segmented . . . . ♀ *Scottocalanus* (fig. 538)
Fifth legs two-segmented . . . . ♀ *Scaphocalanus* (fig. 539)

52. Fifth legs uniramous and asymmetrical. Males only 53
Fifth legs biramous. Males only 54

53. Basal portion of fifth legs enlarged . . . . ♂ *Scolecithricella* (figs. 535, 536)
Basal portion of fifth legs not enlarged . . ♂ *Chiridius* (fig. 534)

54. Second basipod of right fifth leg much swollen proximally . . . . ♂ *Scaphocalanus* (fig. 539)
Second basipod of right fifth leg not thus 55

55. Endopod of left fifth leg shorter than exopod . . . . ♂ *Scottocalanus* (fig. 538)
Endopod of left fifth leg longer than exopod . . . . ♂ *Lophothrix* (fig. 537)

56. Basipod and exopod of second and third legs wider than those of fourth leg . . . . *Clausocalanus* (fig. 540)
Basipod and exopod of fourth leg as wide as in second and third . . 57

57. Distal margin of basipod, and outer marginal spines of exopod, with heavy teeth on second to fourth legs . . . . *Ctenocalanus* (fig. 541)
These structures not thus 58

58. Fifth legs uniramous and asymmetrical. Males only 59
One or both fifth legs biramous. Males only 61

59. Last segment of metasome rounded 60
Last segment of metasome pointed . . . . . . ♂ *Aetideus* (fig. 532)

60. Right fifth leg four-segmented . . . . . . ♂ *Pseudocalanus* (fig. 527)
Right fifth leg three- or five-segmented . . . . ♂ *Microcalanus* (fig. 530)

61. Forehead with a median spine . . . . ♂ *Gaetanus* (fig. 529)
Forehead without a median spine 62

62. Posterior faces of rami of third and fourth legs with spines . . . . ♂ *Scolecithrix* (fig. 524)
Posterior faces of third and fourth legs without spines 63

63. Right fifth leg biramous, left uniramous . . . . ♂ *Euchaeta* (fig. 526)
Both fifth legs biramous 64

64. Endopod of second legs two-segmented 65
Endopod of second legs one-segmented 66

65. Last segment of metasome rounded . . . ♂ *Chirundina* (fig. 531)
Last segment of metasome pointed . . . . ♂ *Gaidius* (fig. 533)

66. Terminal segment of right fifth exopod long, curved, and toothed on its concave margin . . . . ♂ *Euchirella* (fig. 525)
Terminal segment of right fifth exopod short, not curved, and devoid of teeth on inner margin . . . . . . ♂ *Undeuchaeta* (fig. 528)

67. Endopod of first legs with one segment . . . *Eurytemora* (fig. 500)
Endopod of first legs with two segments 68
Endopod of first legs with three segments 71

68. Head with a pair of conspicuous dorsal lenses . . . .
. . . . *Labidocera* (fig. 542)
Head without such lenses 69

69. Exopod of second antenna less than half as long as endopod . .
. . . . *Acartia* (fig. 543)
Exopod of second antenna as long as endopod 70

70. Exopods of first four pairs of legs with three segments . . . .
. . . . *Tortanus* (fig. 544)
Exopods of first four pairs of legs with two segments . . . .
. . . . *Candacia* (fig. 545)

71. With two pairs of lenses on the head . . . . *Anomalocera* (fig. 546)
With one pair of lenses on the head 72

72. Right fifth leg of male without a chela . . *Epilabidocera* (fig. 547)
Right fifth leg of male with a chela . . . . *Pontella* (fig. 548)

73. Metasome much wider than urosome 74
Urosome nearly as wide as the metasome, or tapering into it 79

74. Exopod and endopod of first to fourth legs with three segments . . 75
Same unevenly segmented 78

75. Second antenna with two segments . . . . *Oithona* (fig. 549)
Second antenna with three segments 76
Second antenna with four segments . . . . . . *Sapphirina* (fig. 550)

76. Fifth leg with one segment . . . . *Oncaea* (fig. 551)
Fifth leg vestigial, replaced by one or more setae 77

77. Fifth leg replaced by one seta . . . . *Conaea* (fig. 552)
Fifth leg replaced by two setae . . . . *Oithonina* (fig. 553)

78. Fourth metasome segment sharply pointed . . . .
. . . . *Corycaeus* (fig. 554)
Fourth metasome segment rounded . . . . *Copilia* (fig. 555)

79. Endopod of first legs with two segments 80
Endopod of first legs with three segments 82

80. Exopod of first legs with two segments . . . . *Euterpina* (fig. 556)
Exopod of first legs with three segments 81

81. One or both rami of first legs broadened or modified . . . .
. . . . *Metis* (fig. 557)
Neither ramus of first legs modified . . . . *Macrosetella* (fig. 558)

82. Exopod of first legs with one or two segments . . . .
. . . . *Clytemnestra* (g. 559)
Exopod of first legs with three segments 83

83. Urosome as wide as metasome . . . . *Microsetella* (fig. 560)
Urosome much narrower than metasome . . *Aegisthus* (fig. 561)

Interested persons should consult Davis (1949), Giesbrecht and Schmeil (1898), Herrick and Turner (1895), Kiefer (1929), Lang (1948), Marsh (1929, 1933), Pennak (1953), Vervoort (1946), Ward and Whipple (Marsh, 1918), and Wilson (1932) for further information on the Copepoda.

d. Cirripedia.—All adult cirripedes are marine, or live in brackish waters. All are either sessile (barnacles, etc.), or they are parasites, often extremely degenerate (e.g., *Sacculina*). None of the adults, therefore, are ever members of the plankton. A few species of *Lepas*, however, are found only on the high seas, attached to small pieces of floating wood or other objects, and ecologically these come very close to being plankters, even though they are not strictly so.

On the other hand, all or nearly all of the species have pelagic larval stages. Usually fertilization of the eggs is internal, and the embryos develop in the brood sac of the female to a typical nauplius stage, whereupon they hatch. The nauplius is easily distinguishable from other nauplii by the presence of two shorter or longer spines on the anterolateral corners of the carapace (fig. 562). Other carapace spines may or may not be present. The naupliar appendages are of the usual three kinds. The first antennae are uniramous, and fairly large. The second antennae are larger and biramous. These first two pairs of appendages are the main natatory appendages of the nauplius. The third pair of appendages, smaller than the other two, is the pair of mandibles. However, at this stage the mandibles have no masticatory function. In *Sacculina,* the first antennae of the nauplius are modified to form hooks for grasping the crustacean host.

After a few molts, the nauplius metamorphoses into a second type of larva entirely different in its general appearance and structure. This is the cyprid larva (fig. 563), so called because of its surprisingly close resemblance to the ostracod *Cypris.* There is a bivalved shell covering the entire body and the appendages. The appendages differ somewhat, but not radically, from those of *Cypris* and the other ostracods, and there is a simple eye at the anterior end of the body. The main difference between the barnacle cyprid and the ostracod lies in the presence of a cement gland at the anterior end in the former, by means of which attachment is made to the bottom or other surface at the time of metamorphosis into the adult condition.

Barnacle nauplii are very characteristic meroplankters in marine littoral waters, occurring in vast swarms at times when nearby barnacles are breeding. Cyprids also may occur in swarms, but it is natural that they should be more scattered and fewer in number than the nauplii because between the nauplius stage and the cyprid stage the water currents have had time to accomplish a considerable proportion of their dispersal of the larvae, and enemies and unfavorable environmental conditions have had the opportunity to reduce the numbers considerably.

Interested persons should consult Hoek (1909) for further information on larval stages of the Cirripedia.

e. Cumacea.—The cumaceans (fig. 564) are malacostracans, with the usual eight thoracic segments, and with six abdominal segments. A short carapace covers the head and the first three or four thoracic segments. The compound eyes, if present, are sessile. The Cumacea are normally bottom-dwelling forms, burrowing in the mud or sand. However, they are found so regularly as tychoplankters that it seems advisable to include a very brief mention of them here. Interested persons should consult Lechevalier (1951), Sars (1900), and Zimmer (1927, 1933, 1941) for further information on the Cumacea.

f. Isopoda.—The isopods are malacostracans without any carapace whatsoever. They are almost always flattened dorsoventrally. The thoracic legs are uniramous, and gills are borne on the abdominal appendages. The compound eyes, when they are present, are sessile. A few are terrestrial, and many are found in fresh water, but by far the greatest number of species is marine.

The vast majority of the isopods are benthic or parasitic, some of the latter having a very bizarre form. A few marine species, on the other hand, are characteristic of the plankton in their adult state, and many of the benthic forms may be temporarily swept into the plankton. In addition, the larval stages of many of the parasitic isopods of the suborder Epicaridea and of *Gnathia*, are found in the plankton. Many of these are present in the plankton only as parasites on various zooplankters, and hence cannot truly be considered as plankters themselves. For example, the Microniscus stage of certain unknown genera of the Epicaridea is a very characteristic parasite of planktonic Copepoda in some marine regions. Often the parasite is nearly as large as the copepod, and it is very difficult, when they are common, to ignore them when analyzing plankton collections. As a matter of fact, even though they may not be true plankters, they assume a position of such importance that it would be incorrect to ignore them. As has become clearly evident to the reader long before this, it is impossible to draw lines of demarcation between the plankton and other ecological categories.

So few adult genera are represented as holoplankters that it seems superfluous to construct a key for them. The only two genera that are commonly planktonic are *Munnopsis* and *Eurydice*. Both of these genera have benthic as well as pelagic species. In *Munnopsis* (fig. 565) the abdominal segments are fused together and the coxopodite of the thoracic legs is small and ordinary. In *Eurydice* (fig. 566) the abdominal segments are not fused, and the coxopodites of the thoracic legs are broad, and partially or wholly fused to the sides of the thoracic segments of the body.

Interested persons should consult Tattersall (1911) for further information on pelagic Isopoda.

g. Amphipoda.—The amphipods, or scuds, are very common animals, familiar to nearly everyone. A few are terrestriai, and many species live in fresh water. However, the majority of species is marine. Most are benthic and some are either parasitic or commensal, but a large number of species in salt water are planktonic. The last belong, for the most part, to the suborder Hyperiidea. The borderline between planktonic species and benthic species is unclear because many of the forms that live on the bottom or among plants will leave their more normal habitat for a period and swim in the plankton. In other cases such benthic species may leave the bottom in vast swarms during the breeding season, or at other times. In the following key only the holoplankters among them will be considered.

In many ways the amphipods are similar structurally to the isopods. Usually they are compressed laterally instead of dorsoventrally, however, and the thoracic legs are of more than one kind. The abdominal appendages also are of two kinds, and they are not provided with gills. Instead, simple filamentous gills are found on the thoracic legs. As in the Isopoda, there is no carapace, and the compound eyes are sessile, when they are present. In many of the planktonic species the eyes are especially large, occupying the main portion of the head. A typical amphipod (nonplanktonic) is shown in figure 567, where labels indicate the more important external features.

Artificial key to the more common genera of planktonic Amphipoda

1. Maxilliped with a three- to four-jointed palp 2
   Maxilliped without palp 3
2. Head directed ventrally; first thoracic segment overlapping it . . . . *Cyphocaris* (fig. 577)
   Head and first thoracic segment not thus . . *Eurythenes* (fig. 578)
3. Eye occupying most of the area of the head 4
   Eye small or lacking 9
4. Second segment of fifth and sixth legs greatly widened . . . . *Brachyscelus* (fig. 568)
   Fifth and sixth legs not thus 5
5. Uropods without rami 6
   Uropods with rami 7
6. Fifth leg only with a hand . . . . *Euprimno* (fig. 569)
   Third to sixth legs with hands . . . . *Phrosina* (fig. 570)
7. Fifth leg with a hand . . . . *Phronima* (fig. 571)
   Fifth leg without hand 8
8. Third and fourth legs much shorter than the fifth and sixth . . . . *Themisto* (fig. 572)
   Third and fourth legs not much shorter than the fifth and sixth . . . . *Hyperia* (fig. 573)
9. Head shorter than the first thoracic segment . . . . *Lanceola* (fig. 574)
   Head not thus 10
10. End segment of seventh leg pointed . . . . *Scina* (fig. 575)
    End segment of seventh leg blunt . . . . *Vibilia* (fig. 576)

Interested persons should consult Schellenberg (1927) for further information on the Amphipoda.

h. Mysidacea.—The orders Mysidacea and Euphausiacea were previously lumped together as families of the order Schizopoda, which name refers to their common characteristic of biramous thoracic appendages. The term schizopod still is used frequently as a matter of convenience, for the two orders are superficially very similar to each other in appearance. It has been found, however, that the mysids are considerably more primitive than the euphausiids, and at the present time the usual taxonomic practice is to place the Mysidacea as an order of the division Eucarida. The reasons for this will be made clear below in the discussion of the structure of the two orders.

The mysids are shrimp-like in appearance, and small, the maximum size being about 16 cm. Many forms are considerably less than 1 cm. in length as adults. The carapace, which covers nearly all of the thorax, is fused at most with the first three thoracic segments. The head bears a pair of stalked, movable compound eyes, the usual two pairs of antennae, a pair of mandibles, and two pairs of maxillae. There are eight pairs of biramous thoracic appendages, the first two of which in many forms show a slight modification in the direction of becoming maxillipeds. Most of the species do not have any gills in the thoracic region, using the carapace itself instead for respiration. Some forms, however, have gills (podobranchs) on several of the thoracic appendages. In this case, the carapace covers the gills for the most part. The abdomen typically bears six pairs of appendages. The first five of these, the swimmerets, are usually much reduced in size. The sixth pair is modified as uropods. The uropods, along with the telson, form an efficient swimming organ, the tail fan, as in many other malacostracans. In a great many cases among the mysids there is a conspicuous statocyst in the endopod of each uropod, and the presence of this organ is an easy method of differentiating those mysids which have them from the euphausiids and from the larvae of the decapod suborder Macrura. The telson of all mysids is simple and without furcal rami, though subterminal spines may be present.

Female adult specimens are provided with thoracic oöstegites, which together form a subthoracic marsupium in which the eggs are carried. The eggs of some species hatch as nauplii, but in most or all species the young are set free from the marsupium only after they have attained the general form of the adult.

The mysids are typically members of the hypoplankton,

living just above the bottom. However, a number of species are regularly, or only, found in the pelagic region far from the bottom. Some forms are true benthic species, for they live on, rather than just above, the bottom. A very few even bury themselves in the substratum. They are primarily marine forms, but a few, such as *Mysis relicta,* are to be found in some streams and in certain fresh-water and saline lakes, where their habitat is similar to that described above. Of the marine planktonic forms, some are true open-sea species, but most are neritic, with a relatively localized distribution.

Key to the commoner planktonic genera of the Mysidacea*

1. Gills present on the coxae of the second to seventh thoracic appendages; both male and female pleopods well developed, with none modified in the males; always seven pairs of oöstegites; suborder Lophogastridea 2
   Gills absent on the coxae of the thoracic appendages; female pleopods usually more or less degenerate, and those of the male usually modified; seldom seven pairs of oöstegites; suborder Mysidea 4
2. Second to eighth thoracic appendages of similar structure 3
   Second to fourth thoracic appendages short and subchelate; fifth to seventh long and subchelate; only eighth appendage normal . . . . *Eucopia* (fig. 579)
3. Outer branch of uropods without transverse articulation; tip of telson truncate and armed with strong spines . . . . *Lophogaster* (fig. 580)
   Outer branch of uropods with transverse articulation; tip of telson constricted, then terminally expanded into a crescentic lobe that bears fine spines . . . . *Gnathophausia* (fig. 581)
4. Mandibular palp very elongate; no statocyst in inner branch of uropods 5
   Mandibular palp not elongate; statocyst present, though sometimes reduced 6
5. First and second thoracic appendages uniramous, bearing a lamellose endite on the merus . . . . . . *Petalophthalmus* (fig. 582)
   Only the first thoracic appendage uniramous; lamellose endite only on the merus of the second thoracic appendage . . . . *Hansenomysis* (fig. 583)

*This key was constructed by Dr. A. H. Banner, of the University of Hawaii, for use in this book. Adapted in part from Illig (1930).

6. Seven pairs of oöstegites in the female; statocyst reduced . . . . *Boreomysis* (fig. 584)
   Two to four pairs of oöstegites in female; statocyst well developed 7
7. Outer branch of uropod with transverse articulation 8
   Outer branch of uropod without such an articulation 9
8. Third thoracic appendages about same length as the fourth, with a well-developed terminal claw . . . . . . *Siriella* (fig. 585)
   Third thoracic appendage very elongate, usually twice the length of the fourth; end claw rudimentary . . . . . *Hemisiriella*
9. Outer margin of outer branch of uropod bearing a single to several spines, but without setae 10
   Outer margin of outer branch of uropod bearing numerous setae, but without spines 11
10. Sixth article of posterior thoracic appendages with two to three secondary articulations . . . . *Anchialina* (fig. 586)
    Sixth article of posterior thoracic appendages with five to many secondary articulations . . . . *Gastrosaccus* (fig. 587)
11. Labrum posteriorly rounded; mandible with cutting edge not expanded; sixth article of first thoracic appendages not expanded distally 12
    Labrum expanded posteriorly into a plate of two unequal lobes; mandible with cutting edge expanded; sixth article of first thoracic appendage broad distally, with spines on the free margin . . . . *Mysidella*
12. Third thoracic appendages similar to the following appendages . . 13
    Third thoracic appendages strong, thick, with the sixth article undivided and bearing strong spines; seventh large and bearing a strong end claw . . . . *Heteromysis* (fig. 588)
13. Sixth article of the second to seventh thoracic appendages with one or two articulations, the proximal oblique and the distal vertical 14
    Sixth article of second to seventh thoracic appendages without any oblique articulations, usually with two, three, or more vertical articulations 23
14. Eyes with visual elements either lacking or poorly developed 15
    Eyes with visual elements well developed 18
15. Eyes united into a flat plate without visual elements . . . . *Pseudomma* (fig. 589)
    Eyes at least distally separated; visual elements may be present . . 16
16. Visual elements rudimentary . . . . *Amblyops* (fig. 590)
    Visual elements present, although poorly developed; usually of light color 17
17. Eyes attached to each other at base by a membrane; outer distal angle produced into a digitiform process . . . . *Dactylerythrops* (fig. 591)
    Eyes distinct and separate; outer distal margin rounded . . . . *Dactylamblyops* (fig. 592)

18. Eyes with corneas rounded and entire, not divided into anterior and lateral groups of visual elements 19
Eyes divided into anterior and lateral corneas . . . . *Euchaetomera* (fig. 593)
19. Telson at least as long as broad 20
Telson markedly broader than long . . . . . . *Erythrops* (fig. 594)
20. Telson about as long as broad . . . . *Katerythrops* (fig. 595)
Telson markedly longer than broad 21
21. Telson with spines on lateral margins . . . . *Holmesiella* (fig. 596)
Telson with spines only on truncate tip 22
22. First pleopod of male well developed, with long exopod and short undivided endopod . . . . *Meterythrops* (fig. 597)
First pleopod of male a simple plate without articulations . . . . *Parerythrops* (fig. 598)
23. Second to fifth pleopods of male either all well developed or else all rudimentary 24
At least the first and second pleopods of the male are rudimentary; fourth pleopod usually well developed 29
24. All pleopods of male and female are rudimentary . . . *Mysidetes*
Second to fifth male pleopods well developed 25
25. Molar portion of mandible well developed 26
Molar portion of mandible lacking . . . . . . *Mysidopsis* (fig. 599)
26. Tip of telson rounded . . . . *Leptomysis* (fig. 600)
Tip of telson with a terminal cleft 27
27. Margins of terminal cleft of telson unarmed . . . . *Mysideis* (fig. 601)
Margins of telson with spines or teeth 28
28. Terminal article of palp of second maxilla broader than long . . . . *Doxomysis*
Terminal article of palp of second maxilla longer than broad . . . . *Tenagomysis* (fig. 602)
29. Third pleopods of the male at most only slightly reduced, and usually biramous 30
Third pleopods of male reduced to a simple plate 35
30. Antennal scale bearing setae either along the whole outer edge or along the distal portion only; if the latter is the case, the glabrous portion is not terminated by a tooth 31
Antennal scale either with entire outer margin glabrous or bearing setae only distally; in either case, the glabrous portion terminated by a well-developed tooth 34
31. Telson with terminal margin either truncate and straight or produced convexly 32
Telson with terminal cleft 33

32. Exopod of the fourth pleopod of the male with three articles; sixth article of the second to eighth thoracic appendages with five to eleven secondary articles . . . . . . *Mesopodopsis* (fig. 603)
Exopod of fourth pleopods of the male with five articles; sixth articles of the second to eighth thoracic appendages with three secondary articles . . . . *Stilomysis* (fig. 604)
33. Third pleopods of the male with the exopod rudimentary and knob-like or entirely lacking; fifth pleopod of the male with the protopod and both rami well developed . . . . *Hemimysis* (fig. 605)
Third pleopods of the male with the exopod of four to six articles; fifth pleopods of the male very much reduced and of two articles . . . . *Mysis* (fig. 606)
34. Antennal scale with the tooth that terminates the glabrous portion midway on outer margin; exopod of fourth pleopods of the male with five to six articles, the last two bearing long, strong bristles . . . . *Schistomysis* (fig. 607)
Antennal scale with the tooth that terminates the glabrous portion located terminally; exopod of fourth pleopods of the male with seven articles, only the last bearing a strong terminal spine that is swollen at its base . . . . . . *Praunus* (fig. 608)
35. Telson with posterior margin cleft . . . . *Anisomysis*
Telson with posterior margin either truncate or rounded 36
36. Antennal scale distally acute 37
Antennal scale distally rounded . . . . . . *Acanthomysis* (fig. 609)
37. Exopod of fourth pleopods of the males composed of two articles . . . . *Neomysis* (fig. 610)
Exopod of fourth pleopods of the males composed of three articles . . . . *Proneomysis*

Interested persons should consult Banner (1947, 1948) Illig (1930), and Tattersall (1951) for further information on the Mysidacea.

i. Euphausiacea.—The euphausiids, as mentioned above, are very similar superficially to the mysids. Upon closer observation, however, the two groups are seen to differ from each other in certain essential features. The general body form of both groups is shrimp-like, and the same appendages are present in both. In the euphausiids, as distinguished from the mysids, however, the carapace is fused with all of the thoracic segments, gills (podobranchs) are always present, and they are not covered by the carapace. The telson always has one subterminal spine on either side. In addition, there are no euphausiids with statocysts in the endopod of the uropod, and none of the thoracic appendages are modified to act as maxillipeds. The seventh and, especially, the eighth thoracic legs are nearly always reduced in size. The swimmerets are strongly developed, and the endopods of the first two pairs in the males are modified as copulatory organs.

Typically the euphausiids are luminescent organisms, usually with a single luminescent organ on the stalk of each eye, one on the base of each second and seventh leg, and a single organ on the ventral side of each of the first four abdominal segments. The luminescent organs are relatively highly developed compared with those of the lower forms we have heretofore considered. Each consists of a convex lens beneath which lies the photogenous tissue. The function of the luminescent organs remains hypothetical.

In contrast to the mysids, euphausiid females do not have a marsupium for the protection of the eggs. In a few species the eggs are temporarily attached to the thoracic legs. However, most forms simply shed their eggs into the surrounding water, where subsequent development takes place. As far as known, the egg always hatches into a nauplius phase (fig. 611). There then ensues a complicated life cycle. After three or four naupliar stages, during which the form of the larva changes considerably, the larva metamorphoses into calyptopis phase (fig. 612), in which three pairs of appendages have developed posterior to the mandibles. The thoracic segments have also become clearly differentiated. The telson, which was first formed in a rudimentary fashion in the later nauplii, becomes much more fully developed. The only further appendages that appear during the three calyptopis stages are the pair of uropods. Soon the calyptopis metamorphoses into a furcilia phase (fig. 613), in which the eyes become stalked and movable as in the adult. Other thoracic and abdominal appendages also appear. The antennae remain natatory in the furcilia phase. The furcilia phase gradually changes until the antennae lose their natatory function, whereupon we refer to the developing organism as a postlarva. Many authorities contend that at this time the larva should be called a cyrtopia, but it does not differ in any fundamental structural manner from the adult, and it changes into the adult so gradually that it seems more proper to consider the so-called cyrtopia as the first post-larval stage.

All of the euphausiids are marine, and all appear to be planktonic at all stages of their life history. Often, especially in colder waters, they swarm at the very surface in enormous numbers, sometimes even discoloring the water by their presence. At this time they are well known to fishermen and whalers, who call them krill. Whalers long have realized that the presence of krill is a good indication of good whaling grounds. The good judgment of the whalers has been confirmed by Hjort and Ruud (1929) (see fig. 614), who found that in Davis Strait there was a close correlation between the catch of whales and the abundance of euphausiids. The enormous quantity of planktonic food, primarily euphausiids, consumed by whales is indicated by the results reported by Krogh (1934) and Bigelow (1926). Krogh found that blue whales increase in size from a weight of about 4,400 pounds at birth to a weight of as much as 177,000 pounds two years later, at maturity. All of this increase of weight would necessarily derive from the planktonic food. Bigelow (*loc. cit.*) reported the results obtained by Collett where the stomach of a single whale contained as much as 1,200 liters of *Thysanoëssa*, a euphausiid.

Artificial key to the genera of Euphausiacea*

1. Eighth thoracic legs of reduced size, but of full development . . . . *Bentheuphausia* (fig. 615)
   Eighth thoracic legs with the endopods reduced or absent 2
2. Endopod of sixth thoracic legs with five segments 3
   Endopod of sixth thoracic legs composed of three segments or less. Third thoracic legs elongate; may or may not be chelate . . . . *Stylocheiron* (fig. 616)
3. Endopod of seventh thoracic legs with five segments 4
   Endopod of seventh thoracic legs with two segments or fewer . . . 5
4. Third thoracic legs greatly elongate . . . . *Nematobranchion* (fig. 617)
   Third thoracic legs similar to other anterior legs in length . . . . *Thysanopoda* (fig. 618)
5. Exopod of seventh thoracic legs styliform or absent 6
   Exopod of seventh thoracic legs of normal structure 8
6. Endopod of seventh thoracic legs with two segments . . . . *Nyctiphanes* (fig. 619)
   Endopod of seventh thoracic legs rudimentary 7

* Adapted from mimeographed laboratory notes prepared by Dr. A. H. Banner, of the University of Hawaii.

7. Rostrum tapering to an acute tip . . . . *Euphausia* (fig. 620)
   Rostrum broad and truncate . . . . *Pseudophausia*
8. Antennular peduncle with an upstanding leaflet before the eyes . . . . *Meganyctiphanes* (fig. 621)
   Antennular peduncle without such an upstanding leaflet 9
9. Both second and third thoracic legs elongate . . *Tessarabranchion*
   Third thoracic legs never elongate; second various 10
10. Second thoracic legs either similar in size to the others, or, if elongate, with stiff setae along both margins of the propodus . . . . *Thysanoëssa* (fig. 622)
    Second thoracic legs elongate; distal margin only of propodus with stiff spines . . . . *Nematoscelis* (fig. 623)

Interested persons should consult Banner (1949), Hansen (1915), and Illig (1930) for further information on the Euphausiacea.

j. Decapoda.—The decapods include those Malacostraca with eight pairs of thoracic appendages, the first three of which are considerably modified to form true maxillipeds, and the remainder of which are almost always uniramous (Decapoda with biramous thoracic legs are very unusual, and in most cases the exopod is rudimentary). The uniramous legs are in part chelate, and are used as walking appendages and as organs for grasping or crushing the prey. The head and the thorax are fused together to form the cephalothorax, and both together are covered with a carapace. Gills are located on the thoracic appendages, and on the body wall near their bases, and the gills are covered by special portions of the carapace, the branchiostegites. The abdomen varies greatly in the different groups. Typically it bears a series of swimmerets, the first two of which in the male, as in the Euphausiacea, are modified as copulatory organs. The next to the last abdominal segment typically bears a pair of modified swimmerets, the uropods. These, with the last abdominal segment or telson, form the tail fan, which in many forms is an extremely powerful swimming organ. In the true crabs or Brachyura the abdomen is greatly reduced and folded under the cephalothorax. Here it has no natatory function, but in the female serves to carry the eggs. In the male it bears the copulatory organs.

The decapods average very large compared to other crustacean orders, and because of this fact very few of them are planktonic as adults. If they have adopted a pelagic existence they are usually sufficiently powerful to make headway against the currents, hence are members of the nekton. The vast majority of the species of the order are benthic. The bizarre genus *Lucifer* (fig. 624) is the only genus of the Decapoda that is highly modified for planktonic existence. However, several genera of deep-sea prawns, such as *Gennadas* (fig. 625) and *Sergestes* (fig. 626), are pelagic, yet so weak as swimmers that they take their place in the macroplankton.

In the sea a large number of benthic and nektonic species of decapods have planktonic larval stages. These are of great variety as well as of great biological interest and ecological importance in the meroplankton. No larval stages, however, are planktonic in fresh waters. Fresh-water Decapoda commonly deposit large amounts of yolk in their eggs, with the result that hatching is delayed until after the embryo has passed through stages which in many marine species are spent free-swimming in the plankton. Furthermore, in many fresh-water species the young after they are hatched cling to their mothers until able better to fend for themselves.

The larval stages of marine decapods have been more thoroughly treated by Gurney (1942) than by any other worker, and the same author (1939, 1942) has also published virtually complete bibliographies on larval decapods. Interested persons are referred to these publications for more detailed information than can be given here.

The study of planktonic larvae of the decapods is greatly complicated by the plethora of names that have been given to the several stages of the many decapod species. In fact, larval stages of many forms were first discovered before anyone had even an inkling that any connection existed between the discovered organisms and adult decapods. In many taxonomic works of the past, therefore, larvae were described as though they were adults, and were given generic and specific names. Gurney (1942) lists such larval genera, and finds it necessary to devote two full pages of his book to the task. Some of these invalid generic names have been passed down even to the present as names of specific stages in decapod life histories. Others have been forgotten. The names *zoea* and *megalops*, which originally were generic names given to the larval stages occurring in Brachyura, are still in use. Usually the significance of the former has been broadened to include many other decapod larvae, of other groups, at the same general stage of development. On the other hand, the term *eretmocaris*, although still widely used to refer to macruran larvae with eyes on extremely long stalks, has no real significance because the larvae of some species of a genus will pass through an "eretmocaris" stage, whereas others species of the same genus will not do so. And some species of other genera, not at all closely related, also pass through an "eretmocaris" stage. Still other names, such as *prophylax* or *myto*, are never used any more.

Lucifer spp. and some of the penaeid shrimps leave the egg in the nauplius stage (fig. 627), and the nauplius does not differ in its essentials from that of the Copepoda and the Euphausiacea except that it is lacking in most of the numerous natatory setae of the three pairs of naupliar appendages. Also, the naupliar mouth is sealed shut, and the larvae depend upon stored yolk material for their nutrition, rather than upon an external source of food. Most other decapods also pass through a distinct naupliar stage, clearly recognizable within the egg before hatching.

Gurney (1942) has correctly pointed out the confusion of terminology that reigns in giving names to decapod larvae. He states, for example, (pp.31-32):

> The term "Mysis Stage" can be dropped with advantage. It is an unfortunate term, implying a false relationship, and it is not easily definable. Referring as it does to a stage with biramous legs, it cannot legitimately be applied to a stage with no exopods, although that stage is precisely equivalent to a Mysis stage in other forms. Furthermore, if the term is really used, as it has been, to imply a phylogenetic meaning, it should be applied to a stage in which the thoracic terga are free from the carapace, since this is a much more fundamental feature of the Mysidacea than the possession of exopods.

Gurney would like to divide all preadult decapods into four categories: nauplius, protozoea, zoea, and postlarval. Of these, the nauplius is that stage where the first three pairs of appendages only have been developed, and the carapace is undeveloped; the protozoea is that stage in which the first two or three pairs of maxillipeds have appeared and the carapace is developed, but not yet fused with all the thoracic terga; the zoea is that stage in which the remaining thoracic appendages, and often the abdominal appendages as well, appear; and, finally, the post larval stage is that stage in which the purely adult characters first appear. The mysis stage, then, for example, Gurney would consider as a zoea, and the megalops stage of the Brachyura he would class as postlarval. There is a great deal of merit to Gurney's scheme, but there are also some difficulties, aside from the relatively simple problem of overcoming long-continued usage of the older involved terminology. The main difficulty seems to lie in the fact that in most forms there is no sharp metamorphosis between the stages Gurney would class as protozoeal and those he would class as zoeal. For example, according to the old terminology the Brachyuran larva is hatched from the egg as a "prezoea" (fig. 628), wherein the larva is covered by the embryonic cuticle. This prezoea larva, which would have to be classed as a protozoea according to Gurney's scheme, has no thoracic spines nor long rostrum. The prezoea remains in this stage only a very short time, after which it molts from its embryonic cuticle and emerges, according to the old terminology, as a zoea (fig. 629), in which the carapace is provided with one dorsal and two lateral spines, and with an anterior rostrum. As a rule the first zoeal stage still must be classed, according to Gurney, as a protozoea, for the thoracic legs are not yet developed. Subsequently the larva in question develops thoracic and abdominal appendages, whereupon, the old and the new terminologies agree, it should be called a zoea. It would appear that in many cases, as Gurney himself admits is necessary, it is preferable simply to number the protozoeal-zoeal stages without making any attempt to distinguish between them.

Later there is a sudden metamorphosis, and the zoea is transformed into a megalops (fig. 630). In contrast to the adult, the megalops has a well-developed abdomen, yet the appendages in general have the adult characteristics, and subsequent molts result in a more or less gradual metamorphosis into the adult body form, with its reflexed abdomen.

It is customary to divide the Decapoda, somewhat artificially, into three suborders: the Macrura, with a very well developed abdomen; the Anomura, with an intermediate abdomen; and the Brachyura, with the abdomen relatively very small and folded beneath the cephalothorax. There is considerable uniformity in the larval stages of the Brachyura, which were discussed immediately above. However, in the other two suborders there is great variety. In some of the Anomura, such as *Emerita* and *Albunea* (fig. 631), the larva shows great similarity to the zoeal stages of the Brachyura. Also, except for the extreme length of the spines of the carapace, the larvae of the species of the family Porcellanidae (fig. 632) show at least a superficial resemblance. Even in larval Macrura a few genera show a general brachyuran form. The very peculiar zoea of Sergestes (fig. 633), in which spines of the carapace are present and often highly branched, is an example. Later stages in the development of *Sergestes,* however, do not have the appearance of brachyuran zoeae.

Of the remaining Anomura not discussed above, a number have macruran-like larvae. Space does not permit the description of them here except for the common hermit crabs, the Paguridae. As in the Brachyura, there is no sharp distinction between the protozoea and the zoea. The larvae are easily distinguishable from other similar forms by the condition of the carapace (fig. 634). A rostrum is present, the posterior lateral borders of the carapace bear small spines, and the lateral margins are smooth. The first postlarval stage of at least some of the Paguridae is the peculiar glaucothoë (fig. 635), which is similar to the adult as far as its appendages in general are concerned, but which has a symmetrical abdomen in sharp contrast to the adult.

Among the Macrura the larval stages show considerable similarity. Investigators of the plankton frequently are tempted (and succumb to the temptation) to list them simply as larval shrimps and let it go at that. Such a method may be legitimate at times, but in general it is unsatisfactory for most purposes. Similarly, the classification of these larvae into zoeae (fig. 636) and mysis stages (fig. 637), according to the older terminology, or protozoeae and zoeae according to Gurney's suggestions, remains unsatisfactory. Yet in most parts of the world the life histories of the local macrurans remain unknown in large part, and the recognition of most species is therefore impossible. Gurney (*op. cit.*) has shed considerable light on the problem, and the utilization of his valuable work will permit investigators everywhere, with some difficulty, to identify their material at least to family. In at least some regions, use of his bibliographies (1939, 1942) will also make possible specific identifications. By all means the easiest of the macruran larvae to identify to family are the phyllosoma larvae of the Palinuridae (fig. 638), with their extremely broad, flat, and transparent bodies and long thoracic legs. The mysis stage of the Crangonidae (fig. 639), with the extremely long fifth thoracic legs terminating in long apical spines, is also relatively easy to recognize. The peculiar arrangement of the spines of the carapace makes the recognition of the larvae of *Solenocera* (fig. 640) simple. The eretmocaris larvae (fig. 641), with their long, and often segmented, eyestalks, are easy to determine as a larval form; but, as mentioned above, species of unrelated genera pass through eretmocaris stages, hence the stage has no real taxonomic significance.

k. Stomatopoda.—The stomatopods are relatively large Malacostraca, up to about twelve inches in length. The carapace covers only the first four segments of the thorax and part of the head. The anterior portion of the head, however (and this is a unique situation among those Crustacea that have a carapace), is separated from the remainder of the head by a movable articulation. The abdomen is very broad and bears a number of well-developed swimmerets.

As for the thoracic appendages, there are five pairs of uniramous maxillipeds, of which the second is large and provided with a strong raptorial claw. The appendages corresponding to the thoracic legs of the decapods are three in number; they are small and biramous, and because of their structure cannot be used as walking legs.

All of the stomatopods are benthic in their habits, as a rule living either in holes they have constructed in the sand, or under stones, in cavities of sponges, etc. However, at night they may temporarily leave their shelter and become nekters. None are ever planktonic as adults, but many of them have larvae that are members of the meroplankton. The eggs hatch either as antizoeae or pseudozoeae, these forms not being comparable in any way to, nor derivatives from, the zoea of the decapods. The antizoea (fig. 642) is the first larval stage of *Lysiosquilla* and other genera, and it has a carapace that is so large it covers much of the body of the animal, though it is attached only to the head. There is a large rostrum, and two posterolateral spines on the carapace, in addition to other smaller spines. The thorax and abdomen are not clearly differentiated, but the first five thoracic segments bear simple biramous appendages. The pseudozoea is the first larval stage of *Squilla et al.* (fig. 643), and it is very different from the antizoea. In addition to the rostrum and the two posterolateral spines, there is a pair of anterolateral spines on the carapace, thus giving the carapace an entirely different shape. There are only two thoracic appendages, and they are uniramous. The second of these is very large and raptorial, just as in the adult. In addition, there are four or five pairs of biramous abdominal appendages. Thus the pseudozoea shows many of the features of the adult.

Both the antizoeae and the pseudozoeae develop into modified and somewhat more complex pseudozoea-type larvae, known as erichthus and alima larvae, depending upon relatively insignificant structural differences. All antizoeae and many pseudozoeae develop to become erichthus larvae, whereas the remainder of the pseudozoeae become alima larvae. These more advanced larvae need not be described here in detail (see Giesbrecht, 1910).

l. Diptera.—The majority of all known species of animals belong to the great arthropod class Insecta. However, the class is primarily terrestrial, and all of the forms that are aquatic as adults still depend directly upon the atmosphere for their supply of oxygen. No aquatic adult insects are planktonic.

On the other hand, many of the aquatic larval stages, both of terrestrial and aquatic adults, have divorced themselves from direct dependence upon the atmosphere (stonefly larvae, mayfly larvae, *Chironomus*, *Corethra*, etc.) by various means, and these either obtain their oxygen supply from the surrounding water or, in a few cases, they live anaerobically. Other aquatic larvae (mosquito larvae, etc.) must come to the surface of the water to obtain their supply of oxygen. Practically all of the aquatic insect larvae are benthic or epiphytic. A few of them, such as mosquito larvae among others, may occasionally be found in collections of fresh- or brackish-water plankton, where they have accidentally taken up a temporary existence as tychoplankters. Among all the aquatic larvae of insects, only specimens of the dipteran genus *Corethra* (*Chaoborus,* fig. 644) can be considered as true members of the plankton. *Corethra* larvae are easily recognized by their great transparency, and by the presence of two pairs of internal air sacs (modified tracheae), one towards the posterior end of the animal and the other towards the anterior. These act as organs of flotation and, in addition, perhaps as organs for the storage of oxygen, for the animals often spend considerable time in the hypolimnion in conditions where oxygen is completely lacking. The *Corethra* larvae also have a series of setae on the tail which act as a fin for swimming. Their food appears to be almost entirely Crustacea and aquatic insects.

As mentioned above (p. 111), *Corethra* larvae undertake extensive diurnal vertical migrations, larger specimens actually burying themselves in the bottom muds during the hours of daylight, then populating the entire water mass up to the very surface at night. In the lower regions they have to withstand not only the sparsity, or even total lack, of free oxygen, but also they are at times called upon to withstand considerable quantities of toxic hydrogen sulphide.

m. Acarina.—Like the Insecta, the Arachnoidea, aside from the primitive, benthic Xiphosura of the sea, are primarily terrestrial forms, respiring by means of book lungs or tracheae. A few spiders are aquatic, carrying a bubble of air beneath the water with them for respiratory purposes. Two families of mites are truly aquatic: the Hydrachnidae which live largely in fresh waters, and the Halacaridae, in marine waters near the shore. In both these families, as in most mites, the small size of the body makes special respiratory organs unnecessary. Practically all of the water mites are bottom-living forms, or else they live on and among aquatic plants. They are probably all parasitic, sucking the body juices of larger aquatic animals, at least during their younger stages. Both in shallower, less saline parts of the sea and in fresh waters, mites are more or less regular members of the tychoplankton, but the only planktonic representatives among them are such forms as a few species of *Hygrobates* (fig. 645), in fresh water.

Interested persons should consult Dahl *et al.* (1909), Pennak

(1953), and Ward and Whipple (Wolcott, 1918) for further information on the Hydrachnidae.

18. MOLLUSCA

THE MOLLUSCA are characterized by having a soft, unsegmented body, in which the main viscera are as a general rule (to which there are very few exceptions) covered by a shell gland called the mantle. In the vast majority of species the shell gland secretes an external shell, though in some important groups (e.g., many Cephalopoda) the shell is greatly reduced and often completely internal. It may even be completely lacking, as in the Nudibranchiata, in *Octopus*, and in several planktonic forms.

The molluscs are usually divided into five classes: *(a)* the Amphineura, with their broad, flat foot, reduced head, and (usually) dorsal shell composed of eight overlapping plates; *(b)* the Gastropoda, with a foot similar to the above, but with the head well developed and a shell composed of a single dorsal piece, often twisted into a spire; (*c*) the Scaphopoda, with a hatchet-shaped foot, and a tooth-like univalve dorsal shell which is open at both ends; (*d*) the Pelecypoda, with a hatchet-shaped foot, or with the foot reduced or lacking, and with a bivalved shell closed by one or two strong adductor muscles; and *(e)* the Cephalopoda, in which the foot surrounds the head and forms a series of strong tentacles or arms, and in which almost all of the modern species have a reduced or lacking shell. All except the Pelecypoda have a peculiar tongue-like rasping organ, the radula, in the buccal cavity.

The molluscs are more abundant in the sea, both in number of species and number of individuals involved, than elsewhere, but nevertheless they are widespread both in fresh water (Gastropoda, Pelecypoda) and on land (Gastropoda). In fresh-water habitats practically all forms have eggs rich in yolk, and consequently they have a long developmental period. Fresh-water molluscs usually hatch from the egg in the mature form and with mature habits, and none are planktonic. A partial exception to this rule is to be found in some of the fresh-water clams *(Unio, Anodonta),* where the young hatch as glochidium larvae, which leave the parent's brood chamber and lie on the bottom until a fish approaches, whereupon they attack the fish and become temporary parasites thereon. The glochidium larvae cannot, however, be spoken of as having any sort of real planktonic existence.

In the sea, on the other hand, developmental or immature stages of some species in all of the great classes may be taken regularly or occasionally in the plankton hauls. Except in the Cephalopoda, the embryonic development of many forms results in the production and hatching from the egg of typical trochophore larvae, these in their general structure being indistinguishable from the trochophore larvae of the Annelida, described (p. 222) and illustrated (fig. 470) above. In other species the trochophore stage is passed within the egg membranes, and the larvae hatch at a later stage of development. On one side of the trochophore there develops, according to their nature, either a one-lobed or a two-lobed shell gland, which soon secretes a univalved or bivalved larval shell. Meanwhile, around the mouth region there develops a two-lobed, heavily ciliated extension of the body wall, the velum, which is to be the post-trochophore larval locomotory organ. A foot develops as a protrusion between the mouth and the anus. In this manner the trochophore larva is transformed into a veliger larva. All or nearly all veliger larvae are planktonic, and the individuals maintain their position in the plankton mainly by the action of the velum, though this organ also has other functions, such as food-getting. Veliger larvae of gastropods (fig. 646) and of pelecypods (fig. 647) are extremely characteristic plankters in marine environments, especially in relatively shallow areas near shore where the adults live in great abundance on the bottom. It must not be thought that all forms have planktonic veliger larvae. Many pass through the veliger stage within the egg, and hatch with purely adult characteristics.

Probably a highly modified veliger larva is the echinospira larva (fig. 648), found in temperate and warmer seas, but abundant only in the tropics. In the echinospira larva there is a large, hyaline, noncalcareous shell, the scaphoconch, containing the soft parts, and in most species a second, calcareous shell as well. The scaphoconch does not appear to grow as the animal grows, but maintains its original size throughout until deserted by the growing animal. Echinospirae, which are very striking and beautiful objects in the plankton, are thought to be larvae of the gastropod family Lamellariidae.

Subsequent to the veliger stage, except in shell-less species in which the larval shell is shed at this point, the larvae rid themselves of the velum, and start to secrete the adult shell. Even in those forms with shells, the larval shell usually wears off eventually, for as a rule it is the tiniest tip of the adult shell, and is the oldest portion. Thus the postlarval condition is attained, and at this time the animal adopts adult habits. Occasionally small postlarval forms are encountered as tychoplankters, either having accidentally been washed up into the plankton by strong currents, or not yet having had the opportunity to sink to the bottom after their metamorphosis from the veliger stage.

The cephalopods do not have trochophore or veliger stages in their life history, their yolk-filled eggs hatching as immature individuals with the general characteristics of the adult. In some of the pelagic cephalopods these immature forms are occasionally temporary members of the plankton (see Allan , 1945, and Joubin, 1937).

Of the five molluscan classes, all of the adults of the Amphineura, the Scaphopoda, and the Pelecypoda are benthic in habits, whereas the Cephalopoda are both nektonic and benthic. In the Gastropoda, likewise, the vast majority of species are benthic, but a few small groups of gastropods have adopted a pelagic and planktonic existence. For members of a primarily benthic class of organisms to adopt planktonic habits, obviously necessitates considerable structural modifications. The main modifications have been in the direction of a reduction of the shell and shell gland (in some cases to the point of extinction), and of the development of various special means of flotation. Some of the planktonic gastropods are so bizarre, and differ so greatly from typical gastropods, that for a long time they were placed by themselves in an entirely separate molluscan class. Some authors in the past placed some of them in the class Cephalopoda, as discussed briefly below (p. 264).

The class Gastropoda is usually divided into three orders. The first of these is the order Streptoneura, in which during larval development (i. e., during the metamorphosis from the veliger stage to postlarval, or within the egg in those forms which hatch later) the body of the animal undergoes torsion. In the veliger stage the larvae are bilaterally symmetrical, with the mouth anterior and the anus and mantle cavity gills posterior. During torsion it is as though the foot maintained its original position while the dorsal visceral mass went through a horizontal turn of 180°. This results in bringing the anus and mantle gills forward to lie near the mouth and dorsal to it. At the same time, internally, the whole intestine has been twisted into a figure eight, as has also the nervous system (hence the name Streptoneura). Completely independent of the phenomenon of torsion, there is also in most of the Streptoneura a spiral twisting of the visceral mass and of the mantle which covers it. This is apparently the evolutionary result of the extreme lengthening of the ventrodorsal axis of the animal. The enormously lengthened visceral mass became, in an evolutionary sense, unmanageable, and consequently was folded around itself to form a spire. The very fact of this spiral twisting resulted in the degeneration of many of the organs on one side (usually the left) of the animal. Thus the original bilateral symmetry is completely destroyed both by torsion and by spiral twisting.

The order Opisthobranchia consists of gastropods with the shell reduced or absent and in which during larval development, although torsion takes place, it is followed by a detorsion bringing the animal back to its original form as far as the digestive tract and nervous system are concerned. That the opisthobranchs are derived from streptoneuran ancestors (or both are derived from common ancestors) is indicated not only by this, but also by the fact that the organs on the left side of the body degenerate during development.

The third order, the Pulmonata, does not undergo torsion during embryonic development. The spiral twisting of the visceral mass, however, is as in the Streptoneura. No primary mantle cavity gills are present, though in a few forms secondary gills occur, not homologous with the gills of other forms. Respiration is typically by means of a true, though simple, lung. Gastropods of this order are primarily terrestrial, but many have secondarily taken to water again.

A few adult species both in the Streptoneura and the Opisthobranchia have adopted a planktonic existence, but none of the Pulmonata are planktonic. By far the most commonly encountered adult molluscan plankters are the Pteropoda, especially common being the genera *Limacina, Clio, Creseis, Cavolinia, Cuvierina, Clione,* and *Halopsyche*. Pteropods were at one time considered to constitute a separate class of molluscs by themselves, and because of the lobe-like development of the foot they have in the past been thought by some to be simple cephalopods. Their true position in a natural system of classification has never been satisfactorily settled, though there is general agreement on the basis of internal anatomy that they are opisthobranchs. Cooke (1927) considers them to be a separate suborder of the Opisthobranchia, while Parker, Haswell, and Lowenstein (1940) include them along with other forms among the members of the opisthobranch suborder Tectibranchiata. In the pteropods the foot is produced into two large fins, the functions of which are both locomotion and flotation. The shell is either very thin and delicate or completely lacking. When present it is seldom coiled. Pteropods often occur in vast swarms, to the point of discoloring many square miles of the ocean surface. Some forms, such as *Limacina* and *Clione,* are of great importance as food for whales in northern waters, and all of the common species are important as food for various marine animals. The shelled (thecosomatous) pteropods appear to be largely vegetarian, whereas the unshelled (gymnosomatous) forms are voracious predators.

Thecosomatous pteropods live in such vast numbers in the sea that empty shells raining to the bottom after their death form a significant ingredient of many bottom deposits. In some regions of the South and North Atlantic Oceans in moderate depths (averaging around 2,000 m.) the major ingredient of the deposits is pteropod shells, hence we speak in these cases of pteropod ooze. Such oozes occupy approximately 1,500,000 sq. km. of the bottom in the Atlantic Ocean. They do not appear to occur either in the Pacific or the Indian Oceans, though thecosomatous pteropods occur there commonly in the plankton.

A second group of planktonic gastropods, the Heteropoda, are regularly encountered in warmer seas, but they are practically confined to the tropics and the subtropics except insofar as currents cary them elsewhere. The most common genera are *Atlanta, Carinaria,* and *Pterotrachea.* The heteropods are streptoneurans, as shown by their internal anatomy. As in the case of the pteropods, the foot is modified to form an organ of flotation, usually forming from one to three fins. Often the foot also bears a sucker, and in a few forms it is operculate. There is usually a pair of tentacles, and a pair of large, well-developed eyes. The shell is either present or absent, even the shell gland, or mantle, being absent in *Pterotrachea.* As a whole, the heteropods are very bizarre gastropods, very highly modified for their planktonic existence. Their shells are often found in bottom deposits, but there are no true heteropod oozes.

Besides the pteropods and the heteropods there are a few other specialized planktonic gastropods. The most regularly encountered of these are species of *Janthina*, a thin-shelled streptoneuran whose foot is not modified for flotation, but which relies for maintenance of its position upon a special large float which it secretes by glands in its foot. *Janthina* is a beautiful snail, with a lilac or violet shell and a bright red (variable) body. When irritated it secrets a purple ink into the water. It is largely confined to tropical waters. Another striking gastropod in the tropical plankton is the nudibranch (opisthobranch) genus *Phyllirhoë*. It is of less general importance than the others because of its relative rarity.

Artificial key to the more common genera of planktonic Gastropoda

1. Shell present 2
   Shell absent 9
2. Shell spiral and covers most or all of the body 3
   Shell, if spiral, covers only a small portion of the body 5
3. Large eyes present, foot operculate . . . . *Atlanta* (fig. 649)
   Eyes absent, foot with or without operculum 4
4. Foot modified to form two large fins . . . .
   . . . . *Spiratella (Limacina)* (fig. 650)
   Foot not thus. With large secreted float . . . . . *Janthina* (fig. 651)
5. Shell small and spiral, forming a sort of cap on the soft body
   . . . . *Carinaria* (fig. 652)
   Shell not spiral, covering most of body 6

6. Shell straight, subcylindrical...... *Herse* (*Cuvierina*) (fig. 653)
   Shell conical or with lateral spines or angles.................. 7
7. Shell narrow and conical................ *Creseis* (fig. 654)
   Shell, if conical, very broad anteriorly...................... 8
8. Shell conical or not; if with lateral spines or angles these are towards the anterior end of the shell.......... *Clio* (fig. 655)
   Shell with lateral spines or angles, towards the posterior end of the shell........................... *Cavolina* (fig. 656)
9. Body with a bilateral series of external gills......... *Phyllirhoë*
   Body not thus.. 10
10. Foot a pair of fins towards the anterior end................. 11
    Foot a single ventral fin-like structure... *Pterotrachea* (fig. 657)
11. Body elongate......................... *Clione* (fig. 658)
    Body ovate, rounded behind... *Anopsia* (*Halopsyche*) (fig. 659)

Interested persons should consult Cooke (1927), Simroth (1911, 1913), Thorson (1946), and Tesch (1906, 1949) for further information on planktonic Mollusca.

19. Echinodermata

The Echinodermata are primarily benthic organisms with in general very slow movements, and with a consequent low development of the nervous system. Without exception they are marine. They are highly evolved as far as many of their body systems are concerned, and although most of them are nearly radially symmetrical, they are not at all closely related to the Coelenterata and Ctenophora, with which phyla they were once united in the single phylum Radiata. Echinoderms are triploblastic, with a well-developed coelomic cavity, and for the most part they have a heavy endoskeleton consisting of calcareous plates, rigidly fused together in the Echinoidea (sea urchins, sand dollars, and their relatives), small and widely separated in the Holothuroidea (sea cucumbers), and intermediate in character in the Asteroidea (true starfish), Ophiuroidea (brittle stars), and Crinoidea (sea lilies). Of all the nearly 5,000 existing known species of echinoderms, only three species of sea cucumbers, belonging to two genera, are known that are pelagic. In these slow-moving creatures to have a pelagic existence means to be planktonic. None of the three species is at all common, but *Pelagothuria* is illustrated here (fig. 660) to show the means of flotation whereby the oral tentacles are greatly expanded to form a sort of aquatic parachute to retard sinking.

Although adult echinoderms are almost universally benthic, a very large number of them have planktonic larval stages. It usually has been stated (e. g., see MacBride, 1914) that the typical development of echinoderms passes through pelagic larval stages, as described below. In recent times this concept has been challenged, however. Fell (1945) estimated that the majority of echinoderms have yolky eggs, and either an abbreviated pelagic development with atypical larval stages, or else they skip a larval stage altogether and are hatched or born directly into the benthos. Thorson (1950) came to the conclusion that benthic invertebrates of all kinds, including the echinoderms, very seldom have pelagic larval stages in the Arctic, the Antarctic and in the very deep sea, whereas in temperate regions a larger proportion of the species have such larval stages, and in the tropics most forms have planktotrophic pelagic larvae.

Planktonic echinoderm larvae differ somewhat from class to class, though all of them have the same general structure. They all belong to a general type known as the pluteus type, after the larval genus *Pluteus*, a name at one time used for organisms we now know to be the larvae of echinoids and ophiuroids. The same general type of larva is found in some of the Chordata (see below, p. 272), and it is for this reason that it is now thought that the phyla Echinodermata and Chordata are closely related, whereas a thorough examination of the adults of the two phyla could not possibly lead one to this conclusion.

In its simple form the pluteus-type larva is superficially somewhat similar to the trochophore larva (compare fig. 470 with fig. 661). However, in the pluteus the arrangement of the ciliated band is entirely different. Instead of having a simple prototroch encircling the body anterior to the mouth and a metatroch doing the same between the mouth and the anus, the simple pluteus has a single band, which in essence encircles the mouth. It is transverse anterior to the mouth, bending somewhat further forward at the sides, then on each side it turns to run posteriorly along the side of the larva. Finally, towards the posterior end, each lateral band turns first anteriad, then towards the median ventral surface, where the two join between the mouth and the anus. The apical organs of the trochophore and the pluteus do not differ materially.

Internally also the trochophore and pluteus larvae differ a great deal, although, in general aspect, the digestive tracts are similar in both. In the trochophore the mesoderm buds off from the endoderm as a solid mass of cells in which later there develops a segmentally arranged (Annelida) or nonsegmental (Mollusca) coelomic cavity. In the pluteus, however, the mesoderm is produced from the endoderm by means of segmentally arranged coelomic pouches, where the cavity is present from the beginning and at first is simply a continuation of the enteric cavity (this is the same manner as that in which mesoderm and coelom are produced in lower chordates). Two pairs of coelomic pouches are produced in this manner in the echinoderm pluteus, and a third pair (often this develops only on the left side) is formed by the pinching off of the posterior portion of the first pair of pouches. These middle pouches are called hydrocoels. From the anterior pouch on the left there develops a narrow prolongation, the pore canal, which passes dorsally and opens on the dorsal surface to the outside through a minute opening, the water pore. In some forms the same occurs on the right side as well. Thus the pluteus larva is a bilaterally symmetrical, segmented animal, even though in the Echinodermata it metamorphoses to become a radially symmetrical, unsegmented adult animal. (In many echinoderms, such as the heart urchins and the sea cucumbers, the animal after developing radial symmetry again becomes bilaterally symmetrical by specializations of that side of the animal that is habitually directed forward and of the opposite side that is habitually directed backward during movements. In such cases we have animals in which there was a primitive bilateral symmetry upon which is superimposed a secondary radial symmetry, upon which in turn is superimposed a tertiary bilateral symmetry.) The metamorphosis of the pluteus takes place as follows: the left hydrocoel gradually grows around the esophagus, or around a portion of the stomach, forming five radiating pouches. This left hydrocoel is destined to form the water vascular system of the adult. The main portion of the hydrocoel forms the ring canal, the five branches form the five radial canals, the connection between the hydrocoel and the left anterior coelomic pouch forms the stone canal, the anterior pouch forms the small vesicle beneath the madreporite, and the water pore forms the first pore of the madreporite. After the formation of the ring canal, all the lefthand tissues of the body of the larva, and some of the right side as well, orientate themselves around the ring canal and its five branches, with the consequence that the radially symmetrical adult begins to grow on the left side of the larva, and at its expense. Soon the radial growth dominates, and before long the bilateral larva is completely absorbed.

Except in the very early stages of some, the simple pluteus described above is not produced. Instead there are complications involving mainly the elongation of the ciliated band, its folding into several lobes or arms, and the production of a larval endoskeleton that is by no means homologous with the skeleton of the adult. In each class of the echinoderms where pluteus larvae exist, they were first discovered before it was realized they were not adult organisms, so that they were described by systematists and given generic and specific names. The generic name *Pluteus* was given to the larvae of the Echinoidea and Ophiuroidea, *Bipinnaria* and *Brachiolaria* to larvae of the Asteroidea, and *Auricularia* to the larvae of the Holothuroidea.

The bipinnarian (fig. 662) differs from the simple case described above mainly in the slight prolongation of the ciliated band, and in the separation of its preoral portion to form a ciliated ring anterior to the mouth in addition to the usual transverse band. Just before metamorphosis the bipinnarian develops additional long tentacles, which are not clothed by any portion of the ciliated band. The function of these tentacles is that of providing temporary attachment for the larva during metamorphosis. The larva at this stage is known as a brachiolarian larva (fig. 663).

The auricularian larva (fig. 664) is rather similar to the bipinnarian, but there is no ring of cilia anterior to the mouth. Many auricularia have a few calcareous skeletal plates imbedded in their tissues. The auricularian larva often metamorphoses to form a so-called pupa or doliolarian larva (fig. 665).

This is really a postlarval, bilaterally symmetrical form, which gradually takes on the characteristics of the adult. It bears four or five parallel transverse bands of cilia around its cylindrical body.

The pluteus larvae of the ophiuroids and echinoids differ considerably in superficial appearance from the forms already described, but this is mainly due to the enormous development of the tentacles or arms. These, of course, are formed by the great prolongation of the ciliated band. The arms are not muscular and are not used in locomotion except indirectly through the activity of the cilia. Instead they appear mainly to be organs of flotation. Indeed, they are supported by rigid skeletal spicules, which would make movements impossible.

The plutei of the echinoids and ophiuroids, though similar, are easily distinguishable from each other. They are designated respectively as echinoplutei and ophioplutei. In an ophiopluteus larva (fig. 666) the most posterior pair of arms is well developed and is always directed forward. The skeleton consists of two symmetrical halves. On the other hand, in an echinopluteus larva (fig. 667) the posterior pair of lateral arms is often very small, and is directed laterally or even posteriorly. The skeleton consists of at least four paired parts and one unpaired.

Recently metamorphosed ophiuroids and echinoids are sometimes found in the plankton, not yet having had the opportunity to sink to the bottom subsequent to metamorphosis.

Interested persons should consult Chadwick (1914), Dawydoff (1928), and Mortensen (1901, 1921) for further information on planktonic larvae of the Echinodermata.

20. Chordata

The phylum Chordata differs from all others in the presence at some time in the life history of its individual members of gill slits in the pharynx (except in one genus often considered a primitive enteropneustan), a dorsal hollow nerve cord, and a notochord lying dorsal to the digestive tract and ventral to the nerve cord. The development of the chordate-type central nervous system was a most important evolutionary step, for although the most primitive chordates do not, to say the least, show any intellectual advance over members of many other phyla, the potentialities are present for the greatest advances —advances which have culminated in the evolutionary development of man, who painfully but surely is learning to control his environment for his own needs. The potentialities are present because in the chordates the digestive system no longer passes through the middle of the most delicate and complex portion of the central nervous system, as it does in the more highly developed "invertebrate" phyla.

It is customary to divide the phylum Chordata into four subphyla. The most primitive of these is the subphylum Enteropneusta, in which the adult body is wormlike and divided into three segments—an anterior proboscis, followed by a short collar and a long tapering trunk. The anus is at the posterior end, as in the Annelida. The gill slits are present in the most common species, but are lacking all during the life cycle in *Rhabdopleura.* The notochord is very short and primitive and maintains its primitive direct connection with the digestive tract. The notochord originates in the collar segment, but passes forward into the proboscis.

The subphylum Urochorda is considerably more highly developed, with numerous gill slits in the pharynx and with the notochord large and located in the tail of the animal. However, in most species, after a larval period in which the diagnostic characters listed above are present, the animal becomes sessile and undergoes a marked degenerative metamorphosis, in which the notochord is completely lost and the dorsal hollow nerve cord disappears except for a small rudimentary ganglion. The body becomes saclike, and the animals are hardly recognizable any longer as chordates. But other forms, all of which are planktonic, maintain their primitive condition and fail to degenerate. The Urochorda include the sessile tunicates, the salps, the appendicularians, etc.

The subphylum Cephalochorda, which includes the amphioxi, is clearly segmented, has a notochord for the full length of the body, and has the dorsal hollow nerve cord somewhat expanded at the anterior end to form a primitive brain. There is no sign of any vertebral column.

The fourth subphylum is the Vertebrata, which includes most of the species and all of the higher forms among the

Chordata. The characters are much as in the Cephalochorda except that the brain is more highly developed and there is a cartilagenous or bony vertebral column surrounding the notochord and the nerve cord, which in the higher groups partially or completely obliterates the notochord.

All four subphyla contain species or developmental stages that are planktonic.

a. Enteropneusta.—All adult enteropneustans are benthic, for they are either attached to the bottom or burrow in the mud. All are marine. In *Balanoglossus* and its allies the larval stage, which is called a tornaria after the old larval genus *Tornaria*, is very well known. The tornaria (fig. 668) is a true pluteus-type larva, with all the characteristics of the echinoderm pluteus, as described above (pp. 267-268), even to the production of a hydrocoel, a pore canal, and a water pore on the left side. In some tornarias the ciliated band becomes extremely complicated, but still, in spite of its complications, it remains as a ciliated band around the mouth.

Interested persons should consult MacBride (1914), Stiasny-Wijnhoff and Stiasny (1927), and Stiasny (1941) for further information on tornaria larvae of the Enteropneusta.

b. Urochorda.—The majority of the urochords are sessile as adults (tunicates), but a fair number of them are planktonic (salps, appendicularians, *Pyrosoma*, etc.). All are marine or else they live in brackish waters of marine origin. Many of the sessile forms, however, have a planktonic larval stage, and some also have a planktonic egg (fig. 669). The planktonic larva of the sessile tunicates (fig. 670) shows the chordate affinities of the group very well. The whole larva appears superficially somewhat similar to a tiny tadpole, and is referred to as such by some workers. Anteriorly there is a broadened portion, the "head" of the larva, where the most essential systems of the body are located. The mouth is near the anterior extremity of this "head." It leads into a pharynx, in which there is at least one pair of gill slits. Surrounding the pharynx is a relatively small atrial cavity, which opens to the exterior through an atrial pore. The pharynx is followed by a short esophagus, a stomach, and a curved intestine. Dorsal to the digestive tract is the wide anterior portion of the dorsal hollow central nervous system, and also the single eye. In addition, the "head" bears an adhesive gland at the anterior end, the function of which is to fasten the animal to the substratum at the time of metamorphosis. The tunic, which by no means is as well developed as in the adults, invests the body closely. It is not underlaid by a large body cavity as in the adults, except in the region of the pharynx, where it covers the atrial cavity.

Posteriorly on the "head" there is a long, well-developed, finned tail, whose function is locomotion. Through the middle of the tail there runs the conspicuous notochord, which consists of a closed cylinder of special cells surrounding a gelatinous core. Above the notochord lies the dorsal nerve cord, which is continuous with the brain mentioned above.

Larval tunicates swim very briefly in the plankton, then settle down on some solid object, attach themselves by their adhesive glands, and undergo a remarkable degenerative metamorphosis. The tail, with its notochord and nerve cord, is completely reabsorbed, while the "head" is transformed into the entire body of the adult tunicate by the enlargement of the pharynx, the great increase in the complexity and number of the gill slits, the degeneration of the anterior portion of the nerve cord to form the adult ganglion or brain, and the great enlargement of the atrium and of the enveloping tunic.

Interested persons should consult Lohmann (1911), MacBride (1914), and Parker, Haswell, and Forster-Cooper (1940) for further information on larval tunicates.

The urochord order Larvacea consists of a group of planktonic forms whose structure is surprisingly like that of the larval tunicates described immediately above. The order includes but a single family, the Appendiculariidae. It has been thought that the appendicularians constitute an example of paedogenesis, in which the formation of gonads has occurred in a developmental stage corresponding to the larva of ancestral forms, with the result that the ancestral sessile condition has been lost and the organisms have adopted a pelagic existence. At any rate, the appendicularians do not differ in any important respect from larval tunicates, except in the presence of an anus, sharper differentation of the "head" from the tail, the higher development of the nervous system, the presence of a series of ganglia in the tail portion of the nerve cord, and the lack of a tunic closely investing the body. The animal, however, as described in some detail above (pp. 128-129), is able to secrete a "house" whose functions are filter feeding and locomotion. This "house" is frequently discarded and a new one quickly secreted. Some authorities have considered the "house" to represent the tunic, but it hardly seems that such a temporary secretion could be homologous with the tunic of the ascidian larva still less with the complicated tunic of the adult ascidians.

The urochord order Thaliacea includes a number of peculiar species, all of which, if we except the genus *Octacnemus* (which has very uncertain affinities), are planktonic. The body is extremely transparent and is roughly cylindrical, with an opening at or very near each end. One of these openings is the wide oral aperture, which leads directly into the pharynx. The pharynx bears two to several gill slits, the form of which is typical of urochords; but, unlike their location in the Ascidiacea (tunicates), most of these slits are on the posterior surface of the pharynx. Water passing into the pharynx through the oral aperture goes through the gill slits into a large atrium behind the pharynx, and finally out through the atrial aperture at the posterior end. Food filtered out of the water entering the pharynx eventually gets into the remainder of the digestive tract, which consists of a short esophagus, a small stomach, and a small intestine, the anus of which opens into the atrial cavity near the pharynx. Gonads in sexual forms are located near the intestine. The tunic, which surrounds the whole of the animal and gives the general shape to the body, contains conspicuous muscles arranged transversely in complete rings (family Doliolidae) or partial rings (family Salpidae). The first and last rings of muscle can serve as sphincters for the oral and atrial apertures respectively, while the several rings together are used in the locomotory activities of the animal. Locomotion is by jet propulsion. Rhythmic contraction of the muscle rings forces water out of the atrial aperture at sufficient speed to drive the animal through the water in a forward direction.

The life history differs in the Thaliacea as between the Doliolidae and the Salpidae. In the former there is a tailed larva, similar in many respects to that of the tunicates discussed above. However, the tail, which is ventral, is relatively short, and the "head" shows distinct circular muscle bands. On the tail end of the larva, but on the dorsal surface, there is a short fleshy process, the cadophore, and on the ventral surface of the animal, nearer to the tail than to the oral aperture, there is a much smaller process, the ventral stolon.

As the larva metamorphoses, the most conspicuous results are the complete loss of the tail and the exaggeration of the adult characters already present in the "head." No gonads are produced, however, for this particular generation reproduces only asexually. When mature this generation, which is called the first asexual stage or "nurse," produces small buds of cells on the ventral stolon. The buds break free from the ventral stolon and migrate over the surface of the tunic to the cadophore. Here they become attached (after further growth and division) in such a manner that the cadophore, which meanwhile becomes more elongate, eventually bears two lateral and one median row of the buds. While all of this is happening the parent also changes by increasing the strength and the thickness of the muscle bands, increasing the development of the nervous system, and decreasing the structure and function of the digestive system. At the same time the buds develop into zooids, so that the whole comes to have the character of a colony, in which the parent person is the propulsive and sensory individual and the three rows of zooids become the nutritive and respiratory persons. The whole is somewhat comparable, though there is considerably less complexity, to the arrangement of the persons in the siphonophores among the coelenterates.

All of the lateral zooids remain as nutritive and respiratory persons of the colony, and the same is true of many of the median zooids as well. However, some of the latter break loose from the colony and become independent free-swimming individuals, carrying with them some of the stalk by which they were attached. These individuals are called phorozooids, and they are the second asexual stage, for they produce buds on the stalk. The buds develop to form sexual zooids. Finally these break loose from the phorozooid, become mature, and produce eggs and sperm. Fertilized eggs develop into tailed larvae, and the cycle begins anew. The general appearance of the sexual and the two asexual stages differs somewhat within the confines of a single species, with the result that earlier systematists, who remained unaware of the complexities of the extraordinarily unusual life history, gave a number of specific names to one and the same organism.

In the Salpidae there is no tailed larva in the life cycle. Instead the zygote develops within the body of the parent. The offspring is similar to its parent except that no gonads are developed. Eventually it is set free as an independent asexual zooid. A stolon is formed on these individuals, on which a number of buds develop. The buds become zooids and thus a colonial form is produced. Each zooid eventually develops to take on the form of the sexual individuals. They then break off, first in groups and then as individuals, and undergo sexual reproduction, and the cycle begins anew.

Of the order Ascidiacea, which constitutes the vast majority of the Urochorda, practically all are sessile forms. A notable exception is to be found in *Pyrosoma,* a genus of especially brilliant luminescent forms of the tropics. *Pyrosoma* is a colonial tunicate (Ascidiae compositae)* forming a hollow cylindrical colony with a common central cavity into which all the atrial apertures of the colony empty. This cavity opens to the outside at either one or both ends of the colony. Individual colonies may be from two inches or less to several feet in length. No tailed larval stage is present.

Artificial key to the more common genera of adult planktonic Urochorda

1. Tail and notochord present in adult stage 2
   Tail and notochord absent in adult, present or absent in immature stage 4
2. "Head" greatly elongate . . . . *Fritillaria* (fig. 671)
   "Head" not thus 3

*Sometimes considered as a separate order, Lucida.

3. Tail gradually diminishing posteriorly, ending in a single point . . . . *Oikopleura* (fig. 672)
   Tail broad posteriorly, bifid terminally . . . . *Appendicularia* (fig. 673)
4. Colonial species, forming hollow cylindrical colonies, often of considerable length . . . . *Pyrosoma* (fig. 674)
   Simple or colonial species, not as above 5
5. Transverse muscle bands forming complete rings 6
   Transverse muscle bands forming interrupted rings . . . . *Salpa* (fig. 675)
6. Musculature strongly developed . . . . *Doliolum* (fig. 676)
   Musculature not strongly developed . . . . . . *Doliopsis* (fig. 677)

Interested persons should consult Apstein (1901), Borgert (1894, 1901), Ihle (1927), Lohmann (1901, 1911), and Metcalf (1919) for further information on the planktonic Urochorda, exclusive of the larval tunicates.

c. Cephalochorda.—The amphioxi are all benthic as adults, spending their time buried in the sand. However, immature stages (fig. 678) comparable to the adults in structure are encountered upon rare occasions as members of the hypoplankton. They are too uncommonly found to merit further discussion here.

d. Vertebrata.—The vertebrates are mostly relatively large animals, and many are terrestrial. Of the classes that are aquatic, the adults are almost universally of such a size and of such activity that they do not enter into the plankton community. The sunfish or headfish, *Mola mola* (fig. 679), might be considered to be an outstanding exception to this statement. This filter-feeding species, which also feeds as a predator upon jellyfish, is not usually considered as planktonic, though its sluggish activities easily bring it within the boundaries of the plankton world, for it is carried willy-nilly by the currents. If we do consider it planktonic it is by all means the giant of all plankters, for individuals may reach a length of as much as eight feet and a weight of twelve hundred pounds.

Aside from adult fish, however, many marine species lay floating eggs that remain as members of the meroplankton for a shorter or longer period of time. Many of these eggs are provided with droplets of oil or with large water-filled spaces under the membrane as flotation aids. In particular localities the pelagic eggs of certain local species are recognizable, as for example the egg of the California sardine, *Sardina caerulea,* and of the anchovy, *Engraulis mordax,* off the coast of Southern California. Particular eggs can be identified by the size, the shape of the membrane, the diameter of the egg proper compared with that of the membrane, the number and distribution of oil droplets, the presence or absence of reticulation of the membrane, etc. Our present knowledge of fish life histories, however, especially of noncommercial species, is extremely deficient and, furthermore, the eggs of different species often are very similar in size, shape, and structure, so that it is not feasible to attempt to construct here any universal scheme by means of which pelagic fish eggs obtained in all the various marine waters of the world can be identified. Interested persons will have to depend upon local studies, if any, of the particular water body with which they are concerned, for such information as they may need. An example of such a study is that of Ehrenbaum (1905, 1909), which, however, remains very incomplete because of lack of information on the life histories of all the fish occurring in the region under consideration.

No fresh-water fish are known to date that have pelagic fish eggs, though there is no *a priori* reason to say definitely that no such fish will be discovered in the future. A number of fresh-water species are closely related to marine species that have such pelagic eggs.

In the marine fish with planktonic eggs, and in large numbers of those with sessile or demersal eggs as well, the fry hatch and take up a temporary position in the plankton. Immediately after hatching, these small fry are usually so generalized in structure and so lacking in adult characteristics that it is next to impossible, except with particular species in particular localities, to identify them. An exception to this is the leptocephalus larva of the eel (fig. 680), the life history of which was given above (pp. 44-45). Another exception is the fry of the headfish, *Mola mola* (fig. 681). Older fry, as they develop the characteristics of the adults, become more and more recognizable. By this time they have as a rule left the plankton, however, and become members of the nekton or of the benthos, as the case may be. Naturally, here as elsewhere, the distinction between plankton, nekton, and benthos is a tenuous one, and it is impossible to draw sharp lines of demarcation between them.

In fresh waters there are a number of larval fish that are obtained regularly in plankton collections. Bodies of fresh water usually contain fewer species of fish than any particular local area of the sea, and therefore the determination of the species of planktonic fish fry is greatly simplified. On the other hand, it is as difficult to generalize about the fish larvae of the fresh waters of the world as a whole as it is to generalize about marine pelagic fish eggs or larvae, and interested students of both marine and fresh-water environments will have to rely upon any local studies for aid in determinations. An example of such a study for marine fish larvae is the previously cited work of Ehrenbaum (1905, 1909).

GLOSSARY OF TECHNICAL TERMS NOT DEFINED IN THE TEXT

¶For references to definitions or clarifications of other terms, please consult the index.

A

ABORAL—In radially symmetrical and biradially symmetrical animals, the portion of the body opposite to the mouth—in attached forms, the attached end.

ACICULUM—In the Polychaeta (Annelida), a large seta nearly completely buried in the fleshy parapodium, in distinction to the ordinary setae (*q. v.*).

ADDUCTOR MUSCLE—In the bivalved molluscs, the ostracods, etc., a muscle whose function is to close a bivalved shell.

AGGLOMERATE—A crowded cluster of objects or structures.

ALVEOLUS—A pit or a cavity in a surface, as in the "shell" of the Tintinnidiidae.

ANISOGAMOUS—Not isogamous (*q. v.*), hence with a differentiation of gametes. The term usually signifies a condition in which the gametes are not truly heterogamous (*q. v.*) but show only a slight differentiation.

ANTAPEX—The end opposite the apex, hence the base.

ANTENNAL SCALE—The exopodite of the second antenna in the Mysidacea.

ANTERIAD—Toward the anterior end.

APERTURE—Any opening through an object—mainly applied to relatively large openings.

APOPHYSIS—In the Radiolaria, a horizontal outgrowth at right angles to the longitudinal axis of one of the radial spines.

ARCUATE—Arched or curved.

AREOLATION—Minute spaces surrounded by solid or liquid material, forming a sort of a network, or meshwork, as in the "shells" of certain Tintinnidiidae.

ARTICLE—In the Arthropoda, one of the primary or secondary segments of one of the jointed appendages.

ASPINAL—Without spines, as the aspinal pores of certain of the Radiolaria, where the orifice of the pore is not armed with spines.

AUTOTROPHIC—With reference to the nutrition type of organisms, those nutrition types in which the only food required is inorganic, and in which all organic materials are built up from inorganic by synthetic means. The two autotrophic nutritions are holophytic nutrition (*q. v.*) and chemotrophic nutrition (*q. v.*).

B

BASIPOD—In the Arthropoda (primarily in the Crustacea), the basal, attached portion of the jointed appendage. It usually consists of one or two segments, and terminates, in biramous appendages (*q. v.*) at the place where the appendage branches. In uniramous appendages (*q. v.*) it terminates at a homologous location.

BIFLAGELLATED—With two flagella (*q. v.*).

BIFURCATE—Divided into two branches. Forked.

BINARY FISSION—Asexual reproduction in which an organism divides into two approximately equal parts, each of which then continues life as a whole organism.

BIRAMOUS—In the appendages of the Crustacea and other arthropods, the condition in which there is a basal attached portion, the basipod (*q. v.*), and two terminal, segmented branches, namely the exopod (*q. v.*) and the endopod (*q. v.*).

BOOK LUNG—In certain of the Arachnoidea, an external respiratory device in which there is a sac-like lung, at the base of which lies a series of many leaf-like structures or laminae, all lying flat against each other as in the pages of a book.

BUDDING—Asexual reproduction in which the organism divides into two unequal portions, each of which continues life as a whole organism.

Buffer Salts—Salts which, in water solution, are able to react with and thus remove hydrogen and hydroxide ions. This retards changes of acidity or alkalinity of the solution when acids or bases are added. In the presence of such buffers (as in sea water or salty lakes) there is greater constancy of hydrogen ion concentration (thus of pH) than in fresh water, in which such salts are deficient.

C

Carapace—In the Crustacea, a fold of the skin and exoskeleton of the dorsal body wall, which covers the head and at least a portion of the thorax, usually forming a lateral shield that protects the sides of the body.

Carotinoid—A yellow or orange pigment related chemically to carotene, the coloring matter of carrots.

Caudal—Refers to the tail. In the Copepoda this term is synonymous with the term "furcal".

Centric—In the diatoms (Bacillariaceae), any species belonging to the order Centrales, hence with the ornamentation of the valves arranged radially symmetrically around a central point.

Cephalic—Pertaining to the head.

Chela—In the Crustacea, the pincer on the end of certain appendages, in which the terminal segment of the appendage acts as one movable jaw of the pincers, and the prolonged terminus of the penultimate segment acts as the other, immovable jaw.

Chemotrophic Nutrition—Autotrophic nutrition (*q. v.*), in which energy for organic synthesis is derived from the oxidation of inorganic materials, and in which there is no chlorophyll. This nutrition type is found in certain bacteria, and in a very few other of the lower organisms.

Chitin—An impervious and chemically stable substance synthesized in many animals as an external covering material. It is somewhat similar to cellulose chemically, being a polymer of glucosamine, which is an amino derivative of glucose. Cellulose is a polymer of glucose itself.

CHLOROPLAST—In plant cells, a cytoplasmic cell structure (plastid) containing the green coloring matter, chlorophyll. The seat of photosynthesis in holophytic nutrition (*q. v.*).

CHROMATIN—Material, usually within the nucleus of the cell, which stains deeply with certain particular dyes, such as hematoxylin.

CHROMATOPHORE—In plant cells, a cytoplasmic cell structure (plastid), which acts as the seat of photosynthesis in holophytic nutrition, but whose color is other than green (i.e., yellow, brown, etc.).

CILIUM—A short, vibratile cytoplasmic structure protruding from the surface of a cell.

CIRCULAR CANAL—In the jellyfish, a tube of the gastrovascular cavity lying toward the periphery of the animal and encircling the entire disc.

CIRRUS—A slender, thread-like appendage or tentacle. In the Protozoa, a cone-shaped fused group of cilia.

CLAVATE—Shaped like a club.

COCCOLITH—The tiny calcareous plates and other concretions in the Coccolithophoridae.

COELOM—A body cavity lined on both sides by mesoderm, hence a sizable split in the mesoderm.

COLLARETTE—In the Chaetognatha, a thickened area of the ectoderm in the neck region of the body.

COMMENSALISM—An ecological association between two organisms, whereby one of the pair benefits by the relationship, and the other is neither harmed nor benefited.

COMPLETE DIGESTIVE TRACT—An extracellular digestive tract in which there is an anus.

CONJUGATON—In the Protozoa, sexual reproduction in which there is a temporary fusion of the cytoplasms, and an exchange and fusion of nuclear material, followed by separation of the individuals involved.

COPEPODID STAGE—In the developmental history of the Copepoda, those immature stages in which the general form of the body is similar to that of the adult.

CORONAL PORES—In certain Radiolaria, a circle of smaller pores, surrounding the parmal pores. (*q. v.*).

CORTICAL SHELL—In the Radiolaria, that portion of the shell forming an outer framework.

COXA—In the Crustacea, the basal segment of the jointed appendages. Hence the first segment of the basipod.

CRUCIATE, CRUCIFORM—In the form of a cross.

CYST—A resistant wall or shell secreted around an organism, spore or egg to protect it against the environment during periods unfavorable to ordinary life activities.

D

DICHOTOMOUSLY—Provided with two parts or branches.

DIGITIFORM—In the shape of a finger.

DIPLOBLASTIC—With two layers of cells, namely the endodermal lining of the digestive cavity and the ectodermal covering on the outside: i.e., without an intermediate mesodermal layer of cells. This is the characteristic construction of the Coelenterata.

DISTAL—The portion of an appendage or other structure farthest from the point of attachment.

E

ENDITE—On the appendages of the Crustacea, an inner lobe on certain of the basal segments.

ENDOPOD—The inner branch or ramus of a biramous crustacean appendage. The term is also used to indicate the homologous part in uniramous appendages.

ENDOSKELETON—A skeleton produced within the flesh of an animal, and which typically remains imbedded within the flesh.

ENTOMOSTRACA—At one time the Crustacea were divided taxonomically into two sub-classes, the Entomostraca and the Malacostraca. Now this system has been shown to be artificial. However, the term "entomostraca" has been retained as a convenience to indicate Crustacea of all subclasses except the Malacostraca, though it has no taxonomic validity.

Epiphyte—An organism growing attached upon the exterior of a plant. The epiphyte may be parasitic, commensal, or symbiotic.

Epizoon—An organism growing attached upon the external surface of an animal. The epizoon may be parasitic, commensal, or symbiotic.

Exopod—The outer branch or ramus of a biramous Crustacean appendage.

Exoskeleton—A skeleton secreted by the external or ectodermal cells of an organism, or in one-celled organisms by the external surface of the cell. The term is also used to designate homologous skeletons, such as the quill of the squid, etc., even though these actually may be surrounded by fleshy structures.

Extracapsular Cytoplasm—In the Radiolaria, that cytoplasm lying outside of the central capsule.

Exumbrella—In the jellyfish, the aboral, convex surface.

F

Fascicle—A bundle or cluster of thread-like or rod-like structures bound together or lying parallel to each other.

Fenestra—A small opening, as in a shell or case.

Ferruginous—With the color of iron rust, as in the shells or tests of many of the Foraminifera.

Filament—A thread, or a thread-like structure.

Filiform—In the shape of a thread.

Flagellum—In the Protozoa and elsewhere, a long, thread-like, contractile structure projecting from the surface of the cells, and used in locomotion, etc.

G

Geniculate—With a knee-like bend, as in the male first antennae of certain copepods.

Glabrous—With a smooth surface, not covered with hairs, bristles, or spines.

H

Helix—A coiled spire, as in the shell of the Gastropoda.

Heterogamous—With the sexual reproductive cells differentiated into two very distinct types, namely the small motile male cells, or sperm, and the large immobile female cells, or ova. Thus, with the sexes differentiated.

Holophytic Nutrition—The nutrition of typical green plants. An autotrophic type of nutrition (*q. v.*) in which organic material is synthesized from inorganic by photosynthesis, utilizing light energy (usually from the sun) in the presence of chlorophyll.

Holozoic Nutrition—The nutrition of typical animals. Organic food, derived ultimately from plants, is necessary, and this food is taken into a digestive cavity in order to prepare it for absorption into the body.

Hyaline—Transparent or semitransparent, and with a gelatinous or glassy consistency.

Hydroecium—In the siphonophores, the extended portion of a swimming bell, covering over the proximal part of the stem of the colony.

Hydrostatic Mechanism—A mechanism for maintaining a particular vertical position or orientation within the water mass.

I

Incomplete Digestive Tract—A digestive tract in which there is a mouth and digestive cavity, but in which the anus is lacking, as in the Coelenterata, Platyhelminthes, etc.

Interradial—In the jellyfish, the portion of the disc lying midway between the primary (perradial) canals of the gastrovascular cavity.

Isogamete—Isogamous sexual reproductive cells.

Isogamous—With the sexual reproductive cells all alike. Hence, sexual reproduction without differentiation of males and females.

L

LAMELLOSE—Flat and shaped like a leaf.

LAMINA—A leaf-like structure or object.

LAMINATE—Forming laminae.

LINEAR—Forming a line, or an end-to-end arrangement of individuals, cells or other objects.

LITHOCYST.—In the jellyfish, a lithostyle (*q. v.*) partly or wholly imbedded in the tissues of the margin of the disc. Hence, a cyst-like hydrostatic organ.

LITHOSTYLE—The sense-club or tentaculocyst of the jellyfish. A club-shaped organ on the free margin of the disc, containing usually a small calcareous concretion, and with a cushion of haired sensory cells at the base. Found in a modified form in the lithocysts (*q. v.*).

LOPHOPHORE—The horseshoe of tentacles, and their supporting structures, surrounding the mouth in the Bryozoa, Brachiopoda, and Phoronidea.

LORICA—The protective case, or shell, fitting loosely over the outside of such animals as the Tintinnidiidae, etc.

LUNATE—In the shape of a crescent.

M

MACRONUCLEUS—In the Infusoria (Protozoa); the larger or vegetative type of nucleus found within the cell.

MANUBRIUM—In the jellyfish, a protrusion on the sub-umbrellar surface, at the end of which lies the mouth.

MASTICATORY—Refers to the process of chewing. Hence, any structure involved in chewing is a masticatory structure.

MEDULLARY SHELL—In the Radiolaria, that portion of the shell forming an inner framework.

MEMBRANELLE—A group of cilia fused together to form a short vibratile plate.

MERIDIONAL—Toward the meridian. Along any line dividing a sphere into two equal halves.

MERUS—In the Mysidacea, the third segment of the endopod of the thoracic legs.

Micronucleus—In the Infusoria (Protozoa), the smaller or reproductive type of nucleus found within the cell.

Motorium—In the Protozoa, the nervous center to be found in some of the more highly developed types.

Mucinoid, or Mucoid—Similar to mucus, or to mucin, which is the chief ingredient of mucus.

Myofibril—An intracellular contractile fibril, in the form of plasmagel.

N

Natatory—With the function of swimming.

Nectophore—Among the siphonophores, a medusoid person specialized for the function of swimming. A swimming bell.

Nematocyst—The functional portion of the stinging cell (cnidoblast) in the Coelenterata. A cyst within the stinging cell, containing a coiled thread, which is exploded upon suitable stimulation, with the consequent eversion of the thread. Used for predation and for protection.

Notochord—A structure characteristic of the phylum Chordata. It is a stiff rod of turgid, vacuolated cells extending longitudinally between the digestive tract and the dorsal nerve cord, and is used as a skeletal support for the body.

Nuchal—Pertaining to the region of the neck.

O

Ocellus—A simple eye, consisting usually of a pigment spot or area, a light-sensitive region, and a simple lens. Widely occurring among invertebrates, either as the only photoreceptor, or supplemental to more complicated organs.

Oöstegite—In the Mysidacea, a large inner expansion of the coxa of several of the thoracic appendages, the total of the oöstegites forming a brood pouch within which the embryos remain until a relatively late stage of development.

Ooze—Soft, sand-like deposits on the floor of the ocean, composed largely of the shells or cell walls of dead planktonic organisms. The most important such deposits in the oceans are Globigerina ooze (foraminiferan ooze), radiolarian ooze, diatom ooze, and pteropod ooze, depending upon the nature of the organisms which contribute most heavily to the deposits.

Operculum—Among the gastropods, a shelly structure secreted by the foot and used to close the opening of the main portion of the shell when the animal has withdrawn its soft parts for protection.

Orifice—Any relatively large opening, as in a shell.

Otoporpae—In certain of the jellyfish, tracts of nematocysts (*q.v.*) on the exumbrellar surface above the lithostyles (*q.v.*).

Ovate—Egg-shaped. Oval.

P

Palp—In the Crustacea and other Arthropoda, a sensory portion of certain appendages in the region of the mouth (mandibles, maxillae, etc.). The main portion of the appendage is used for other functions, associated with food gathering, chewing, etc.

Parmal Pores—In certain Radiolaria, the primary pores through the shell, surrounding each radial spine.

Parthenogenesis—The development of an egg without the benefit of the entry of (fertilization by) a sperm.

Pectin—A complex carbohydrate produced in plants. Pectin forms a gel when combined with suitable quantities of sugar and organic acids.

Pedalium—In certain of the jellyfish, an expanded gelatinous basal portion of a marginal tentacle or marginal sense organ.

Peduncle—In certain jellyfish, an extension of the center of the subumbrellar surface, on which the manubrium is borne.

PENNATE—In the diatoms (Bacillariaceae), any species belonging to the order Pennales, hence with the ornamentation of the valves bilaterally or asymmetrically arranged around a longitudinal axis.

PERFORATE—Provided with holes or pores.

PERIPHERY—The edges or borders of an object, or the surface of an organism.

PERRADIAL—In the jellyfish, the portion of the disc in which the primary radial canals lie.

PERSON—In colonial coelenterates, colonial urochords, and other colonial forms, one of the individuals of the colony.

PHOTOGENOUS—Pertaining to that which can generate light, as photogenous tissues in luminescent organisms.

PHOTORECEPTOR—Any sense organ specialized for the reception of light stimuli. An eye or ocellus.

PLEOPOD—In the Malacostraca among the Crustacea, the swimmerets, located on the abdominal segments.

PODOBRANCH—Among certain of the Crustacea, a gill attached to the basal segment of one of the thoracic legs.

PROPODUS—In the Malacostraca among the Crustacea, the next to the last segment of the endopod of the thoracic appendages.

PROTISTA—Those organisms, both plant and animal, that consist of a single cell, or of several to many cells essentially alike in structure and function. Thus, the Protozoa plus the Protophyta.

PROTOPLAST—The mass of protoplasm within the cell wall.

PROXIMAL—The portion of an appendage or other structure nearest to the point of attachment.

PSEUDOCHITINOUS—Composed of a substance with a superficial similarity to chitin (*q. v.*), but which chemically is not the same as this substance.

PSEUDOCOELOM—A well-developed body cavity which is not a true coelomic cavity. The pseudocoelom is a fluid-filled space lying between the endoderm of the digestive tract and the mesodermal layers of the body.

Pseudopodium—A temporary cytoplasmic protrusion of the periphery of a cell, as occurs in the Sarcodina among the Protozoa, and in many other cells in most of the animal phyla.

Pyrenoid—A structure in plant cells within the chloroplasts and other plastids, associated with the synthesis of starch.

R

Radial Canal—In the jellyfish, tubes of the gastrovascular cavity that pass from the central stomach toward the periphery of the disc. In simple cases there are but four of these canals, the perradial canals, but in more complicated cases there are in addition interradial and adradial canals, many of which may be complexly branched.

Ramus—One of the branches of the biramous crustacean appendage.

Raptorial—Pertaining to structures associated with seizing the prey or enemy, as in certain appendages of the Stomatopoda, where the terminal segment of the appendage folds back strongly upon the spiny penultimate segment to form an efficient seizing organ.

Reticulum—A network.

Rhopalia—In the jellyfish, marginal sense organs, which are reduced tentacles containing endodermal statoliths (*q.v.*).

Rostrum—In the Crustacea, a prolongation of the anterior portion of the carapace beyond or below the head.

S

Saprophytic Nutrition—The nutrition type of most bacteria, and of many non-green plants. A certain amount of organic food is required, but this is absorbed through the body surface. In those cases where any digestion is required before such absorption takes place, this digestion takes place outside the body of the organism, and not in a special digestive cavity. Saprophytic nutrition differs from the somewhat similar saprozoic nutrition (occurring in certain parasitic animals, such as the tapeworms) in that the subsequent metabolism of the organic food is of a different type than occurs in animals.

SCAPULET—In certain of the jellyfish, outgrowths of the oral arms near the bell. These outgrowths bear supplementary mouths.

SESSILE—With reference to organisms, those that are attached permanently to the substratum. With reference to the compound eyes of certain Crustacea, those that are not borne on stalks, but instead grow close to the surface of the head.

SETAE—In the annelid worms, the Chaetognatha, etc., solid, bristle-like secretions which serve as an aid in locomotion, in flotation, in predation, and in protection. In the Arthropoda, long, relatively wide (as contrasted to hairs) bristles, which are attached to the appendages, etc., by means of a joint. Arthropod setae have a core of living cells.

SIGMOID—In the shape of an S.

SILICIOUS—Composed of silicon dioxide or other silicon compounds. Glassy.

SIPHONOGLYPH—In certain of the Coelenterata, a ciliated canal commencing at one end of the mouth and coursing along the side of the gullet. It serves as a means for the passage of a current of water in and out of the gastrovascular cavity during those times when the animal is in a contracted condition.

SOMATOCYST—In many siphonophores, that portion of the gastrovascular canal lying in front of the growing (proximal) portion of the stem.

SPERMATOPHORE—A packet of sperm within a more or less complicated case that has been secreted by the male, as in the Copepoda, the Cephalopoda, etc.

SPICULATE—Composed of spicules.

SPINOSE—Spiny, or bearing spines.

STATOLITH—In the jellyfish, a small calcareous concretion within the lithocyst (*q. v.*).

STATOCYST—The same as lithocyst (*q. v.*).

STELLATE—In the form of a star, or similar to a star in shape.

STYLIFORM—With the form of a stiletto.

SUBCHELATE—With a poorly developed chela (*q. v.*).

SUBQUADRATE—Approximately square, but not quite so.

SUBUMBRELLA—In the jellyfish, that portion of the body lying on the oral surface, between the margin (periphery) and the manubrium.

SWIMMERET—In the higher Crustacea, a series of rather simple, usually biramous, appendages on the abdominal segments.

SYMBIOSIS—An ecological association between two organisms whereby both the organisms are benefited by the relationship.

SYNCYTIUM—A mass of cytoplasm containing two to many nuclei without corresponding divisions of the cytoplasm. A multinuclear cell.

T

TELSON—The last segment of the abdomen in the higher Crustacea.

TENTACULOCYST—The same as lithostyle (*q. v.*).

TENTILLA—Lateral branches of the contractile tentacle of the feeding persons (gastrozoids) in the Siphonophora.

TERGUM—Among the decapod Crustacea, the skeletal plate covering the dorsal portion of each free segment.

THECA—The case, lorica, or shell surrounding an organism.

TORSION—In many of the Gastropods, during larval development, a process by which the visceral mass (on the dorsal side of the animal) turns through 180°, while at the same time the ventral foot remains as it was. This results in twisting the digestive tract and the nervous system into a figure 8, and brings the anus to lie at the anterior end of the animal, dorsal to the mouth.

TRACHEAE—Air tubes with annular or spiral cuticular thickenings, which serve as the external respiratory mechanism in most of the terrestrial Arthropoda.

TRICORNUATE—With three horns or divisions—trifid.

Triploblastic—With three layers of cells. With a layer of mesodermal cells lying between the external covering of ectodermal cells, and the layer of endodermal cells lining the digestive cavity.

Truncate —With the end flat, or squared off.

Tubulus—In certain of the Radiolaria, a radial, cylindrical extension of the skeleton having considerable width.

U

Unicornuate—With one horn, hence unbranched.

Uniramous—An arthropod appendage in which either the endopod or (usually) the exopod is lacking. Thus the appendage is unbranched.

Univalve —With one valve or shell, as in the Gastropoda.

Uropod—In the higher Crustacea, one of the swimmerets, specialized in such a manner that its parts are very broad and flat. In the Decapoda and others, the uropods and the telson together make up a powerful tail fan, used in rapid backward-swimming movements.

V

Vacuole—A small fluid-filled space within the protoplasm of a cell.

Z

Zoogamete—Among plants, a sexual reproductive cell that is flagellated and able to swim actively through fluid media.

Zoospore—Among plants, a spore that is flagellated and able to swim actively through fluid media.

Zygote—The resultant of the fusion of two gametes. The fertilized egg.

LITERATURE CITED

ALLAN, JOYCE. 1945. Planktonic cephalopod larvae from the eastern Australian coast. *Rec. Austral. Mus.,* 21(6):317-350.

ALLEN, W. E. 1939. Surface distribution of marine plankton diatoms in the Panama region in 1933. *Bull. Scripps Inst. Oc., Tech. Ser.,* 4(7):181–196.

ALLEN, W. E. 1946. "Red water" in La Jolla Bay in 1945. *Trans. Amer. Micr. Soc.,* 65 (2):149-153.

AMERICAN GUIDE SERIES. 1939. Florida, a guide to the southernmost state. Fed. Works Project. Oxford Univ. Press, New York. Pp. i–xxvi, 1–600.

APSTEIN, C. 1901. Salpidae, Salpen. *Nord. Plank.,* 3:5–10.

BALL, ROBERT C. 1948. Fertilization of natural lakes in Michigan. *Trans. Amer. Fisheries Soc.,* 78:145–155.

BALL, R. C. and H. A. TANNER. 1951. The biological effects of fertilizer on a warm-water lake. *Mich. State College Agric. Exper. Sta. Tech. Bull.,* 223:1–32.

BAINBRIDGE, ROBERT. 1949. Movement of zooplankton in diatom gradients. *Nature,* 163:910–912.

BANNER, ALBERT H. 1947. A taxonomic study of the Mysidacea and Euphausiacea (Crustacea) of the northeastern Pacific. Part 1: Mysidacea, from family Lophogastridae through tribe Erythropini. *Trans. Roy. Canad. Inst.,* 26:345–399, pls. 1–9.

BANNER, ALBERT H. 1948. A taxonomic study of the Mysidacea and Euphausiacea (Crustacea) of the northeastern Pacific. Part 2: Mysidacea, from tribe Mysini through subfamily Mysidellinae. *Trans. Roy. Canad. Inst.,* 27:65–111, pls. 1–7.

BANNER, ALBERT H. 1949. A taxonomic study of the Mysidacea and Euphausiacea (Crustacea) of the northeastern Pacific. Part 3: Euphausiacea. *Trans. Roy. Canad. Inst.,* 28:1–62, pls. 1–4.

BARNES, C. A., and T. G. THOMPSON. 1938. Physical and chemical investigations in Bering Sea and portions of the North Pacific Ocean. *Univ. of Wash. Publ. in Oc.,* 3(2): 35-79, appendix 1-164.

BIGELOW, HENRY B. 1911. Albatross Siphonophorae. *Mem. Mus. Comp. Zool.,* 38(2):173–401, pls. 1–32.

BIGELOW, HENRY B. 1926. Plankton of the off shore waters of the Gulf of Maine. *Bull. U. S. Bur. Fish.,* 968:1-509.

BIGELOW, H. B., and M. SEARS. 1937. Siphonophorae. *Rept. Danish Oceanog. Expeds. 1908–10 to the Medit. and Adj. Seas,* 2(Biol. H 2):1–144.

BIRGE, EDWARD A. 1918. The water fleas (Cladocera). In Ward and Whipple, *Fresh-water biology,* pp.676–740.

BIRGE, E. A., and C. JUDAY. 1922. The inland lakes of Wisconsin. The plankton. 1: Its quantity and chemical composition. *Bull. Wisc. Geol. Nat. Hist. Surv.,* 64 (Sci. Ser. No. 13):i-ix, 1-222.

BOGOROV, V. G. 1946*a*. Peculiarities of diurnal vertical migrations of zooplankton in polar seas. *Jour. Mar. Res.,* 6(1):25-32.

BOGOROV, V. G. 1946*b*. Zooplankton collected by the "Sedov" Expedition 1937–1939. *Trudy dreifuiushchei ekspeditsii glavsevmorputi na ledokol'nom parokhode "G. Sedov" 1937-1940 gg,* pp. 336-370. (In Russian, English summary.)

BÖHM, ANTON. 1931. Distribution and variability of *Ceratium* in the northern and western Pacific. *Bernice P. Bishop Mus. Bull.,* 87:1–46.

BORGERT, A. 1894. Die Thaliacea der Plankton-Expedition. *Ergebn. Plank.-Exped.,* 2:1-68, pls. 5-8.

BORGERT, A. 1901. Die nordischen Dolioliden. *Nord. Plank.,* 3:1–4.

BOYER, C. S. 1927. Synopsis of the North American Diatomaceae. Part I, *Proc. Acad. Nat. Sci. Phila.,* 78 (Suppl.): 1-228. Part II, *ibid.,* 79 (Suppl.):229–583.

BROOKS, JOHN L. 1946. Cyclomorphosis in *Daphnia.* I: An analysis of *D. retrocurva* and *D. galeata. Ecol. Monographs,* 16(4):409–447.

BRUNEL, J., G. W. PRESCOTT, L. H. TIFFANY, *et al.* 1950. The culturing of algae. Charles F. Kettering Foundation. Pp. I-IX, 1-114.

CAMPBELL, ARTHUR S. 1942. The oceanic Tintinnoina of the plankton gathered during the last cruise of the "Carnegie". *Publ. Carnegie Inst. Wash.,* 537:i-v, 1–163.

CANNON, H. G. 1928. On the feeding mechanism of the copepods, *Calanus finmarchicus,* and *Diaptomus gracilis. Brit. Jour. Exper.Biol.,* 6:131–144.

CHADWICK, H. C. 1914. Echinoderm larvae. *Liverpool Mar. Biol.Comm. Mem.,* 22:1–32, pls. I–IX.

CHANDLER, DAVID C. 1940. Limnological studies of western Lake Erie. I: Plankton and certain physical-chemical data of the Bass Islands region, from September, 1938, to November, 1939. *OhioJour. Sci.,* 40(6):291–336.

CHUN, C. 1880. Die Ctenophoren des Golfes von Neapel und der angrenzenden Meeresabschnitte. *Fauna und Flora des Golfes von Neapel,* 1:i-xvii, 1–313, pls. 1–18.

CHUN, C. 1898. Die Ctenophoren der Plankton-Expedition. *Ergebn. Plank.-Exped.,* 2Ka:1–32.

CLARKE, G. L. 1934*a.* The role of copepods in the economy of the sea. *5th Pac. Sci. Congress,* Vancouver, B. C., A5. 5. 2017–2021.

CLARKE, G. L. 1934*b.* Factors affecting vertical distribution of copepods. *Ecol.Monographs,* 4:530–540.

CLARKE, G. L. 1939*a.* The relation between diatoms and copepods as a factor in the productivity of the sea. *Quart. Rev. Biol.,* 14(1):60–64.

CLARKE, G. L. 1939*b.* Plankton as a food source for man. *Science,* 89(2322):602–603.

CLARKE, G. L. 1949. The nutritional value of marine zooplankton with a consideration of its use as an emergency food. *Ecology,* 29(1):54–71.

CLARKE, G. L., and S. S. GELLIS. 1935. The nutrition of copepods in relation to the food cycle of the sea. *Biol. Bull.,* 68(2):231–246.

COBB, N. A. 1918. Free-living nematodes. In Ward and Whipple, *Fresh-water biology.* pp. 459–505.

COE, WESLEY R. 1926. The pelagic nemerteans. *Mem. Mus. Comp. Zool.*, 49:1–244.

COLLIN, A., H. DIEFFENBACH, R. SACHSE, and M. VOIGT. 1912. Rotatoria und Gastrotricha. *Die Süsswasserfauna Deutschlands*, 14;1–273.

CONN, H. W., and C. H. EDMONDSON. 1918. Flagellate and ciliate Protozoa (Mastigophora et Infusoria). In Ward and Whipple, *Fresh-water biology*, pp. 238–300.

COOK, A. H. 1945. Algal pigments and their significance. *Biol. Rev.*, 20(3):115–132.

COOKE, A. H. 1927. Mollusca. In *Cambridge Natural History*, Vol. 3. Macmillan and Co., Ltd., London. Pp. i–xii, 1–459.

COWLES, R. P. 1930. A biological study of the offshore waters of Chesapeake Bay. *U. S. Bur. Fish., Fish. Doc.*, 1091:277–381.

CUPP, EASTER E. 1943. Marine plankton diatoms of the west coast of North America. *Bull. Scripps Inst. Oc., Tech. Ser.*, 5(1):1–238.

CUSHMAN, JOSEPH A. 1910. A monograph of the Foraminifera of the North Pacific Ocean. Part 1: Astrorhizidae and Lituolidae. *Bull. U. S. Nat. Mus.*, 71(1):i–xiv, 1–134.

CUSHMAN, JOSEPH A. 1911. A monograph of the Foraminifera of the North Pacific Ocean. Part 2: Textulariidae. *Bull. U. S. Nat. Mus.*, 71(2):i–xiii, 1–108.

CUSHMAN, JOSEPH A. 1913. A monograph of the Foraminifera of the North Pacific Ocean. Part 3: Lagenidae. *Bull. U. S. Nat. Mus.*, 71(3):i–ix, 1–125.

CUSHMAN, JOSEPH A. 1914. A monograph of the Foraminifera of the North Pacific Ocean. Part 4: Chilostomellidae, Globigerinidae, Nummulitidae. *Bull. U. S. Nat. Mus.*, 71(4):i–vi, 1–46.

CUSHMAN, JOSEPH A. 1915. A monograph of the Foraminifera of the North Pacific Ocean. Part 5: Rotaliidae. *Bull. U. S. Nat. Mus.*, 71(5):i–vii, 1–87.

CUSHMAN, JOSEPH A. 1917. A monograph of the Foraminifera of the North Pacific Ocean Part. 6: Miliolidae. *Bull. U. S. Nat. Mus.*, 71(6):i–vii, 1–108.

CUSHMAN, JOSEPH A. 1918. The Foraminifera of the Atlantic Ocean. Part 1: Astrorhizidae. *Bull. U. S. Nat. Mus.*, 104(1):i–vii, 1–111.

CUSHMAN, JOSEPH A. 1920. The Foraminifera of the Atlantic Ocean. Part 2: Lituolidae. *Bull. U. S. Nat. Mus.*, 104(2):i–vii, 1–111.

CUSHMAN, JOSEPH A. 1922. The Foraminifera of the Atlantic Ocean. Part 3: Textulariidae. *Bull. U. S. Nat. Mus.*, 104(3):i–viii, 1–149.

CUSHMAN, JOSEPH A. 1923. The Foraminifera of the Atlantic Ocean. Part 4: Lagenidae. *Bull. U. S. Nat. Mus.*, 104(4):i–x, 1–228.

CUSHMAN, JOSEPH A. 1924. The Foraminifera of the Atlantic Ocean. Part 5: Chilostomellidae and Globigerinidae. *Bull. U. S. Nat. Mus.*, 104(5):i–v, 1–55.

CUSHMAN, JOSEPH A. 1928. Foraminifera, their classification and economic use. Cushman Laboratory, Sharon, Pa., pp. 1–401.

DAHL, F., F. KOENIKE, and A. BRAUER. 1909. Araneae, Acarina, und Tardigrada. *Die Süsswasserfauna Deutschlands*, 14:1–273.

DAMAS, D.1905. Notes biologiques sur les copépodes de la mer Norvégienne. *Conseil Perm. Internat. p. l'Explor. de la Mer, Publ. de Circ.*, 22:1–23.

DAVIS, CHARLES C. 1943. The larval stages of the calanoid copepod, *Eurytemora hirundoides* (Nordquist). *Publ. Chesapeake Biol. Lab.*, 58:1–52.

DAVIS, CHARLES C. 1948. The effect of industrial copperas pollution upon the plankton in the Baltimore Harbor area. *Publ. Chesapeake Biol. Lab.*, 72:1–12.

DAVIS, CHARLES C. 1949. The pelagic Copepoda of the northeastern Pacific Ocean. *Univ. of Wash. Publ. in Biol.*, 14:1–118.

DAVIS, JOHN H. 1940. The ecology and geological role of mangroves in Florida. *Carnegie Inst. Wash. Publ.*, 517:303–412.

DAWYDOFF, C. 1928. Traité d'embryologie comparée des invertébrés. Masson et Cie., Paris. Pp. i-xiv, 1-930.

DITLEVSEN, H. 1926. Freeliving nematodes. *Danish Ingolf-Exped.*, 4(6):1–42, pls. 1–15.

DITTMAR, W. 1884. Report on researches into the composition of ocean water, collected by H.M.S. *Challenger. Challenger Rept., Phys. and Chem.*, 1:1–251.

DUBOIS, R. 1928. Lumière (production de la) ou Biophotogénèse. In C. Richet, *Dictionnaire de Physiologie.* 10:277–394.

DÜRKEN, BERNHARD. 1932. Experimental analysis of development. W. W. Norton and Co., Inc., New York. Pp. 1–288.

EHRENBAUM, E. 1905. Eier und Larven von Fischen. *Nord. Plank.*, 1:1–216.

EHRENBAUM, E. 1909. Eier und Larven von Fischen. *Nord. Plank.*, 1:217–414.

EINARSSON, HERMANN. 1948. Echinoderma. *The Zool. of Iceland.* 4(70):1–67.

EKMAN, SVEN. 1935. Tiergeographie des Meeres. Leipzig. Pp. i–xii, 1–542.

ELMORE, C. J. 1922. The diatoms (Bacillarioideae) of Nebraska. *Univ. of Nebraska Studies*, 21:22–214, pls. 1–23.

FELL, H. BARRACLOUGH. 1945. A revision of the current theory of echinoderm development. *Trans. Roy. Soc. New Zealand*, 75(2):73–101.

FISH, CHARLES J. 1935. Marine biology and paleoecology. *Jour. Paleontol.*, 9(1):92–100.

FITSCH, C. P., L. M. BISHOP, W. L. BOYD, R. A. GORTNER, C. F. ROGERS, and J. E. TILDEN 1934. "Water bloom" as a cause of poisoning in domestic animals. *Cornell Veterinarian*, 24:30–39.

FRITSCH, F. E. 1935. The structure and reproduction of the algae, Vol. 1. Cambridge University Press. Pp. i–xvii, 1–791.

FRITSCH, F. E. 1945. The structure and reproduction of the algae, Vol. 2. Cambridge University Press. Pp. i–xiv, 1–939.

FULLER, J. L., and G. L. CLARKE, 1936. Further experiments on the feeding of *Calanus finmarchicus. Biol. Bull.*, 70(2):308–320.

GALADZIEV, M., and E. MALM. 1929. L'influence de quelques facteurs physicochimiques sur les Protozoa marins. *Compt. Rend. Acad. Sci. U. R. S.* S., Sér. A, 1929 (18):433–436.

GALLOWAY, JESSE JAMES. 1933. Manual of the Foraminifera. Principia Press, Bloomington, Ind. Pp.i-xii, 1–483.

GEITLER, L. 1925. Cyanophyceae. In A. Pascher, *Die Süsswasserflora Deutschlands, Österreichs, und der Schweiz, 12:1–450.*

GEITLER, L. 1930–31. Cyanophyceae. In L. Rabenhorst, *Kryptogamen-Flora von Deutschland, Österreich, und der Schweiz,* 14:1–464.

GIESBRECHT, W. 1910. Stomatopoden. *Fauna und Flora des Golfes von Neapel,* 33:1–239, pls. 1–11.

GIESBRECHT, W., and O. SCHMEIL. 1898. Copepoda. I: Gymnoplea. *Das Tierreich,* 6:1–169.

GILLBRIGHT, MAX. 1952. Untersuchungen zur Produktionsbiologie des Planktons in der Kieler Bucht. *Kieler Meeresforschungen,* 8(2): 173-191.

GOODEY, T. 1951. Soil and fresh-water Nematodes. A monograph. John Wiley & Sons, Inc., New York. Pp. 1-389.

GRAFF, L. V. 1905. Turbellaria. I: Acoela. *Das Tierreich,* 23:1–35.

GRAFF, L. V. 1913. Turbellaria. II: Rhabdocoelida und Alloeocoela. *Das Tierreich,* 35:1–484.

GRAHAM, HERBERT W. 1941. Plankton production in relation to character of water in the open Pacific. *Jour. Mar. Res.,* 4(3):189–197.

GRAHAM, HERBERT W. 1942. Studies in the morphology, taxonomy, and ecology of the Peridiniales. *Publ. Carnegie Inst. Wash.,* 542:i–vii, 1–129.

GRAHAM, HERBERT W. 1943. Chlorophyll-content of marine plankton. *Jour. Mar. Res.,* 5(2):153–160.

GRAHAM, H. W., and N. BRONIKOVSKY. 1944. The genus *Ceratium* in the Pacific and North Atlantic Oceans. *Publ. Carnegie Inst. Wash.,* 565:i–vii, 1–209.

GRAN, H. H. 1931. On the conditions for the production of plankton in the sea. *Conseil Perm. Internat. pour l'Explor. de la Mer, Rapp. et Proc.-Verb.,* 75:37–46.

GRAN, H. H., and T. BRAARUD. A quantitative study of the phytoplankton in the Bay of Fundy and the Gulf of Maine (including observations on hydrography, chemistry, and turbidity). *Jour. Biol.Bd. Canada,* 1(5):279–467.

GRAVELY, F. H. 1909. Polychaet larvae. *Liverpool Mar. Biol. Comm. Mem.,* 19:1–79, pls. I–IV.

GROOM, THEODORE T. 1894. The life history of the rock barnacle, *Balanus.* Part I. *Jour. Mar. Zool. and Micr.,* 1:81–96.

GROOM, THEODORE T. 1895. The life history of the rock barnacle, *Balanus.* Part II. *Jour. Mar. Zool. and Micr.,* 2:1–5.

GROSS, F. 1934. Zur Biologie und Entwicklungsgeschichte von *Noctiluca miliaris. Arch. Protistenk.* 83:178–196.

GROSS, F. 1937. Notes on the culture of some marine plankton organisms. *Jour. Mar. Biol. Assoc.,* 21:753–768.

GUNTER, G., F. G. W. SMITH, and R. H. WILLIAMS. 1947. Mass mortality of marine animals on the lower west coast of Florida, November, 1946–January, 1947. *Science,* 105(2723):256–257.

GUNTER, G., R. H. WILLIAMS, C. C. DAVIS, and F. G. W. SMITH. 1948. Catastrophic mass mortality of marine animals and coincident phytoplankton bloom on the west coast of Florida, November, 1946 to May, 1947. *Ecol. Monographs,* 18(3):310–324.

GURNEY, ROBERT. 1939. Bibliography of the larvae of decapod Crustacea. Ray Soc., London. Pp. i–viii, 1–123.

GURNEY, ROBERT. 1942. Larvae of decapod Crustacea. Ray Soc., London. Pp. i–viii, 1–306.

HAECKEL, ERNST. 1887. Report on the Radiolaria collected by H. M. S. "Challenger" during the years 1873–76.*Rept. Sci. Res. "Challenger,"* 18(1):i–clxxxviii, 1–888; 18(2):889–1803; 18*a:* pls. 1–140.

HAECKER, V. 1908. Tiefseeradiolarien. *Wiss. Ergebn. Deutsch. Tiefsee-Exped. a. d. Damfper "Valdivia,"* 1898–1899, 14:477–706, pls. 86–87, 2 charts.

HANSEN, H. J. 1915. The Crustacea Euphausiacea of the United States National Museum. *Proc. U. S. Nat. Mus.*, 48:59–114.

HANUŠKA, LADISLAV. 1949. Hydrobiologie de l'écluse à Vrané sur la Vltava I. (Contribution au jugement biologique des eaux superficielles.) (In Czeck, French summary.) *Vestník Československé Zoologické Společnosti.* 13:69–93.

HARDY, A. C. 1941. Plankton as a source of food. *Nature,* 147:695–696.

HARDY, A. C. and R. BAINBRIDGE. 1951. Vertical migration of plankton animals. *Nature.* 168:327–328.

HARDY, A. C., and E. R. GUNTHER. 1935. The plankton of the South Georgia whaling grounds and adjacent waters, 1926–1927. *Discovery Repts.*, 11:1–456.

HARDY, A. C., C. E. LUCAS, G. T. D. HENDERSON, and J. H. FRASER. 1936. The ecological relations between the herring and the plankton investigated with the plankton indicator. *Jour. Mar. Biol. Assoc.*, 21(1):147–291.

HARRING, H. K. 1913. Synopsis of the Rotatoria. *Bull. U. S. Nat. Mus.*, 81:1–226.

HARTLAUB, C. L. 1907–15. Craspedota Medusen. *Nord. Plank.,* 6(12):1–539.

HARVEY, E. NEWTON. 1940. Living Light. Princeton University Press. Pp. 1–328.

HARVEY, E. NEWTON. 1952. Bioluminescence. Academic Press, Inc. New York. Pp. 1–649.

HARVEY, H. W. 1933. On the rate of diatom growth. *Jour. Mar. Biol. Assoc.*, 19:253–276.

HARVEY, H. W. 1937. Notes on selective feeding by *Calanus. Jour. Mar. Biol. Assoc.*, 22:97–100.

HARVEY, H. W., L. H. N. COOPER, M. V. LEBOUR, and F. S. RUSSELL. 1935. Plankton production and its control. *Jour, Mar. Biol. Assoc.*, 20:407–442.

HEDGPETH, JOEL W. 1951. The classification of estuarine and brackish waters and the hydrographic climate. In H. S. Ladd *et al.*, *Report of the Committee on a Treatise on Marine Ecology and Paleoecolgy* 1950–1951, pp. 49–56.

HERRICK, C. L., and C. H. TURNER. 1895. Synopsis of the Entomostraca of Minnesota, *Geol. and Nat. Hist. Surv. Minn., Zool. Ser.*, 2:1–525.

HERZ, L. E. 1933. The morphology of the later stages of *Balanus crenatus* Brugiere. *Biol. Bull.*, 64:432–442.

HEYERDAHL, THOR. 1950. Kon-Tiki. Rand McNally & Co., Chicago. Pp. 1–304.

HJORT, J. and J. T. RUUD. 1929. Whaling and fishing in the North Atlantic. *Conseil Perm. Internat. pour l'Explor. de la Mer, Rapp. et Proc.-Verb.*, 56(1):1–123.

HOEK, P. P. C. 1909. Die Cirripedien des nordischen Plankton. *Nord. Plank.*, 8:265–332.

HUBER-PESTALOZZI, G. 1923. Experimentelle Studien über Entwicklung und Formgestaltung bei *Ceratium hirundinella. Verhand. Internat. Vereinig. Theor. u. Angewandte Limnologie,* 1:15–19.

HUBER-PESTALOZZI, G. 1938–41. Das Phytoplankton des Süsswassers. Systematik und Biologie. 1. Teil in *Die Binnengewässer*, Stuttgart, 1938, pp. 1–342, pls.3–66. 2. Teil, *ibid.*, 1941, pp. 1-365.

HUSTEDT, FRIEDRICH. 1930. Bacillariophyta. (Diatomeae). In *Süsswasserflora Mitteleuropas*, 10:i–viii, 1–464.

HYMAN, LIBBIE H. 1940. The invertebrates: Protozoa through Ctenophora. McGraw-Hill Book Co., New York. Pp.1–720.

IHLE, J. E. W. 1927. Thaliacea. *Die Tierwelt der Nord- und Ostsee*, 12(9):21–48.

ILLIG, G. 1930. Die Schizopoden der deutschen Tiefsee-Expedition. *Wiss. Ergebn. Deutsch. Tiefsee-Exped. a. d. Dampfer "Valdivia," 1898–1899,* 22(6):379–625.

JAKUBOWA, L., and E. MALM. 1931. Die Beziehung einiger Benthos-Formen des Schwarzen Meeres zum Medium. *Biol. Zentralbl.*, 51(3):105–116.

JENKIN, PENELOPE M. 1937. Oxygen production by the diatom *Coscinodiscus excentricus* Ehr. in relation to submarine illumination in the English Channel. *Jour. Mar. Biol. Assoc.*, 22(1):301–343.

JENNINGS, H. S. 1901. Synopsis of North American invertebrates. XVII: The Rotatoria. *Amer. Nat.*, 35:725–777.

JENNINGS,H. S. 1918. The wheel animalcules (Rotatoria). In Ward and Whipple, *Fresh-water biology*, pp. 553-620.

JESPERSEN, POUL. 1935. Quantative investigation on the distribution of macroplankton in different oceanic regions. *Carlsberg Foundation's Oceanog. Exped. Round the World 1928–30. Dana-Rept.*, 7:1–44.

JESPERSEN, POUL. 1939. Investigations on the copepod fauna in east Greenland waters. *Meddelelser om Grönland*, 119(9):1–106.

JESPERSEN, POUL. 1940. Investigations on the quantity and distribution of zooplankton in Icelandic waters. *Meddelelser Komm. Danmarks Fiskeri- og Havunders., Serie Plankton*, 3(5):1–77.

JESPERSEN, P., and M. F. S. RUSSELL. 1939–52. Fiches d'identification du zooplancton. Nos. 1–49. *Conseil. Perm. Internat. pour l'Explor. de la Mer.*

JOHNSON, MARTIN W. 1939. The correlation of water movements and dispersal of pelagic larval stages of certain littoral animals, especially the sand crab, *Emerita. Jour. Mar. Res.*, 2:236–245.

JOHNSTONE, J. 1908. Conditions of life in the sea. Cambridge University Press. Pp. i–xiv, 1–332.

JOHNSTONE, J., A. SCOTT, and H. C. CHADWICK. 1924. The marine plankton. University Press, Liverpool. Pp. 1–194.

JOUBIN, LOUIS. 1937. Les Octopodes de la croisière du "Dana" 1921–22. *Carlsberg Foundation's Oceanog. Exped. Round the World* 1928–30. *Dana-Rept.*, 11:1–49.

JUDAY, CHANCEY. 1943. The utilization of aquatic food resources. *Science*, 97(2525):456–458.

KEILHACK, L. 1910. Phyllopoda. In Brauer's *Die Süsswasserfauna Deutschlands*, Part 10, pp. 1–112.

KIEFER, F. 1929. Crustacea Copepoda. II. Cyclopoida Gnathostoma. *Das Tierreich*, 53:i–xvi, 1–102.

KIKUCHI, KENZO. 1930. Diurnal migration of plankton Crustacea. *Quart. Rev. Biol.*, 5:189–206.

KINCAID, TREVOR. 1942. The biotic and economic relations of the plankton. *Calif. Fish and Game*, 28(4):210–215.

KISELEV, I. A. 1950. Pantsyrnye zhgutikonostsy (Dinoflagellata) morei i presnykh vod SSSR. (In Russian). *Committee on the Fauna of the USSR, Zoological Institute, Academy of Science, USSR.* 33:1–280.

KISER, RUFUS W. 1950. A revision of the North American species of the Cladoceran genus *Daphnia*. Edwards Bros., Inc., Ann Arbor. Pp. 1–65.

KNOBLOCH, IRVING W. 1948. Readings in biological science. Appleton-Century-Crofts, Inc., New York. Pp. i–xiii, 1–449.

KOFOID, C. A., and A. S. CAMPBELL. 1929. A conspectus of the marine and fresh-water Ciliata belonging to the sub-order Tintinnoinea, with descriptions of new species principally from the Agassiz Expedition to the eastern tropical Pacific 1904–1905. *Univ. of Calif. Publ. in Zool.,* 34(1):1–404.

KOFOID, C. A., and A. S. CAMPBELL. 1939. Reports on the scientific results of the expedition to the eastern tropical Pacific in charge of Alexander Agassiz, by the U. S. Fish Commission steamer "Albatross," from October, 1904 to March, 1905, Lieut.-Commander L. M. Garrett, U. S. N., commanding. XXXVII. The Ciliata: The Tintinnoinea. *Bull. Mus. Comp. Zool.,* 84:1–473, pls. 1–36.

KOFOID, C. A., and O. SWEZY. 1921. The free-living unarmored Dinoflagellata. *Univ. of Calif. Mem.,* 5:1–538.

KRAMP, P. L. 1919. Medusae. Part I: Leptomedusae. *Repts. Danish Ingolf-Exped.,* 5(8):1–111, pls. 1–5.

KRAMP, P. L. 1926. Medusae. Part II: Anthomedusae. *Repts. Danish Ingolf-Exped.,* 5(10):1–102, pls. 1–2.

KRAMP, P. L. 1947. Medusae. Part III: Trachylina and Scyphozoa, with zoo-geographical remarks on all the medusae of the North Atlantic. *Repts. Danish Ingolf-Exped.,* 5(4):1–66, pls. 1–6.

KREPS, E., and N. VERJBINSKAYA. 1930. Seasonal changes in the phosphate and nitrate content and in hydrogen ion concentration in the Barents Sea. *Conseil Perm. Internat. p. l'Explor. de la Mer, Jour. du Conseil.* 5(3): 329–346.

KŘÍŽENECKÝ, J. 1925. Untersuchungen über die Assimilationsfähigkeit der Wassertiere für im Wasser gelöste Nahrstoffe. *Biologia Generalis.* 1:79–149.

KROGH, A. 1931. Dissolved sustances as food of aquatic organisms. *Conseil Perm. Internat. p. l'Explor. de la Mer, Rapp. et Proc.-Verb.* 75:7-26.

KROGH, A. 1934. Physiology of the blue whale. *Nature.* 133: 635–637.

KROGH, A. 1939. Osmotic regulation in aquatic animals. Cambridge University Press, Cambridge. 242 pp.

KROGH, A., and E. BERG. 1931. Über die chemische Zusammensetzung des Phytoplanktons aus dem Frederiksborg-Schlossee und ihre Bedeutung für die Maxima der Cladoceren. *Int. Rev. d. ges. Hydrobiol.u. Hydrog.*, 25:204-218.

KRUMBACH, THILO. 1927. Ctenophora. *Die Tierwelt der Nord- und Ostsee,* 2(2):1–12.

KUDO, ROKSABRO. 1946. Protozoology. 3rd ed. C. C. Thomas, Springfield, Ill. Pp. 1–778.

KUENEN, PH. H. 1941. Geochemical calculations concerning the total mass of sediments in the earth. *Amer. Jour. Sci.*, 39:161–190.

KUENEN, PH. H. 1950. Marine geology. John Wiley and Sons, Inc., New York. Pp. i–x, 1–568.

KUHL, W. 1938. Chaetognatha. In Bronn's *Klassen und Ordnungen des Tierreichs,* Vol. 4, Part 4, Bk. 2, Sec. 1, pp. i–viii, 1–226.

LACKEY, JAMES B. 1949. Plankton as related to nuisance conditions in surface water. In *Limnological aspects of water supply and water disposal.* A. A. A. S. Publ., pp. 56–63.

LANG, KARL. 1948. Monographie der Harpacticiden. Håken Ohlssons Boktryckeri, Lund. Vol. I, pp. 1–896, Vol. II, pp. 897–1683.

LEBOUR, MARIE V. 1923. The food of plankton organisms. II. *Jour Mar. Biol. Assoc.,* 13(1):70–92.

LEBOUR, MARIE V. 1925. The dinoflagellates of northern seas. Marine Biol. Lab., Plymouth. Pp. 1–172.

LEBOUR, MARIE V. 1926. A general survey of larval euphausiids, with a scheme for their identification. *Jour. Mar. Biol. Assoc.*, 14(2):519–527.

LEBOUR, MARIE V. 1930. The planktonic diatoms of northern seas. Ray Soc., London. Pp. i–ix, 1–244.

LECHEVALIER, PAUL. 1951. Cumacés. *Faune de France.* 54:1–136.

LEFEVRE, M., M. NISBET and E. JAKOB. 1949. Action des substances excrétées en culture, par certaines espèces d'Algues, sur le métabolisme d'autres espèces d'Algues. *Verhand. Internat. Vereinig. theor. u. angewandte Limnologie.* 10:259–264.

LENS, A. D., and T. VAN RIEMSDIJK. 1908. The Siphonophora of the Siboga Expedition. *Siboga-Exped. Monogr.,* 38:1–130, 24 pls.

LEWIS, RALPH C. 1929. The food habits of the California sardine in relation to the seasonal distribution of the microplankton. *Bull. Scripps Inst. Oc., Tech. Ser.,* 2(3): 155–180.

LOHMANN, H. 1901. Die Appendicularien. *Nord. Plank.,* 3:11–21.

LOHMANN, H. 1910. Die Cyphonautes der nordisches Plankton. *Nord. Plank.,* 9:31–40.

LOHMANN, H. 1911*a*. Die Appendicularien. Nachtrag. *Nord. Plank.,* 3:23–29.

LOHMANN, H. 1911*b*. Die Ascidienlarven des nordischen Plankton. *Nord. Plank.,* 3:31–47.

LOHMANN, H. 1920. Die Bevölkerung des Ozeans mit Plankton nach den Ergebnissen der Zentrifugenfänge während der Ausreise der "Deutschland" 1911. Zugleich ein Beitrag zur Biologie des Atlantischen Ozeans. *Arch. Biontol.,* 4(3):1–617, pls. 1–16.

LUND, E. J. 1936. Some facts relating to the occurrence of dead and dying fish on the Texas Coast during June, July, and August, 1935. *Ann. Rept., Texas Game, Fish, and Oyster Comm.,* 1934-1935:47–50.

LUND, J. W. G. 1949. The dynamics of diatom outbursts, with special reference to *Asterionella. Proc. Internat. Assoc. Theoretical and Applied Limnol.,* 10:275–276.

LUND, J. W. G. 1950. Studies on *Asterionella formosa* Hass. *Jour. Ecol.,* 38(1):1–36.

MACBRIDE, ERNEST W. 1914. Textbook of embryology. Vol. 1: Invertebrata. Macmillan and Co., Ltd., London. Pp. i–xxxii, 1–692.

MARSH, C. DWIGHT. 1918. Copepoda. In Ward and Whipple, *Fresh-water biology*, pp. 741–789.

MARSH, C. DWIGHT. 1929. Distribution and key to the North American copepods of the genus *Diaptomus*, with the description of a new species. *Proc. U. S. Nat. Mus.*, 75:1–27.

MARSH, C. DWIGHT. 1933. Synopsis of the calanoid crustaceans, exclusive of the Diaptomidae, found in fresh and brackish waters, chiefly of North America. *Proc. U. S. Nat. Mus.*, 82(18):1–58, pls. 1–24.

MARSHALL, S. M. 1933. The production of microplankton in the Great Barrier Reef Region. *Brit. Mus. Great Barrier Reef Exped., 1928–29, Sci. Repts.*, 2(5): 111-157.

MAYER, ALFRED G. 1910. Medusae of the World. *Publ. Carnegie Inst. Wash., 109.* 1:i–xv, 1–230; 2:i–xv, 231–498; 3:i–iv, 499–735.

MAYER, ALFRED G. 1912. Ctenophores of the Atlantic coast of North America. *Publ. Carnegie Inst. Wash.*, 162:1-58.

MEEHEAN, O. LLOYD. 1934. The role of fertilizers in pond-fish production. II: Some ecological aspects. *Trans. Amer. Fisheries Soc.* 64:151–154.

METCALF, M. M. 1919. The Salpidae: a taxonomic study. *Bull. U. S. Nat. Mus.* 100,2(2):1–193.

MICHAEL, ELLIS L. 1911. Classification and vertical distribution of the Chaetognatha of the San Diego region. *Univ. of Calif. Publ. Zool.*, 8(3):21–186.

MIYAJIMA, M. 1934. La question de "l'eau rouge". Un péril pour les huitres perlières. *Bull. Soc. Gén. d'Aquiculture et de Pèche*, 41:97–110.

MORI, TAKAMOCHI. 1937. The pelagic Copepoda from the neighboring waters of Japan. Yokendo Co., Tokyo. Pp. 1–150, pls. 1–80.

MORTENSEN, THEODOR. 1901. Die Echinodermen-Larven. *Nord. Plank.*, 9:1–30.

MORTENSEN, THEODOR. 1921. Studies of the development and larval forms of echinoderms. G. E. C. Gad., Copenhagen. Pp. 1–266.

MOSER, FANNY. 1903. Die Ctenophoren der Siboga-Expedition. *Siboga-Exped. Monogr.*, 12:1–34, pls. 1–4.

MOSER, FANNY. 1909. Die Ctenophoren der deutschen Südpolar-Expedition. *Deut. Südpolar-Exped.*, 11(Zool.3):115–192.

MOSER, FANNY. 1925. Die Siphonophoren der deutschen Südpolar-Expedition, 1901–1902. *Deut. Südpolar-Exped.*, 17 (Zool.9):1–541, pls. 1–33.

MÜLLER, G. W. 1894. Die Ostracoden des Golfes von Neapel. *Fauna und Flora des Golfes von Neapel*, 21:1–404, pls. 1–40.

MÜLLER, G. W. 1900. Deutschlands Süsswasser-Ostracoden. *Zoologica*, 30:1–112.

NAUMANN, E. 1923. Spezielle Untersuchungen über die Ernährungsbiologie des tierischen Limonoplanktons. II. Über den Nahrungswerb und die natürliche Nahrung der Copepoden und der Rotiferen des Limnoplanktons. *Lunds Univ. Årsskr. n.f.*, Avd. 2. ,19:3–17.

NEEDHAM, J. G., and P. R. NEEDHAM. 1930. A guide to the study of fresh-water biology. C. C. Thomas, Springfield, Ill. Pp. 1–88.

NEUMANN, GÜNTHER. 1913. Tunicata (Manteltiere). In Bronn's *Klassen und Ordnungen des Tierreichs*. 3(Suppl. 2): 145–182, 4 pls.

NICHOLLS, A. G. 1933. On the biology of *Calanus finmarchicus*. III. Vertical distribution and diurnal migration in the Clyde Sea area. *Jour. Mar. Biol. Assoc.*, 19:139–164.

OLIVE, EDGAR W. 1918. Blue-green algae (Cyanophyceae). In Ward and Whipple, *Fresh-water biology*, pp. 100–114.

OLIVEIRA, LEJEUNE P. H. de. 1948. Estudo hidrobiológica das lagôas de Piratininga e Itaipú. *Mem. Inst. Oswaldo Cruz*, 46(4):673-718.

OLSON, R. A., R. F. BRUST, and W. L. TRESSLER. 1941. Studies of the effects of industrial pollution in the lower Patapsco River area. 1: Curtis Bay area, 1941. *Publ. Chesapeake Biol. Lab.*, 43:1–40.

OLSON, THEODORE A. 1951. Toxic Plankton. *Proc. Inservice Training Course in Water Works Problems, Feb. 15-16* (Univ. Mich., School of Pub. Health). Pp. 86–95.

OLSON, THEODORE A. 1952. Toxic Plankton. *Water and Sewage Works.* 99(2):75–77.

OLTMANNS, FRIEDR. 1922–23. Morphologie und Biologie der Algen. Vol. 1:i–vi, 1–459; Vol. 2:i–iv, 1–439; Vol. 3:i–vi, 1–459. Jena.

OSTWALD, W. 1902. Zur Theorie des Planktons. *Biol. Zentralbl.* 22:596–605, 609–638.

PARKER, T. J., W. A. HASWELL, and O. LOWENSTEIN. 1940. A textbook of zoology. Vol. 1, Revised ed. Macmillan and Co., Ltd., London. Pp. i–xxxiii, 1–770.

PARKER, T. J., W. A. HASWELL, and C. FORSTER-COOPER. 1940. A textbook of zoology. Vol. 2. Revised ed. Macmillan and Co., Ltd., London. Pp. i-xxiii, 1–758.

PASCHER, A. 1913. Chrysomonadinae. In *Die Süsswasserflora Deutschlands, Österreichs, und der Schweiz*, Heft 2, Flagellata, 2:7–95.

PASCHER, A. 1927. Volvocales. In *Die Süsswasserflora Deutschlands, Österreichs, und der Schweiz,* 4:1–506.

PEARSALL, W. H. 1932. Phytoplankton in the English lakes. II: The composition of the phytoplankton in relation to dissolved substances. *Jour. Ecol.*, 20:241–262.

PENNAK, ROBERT W. 1946. The dynamics of fresh-water plankton populations. *Ecol. Monographs*, 16:340–355.

PENNAK, ROBERT W. 1953. Fresh-water invertebrates of the United States. Ronald Press, N. Y. Pp. i–ix, 1–769.

PETERS, R. A. 1921. The substances needed for the growth of a pure culture of *Colpidium colpoda. Jour. Physiol.*, 55:1–32.

PHIFER, L. D. 1934. Phytoplankton of East Sound, Washington, February to November, 1932. *Univ. of Wash. Publ. in Oc.,* 1(4):97–110.

PRATJE, A. 1921. *Noctiluca miliaris* Suriray. Beiträge zur Morphologie, Physiologie, und Cytologie, I: Morphologie und Physiologie (Beobachtungen an der lebenden Zelle). *Arch. Protistenk.* 42:1–98.

PRATT, H. S. 1935. A manual of the common invertebrate animals exclusive of insects. The Blakiston Co., Philadelphia. Pp. i–xviii, 1–854.

PRESCOTT, G. W. 1939. Some relationships of the phytoplankton to limnology and aquatic biology. In *Publ. A. A. A. S., No. 10* (Science Press). Pp. 65–78.

PRESCOTT, G. W. 1948. Objectional algae with reference to the killing of fish and other animals. *Hydrobiologia,* 1(1) :1-13.

PRESCOTT, G. W. 1951. Algae of the central Great Lakes region. Cranbrook Press, Bloomfield Hills, Mich. Pp. 1–946.

PROSSER, C. LADD *et al.* 1950. Comparative animal physiology. W. B. Saunders Co., Philadelphia. Pp. 1–888.

PÜTTER, A. 1909. Die Ernährung der Wassertiere und der Stoffhaushalt der Gewässer. Fischer, Jena. Pp. 1–168.

REDFIELD, A. C. 1941. The effect of the circulation of water on the distribution of the calanoid community in the Gulf of Maine. *Biol. Bull.,* 80:86–110.

REIBISCH, J. 1905. Anneliden. *Nord. Plank.,*7:1–10.

RICE, THEODORE R. 1954. Biotic influences affecting population growth of planktonic algae. Fish. Bull., U. S. Fish and Wildlife Serv. 54:226–245.

RILEY, G. A., H. STOMMEL, and D. F. BUMPUS. 1949. Quantitative ecology of the plankton of the western North Atlantic. *Bull. Bingham Oc. Coll.,* 12(3):1–169.

RUSSELL, F. S. 1934. The zooplankton. III. A comparison of the abundance of zooplankton in the Barrier Reef Lagoon with that of some regions in northern European waters. *Brit. Mus. Great Barrier Reef Exped., 1928–29, Sci. Repts.,* 2(6):176-201.

RUSSELL, F. S. 1935. A review of some aspects of zooplankton research. *Conseil Perm. Internat. p. l'Explor. de la Mer, Rapp. et Proc.-Verb.,* 95:3–30.

RUSSELL, F. S., and C. M. YONGE. 1928. The seas. Frederick Warne and Co., New York. Pp. i-xiii ,1-378.

RYLOV, W. M. 1935. Das Zooplankton der Binnengewässer. Einführung in die Systematik und Ökologie des tierischen Limnoplanktons mit besonderer Berücksichtigung der Gewässer Mitteleuropas. *Die Binnengewässer*, 15:i-ix, 1–272.

SARS, G. O. 1900. An account of the Crustacea of Norway. III. Cumacea. Bergen. Pp. i-x, 1–115, pls. 1–72.

SAWYER, C. N., J. B. LACKEY, and A. T. LENZ. 1943. Investigation of the odor nuisance occurring in the Madison lakes, particularly Lakes Monona, Waubesa, Kegonsa from July, 1942 to July, 1943. *Rept. for the Governor's Comm. State of Wisconsin* (mimeographed).

SAWYER, C. N., J. B. LACKEY, and A. T. LENZ. 1944. Investigation of the odor nuisance occurring in the Madison lakes, particularly Lakes Monona, Waubesa, Kegonsa, from July, 1943 to July, 1944. *Rept. for the Governor's Comm., State of Wisconsin.* (mimeographed).

SCHELLENBERG, A. 1927. Amphipoda des nordischen Plankton. *Nord. Plank.*, 6:589–722.

SCHILLER, JOSEF. 1926. Über Fortpflanzung, geissellose Gattungen und die Nomenklatur der Coccolithophoraceen nebst Mitteilung über Copulation bei *Dinobryon. Arch. Protistk.*, 53(2):326–342.

SCHILLER, JOSEF. 1930. Coccolithineae. In Rabenhorst's *Kryptogamen-Flora von Deutschland, Österreich, und der Schweiz*, 10(2):89–267.

SCHMIDT, G. A. 1937. Bau und Entwicklung der *Pilidium* von *Cerebratulus pantherinus* und *marginatus* und die Frage der morphologischen Merkmale der Pilidien. *Zool. Jahrb., Abt. Anat.*, 62(4):423–448.

SCHMIDT, JOHANNES. 1909. The distribution of pelagic fry and the spawning regions of gadoids in the North Atlantic from Iceland to Spain, based chiefly on Danish investigations. *Conseil Perm. Internat. p. l'Explor. de la Mer, Rapp. et Proc.-Verb.*, 10(4):1–229.

SCHMIDT, JOHANNES. 1925. The breeding places of the eel. *Smithsonian Inst. Ann. Rept. for 1924,* pp. 279–316.

SCHOENBORN, HENRY W. 1946. Studies on the nutrition of colorless euglenoid flagellates. II: Growth of *Astasia* in an inorganic medium. *Physiol. Zool.,* 19(4):430–442.

SCHULTZ, PAUL. 1928. Beiträge zur Kenntnis fossiler und rezenter Silicoflagellaten. *Bot. Arch.,* 21(2):225-292.

SCOFIELD, EUGENE C. 1934. Early life history of the California sardine (*Sardina caerulea*), with special reference to distribution of eggs and larvae. *Calif. Div. Fish and Game Fish Bull.,* 41:1–48.

SEARS, MARY. 1941. Notes on the phytoplankton on Georges Bank in 1940. *Jour. Mar. Res.,*4(3):247–257.

SHARPE, R. W. 1918. The Ostracoda. In Ward and Whipple, *Fresh-water biology,* pp. 790–827.

SHROPSHIRE, R. G. 1944. Plankton harvesting. *Jour. Mar. Res.,* 5(3):185–188.

SIMROTH, H. 1911. Die Gastropoden des nordischen Planktons. *Nord. Plank.,* 5:1–35.

SIMROTH, H. 1913. Die Acephalen des nordischen Planktons. *Nord. Plank.,* 5:36–55.

SKOGSBORG, TAGE. 1920. Studies on the marine Ostracoda. I: Cypridinids, halocyprids, and polycopids. *Zool. Bidrag från Uppsala,* Suppl. Bd. I:1–787.

SKOGSBORG, TAGE. 1928. Studies on the marine Ostracoda. II: External morphology of the genus *Cythereis* and descriptions of 21 new species. *Occ. Papers. Calif. Acad. Sci.* 15:1–155.

SLEGGS, G. F. 1927. Marine phytoplankton in the region of La Jolla, California, during the summer of 1924. *Bull. Scripps Inst. Oc., Tech. Ser.,* 1(9):93–117.

SMITH, F. G. W., R. H. WILLIAMS, and C. C. DAVIS. 1950. An ecological survey of the tropical waters adjacent to Miami. *Ecology.* 31((1):119–146.

SMITH, G. M. 1950. The fresh-water algae of the United States. 2nd Ed. McGraw-Hill Book Co., New York. Pp. i-vii, 1-719.

SNOW, JULIA W. 1918. The fresh-water algae (excluding the bluegreen algae). In Ward andWhipple, *Fresh-water biology,* pp. 115-177.

STEEMAN NIELSEN, E. 1935. The production of phytoplankton at the Faroë Isles, Iceland, East Greenland, and in the waters around. *Meddelelser Komm. Danmarks Fiskeri- og Havunders., Serie Plankton,* 3(1):1–92.

STEEMAN NIELSEN, E. 1937. On the relation between the quantities of phytoplankton and zooplankton in the sea. *Jour. du Conseil Perm. Internat. p. l'Explor. de la Mer,* 12:147–154.

STEUER, ADOLPH. 1910. Planktonkunde. B. G. Teubner, Berlin. Pp. 1-723.

STEUER, ADOLPH. 1911. Leitfaden der Planktonkunde. B. G. Teubner, Leipzig. Pp. 1–382.

STIASNY, GUSTAV. 1941. Über Tornarien. *Zool. Jahrb., Abt. Syst., Ök. u. Geog.,* 74(5/6):361–374.

STIASNY-WIJNHOFF, G., and G. STIASNY. 1927. Die Tornarien. Kritik der Beschreibungen und Vergleich sämmtlicher bekannter Enteropneustenlarven. *Erg. Fortsch. Zool.,* 7:38–208.

SVERDRUP, H. U., and W. E. ALLEN. 1939. Distribution of diatoms in relation to the character of water masses off Southern California. *Jour. Mar. Res.,* 2:131–144.

SVERDRUP, H. U., and R. H. FLEMING. 1941. The waters off the coast of Southern California, March to July, 1937. *Bull. Scripps. Inst. Oc.,* 4(10):261–378.

SVERDRUP, H. U., M. W. JOHNSON, and R. H. FLEMING. 1946. The oceans. Prentice-Hall, Inc., New York. Pp. i–x, 1–1087.

SWINGLE, H. S., and E. V. SMITH. 1939. Fertilizers for increasing the natural food for fish in ponds. *Trans. Amer. Fisheries Soc.* 68:126–134.

TATTERSALL, W. M. 1911. Die nordischen Isopoden. *Nord. Plank.,* 6:181–314.

TATTERSALL, W. M. 1951. A review of the Mysidacea of the United States National Museum. *Bull. U. S. Nat. Mus.,* 201:1–292.

TESCH, J. J. 1906. Die Heteropoden der Siboga-Expedition. 51:1–112, pls. 1–14.

TESCH, J. J. 1949. Heteropoda. *Carlsberg Foundation's Oceanog. Exped. Round the World 1928–30. Dana-Rept.*, 34:1–53.

THOMPSON, J. M. 1947. The Chaetognatha of Southeastern Australia. Bull. 222, Div. of Fish, Rept. 14. *Austral. Council Sci. Indust. Research,* pp. 1–43.

THORSON, GUNNAR. 1946. Reproduction and larval development of Danish marine bottom invertebrates, with special reference to the planktonic larvae in the Sound (Øresund). *Meddelelser Komm. Danmarks Fiskeri- og Havunders., Serie Plankton,* 4(1):1–523.

THORSON, GUNNAR. 1950. Reproductive and larval ecology of marine bottom invertebrates. *Biol. Reviews,* 25(1): 1–45.

TILDEN, JOSEPHINE. 1910. Minnesota algae. Vol. 1: The Myxophyceae of North America and adjacent regions. *Rept. Minn. Surv., Bot. Ser.,* 8:i-iv, 1–328, pls. 1–20.

VANHÖFFEN, E. 1906. Siphonophoren. *Nord. Plank.,* 11:9–39.

VERDUIN, JACOB. 1951. Comparison of spring diatom crops of western Lake Erie in 1949 and 1950. *Ecology,* 32(4):662–668.

VERVOORT, W. 1946. The Copepoda of the Snellius Expedition. I. E. J. Brill, Leiden. Pp. 1–181.

WARD, H. B., and G. C. WHIPPLE. 1918. Fresh-water biology. John Wiley and Sons, Inc., New York. Pp. i-ix, 1–1111.

WATERMAN. T. H., R. F. NUNNEMACHER, F. A. CHACE, and G. L. CLARKE. 1939. Diuranl vertical migrations of deep-water plankton. *Biol. Bull.,* 76(2): 256–279.

WAWRIK, FRIEDERICKE. 1952. Beziehungen zwischen Kieselsäurehaushalt und Diatomeenblüte in der Seebachlacke bei Kienberg-Gaming, N.-Ö. *Osterreich. Bot. Zeitschr.* 99(2–3):286–294.

WELCH, PAUL S. 1935. Limnology. McGraw-Hill Book Co., New York. Pp. i–xiv, 1–471.

WELCH, PAUL S. 1948. Limnological methods. The Blakiston Co., Philadelphia. Pp. 1–381.

WELCH, PAUL S. 1952. Limnology. 2d ed. McGraw-Hill Book Co., New York. Pp. i-xi, 1–538.

WELSH, J. H. 1933. Light intensity and the extent of activity of locomotor muscles as opposed to cilia. *Biol. Bull.*, 65: 168–174.

WESENBERG-LUND, C. 1900. Von dem Abhängigkeitsverhältnisse zwischen dem Bau der Planktonorganismen und dem spezifischen Gewicht des Süsswassers. *Biol. Zentralbl.*, 20:606–619; 644–656.

WESENBERG-LUND, C. 1926. Contributions to the biology and morphology of the genus *Daphnia* with some remarks on heredity. *Mém. Acad. R. Sci. et Let. Danemark*, 8me Sér., 11:91–250.

WIGGLESWORTH, V. B. 1931. The respiration of insects. *Biol. Rev.*, 6:181–220.

WILSON, CHARLES B. 1932. The copepods of the Woods Hole region, Massachusetts. *Bull. U. S. Nat. Mus.*, 158:1–635.

WILSON, CHARLES B. 1942. The copepods of the plankton gathered during the last cruise of the "Carnegie." *Publ. Carnegie Inst. Wash.*, 536:i–v, 1–237.

WIMPENNY, R. S. 1936. The size of diatoms. I: The diameter variation of *Rhizosolenia styliformis* Brightw. and *R. alata* Brightw. in particular and of pelagic marine diatoms in general. *Jour. Mar. Biol. Assoc.*, 21(1):29–60.

WOHLSCHLAG, D. E., and A. D. HASLER. 1951. Some quantitative aspects of algal growth in Lake Mendota. *Ecology*, 32(4):581–593.

WOLCOTT, ROBERT H. 1918. The water mites (Hydracarina). In Ward and Whipple, *Fresh-water biology*, pp. 851–875.

WOLLE, FRANCIS. 1887. Fresh-water algae of the United States. Bethlehem, Pa. Pp. 1–364, pls. 1–210.

WOLLE, FRANCIS, 1892. Desmids of the United States. Moravian Publishing Office, Bethlehem, Pa. Pp. i–xiv, 1–182, pls. 1–64.

WOLLE, FRANCIS. 1894. Diatomaceae of North America. Comenius Press, Bethlehem, Pa. Pp. 1–48, pls. 1–112.

WOLTERECK, R. 1928. Über die Spezifität des Lebensraumes, der Nahrung, und der Körperformen bei pelagischen Cladoceren und über "Ökologische Gestalt-Systeme." *Biol. Zentralbl.*, 48:521–551.

ZIMMER, C. 1927*a*. Mysidacea. In Kükenthal and Krumbach's *Handbuch der Zoologie*, 3[1]:607–650.

ZIMMER, C. 1927*b*. Cumacea. In Kükenthal-Krumbach's *Handbuch der Zoologie*, 3[1]:651–682.

ZIMMER, C. 1933. Cumacea. *Die Tierwelt der Nord- und Ostsee*, 10(23):70–120.

ZIMMER, C. 1941. Cumacea. *Klassen u. Ord. d. Tierreichs*, 5(1):1–222.

ZOBELL, CLAUDE E. 1946. Marine microbiology. Chronica Botanica, Waltham, Mass. Pp. i–xv, 1–240.

DESCRIPTIONS OF FIGURES IN PLATES

Fig. 49. *Merismopedia* sp.

Fig. 50. (A) *Holopedium irregulare* Lagerh. (B-C) *H. geminatum* Lagerh. From Smith, 1933, after Lagerheim. By permission from *The Fresh-water Algae of the United States,* 1st ed., by G. M. Smith. Copyright, 1933. McGraw-Hill Book Co., Inc.

Fig. 51. *Gomphosphaeria aponina* Kuetz. From Tilden, 1910, after West.

Fig. 52. *Coelosphaerium kuetzingianum* Naegeli.

Fig. 53. *Microcystis aeruginosa* Kuetzing. Portion of a colony.

Fig. 54. *Aphanocapsa grevillei* (Hass.). From Tilden, 1910, after West.

Fig. 55. *Chroococcus giganteus* W. West.

Fig. 56. *Gloeothece rupestris* (Lyngb.). Redrawn from Tilden, 1910, after Cooke.

Fig. 57. *Aphanothece stagnina* (Spreng.). From Tilden, 1910, after Lemmermann.

Fig. 58. *Spirulina* sp.

Fig. 59. *Nodularia hawaiiensis* Tilden. From Tilden, 1910.

Fig. 60. *Anabaena* sp.

Fig. 61. *Gloeotrichia echinulata* (J.Smith) Richter. Redrawn from Tiffany.

Fig. 62. *Nostoc commune* Vauch. From Tilden, 1910, after Hansgirg.

Fig. 63. *Aphanizomenon flosaquae* (L.) Ralfs.

Fig. 64. *Lyngbya confervoides* C. Ag. From Tilden, 1910, after Gomont.

Fig. 65. *Skujaella Thiebauti* (Gom.) J. de Toni.

Fig. 66. *Oscillatoria limosa* Kuetzing.

Fig. 67. *Planktoniella sol* (Wall.) Schutt. From Hustedt.

Fig. 68. *Cyclotella stelligera* Cleve and Grunow. From Hustedt.

Fig. 69. *Coscinodiscus oculus iridis* Ehr. From Hustedt.

Fig. 70. *Stephanodiscus* sp.

Fig. 71. *Coscinosira Oestrupi* Ostenfeld. From Hustedt.

Fig. 72. *Thalassiosira* sp.

Fig. 73. *Cerataulina Bergonii* Peragallo. From Hustedt.

Fig. 74. *Biddulphia aurita* (Lyngb.) Brébisson and Godey. From Hustedt.

Fig. 75. *Chaetoceros decipiens* Cleve.

Fig. 76. *Corethron hystrix* Hensen. From Hustedt.

Fig. 77. *Bacteriastrum delicatulum* Cleve. From Hustedt.

Fig. 78. *Rhizosolenia* spp.

Fig. 79. *Ditylum Brightwellii* (West) Grunow. From Hustedt.

Fig. 80. *Lauderia borealis* Gran. From Hustedt, after Gran.

Fig. 81. *Hemiaulus Hauckii* Grunow. From Hustedt.

Fig. 82. *Eucampia zoodiacus* Ehrenberg. From Hustedt.

Fig. 83. *Melosira* sp.

Fig. 84. *Stephanopyxis turris* (Grev. et Arn.) Ralfs. From Hustedt.

Fig. 85. *Skeletonema costatum* (Grev.) Cleve.

Fig. 86. *Asterionella* sp.

Fig. 87. *Striatella interrupta* (Ehr.) Heiberg. From Hustedt.

Fig. 88. *Campylodiscus cribrosus* W. Smith. From Ward and Whipple (Snow), after Smith. Reproduced by permission from *Fresh-water Biology* by H. B. Ward and G. C. Whipple. John Wiley & Sons, Inc., 1918.

Fig. 89. A. *Fragillaria crotonensis* Kitton. From Hustedt. B. *F.* sp.

Fig. 90. *Thalassiothrix nitzschioides* (Grunow). Redrawn from Hustedt.

Fig. 91. *Tabellaria* sp.

Fig. 92. *Synedra Utermohlii* Hustedt. From Hustedt.

Fig. 93. *Diatoma anceps* (Ehr.) Kirchner. From Hustedt.

Fig. 94. *Pleurosigma* sp.

Fig. 95. *Navicula* sp.

Fig. 96. *Nitzschia Brebissonii* W. Smith. From Smith, 1933. By permission from *The Fresh-water Algae of the United States,* 1st ed., by G. M. Smith. Copyright, 1933. McGraw-Hill Book Company, Inc.

Fig. 97. *Surirella* sp. Girdle view.

Fig. 98. *Amphora ovalis* Kuetzing. From Ward and Whipple (Snow), 1918. Reproduced by permission from *Freshwater Biology* by H. B. Ward and G. C. Whipple. John Wiley & Sons, Inc., 1918.

Fig. 99. *Netrium digitus* (Ehr.) Itz. and Rothe. From Smith. By permission from *The Fresh-water Algae of the United States,* 1st ed., by G. M. Smith. Copyright, 1933. McGraw-Hill Book Co., Inc.

Fig. 100. *Closterium* sp.

Fig. 101. (A) *Pleurotaenium nodosum* (Bailey) Lund. (B) P. *Ehrenbergii* (Bréb.) DBy. (C) P. *truncatum* (Bréb.) Näg. From Smith. By permission from *The Fresh-water Algae of the United States,* 1st ed., by G. M. Smith. Copyright, 1933. McGraw-Hill Book Company, Inc.

Fig. 102. (A) *Cosmarium reniforme* (Ralfs) Arch. (B) *C. granatum* Bréb. From West and Fritsch. By permission from *A Treatise on the British Freshwater Algae* (revised ed.) by G. S. West and F. E. Fritsch. Copyright, 1927. Cambridge: The University Press.

Fig. 103. (A) *Micrasterias americana* (Ehr.) Ralfs. (B) *M. apiculata* (Ehr.) Menegh. (C) *M. radiata* Hass. From Smith. By permission from *The Fresh-Water Algae of the United States,* 1st ed., By G. M. Smith. Copyright, 1933. McGraw-Hill Book Company, Inc.

Fig. 104. *Arthrodesmus Incus* (Bréb.) Hass. Redrawn from West and Fritsch. By permission from *A Treatise on the British Freshwater Algae* (revised ed.) by G. S. West and F. E. Fritsch. Copyright, 1927. Cambridge: The Univerity Press.

Fig. 105. *Xanthidium armatum* (Bréb.) Rabenh. From West and Fritsch. By permission from *A Treatise on the British Freshwater Algae* (revised ed.) by G. S. West and F. E. Fritsch. Copyright, 1927. Cambridge: The University Press.

Fig. 106. *Staurastrum* sp.

Fig. 107. (A) *Hyalotheca mucosa* (Dillw.) Ehr. (B) *H. dissiliens* (Smith) Bréb. From Smith. By permission from *The Fresh-water Algae of the United States,* 1st ed., by G. M. Smith. Copyright, 1933. McGraw-Hill Book Company, Inc.

Fig. 108. *Desmidium Swartzii* Ag. From West and Fritsch. By permission from *A Treatise on the British Freshwater Algae* (revised ed.) by G. S. West and F. E. Fritsch. Copyright 1927. Cambridge: The University Press.

Fig. 109. *Sphaerozosma vertebratum* Ralfs. Redrawn from West and Fritsch. By permission from *A Treatise on the British Freshwater Algae* (revised ed.) by G. S. West and F. E. Fritsch. Copyright, 1927. Cambridge: The University Press.

Fig. 110. (A, B) *Spondylosium moniliforme* Lund. (C) *S. planum* (Wolle) W. and G. S. West. From Smith. By permission from *The Fresh-water Algae of the United States,* 1st ed., by G. M. Smith. Copyright, 1933. McGraw-Hill Book Company, Inc.

Fig. 111. *Geminella mutabilis* (Bréb.) Wille. Redrawn from West and Fritsch. By permission from *A Treatise on the British Freshwater Algae* (revised ed.) by G. S. West and F. E. Fritsch. Copyright, 1927. Cambridge: The University Press.

Fig. 112. *Microspora Willeana* Wittr. From Smith. By permission from *The Fresh-water Algae of the United States,* 1st ed., by G. M. Smith. Copyright, 1933. McGraw-Hill Book Company, Inc.

Fig. 113. *Zygnema pectinatum* (Vauch.) Ag. From Smith. By permission from *The Fresh-water Algae of the United States,* 1st ed., by G. M. Smith. Copyright, 1933. McGraw-Hill Book Company, Inc.

Fig. 114. (A-D) *Mougeotia viridis* (Kütz.) Wittr. (A) vegetative filament; (B) parthenospores; (C, D) zygotes. (E) *M. scalaris* Hass. (F, G) Lateral conjugation in *M. genuflexa* (Dillw.) Ag. Redrawn from Smith. By permission from *The Fresh-water Algae of the United States,* 1st ed., by G. M. Smith. Copyright, 1933. McGraw-Hill Book Company, Inc.

Fig. 115. *Spirogyra pseudocylindrica* Prescott. Redrawn from Prescott.

Fig. 116. *Sirogonium sticticum* (Engl. Bot.) Kütz. From Smith. By permission from *The Fresh-water Algae of the United States,* 1st ed., by G. M. Smith. Copyright, 1933. McGraw-Hill Book Company, Inc.

Fig. 117. *Scenedesmus* sp.

Fig. 118. *Crucigenia lauterbornei* Schmidle. Redrawn from Tiffany, after Smith.

Fig. 119. *Westella botryoides* (West) de Wildemann. Redrawn from Tiffany

Fig. 120. *Dictyosphaerium planktonicum* Tiffany and Alstrom.

Fig. 121. *Planktosphaeria gelatinosa* G. M. Smith. From Smith. By permission from *The Fresh-water Algae of the United States,* 1st ed., by G. M. Smith. Copyright, 1933. McGraw-Hill Book Co., Inc.

Fig. 122. *Tetraspora lacustris* Lemmermann. Redrawn from Tiffany, after Smith.

Fig. 123. *Asterococcus limneticus* G. M. Smith. Redrawn from Tiffany, after Smith.

Fig. 124. *Sphaerocystis Schroeteri* Chod. From Smith. By permission from *The Fresh-water Algae of the United States,* 1st ed., by G. M. Smith. Copyright, 1933. McGraw-Hill Book Company, Inc.

Fig. 125. *Gloeocystis gigas* Kütz.) Lagerheim. Redrawn from Tiffany after Cienkowsky.

Fig. 126. *Kirchneriella lunaris* (Kirchner) Möbius. Redrawn from Tiffany, after Smith.

Fig. 127. *Quadrigula closteroides* (Bohlin) Printz. Redrawn from Tiffany.

Fig. 128. *Elakatrothrix viridis* (Snow) Printz. Redrawn from Tiffany.

Fig. 129. (A) *Nephrocytium Agardhianum* Näg. (B) *N. ecdysiscepanum* W. and G. S. West. (C) *N. limneticum* G. M. Smith. From Smith. By permission from *The Fresh-water Algae of the United States,* 1st ed., by G. M. Smith. Copyright, 1933. McGraw-Hill Book Co., Inc.

Fig. 130. *Oöcystis borgei* Snow.

Fig. 131. *Dimorphococcus lunatus* A. Br. Redrawn from West and Fritsch. By permission from *A Treatise on the British Freshwater Algae* (revised ed.) by G. S. West and F. E. Fritsch. Copyright, 1927. Cambridge: The University Press.

Fig. 132. *Selenastrum bibraianum* Reinsch.

Fig. 133. *Ankistrodesmus falcatus* (Corda) Ralfs. From Smith. By permission from *The Fresh-water Algae of the United States,* 1st ed., by G. M. Smith. Copyright, 1933. McGraw-Hill Book Company, Inc.

Fig. 134. *Closteridium lunula* Reinsch. From Smith, after Reinsch. By permission from *The Fresh-water Algae of the United States,* 1st ed., by G. M. Smith. Copyright, 1933. McGraw-Hill Book Co., Inc.

Fig. 135. *Schroederia setigera* (Schroeder).

Fig. 136. *Eremosphaera viridis* DBy. From Smith. By permission from *The Fresh-water Algae of the United States,* 1st ed., by G. M. Smith. Copyright, 1933. McGraw-Hill Book Company, Inc.

Fig. 137. *Tetrastrum heteracanthum* (Nordstedt) Chodat. Redrawn from Tiffany.

Fig. 138. *Micractinium eriense* Tiffany and Alstrom.

Fig. 139. *Errerella bornhemiensis* Conrad. From Smith. By permission from *The Fresh-water Algae of the United States,* 1st ed., by G. M. Smith. Copyright, 1933. McGraw-Hill Book Company, Inc.

Fig. 140. *Actinastrum hantzchii* Lagerheim.

Fig. 141. *Chlorella variegatus* Beyerinck.

Fig. 142. *Hydrodictyon reticulatum* (L.) Lagerheim.

Fig. 143. *Coelastrum microporum* Nägeli.

Fig. 144. *Pediastrum simplex* Meyen.

Fig. 145. (A) *Tetraëdron tumidulum* (Reinsch). (B) *T. trigonum* (Nägeli).

Fig. 146. *Pachycladon umbrinus* G. M. Smith. From Smith. By permission from *The Fresh-water Algae of the United States,* 1st ed., by G. M. Smith. Copyright, 1933. McGraw-Hill Book Company, Inc.

Fig. 147. *Lagerheimia citriformis* (Snow) G. M. Smith. Redrawn from Tiffany.

Fig. 148. *Golenkinia paucispina* West. Redrawn from Tiffany, after Smith.

Fig. 149. *Franceia droescheri* (Lemmermann) G. M. Smith. Redrawn from Tiffany, after Smith.

Fig. 150. *Polyedriopsis spinulosa* Schmidle. Redrawn from Tiffany, after Smith.

Fig. 151. *Treubaria varia* Tiffany and Alstrom. Redrawn from Tiffany.

Fig. 152. *Botryococcus braunii.* Kützing.

Fig. 153. *Pontosphaera syracusana* Lohmann. Redrawn from Lohmann.

Fig. 154. *Syracosphaera brasiliensis* Lohmann. (A) Entire organism; (B) details of shell aperture. Redrawn from Lohmann.

Fig. 155. *Halopappus vahseli* Lohmann. Redrawn from Lohmann.

Fig. 156. *Petalosphaera grani* Lohmann. Redrawn from Lohmann.

Fig. 157. *Michaelsarsia splendens* Lohmann. Redrawn from Lohmann.

Fig. 158. *Acanthoica acanthifera* Lohmann. Redrawn from Lohmann, after Ostenfeld.

Fig. 159. *Deutschlandia anthos* Lohmann. Redrawn from Lohmann.

Fig. 160. *Scyphosphaera apsteini* Lohmann. Redrawn from Lohmann.

Fig. 161. *Coccolithophora fragilis* Lohmann. Redrawn from Lohmann.

Fig. 162. *Umbilicosphaera mirabilis* Lohmann. Redrawn from Lohmann.

Fig. 163. *Rhabdosphaera hispida* Lohmann. Redrawn from Lohmann.

Fig. 164. *Discosphaera tubifer* Murr and Blackm. Redrawn from Lohmann.

Fig. 165. *Distephanus* sp.

Fig. 166. *Dichtyocha* sp.

Fig. 167. *Ochromonas ludibunda* Pascher. Redrawn from Huber-Pestalozzi, after Pascher.

Fig. 168. *Mallomonas* sp.

Fig. 169. *Chrysococcus tesselatus* Fritsch. Redrawn from Huber-Pestalozzi, after Fritsch.

Fig. 170. *Chromulina ovalis* Klebs. From Smith, after Klebs. By permission from *The Fresh-water Algae of the United States,* 1st ed., by G. M. Smith. Copyright, 1933. McGraw-Hill Book Company, Inc.

Fig. 171 *Euglena viridis* Ehrenberg. FromPratt, after Doflein. By permission from *A Manual of the Common Invertebrate Animals* (revised ed.) by H. S. Pratt. Copyright, 1935. The Blakiston Co.

Fig. 172. *Chlamydomonas globosa* Snow.

Fig. 173. *Platymonas elliptica* G. M. Smith. From Smith. By permission from *The Fresh-water Algae of the United States,* 1st ed., by G. M. Smith. Copyright, 1933. McGraw-Hill Book Company, Inc.

Fig. 174. *Carteria klebsii* (Dangeard) Dill. Redrawn from Tiffany.

Fig. 175. *Uroglena americana* Calkins. From Smith. By permission from *The Fresh-water Algae of the United States,* 1st ed., by G. M. Smith. Copyright, 1933. McGraw-Hill Book Company, Inc.

Fig. 176. *Dinobryon setularia* Ehrenberg.

Fig. 177. *Synura uvella* Ehrenberg. From Pratt, after Conn. By permission from *A Manual of the Common Invertebrate Animals* (revised ed.) by H. S. Pratt. Copyright, 1935. The Blakiston Company.

Fig. 178. *Spondylomorum quaternarium* Ehrenberg. From Smith, after Jacobsen. By permission from *The Fresh-water Algae of the United States,* 1st ed., by G. M. Smith. Copyright, 1933. McGraw-Hill Book Company, Inc.

Fig. 179. *Gonium pectorale* Müller. Redrawn from Tiffany.

Fig. 180. *Platydorina caudata* Kofoid.

Fig. 181. *Volvox aureus* Ehrenberg.

Fig. 182. *Pleodorina californica* Shaw.

Fig. 183. *Pandorina morum* Bory.

Fig. 184. *Eudorina elegans* Ehrenberg.

Fig. 185. *Phacus pleuronectes* (Müller).

Fig. 186. *Trachelomonas schauinslandii* Lemmermann.

Fig. 187. *Gymnodinium lunula* Schütt. Pyrocystis stage. Redrawn from Kofoid and Swezy.

Fig. 188. *Pyrocystis noctiluca.*

Fig. 189. *Prorocentrum* sp.

Fig. 190. *Haplodinium antjoliense* Klebs. Redrawn from Kofoid and Swezy, after Klebs.

Fig. 191 *Noctiluca scintillans* (Macartney). Redrawn from Kofoid and Swezy.

Fig. 192. *Dinophysis* sp.

Fig. 193. *Ceratocorys horrida* Stein. Redrawn from Graham.

Fig. 194. *Gonyaulax spinifera* (?) (Clap. and Lachm).

Fig. 195. *Ceratium tripos* Ehrenberg.

Fig. 196. *Ceratium hirundinella* Müller.

Fig. 197. *Hemidinium nasutum* Stein. Redrawn from Kofoid and Swezy, after Stein.

Fig. 198. *Peridinium* sp.

Fig. 199. *Pouchetia maculata* Kofoid and Swezy. Redrawn from Kofoid and Swezy.

Fig. 200. *Polykrikos kofoidi* Chatton. Redrawn from Kofoid and Swezy, after Kofoid.

Fig. 201. *Cochlodinium strangulatum* Schütt. Redrawn from Kofoid and Swezy, after Schütt.

Fig. 202. *Amphidinium herdmani* Kofoid and Swezy. Redrawn from Kofoid and Swezy, after Herdman.

Fig. 203. *Glenodinium aciculiferum* (Lemmermann).

Fig. 204. *Gymnodinium gracile* Bergh. From Pratt, after Calkins. By permission from *A Manual of the Common Invertebrate Animals* (revised ed.) by H. S. Pratt. Copyright, 1935. The Blakiston Company.

Fig. 205. *Gyrodinium contortum* (Schütt). Redrawn from Kofoid and Swezy, after Schütt.

Fig. 206. *Tretomphalus bulloides* (d'Orbigny). Redrawn from Brady.

Fig. 207. *Chilostomella ovoidea* Reuss. From Cushman.

Fig. 208. *Orbulina universa* d'Orbigny. Redrawn from Brady.

Fig. 209. *Candeina nitida* d'Orbigny. From Cushman.

Fig. 210. *Pulvinulina menardii* d'Orbigny. From Cushman.

Fig. 211. *Pullenia quinqueloba* (Reuss). From Cushman.

Fig. 212. *Globigerina bulloides* d'Orbigny. From Coker, after Murray and Hjort. Reproduced by permission from *This Great and Wide Sea* by R. E. Coker. Copyright, 1947. The University of North Carolina Press.

Fig. 213. *Hastigerina pelagica* d'Orbigny. From Rhumbler.

Fig. 214. *Actinelius primordialis* Haeckel. From Haeckel.

Fig. 215. *Acanthochiasma fusiforme* Haeckel. From Popofsky.

Fig. 216. *Acanthrometron pellucidum* Müller. From Popofsky.

Fig. 217. *Zygacantha rotunda* Popofsky. From Popofsky.

Fig. 218. *Acanthonia nordgaardi* Jörgensen. From Popofsky.

Fig. 219. *Amphilonche diodon* Haeckel. From Haeckel.

Fig. 220. *Lithoptera tetraptera* Haeckel. From Haeckel.

Fig. 221. *Hexalaspis heliodiscus* Haeckel. From Haeckel.

Fig. 222. *Diploconus hexaphyllus* Haeckel. From Haeckel.

Fig. 223. *Belonaspis datura* Haeckel. From Haeckel.

Fig. 224. *Phatnaspis lacunaria* Haeckel. From Haeckel.

Fig. 225. *Sphaerocapsa cruciata* Haeckel. From Haeckel.

Fig. 226. *Dorataspis micropora* Haeckel. From Haeckel.

Fig. 227. *Coscinaspis parmipora* Haeckel. From Haeckel.

Fig. 228. *Pleuraspis horrida* Haeckel. From Haeckel.

Fig. 229. *Phractaspis prototypus* Haeckel. From Haeckel.

Fig. 230. *Lychnaspis longissima* Haeckel. From Haeckel.

Fig. 231. *Tessaraspis arachnoides* Haeckel. From Haeckel.

Fig. 232. *Icosaspis tabulata* Haeckel. From Haeckel.

Fig. 233. *Phractopelta dorataspis* Haeckel. From Haeckel.

Fig. 234. *Dorypelta tessaraspis* Haeckel. From Haeckel.

Fig. 235. *Collozoum vermiforme* Haeckel. From Haeckel.

Fig. 236. *Thalassicolla pelagica* Haeckel. From Haeckel.

Fig. 237. *Thalassoxanthium cervicorne* Haeckel. From Haeckel.

Fig. 238. *Sphaerozoum geminatum* Haeckel. From Haeckel.

Fig. 239. *Cenosphaera compacta* Haeckel. From Haeckel.

Fig. 240. *Carposphaera melitomma* Haeckel. From Haeckel.

Fig. 241. *Collosphaera polygona* Haeckel. From Haeckel.

Fig. 242. *Chaenicosphaera flammabunda* Haeckel. From Haeckel.

Fig. 243. *Siphonosphaera tubulosa* Müller. From Haeckel.

Fig. 244. *Solenosphaera ascensionis* Haeckel. From Haeckel.

Fig. 245. *Stylosphaera melpomene* Haeckel. From Haeckel.

Fig. 246. *Xiphosphaera gaea* Haeckel. From Haeckel.

Fig. 247. *Xiphostylus trochilus* Haeckel. From Haeckel.

Fig. 248. *Hexastylus thaletis* Haeckel. From Haeckel.

Fig. 249. *Hexalonche anaximandri* Haeckel. From Haeckel.

Fig. 250. *Hexacontium sceptrum* Haeckel. From Haeckel.

Fig. 251. *Haliomma circumtextum* Haeckel. From Haeckel.

Fig. 252. *Actinomma denticulatum* Haeckel. From Haeckel.

Fig. 253. *Arachnosphaera dichotoma* Jörgensen. From Schröder.

Fig. 254. *Acanthosphaera clavata* Haeckel. From Haeckel.

Fig. 255. *Panartus tetrathalamus* Haeckel. From Haeckel.

Fig. 256. *Lithatractus jugatus* Haeckel. From Haeckel.

Fig. 257. *Druppatractus hippocampus* Haeckel. From Haeckel.

Fig. 258. *Xiphatractus armadillo* Haeckel. From Haeckel.

Fig. 259. *Heliosestrum medusinum* Haeckel. From Haeckel.

Fig. 260. *Heliodiscus cingillum* Haeckel. From Haeckel.

Fig. 261. *Porodiscus flustrella* Haeckel. From Haeckel.

Fig. 262. *Stylodictya centrospira* Haeckel. From Haeckel.

Fig. 263 *Hymeniastrum euclidis* Haeckel. From Haeckel.

Fig. 264. *Euchitonia echinata* Haeckel. From Haeckel.

Fig. 265. *Larnacantha bicruciata* Haeckel. From Haeckel.

Fig. 266. *Trizonium tricinctum* Haeckel. From Haeckel.

Fig. 267. *Pylonium quadricorne* Haeckel. From Haeckel.

Fig. 268. *Tetrapyle pleuracantha* Haeckel From Haeckel.

Fig. 269. *Octopyle decastyle* Haeckel. From Haeckel.

Fig. 270. *Lithelius spiralis* Haeckel. From Schröder.

Fig. 271. *Phorticium pylonium* Haeckel. From Haeckel.

Fig. 272. *Aulacantha spinosa* Haeckel. One of the radial tubes. From Haeckel.

Fig. 273. *Aulographis candelabrum* Haeckel. A portion of a specimen, showing a small part of the central capsule, the dark phaeodium, the alveolate calymma, tangential peripheral needles, pseudopodia, and radial tubes. Redrawn from Haeckel.

Fig. 274. *Conchopsis navicula* Haeckel. From Haeckel.

Fig. 275. *Coelodendrum furcatissimum* Haeckel. From Haeckel.

Fig. 276. *Gazelletta atlantica* Borgert. (A) entire shell, showing bases only of the spines; (B) the termination of one of the spines. From Borgert.

Fig. 277. *Challengeron wyvillei* Haeckel. From Haeckel.

Fig. 278. *Challengeria tritonis* Haeckel. From Haeckel.

Fig. 279. *Sagoscena castra* Haeckel. Half of a specimen. From Haeckel.

Fig. 280. *Sagosphaera penicilla* Haeckel. A nodal point and its radial spine. From Haeckel.

Fig. 281. *Aulosphaera dendrophora* Haeckel. From Haeckel.

Fig. 282. *Aulonia hexagonia* Haeckel. From Haeckel.

Fig. 283. *Cortina tripus* Haeckel. From Haeckel.

Fig. 284. *Lithocircus tarandus* Haeckel. From Haeckel.

Fig. 285. *Zygocircus triquetris* Haeckel. From Haeckel.

Fig. 286. *Cortiniscus typicus* Haeckel. From Haeckel.

Fig. 287. *Clathrocircus dictyospyris* Haeckel. From Haeckel.

Fig. 288. *Zygostephanus dissocircus* Haeckel. From Haeckel.

Fig. 289. *Eucoronis angulata* Haeckel. From Haeckel.

Fig. 290. *Botryocampe camerata* Haeckel. From Haeckel.

Fig. 291. *Tripospyris eucolpos* Haeckel. From Haeckel.

Fig. 292. *Dictyospyris distoma* Haeckel. From Haeckel.

Fig. 293. *Petalospyris dictyocubus* Haeckel. From Haeckel.

Fig. 294. *Ceratospyris allmersii* Haeckel. From Haeckel.

Fig. 295. *Cornutella annulata* Ehrenberg. From Schröder.

Fig. 296. *Cyrtocalpis sethophora* Haeckel. From Haeckel.

Fig. 297. *Sethoconus galea* Cleve. From Schröder.

Fig. 298. *Dictyocephalus mediterraneus* Haeckel. From Schröder.

Fig. 299. *Sethocorys odysseus* Haeckel. From Haeckel.

Fig. 300. *Lychnocanium pudicum* Haeckel. From Haeckel.

Fig. 301. *Dictyophimus platycephalus* Haeckel. From Haeckel.

Fig. 302. *Lithomelissa decacantha* Haeckel. From Haeckel.

Fig. 303. *Sethophormis pentalactis* Haeckel. From Haeckel.

Fig. 304. *Anthocyrtium anthemis* Haeckel. From Schröder.

Fig. 305. *Carpocanium verecundum* Haeckel. From Haeckel.

Fig. 306. *Calocyclas veneris* Haeckel. From Haeckel.

Fig. 307. *Theocapsa darwinii* Haeckel. From Haeckel.

Fig. 308. *Theocorys veneris* Haeckel. From Haeckel.

Fig. 309. *Theoconus jovis* Haeckel. From Haeckel.

Fig. 310. *Theocalyptra cornuta* Bailey. From Schröder.

Fig. 311. *Podocyrtis surena* Haeckel. From Haeckel.

Fig. 312. *Pterocorys tubulosa* Haeckel. From Haeckel.

Fig. 313. *Pterocanium pyramis* Haeckel. From Haeckel.

Fig. 314. *Eucyrtidium bütschlii* Haeckel. From Haeckel.

Fig. 315. *Lithocampe diploconus* Haeckel. From Haeckel.

Fig. 316. *Lithostrobus cornutus* Haeckel. From Haeckel.

Fig. 317. *Lithomitra nodosaria* Haeckel. From Haeckel.

Fig. 318. *Eutintinnus* sp.

Fig. 319. *Dictyocysta mexicana* Kofoid and Campbell. From Kofoid and Campbell.

Fig. 320. *Rhabdonella lohmanni* Kofoid and Campbell. From Kofoid and Campbell.

Fig. 321. *Salpingella* sp.

Fig. 322. *Codonellopsis abconica* Kofoid and Campbell. From Kofoid and Campbell.

Fig. 323. *Helicostomella* sp.

Fig. 324. *Stenosemella expansa* (Wailes). From Kofoid and Campbell.

Fig. 325. *Coxliella cymatiocoides* Kofoid and Campbell. From Kofoid and Campbell.

Fig. 326. *Codonella cratera* (Leidy).

Fig. 327. *Tintinnopsis* sp.

Fig. 328. *Ptychocylis urnula* (Clap. and Lachm). From Kofoid and Campbell, after Brandt.

Fig. 329. *Epiplocylis lineata* Kofoid and Campbell. From Kofoid and Campbell.

Fig. 330. *Cyttarocylis conica* Brandt. From Kofoid and Campbell, after Brandt.

Fig. 331. *Parafavella ventricosa* (Jörgensen). From Kofoid and Campbell, after Jörgensen.

Fig. 332. *Xystonellopsis abbreviata* Kofoid and Campbell. From Kofoid and Campbell.

Fig. 333. *Parundella longa* Jörgensen. From Kofoid and Campbell, after Jörgensen.

Fig. 334. *Favella franciscana* Kofoid and Campbell. From Kofoid and Campbell.

Fig. 335. *Proplectella* sp.

Fig. 336. *Undella hyalina* Daday. From Kofoid and Campbell, after Daday.

Fig. 337. Amphiblastula of *Sycandra raphonus.* From Dawydoff, after Schultze. Reproduced by permission from *Traité d'embryologie comparée des invertébrés* by C. Dawydoff. Copyright, 1928. Masson et Cie. (Paris).

Fig. 338. A gemmule of *Spongilla fluviatilis.* From Korschelt and Heider, after Vejdovsky. Reproduced by permission from *Textbook of the Embryology of Invertebrates,* Vol. 1, by E. Korschelt and K. Heider. Copyright, 1895. The Macmillan Company, New York, and Macmillan & Co., Ltd. (London).

Fig. 339. A planula larva from the plankton off the coast of Florida.

Fig. 340. An ephyra larva of a scyphozoan.

Fig. 341. *Solmaris corona* Haeckel. From Hartlaub, after Haeckel.

Fig. 342. *Pegantha clara* Bigelow. From Mayer.

Fig. 343. *Aegina citrea* Eschscholtz. Redrawn from Mayer, after Maas.

Fig. 344. *Aeginopsis laurentii* Brandt. From Mayer, after Brandt.

Fig. 345. *Aeginura incisa* Mayer. Redrawn from Mayer.

Fig. 346. *Cunina prolifera* Gegenbauer. From Mayer.

Fig. 347. *Solmissus albescens* Haeckel. From Mayer.

Fig. 348. *Liriope exigua* (Quoy and Gaimard). From Pratt, after Mayer. By permission from *A Manual of the Common Invertebrate Animals* (revised ed.) by H. S. Pratt. Copyright, 1935. The Blakiston Co.

Fig. 349. *Gonionemus* sp.

Fig. 350. *Cubaia aphrodite* Mayer. Redrawn from Mayer.

Fig. 351. *Olindias tenuis* Browne. Redrawn from Mayer.

Fig. 352. *Aglantha digitalis (Haeckel)* From Pratt, after Hargitt. By permission from *A Manual of the Common Invertebrate Animals* (revised ed.) by H. S. Pratt. Copyright, 1935. The Blakiston Company.

Fig. 353. *Aglaura hemistoma* Péron and Lesueur. Redrawn from Mayer.

Fig. 354. *Homoeonema typicum* Maas. Redrawn from Mayer, after Maas.

Fig. 355. *Pantachogon rubrum* Vanhöffen. Redrawn from Mayer, after Maas.

Fig. 356. *Botrynema ellinorae* Cl. Hartlaub. From H. Hartlaub.

Fig. 357. *Halicreas alba* (Vanhöffen). Redrawn from Mayer, after Vanhöffen.

Fig. 358. *Rhopalonema clavigerum* (Haeckel). Redrawn from Mayer.

Fig. 359. *Melicertum campanula* L. Agassiz. Redrawn from Mayer.

Fig. 360. *Laodicea undulata* (Forbes and Goodsir). (A) Aboral view of entire animal. (B) A portion of the margin of the bell, showing (to the left) a cirrus. Redrawn from Kramp.

Fig. 361. *Chromatonema rubrum* (Fewkes). Redrawn from Kramp.

Fig. 362. *Polyorchis penicillata* A. Agassiz. Redrawn from Mayer, after Fewkes.

Fig. 363. *Dipleurosoma collapsa* Mayer. Redrawn from Mayer.

Fig. 364. *Staurophora mertensii* Brandt. Redrawn from Mayer.

Fig. 365. *Eucheilota duodecimalis* A. Agassiz. From Pratt after Mayer. By permission from *A Manual of the Common Invertebrate Animals* (revised ed.) by H. S. Pratt. Copyright, 1935. The Blakiston Company.

Fig. 366. *Obelia* sp.

Fig. 367. *Tiaropsis diademata* Agassiz. From Pratt, after Mayer. By permission from *A Manual of the Common Invertebrate Animals* (revised ed.) by H. S. Pratt. Copyright, 1935. The Blakiston Company.

Fig. 368. *Clytia johnstoni* (Alder). From Pratt, after Mayer. By permission from *A Manual of the Common Invertebrate Animals* (revised ed.) by H. S. Pratt. Copyright, 1935. The Blakiston Company.

Fig. 369. *Phialidium languidum* Haeckel. Redrawn from Mayer.

Fig. 370. *Tima formosa* Agassiz. From Pratt, after Hargitt. By permission from *A Manual of the Common Invertebrate Animals* (revised ed.) by H. S. Pratt. Copyright, 1935. The Blakiston Company.

Fig. 371. *Eirene pyramidalis* Mayer. Redrawn from Mayer.

Fig. 372. *Halopsis ocellata* A. Agassiz. Redrawn from Kramp.

Fig. 373. *Eutima mira.* McCrady. From Pratt, after Hargitt. By permission from *A Manual of the Common Invertebrate Animals* (revised ed.) by H. S. Pratt. Copyright, 1935. The Blakiston Company.

Fig. 374. *Zygodactyla grönlandica* (Péron and Lesueur). From Pratt, after Mayer. By permission from *A Manual of the Common Invertebrate Animals* (revised ed.) by H. S. Pratt. Copyright, 1935. The Blakiston Company.

Fig. 375. *Aequorea tenuis* A. Agassiz. From Pratt, after Mayer. By permission from *A Manual of the Common Invertebrate Animals* (revised ed.) by H. S. Pratt. Copyright, 1935. The Blakiston Company.

Fig. 376. *Cladonema perkinsii* Mayer. Redrawn from Mayer, after Perkins.

Fig. 377. *Zanclea costata* Gegenbauer. Redrawn from Mayer.

Fig. 378. *Hybocodon prolifer* A. Agassiz. Redrawn from Mayer.

Fig. 379. *Dipurena strangulata* Haeckel. Redrawn from Mayer.

Fig. 380. *Steenstrupia nutans* M. Sars. Redrawn from Mayer, after Hartlaub.

Fig. 381. *Pennaria tiarella* McGrady. Redrawn from Mayer.

Fig. 382. *Ectopleura ochracea* A. Agassiz. From Pratt, after Hargitt. By permission from *A Manual of the Common Invertebrate Animals* (revised ed.) by H. S. Pratt. Copyright, 1935. The Blakiston Co.

Fig. 383. *Sarsia princeps* Haeckel. Redrawn from Mayer, after Hartlaub.

Fig. 384. *Stomotoca dinema* L. Agassiz. Redrawn from Mayer.

Fig. 385. *Amphinema dinema* (Péron and Lesueur). From Hartlaub.

Fig. 386. *Pandea violacea* Agassiz and Mayer. Redrawn from Mayer.

Fig. 387. *Halitholus pauper* Hartlaub. From Hartlaub.

Fig. 388. *Leuckartiara octona* (Fleming). From Hartlaub, after Haeckel.

Fig. 389. *Catablema campanula* Haeckel. From Hartlaub, after Haeckel.

Fig. 390. *Rathkea octopunctata* (M. Sars). Redrawn from Mayer.

Fig. 391. *Nemopsis bachei* Agassiz. From Pratt, after Mayer. By permission from *A Manual of the Common Invertebrate Animals* (revised ed.) by H. S. Pratt. Copyright, 1935. The Blakiston Company.

Fig. 392. *Bougainvillia superciliaris* L. Agassiz. Redrawn from Mayer.

Fig. 393. *Lizzia blondina* Fewkes. From Hartlaub, after Linko.

Fig. 394. *Cytaeis atlantica* (Steenstrup). Redrawn from Mayer, after Haeckel.

Fig. 395. *Podocoryne carnea* Sars. From Pratt, after Mayer. By permission from *A Manual of the Common Invertebrate Animals* (revised ed.) by H. S. Pratt. Copyright, 1935. The Blakiston Company.

Fig. 396. *Turritopsis nutricula* McCrady. From Pratt, after Mayer. By permission from *A Manual of the Common Invertebrate Animals* (revised ed.) by H. S. Pratt. Copyright, 1935. The Blakiston Company.

Fig. 397. *Carybdea xaymacana* Conant. Two of the tentacles are not shown. Redrawn from Mayer.

Fig. 398. *Periphylla hyacinthina* Steen. From Pratt, after Mayer. By permission from *A Manual of the Common Invertebrate Animals* (revised ed.) by H. S. Pratt. Copyright, 1935. The Blakiston Co.

Fig. 399. *Linuche unguiculata* Eschscholtz. (A) Side view of entire animal. (B) Oral view of a portion of the disc. Redrawn from Mayer.

Fig. 400. *Nausithoë punctata* Kölliker. From Pratt, after Mayer. By permission from *A Manual of the Common Invertebrate Animals* (revised ed.) by H. S. Pratt. Copyright, 1935. The Blakiston Co.

Fig. 401. *Aurelia aurita* (Linnaeus). From Mayer.

Fig. 402. *Cyanea capillata* (Linnaeus). Two of the oral palps have been cut away to expose the underlying parts. Redrawn from Mayer.

Fig. 403. *Phacellophora sicula* Haeckel. From Mayer.

Fig. 404. *Discomedusa philippina* Mayer. From Mayer.

Fig. 405. *Pelagia cyanella* Péron and Lesueur. From Pratt, after Mayer. By permission from *A Manual of the Common Invertebrate Animals* (revised ed.) by H. S. Pratt. Copyright, 1935. The Blakiston Co.

Fig. 406. *Chrysaora melanaster* Brandt. Redrawn from Mayer, after Brandt, after Vanhöffen.

Fig. 407. *Dactylometra quinquecirrha* L. Agassiz. Redrawn from Mayer.

Fig. 408. *Stomolophus meleagris* Agassiz. From Pratt, after Mayer. By permission from *A Manual of the Common Invertebrate Animals* (revised ed.) by H. S. Pratt. Copyright, 1935. The Blakiston Co.

Fig. 409. *Rhizostoma pulmo* Agassiz. Redrawn from Mayer.

Fig. 410. *Rhopilema verrillii* (Fewkes). Redrawn from Mayer.

Fig. 411. *Cephea cephea* (Forskål). From Mayer, after Kishinouye.

Fig. 412. *Physalia pelagica* Bosc. From Pratt, after Lankester. By permission from *A Manual of the Common Invertebrate Animals* (revised ed.) by H. S. Pratt. Copyright, 1935. The Blakiston Co.

Fig. 413. *Velella spirans* Forskål. Redrawn from Vanhöffen, after Agassiz.

Fig. 414. *Physophora hydrostatica* Forskål. From Vanhöffen, after Sars.

Fig. 415. *Agalma* sp. From Hyman, after Mayer. By permission from *The Invertebrates,* Vol. 1, by Libbie H. Hyman. Copyright, 1940. McGraw-Hill Book Company, Inc.

Fig. 416. *Stephanomia amphitridis* Péron and Lesueur. A (unicornuate) tentillum of an adult. Redrawn from Bigelow and Sears.

Fig. 417. *Vogtia glabra* Bigelow. Dorsal view of a nectophore. Redrawn from Bigelow and Sears.

Fig. 418. *Hippopodius hippopus* Forskål. Dorsal view of a nectophore. Redrawn from Bigelow and Sears.

Fig. 419. *Sulculeolaria monoica* (Chun). Redrawn from Bigelow.

Fig. 420. *Galetta australis* (Quoy and Gaimard). Redrawn from Bigelow.

Fig. 421. *Abyla leuckartii* Quoy and Gaimard. Lateral view of a colony. Redrawn from Bigelow.

Fig. 422. *Abylopsis eschscholtzii* (Huxley). Lateral view of the superior nectophore. Redrawn from Bigelow.

Fig. 423. *Bassia appendiculata* (Eschscholtz). Oblique dorsal view of the superior nectophore, showing the short dorsal ridge. Modified from Bigelow.

Fig. 424. *Enneagonum hyalinum* Quoy and Gaimard. Redrawn from Bigelow and Sears.

Fig. 425. *Diphyes dispar* Chamisso and Eysenhardt. Redrawn from Bigelow.

Fig. 426. *Eudoxoides spiralis* (Bigelow). Redrawn from Bigelow.

Fig. 427. *Muggiaea atlantica* Cunningham. From Vanhöffen, after Cunningham.

Fig. 428. *Lensia conoidea* Keferstein and Ehlers. Redrawn from Bigelow and Sears.

Fig. 429. Cydippid larva of a ctenophore (*Mnemiopsis leidyi* A. Agassiz). From Mayer.

Fig. 430. *Mertensia ovum* Fabricius. From Vanhöffen.

Fig. 431. *Pleurobrachia brunnea* Mayer. From Mayer.

Fig. 432. *Cestum veneris* Lesueur. From Vanhöffen, after Chun.

Fig. 433. *Leucothea ochracea* Mayer. Redrawn from Mayer.

Fig. 434. *Ocyropsis crystallina* (Rang). This is a half-grown specimen. Redrawn from Mayer.

Fig. 435. *Mnemiopsis leidyi* A. Agassiz. From Pratt, after Mayer. By permission from *A Manual of the Common Invertebrate Animals* (revised ed.) by H. S. Pratt. Copyright, 1935. The Blakiston Co.

Fig. 436. *Bolinopsis infundibulum* (O. F. Müller). From Pratt, after Mayer. By permission from *A Manual of the Common Invertebrate Animals* (revised ed.) by H. S. Pratt. Copyright, 1935. The Blakiston Co.

Fig. 437. Müller's larva of *Yungia*. From Dawydoff, after Lang. Reproduced by permission from *Traité d'embryologie comparée des invertébrés* by C. Dawydoff. Copyright, 1928. Masson et Cie. (Paris).

Fig. 438. Pilidium larva of a nemertean. Redrawn from Thorson.

Fig. 439. *Nectonemertes mirabilis* Ver. From Pratt, after Coe. By permission from *A Manual of the Common Invertebrate Animals* (revised ed.) by H. S. Pratt. Copyright, 1935. The Blakiston Company.

Fig. 440. *Collotheca polyphemus* From Remane, after Harring.

Fig. 441. *Conochilus unicornis* Rousselet. The usual colonial condition. From Ward and Whipple, after Weber. Reproduced by permission from *Freshwater Biology* by H. B. Ward and G. C. Whipple. John Wiley & Sons, Inc., 1918.

Fig. 442. *Conochilus unicornis* Rousselet. An isolated individual. From Ward and Whipple, after Weber. Reproduced by permission from *Freshwater Biology* by H. B. Ward and G. C. Whipple. John Wiley & Sons, Inc., 1918.

Fig. 443. *Conochiloides natans* Seligo. From Ward and Whipple, after Hlava. Reproduced by permission from *Freshwater Biology* by H. B. Ward and G. C. Whipple. John Wiley & Sons, Inc., 1918.

Fig. 444. *Trichocerca longiseta* (?) Shrank.

Fig. 445. *Synchaeta* sp.

Fig. 446. *Ploesoma hudsoni* Imhof. From Ward and Whipple after Wierzejski and Zacharias. Reproduced by permission from *Freshwater Biology* by H. B. Ward and G. C. Whipple. John Wiley & Sons, Inc., 1918.

Fig. 447. *Gastropus hyptopus* Ehrenberg. From Ward and Whipple, after Hudson and Gosse. Reproduced by permission from *Freshwater Biology* by H. B. Ward and G. C. Whipple. John Wiley & Sons, Inc., 1918.

Fig. 448. *Brachionus calyciflorus* Pallas. From Remane, after Beauchamp.

Fig. 449. *Euchlanis macrura* Ehrenberg. From Ward and Whipple, after Weber. Reproduced by permission from *Fresh-water Biology* by H. B. Ward and G. C. Whipple. John Wiley & Sons, Inc., 1918.

Fig. 450. *Pedalion oxyurus* Zernov. From Remane, after Hauer.

Fig. 451. *Filinia longiseta* (Ehrenberg).

Fig. 452. *Asplanchna priodonta* Gosse. From Pratt. By permission from *A Manual of the Common Invertebrate Animals* (revised ed.) by H. S. Pratt. Copyright, 1935. The Blakiston Company.

Fig. 453. *Keratella cochlearis* (Gosse).

Fig. 454. *Kellicottia striata* (Müller). From Lauterborn, after Levander.

Fig. 455. *Notholca longispina* Kellicott. From Remane, after Hudson and Gosse.

Fig. 456. *Chromogaster ovalis* (Bergendal). From Ward and Whipple, after Weber. Reproduced by permission from *Freshwater Biology* by H. B. Ward and G. C. Whipple. John Wiley & Sons, Inc., 1918.

Fig. 457. *Pompholyx complanata* Gosse. From Ward and Whipple, after Hudson and Gosse. Reproduced by permission from *Freshwater Biology* by H. B. Ward and G. C. Whipple. John Wiley & Sons, Inc., 1918.

Fig. 458. *Polyarthra* sp. Entire animal.

Fig. 459. *Polyarthra* sp. A single spine, broadside view.

Fig. 460. Statoblasts of *Pectinatella magnifica* Leidy.

Fig. 461. A cyphonautes larva of a bryozoan.

Fig. 462. The pelagic developmental stage of *Lingula* sp.

Fig. 463. The actinotroch larva of *Phoronis mülleri* Longchamps. Redrawn from Thorson.

Fig. 464. *Pterosagitta draco* Krohn. From Kuhl, after Hertwig.

Fig. 465. *Spadella cephaloptera* Busch. From Kuhl, after Hertwig.

Fig. 466. *Sagitta enflata* Grassi.

Fig. 467. *Krohnitta subtilis* Grassi. From Kuhl, after Ritter-Zàhony.

Fig. 468. *Eukrohnia hamata* Möbius. From Kuhl, after Ritter-Zàhony.

Fig. 469. *Heterokrohnia mirabilis* Ritter-Zàhony. From Kuhl, after Ritter-Zàhony.

Fig. 470. The trochophore larva of *Phyllodoce grönlandica* Ørsted. Redrawn from Thorson.

Fig. 471. The polytroch larva of *Ophryotrocha puerilis*. From Dawydoff, after Claparède and Metschnikoff. Reproduced by permission from *Traité d'embryologie comparée des invertébrés* by C. Dawydoff. Copyright, 1928. Masson et Cie. (Paris).

Fig. 472. The mitraria larva of *Myriochele danielsseni* Hansen. Redrawn from Thorson.

Fig. 473. The nectochaete larva of *Nereis* sp. From Parker, Haswell, and Lowenstein, after E. B. Wilson. Redrawn by permission from *A Textbook* of *Zoology,* Vol 1, 6th ed., by T. J. Parker, W. A. Haswell, and O. Lowenstein. Copyright, 1940. Macmillan & Co., Ltd. (London).

Fig. 474. An old spionid larva. Redrawn from Thorson.

Fig. 475. A rostraria larva of a polychaete. From Dawydoff, after Haecker. Reproduced by permission from *Traité d'embryologie comparée des invertébrés* by C. Dawydoff. Copyright, 1928. Masson et Cie. (Paris).

Fig. 476. *Callizona angelini* (Kimberg). (A) Anterior end of an adult. (B) A single parapodium from near the middle of the body. From Reibisch, after Apstein.

Fig. 477. *Greeffia celox* (Greeff). A single parapodium. From Reibisch.

Fig. 478. *Pelagobia longecirrata* Greef. (A) The anterior end of an adult specimen. (B) A single parapodium. From Reibisch.

Fig. 479. *Tomopteris* sp.

Fig. 480. *Holopedium* sp. Drawn by R. W. Kiser.

Fig. 481. *Sida* sp. Drawn by R. W. Kiser.

Fig. 482. *Diaphanosoma* sp. Drawn by R. W. Kiser.

Fig. 483. *Chydorus* sp. Drawn by R. W. Kiser.

Fig. 484. *Acroperus* sp. Drawn by R.W. Kiser.

Fig. 485. *Bosmina* sp. Drawn by R. W. Kiser.

Fig. 486. *Daphnia longispina* (O. F. Müller). Drawn by R. W. Kiser.

Fig. 487. *Ceriodaphnia* sp. Drawn by R. W. Kiser.

Fig. 488. *Scapholeberis* sp. Drawn by R.W. Kiser.

Fig. 489. *Simocephalus* sp. Drawn by R. W. Kiser.

Fig. 490. *Leptodora kindtii* (Focke). Drawn by R. W. Kiser.

Fig. 491. *Evadne* sp. Drawn by R. W. Kiser.

Fig. 492. *Podon* sp. Drawn by R. W. Kiser.

Fig. 493. *Polyphemus pediculus* (Linnaeus). Drawn by R. W. Kiser.

Fig. 494. *Notodromas monacha* (O. F. Müller). From Ward and Whipple. Reproduced by permission from *Freshwater Biology* by H. B. Ward and G. C. Whipple. John Wiley & Sons, Inc., 1918.

Fig. 495. *Cypris incongruens* Ramdohr. From Ward and Whipple. Reproduced by permission from *Freshwater Biology* by H. B. Ward and G. C. Whipple. John Wiley & Sons, Inc., 1918.

Fig. 496. *Conchoecia* sp.

Fig. 497. Copepod nauplius, usual type.

Fig. 498. Copepod nauplius, *Longipedia* type. Redrawn from Gurney.

Fig. 499. *Epischura lacustris* Forbes. (A) ♂. Fifth pair of legs. (B) ♀. Fifth leg. Redrawn from Marsh.

Fig. 500. *Eurytemora hirundoides* (Nordquist). (A) ♀. Fifth leg. (B) ♂. Left fifth leg. (C) ♂. Right fifth leg. From Davis, 1943.

Fig. 501. *Limnocalanus grimaldii* (de Guerne). (A) ♀. Fifth leg. (B) ♂. Right fifth leg. Redrawn from Marsh.

Fig. 502. *Diaptomus oregonensis* Lilljeborg. (A) ♀. Dorsal view. (B) ♀. Fifth leg. (C) ♂. Fifth pair of legs. From Pratt, after Pearse. By permission from *A Manual of the Common Invertebrate Animals* (revised ed.) by H. S. Pratt. Copyright, 1935. The Blakiston Company.

Fig. 503. *Pseudodiaptomus coronatus* Williams. (A) ♀. Fifth leg. Redrawn from Wilson. (B) ♂. Fifth pair of legs. From Davis.

Fig. 504. *Microcyclops varicans* (Sars). *a,* ♂. Dorsal view. *b,* ♂. First antenna. *c,* ♂. Fifth and sixth legs. *d,* ♀. Fifth leg. From Wilson.

Fig. 505. *Eucyclops serrulatus* (Fischer). (1) Fifth leg. (2) Furcal rami and anal segment. (3) Genital segment. From Pratt, taken from *Süsswasserfauna Deutschlands.* By permission from *A Manual of the Common Invertebrate Animals* (revised ed.) by H. S. Pratt. Copyright, 1935. The Blakiston Company.

Fig. 506. *Macrocyclops albidus* (Jurine). (1) Furcal rami and anal segment. (2) Fifth leg. (3) Genital segment. From Pratt, taken from *Süsswasserfauna Deutschlands.* By permission from *A Manual of the Common Invertebrate Animals* (revised ed.) by H. S. Pratt. Copyright, 1935. The Blakiston Company.

Fig. 507. *Mesocyclops leuckarti* (Claus). Fifth leg.

Fig. 508. *Cyclops americanus* Marsh. Fifth leg.

Fig. 509. *Monstrilla reticulata* Davis. ♀. Lateral view. From Davis, 1949.

Fig. 510. *Mormonilla polaris* G. O. Sars. (A) ♀. Dorsal view. (B) First leg. Redrawn from van Breemen.

Fig. 511. *Pleuromamma xiphias* (Giesbrecht). ♀. Lateral view.

Fig. 512. *Metridia lucens* Boeck. ♀. Second leg.

Fig. 513. *Undinula vulgaris* (Dana). *a.* second segment of second exopod. *b,* ♂. Fifth pair of legs. *c,* ♀. Fifth leg. From Wilson.

Fig. 514. *Calanus finmarchicus* (Gunner). (A) ♀. Lateral view. (B) ♀. Fifth leg.

Fig. 515. *Heterorhabdus* sp. ♀. Ventral view of urosome.

Fig. 516. *Lucicutia grandis* (Giesbrecht). *a,* ♀. Dorsal view. *b,* ♀. Fifth leg. *c,* ♂. Fifth pair of legs. From Wilson.

Fig. 517. *Centropages furcatus* (Dana). (A) ♀. Fifth leg. (B) ♂. Right fifth leg. (C) ♂. Left fifth leg.

Fig. 518. *Tortanus discaudatus* Giesbrecht. ♂. Dorsal view of urosome. Redrawn from Mori.

Fig. 519. *Calocalanus pavo* (Dana). ♀. Dorsal view. Redrawn from van Breemen, after Giesbrecht.

Fig. 520. *Paracalanus parvus* (Claus). ♂. Fourth leg. Redrawn from Mori.

Fig. 521. *Rhincalanus nasutus* Giesbrecht. *a,* ♀. Dorsal view. *b,* ♀. Fifth leg. *c,* ♂. Fifth pair of legs. From Wilson.

Fig. 522. *Eucalanus bungii* Giesbrecht. ♀. Dorsal view.

Fig. 523. *Mecynocera clausi* Thompson. ♀. Dorsal view. From Wilson, after Wheeler.

Fig. 524. *Scolecithrix danae* (Lubbock). ♂. Fifth pair of legs. Redrawn from Wilson, after Giesbrecht.

Fig. 525. *Euchirella pulchra* (Claus). (A) ♀. First basal segment of fourth leg. (B) ♂. Fifth pair of legs.

Fig. 526. *Euchaeta marina* (Prestandrea). (A) ♀ Lateral view. (B) ♂. Left fifth leg, grasping a spermatophore. (C) ♂. Right fifth leg.

Fig. 527. *Pseudocalanus minutus* (Krøyer). (A) ♀. Lateral view. (B) ♂. Left fifth leg. (C) ♂. Right fifth leg.

Fig. 528. *Undeuchaeta minor* Giesbrecht. *a,* ♀. Dorsal view. *b,* ♀. Lateral view of the urosome. *c,* ♂. Fifth pair of legs. From Wilson, *a* after Sars, *b* and *c* after Giesbrecht.

Fig. 529. (A) *Gaetanus armiger* Giesbrecht. ♀. Lateral view. Original. (B) *G. kruppii* Giesbrecht. ♂. Fifth pair of legs. Redrawn from Wilson.

Fig. 530. *Microcalanus pusillus* Sars. ♀. Ventral view of the urosome.

Fig. 531. *Chirundina streetsii* Giesbrecht. (A) ♀. Ventral view of the urosome. (B) ♂. Left fifth leg. (C) ♂. Right fifth leg. Redrawn from Esterly.

Fig. 532. *Aetideus armatus* (Boeck). (A) ♀. Lateral view. Redrawn from Esterly. (B) ♂. Left (only) fifth leg. Redrawn from Wilson, after Sars.

Fig. 533. *Gaidius pungens* Giesbrecht. Second antenna.

Fig. 534. *Chiridius armatus* (Boeck). ♂. Fifth pair of legs. From Wilson.

Fig. 535. *Scolecithricella ovata* (Farran). *a*, ♀. Dorsal view *b*, ♀. Fifth leg. From Wilson.

Fig. 536. *Scolecithricella minor* (Brady). *a*, ♂. Dorsal view. *b*, ♂ Fifth pair of legs. From Wilson, after Sars.

Fig. 537. *Lophothrix frontalis* Giesbrecht. ♀. Fifth leg.

Fig. 538. *Scottocalanus persecans* (Giesbrecht). *a*, ♂. Dorsal view. *b*, ♂. Fifth pair of legs. *c*, ♀. Fifth pair of legs. From Wilson, *a* after Scott, *b* after Giesbrecht.

Fig. 539. *Scaphocalanus magnus* (Scott). *a*, ♀. Fifth pair of legs. *b*, ♂. Fifth pair of legs. *c*, ♀. Dorsal view. From Wilson.

Fig. 540. *Clausocalanus arcuicornis* (Dana). (A) Fourth leg. (B) Third leg.

Fig. 541. *Ctenocalanus longicornis* Mori. ♀. Third segment of the exopod of the third leg. Redrawn from Mori.

Fig. 542. *Labidocera scotti*. (A) ♀. Dorsal view of the urosome and the terminal segments of the metasome. (B) ♂. Dorsal view. (C) ♂. Grasping antenna. (D) ♀. Fifth pair of legs. (E) ♂. Left fifth leg. (F) ♂. Right fifth leg.

Fig. 543. (A) *Acartia floridana* Davis. ♀. Fifth leg. (B) *A. floridana* Davis. ♂. Fifth pair of legs. From Davis, 1948. (C) *A. longiremis* (Lilljeborg). Second antenna.

Fig. 544. *Tortanus discaudatus* (Thompson and Scott). (A) ♀. Dorsal view. (B) ♀. Fourth leg.

Fig. 545. *Candacia pachydactyla* (Dana). (A) ♀. First leg. (B) ♀. Fifth leg.

Fig. 546. *Anomalocera patersonii* Templeton. *a*, ♀. Dorsal view. *b*, ♂. Dorsal view. *c*, ♂. Grasping antenna. *d*, ♂. Fifth pair of legs. *e*, ♀. Fifth leg (somewhat abnormal). From Wilson, after Wheeler.

Fig. 547. *Epilabidocera amphitrites* (McMurrich). ♂. Right fifth leg.

Fig. 548. *Pontella spinicauda* Mori. ♂. Right fifth leg. Redrawn from Mori.

Fig. 549. *Oithona helgolandica* Claus. Second antenna.

Fig. 550. *Sapphirina gemma* Dana. *a*, ♀. Dorsal view. *b*, ♂. Second antenna. *c*, ♀. Fourth leg. *d*, ♂. Dorsal view. From Wilson, *c* after Wheeler.

Fig. 551. *Oncaea venusta* Philippi. *a*, ♀. Dorsal view. *b*, ♂. Dorsal view. *c*, ♀. Second antenna. *d*, ♀. Fourth leg. From Wheeler.

Fig. 552. *Conaea gracilis* (Dana). (A) ♀. Dorsal view. (B) Second antenna. Redrawn from Wilson.

Fig. 553. *Oithonina nana* (Giesbrecht). ♀. Dorsal and lateral views. Drawn from photograph.

Fig. 554. *Corycaeus* sp. ♀. Lateral view showing two spermatophores attached to the genital segment. Drawn from photograph.

Fig. 555. *Copilia mirabilis* Dana. (A) ♀. Dorsal view. (B) ♂. Dorsal view.

Fig. 556. *Euterpina acutifrons* (Dana). ♀. Lateral view. Drawn from photograph.

Fig. 557. *Metis jousseaumei* (Richard). *a*, Mandibular palp. *b*, ♀. Lateral view. *c*, a pair of maxillipeds. *d*, rostrum. *e*, ♀. Fifth leg. *f*, ♀. First leg. *g*, Furcal rami and the anal segment. *h*, ♀. Second antenna. *i*, ♀. First antenna. From Wilson, after Sharpe.

Fig. 558. *Macrosetella gracilis* (Dana). ♀. Lateral view.

Fig. 559. *Clytemnestra rostrata* (Brady). *a*, ♀. Dorsal view. *b*, ♀. Fifth leg. *c*, ♂. Fifth leg. From Wilson.

Fig. 560. *Microsetella rosea* (Dana). ♀. Lateral view. Drawn from photograph.

Fig. 561. *Aegisthus mucronatus* Giesbrecht. ♀. Lateral view, showing only the base of the long furcal setae. (Cf. fig. *12A*.)

Fig. 562. Nauplius of *Balanus perforatus*. Redrawn from Hoek, after Groom.

Fig. 563. A cypris larva of a barnacle. Drawn from photograph.

Fig. 564. *Diastylis bispinosus* (Stimpson). From Pratt, after Paulmeier. By permission from *A Manual of the Common Invertebrate Animals* (revised ed.) by H. S. Pratt. Copyright, 1935. The Blakiston Company.

Fig. 565. *Munnopsis typica* M. Sars. From Richardson, after Harger.

Fig. 566. *Eurydice pulchra* Leach. From Tattersall, after Sars.

Fig. 567. *Gammarus fasciatus* Say. ♂. A nonplanktonic amphipod, showing the structure and terminology of the parts. From Clemens.

Fig. 568. *Brachyscelus crusculum* Bate. From Schellenberg, after Stebbing and Senna.

Fig. 569. *Euprimno macropus* (Guérin). ♂. From Schellenberg, after Bovallius.

Fig. 570. *Phrosina semilunata* Risso. (A) ♀. Lateral view. (B) Fourth pereiopod. From Schellenberg, after Stebbing and Bovallius.

Fig. 571. *Phronima sedentaria* Forskål. From Schellenberg, after Vosseler.

Fig. 572. *Themisto libellula* (Mandt). From Schellenberg, after Sars.

Fig. 573. *Hyperia galba* (Mont.). From Schellenberg, after Sars.

Fig. 574. *Lanceola serrata* Bovallius. ♀. From Schellenberg, after Bovallius.

Fig. 575. *Scina rattrayi* Stebbing. ♀. From Schellenberg, after Stebbing and Vosseler.

Fig. 576. *Vibilia armata* Bovallius. From Schellenberg, after Bovallius.

Fig. 577. *Cyphocaris anonyx* Boeck. From Schellenberg, after Stebbing.

Fig. 578. *Eurythenes gryllus* Mandt. ♀. From Schellenberg, after Sars.

Fig. 579. *Eucopia sculpticauda* Faxon. From Zimmer.

Fig. 580. *Lophogaster typicus* M. Sars. (A) Lateral view. (B) Telson and uropods. From Zimmer.

Fig. 581. *Gnathophausia gigas* Willemoës-Suhm. Telson and left uropod. From Zimmer.

Fig. 582. *Petalophthalmus armiger* Willemoës-Suhm. From Zimmer.

Fig. 583. *Hansenomysis fyllae* (Hansen). Telson and left uropod, dorsal view. From Nouvel.

Fig. 584. *Boreomysis scyphops* G. O. Sars. (A) ♀. Lateral view. (B) Telson and left uropod. From Zimmer.

Fig. 585. *Siriella norvegica* G. O. Sars. From Zimmer.

Fig. 586. *Anchialina typica* (Krøyer). From Kükenthal-Krumbach, after G. O. Sars. Reproduced by permission from *Handbuch der Zoologie* by W. Kükenthal and T. Krumbach. Copyright; 1927. Walter de Gruyter & Co. (Berlin).

Fig. 587. *Gastrosaccus sanctus* (van Beneden). (A) ♀. Lateral view. (B) Uropod. From Zimmer.

Fig. 588. *Heteromysis formosa* S. J. Smith. From Zimmer.

Fig. 589. *Pseudomma roseum* G. O. Sars. Anterior end of the body. From Zimmer.

Fig. 590. *Amblyops crozeti* G. O. Sars. Dorsal view of the body. From Zimmer.

Fig. 591. *Dactylerythrops dactylops* Holt and Tattersall. ♀. Dorsal view of the body. From Zimmer.

Fig. 592. *Dactylamblyops thaumatops* Tattersall. Dorsal view. From Nouvel.

Fig. 593. *Euchaetomera fowleri* (Holt and Tattersall). ♂. Dorsal view of the body. From Zimmer.

Fig. 594. *Erythrops abyssorum* G. O. Sars. Telson. From Zimmer.

Fig. 595. *Katerythrops oceanae* Holt and Tattersall. Telson. From Zimmer.

Fig. 596. *Holmesiella anomala* Ortmann. Redrawn from Ortmann.

Fig. 597. *Meterythrops robusta* (S. J. Smith). (A) ♂. Lateral view. (B) Telson. From Zimmer.

Fig. 598. *Parerythrops spectabilis* G. O. Sars. ♂. Lateral view. From Zimmer.

Fig. 599. *Mysidopsis didelphys* (Norman). ♂. Lateral view. From Zimmer.

Fig. 600. *Leptomysis mediterranea* G. O. Sars. Telson. From Zimmer.

Fig. 601. *Mysideis insignis* (G. O. Sars). Telson. From Zimmer.

Fig. 602. *Tenagomysis atlantica* Nouvel. Telson. From Nouvel.

Fig. 603. *Mesopodopsis slabberi* (van Beneden). Fourth pleopod. From Nouvel.

Fig. 604. *Stilomysis grandis* (Goës). ♂. Lateral view. From Zimmer.

Fig. 605. *Hemimysis abyssicola* G. O. Sars ♂. Third pleopods. From Zimmer.

Fig. 606. *Mysis oculata* (Fabricius). ♂. Lateral view. From Zimmer.

Fig. 607. *Schistomysis ornata* (G. O. Sars). (A) Basal portion of the second antenna. (B) ♂. Fourth pleopod. From Zimmer.

Fig. 608. *Praunus flexuosus* (Müller). (A) Basal portion of the second antenna. (B) ♂. Fourth pleopods. From Zimmer.

Fig. 609. *Acanthomysis sculpta* (Tattersall). Basal portion of the second antenna. From Banner.

Fig. 610. *Neomysis vulgaris* (J. V. Thompson). Telson. From Zimmer.

Fig. 611. (A) Young nauplius larva (nauplius stage 2). (B) Older nauplius larva (nauplius stage 3) of *Euphausia brevis*. From Gurney.

Fig. 612. Calyptopis larva of *Euphausia brevis*. From Gurney.

Fig. 613. Furcilia stage of *Euphausia brevis*. Dorsal and lateral views. From Gurney.

Fig. 614. The interrelationship between the number of whales captured and the abundance of euphausiids in the region of Davis Strait. From Sverdrup, Johnson, and Fleming, after Hjort and Ruud. Redrawn by permission from *The Oceans* by H. U. Sverdrup, Martin W. Johnson, and Richard H. Fleming. Copyright, 1942, by Prentice-Hall, Inc.

Fig. 615. *Bentheuphausia amblyops* (G. O. Sars). Lateral view. From Zimmer.

Fig. 616. *Stylocheiron longicorne* G. O. Sars. Lateral view. From Zimmer.

Fig. 617. *Nematobranchion boopis* (Calman). Lateral view. From Zimmer.

Fig. 618. *Thysanopoda acutifrons* Holt and Tattersall. Lateral view. From Zimmer.

Fig. 619. *Nyctiphanes australis* G. O. Sars. Seventh thoracic leg. From Kükenthal-Krumbach, after G. O. Sars. Reproduced by permission from *Handbuch der Zoologie* by W. Kükenthal and T. Krumbach. Copyright, 1927. Walter de Gruyter & Co. (Berlin).

Fig. 620. *Euphausia luscens* Hansen. Anterior portion of the body from a dorsal view. From Zimmer.

Fig. 621. *Meganyctiphanes norvegica* (M. Sars). Lateral view. From Zimmer.

Fig. 622. *Thysanoëssa gregaria* G. O. Sars. Lateral view. From Zimmer.

Fig. 623. *Nematoscelis megalops* G. O. Sars. Lateral view. From Zimmer.

Fig. 624. *Lucifer* sp. from the Gulf of Mexico.

Fig. 625. *Gennadas elegans* (S. I. Smith). Lateral view. From Boone. Reproduced from the Bulletin of the *Vanderbilt Marine Museum*, Vol. 3. Copyright, 1930, by Lee Boone.

Fig. 626. *Sergestes inous* Faxon. Lateral view. From Faxon.

Fig. 627. Nauplius stage of *Penaeus*. From Dawydoff, after Müller. Reproduced by permission from *Traité d'embryologie comparée des invertébrés* by C. Dawydoff. Copyright, 1928. Masson et Cie. (Paris).

Fig. 628. Prezoea (hatching) stage of the brachyuran *Elamena mathei*. From Gurney.

Fig. 629. Zoea larva of a brachyuran. Drawn from life by S. M. Davis.

Fig. 630. Megalops stage of a brachyuran. Dorsal view.

Fig. 631. Zoea larva of *Albunea* sp. From a specimen collected in the Bahamas.

Fig. 632. Zoea larva of a porcellanid crab.

Fig. 633. Zoea larva of *Sergestes cornutus*. From Gurney.

Fig. 634. Pagurid zoea from Biscayne Bay, Florida. Drawn from photograph.

Fig. 635. *Glaucothoë peronii*. From Gurney.

Fig. 636. Zoea stage of *Pandalus stenolepis*. From Gurney.

Fig. 637. Mysis stage of *Parapenaeus longirostris*. From Gurney.

Fig. 638. Phyllosoma larva of *Palinurus* sp. From Dawydoff, after Claus. Reproduced by permission from *Traité d'embryologie comparée des invertébrés* by C. Dawydoff. Copyright, 1928. Masson et Cie. (Paris).

Fig. 639. Mysis stage of the crangonid *Athanus nitescens*. From Gurney.

Fig. 640. Mysis stage of *Solenocera* sp.

Fig. 641. Eretmocaris larva of a macruran, from the Florida Current. The exopods of the thoracic appendages are missing.

Fig. 642. Antizoea stage of *Lysiosquilla eusebia*. From Gurney.

Fig. 643. Pseudozoea stage of a stomatopod.

Fig. 644. A larva of *Corethra*, from a specimen taken in Lake Apopka, Florida.

Fig. 645. *Hygrobates longipalpis* (Hermann). From Pratt, after Wolcott. By permission from *A Manual of the Common Invertebrate Animals* (revised ed.) by H. S. Pratt. Copyright, 1935. The Blakiston Company.

Fig. 646. (A) Veliger larva of *Eulimella nitidissima* (Montagu), showing the expanded velum. Redrawn from Thorson. (B) A preserved gastropod veliger taken from the plankton, showing the usual retraction of the soft parts in preserved material. Drawn from photograph.

Fig. 647. A pelecypod veliger larva. Drawn from photograph.

Fig. 648. An echinospira larva taken from the Gulf of Mexico.

Fig. 649. *Atlanta peronii*, expanded. From Parker, Haswell, and Lowenstein. Reproduced by permission from *A Textbook of Zoology*, Vol. 1, 6th ed., by T. J. Parker, W. A. Haswell, and O. Lowenstein. Copyright, 1940. Mcmillan & Co., Ltd. (London).

Fig. 650. *Spiratella* (*Limacina*) sp. from the northeastern Pacific Ocean.

Fig. 651. *Janthina fragilis* (Linnaeus). From Cooke, after Quoy and Gaimard. Reproduced by permission from the *Cambridge Natural History*, Vol. 3. Copyright 1895. Macmillan & Co., Ltd. (London).

Fig. 652. *Carinaria mediterranea* Péron and Lesueur.

Fig. 653. *Herse (Cuvierina) columnella* Rang. From Coker, after Souyelet. Reproduced by permission from *This Great and Wide Sea* by R. E. Coker. Copyright, 1947. The University of North Carolina Press.

Fig. 654. *Creseis virgula* Rang. From Coker, after Souyelet. Reproduced by permission from *This Great and Wide Sea* by R. E. Coker. Copyright, 1947. The University of North Carolina Press.

Fig. 655. *Clio acicula.* From Lancester, after Souyelet.

Fig. 656. *Cavolina trispinosa* (Lesueur). From Pratt, after Dall. By permission from *A Manual of the Common Invertebrate Animals* (revised ed.) by H. S. Pratt. Copyright, 1935. The Blakiston Company.

Fig. 657. *Pterotrachea scutata.* Redrawn from Parker, Haswell, and Lowenstein. Reproduced by permission from *A Textbook of Zoology*, Vol. 1, 6th ed., by T. J. Parker, W. A. Haswell, and O. Lowenstein. Copyright, 1940. Macmillan & Co., Ltd. (London).

Fig. 658. *Clione kinkaidi.*

Fig. 659. *Anopsia* (*Halopsyche*) *gaudichaudi* Souleyet. From Cooke, after Souleyet. Reproduced by permission from the *Cambridge Natural History,* Vol. 3. Copyright, 1895. Macmillan & Co., Ltd. (London).

Fig. 660. *Pelagothuria natans.* From Borradaile and Potts. Reproduced from *The Invertebrata* by L. A. Borradaile and F. A. Potts, by permission of the Cambridge University Press. Copyright, 1932.

Fig. 661. Diagram of a primitive pluteus-type larva.

Fig. 662. *"Bipinnaria metschnikoffi"* Mortensen. From Mortensen.

Fig. 663. A (late) brachiolarian larva in the act of attaching itself to the substratum. From MacBride, after Müller. Reproduced by permission from the *Cambridge Natural History,* Vol. 1, 1st ed. Copyright, 1906. Macmillan & Co., Ltd. (London).

Fig. 664. Auricularian larva of *Synapta digitata.* From Mortensen.

Fig. 665. Pupa stage of *Synapta digitata.* Redrawn from Korschelt and Heider, after Semon. Reproduced by permission from *Textbook of the Embryology of Invertebrates,* Vol. 1, by E. Korschelt and K. Heider. Copyright, 1895. The Macmillan Company, New York, and Macmillan & Co., Ltd. (London).

Fig. 666. Ophiopluteus larva. Drawn from photograph.

Fig. 667. Echinopluteus larva of a spatangid. From Dawydoff, after Lang. Reproduced by permission from *Traité d'embryologie comparée des invertébrés* by C. Dawydoff. Copyright, 1928. Masson et Cie. (Paris).

Fig. 668. (A) Lateral view of an early tornaria larva of *Balanoglossus clavigerus* Delle Chiaje. Redrawn from Stiasney. (B) Ventral view of a later tornaria (*Tornaria mülleri*) of *B. clavigerus* Delle Chiaje. Redrawn from Stiasny. (C) *Tornaria snelliusi* Stiasny. From Stiasny.

Fig. 669. Pelagic egg of a tunicate, from Chokaloskee Bay, Florida. Drawn from photograph.

Fig. 670. Larval tunicate from Chokaloskee Bay, Florida. Drawn from photograph.

Fig. 671. *Fritillaria borealis* Quoy and Gaimard. From Lohmann.

Fig. 672. *Oikopleura* sp. Drawn from photograph.

Fig. 673. *Appendicularia sicula* Fol. Redrawn from Lohmann.

Fig. 674. *Pyrosoma atlanticum* Péron. Redrawn from Pratt, after Cambridge Natural History. By permission from *A Manual of the Common Invertebrate Animals* (revised ed.) by H. S. Pratt. Copyright, 1935. The Blakiston Company.

Fig. 675. *Salpa democratica* Forskål. Solitary form. From Pratt, after Vogt and Jung. By permission from *A Manual of the Common Invertebrate Animals* (revised ed.) by H. S. Pratt. Copyright, 1935. The Blakiston Company.

Fig. 676. *Doliolom krohni* Herdman. From Borgert.

Fig. 677. *Doliopsis rubescens* Vogt. Sexual form, lateral view. From Neumann, after Kowalevsky and Barrois.

Fig. 678. "*Amphioxides pelagicus*" Gill. The immature form of an unkown cephalochord from the Gulf of Mexico.

Fig. 679. *Mola mola* (Linnaeus). From MacGinitie and MacGinitie. By permission from *Natural History of Marine Animals* by G. E. MacGinitie and N. MacGinitie. Copyright, 1949. McGraw-Hill Book Company, Inc.

Fig. 680. Leptocephalus larva of *Anguilla vulgaris* Turt. (=*Leptocephalus brevirostris* Kaup.). From Ehrenbaum.

Fig. 681. Newly hatched fry of *Mola mola* (Linnaeus) (specimen kindly loaned by Dr. A. H. Banner of the University of Hawaii).

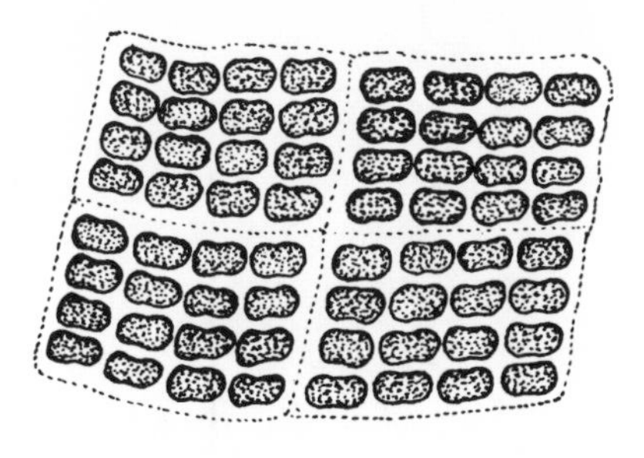

FIGURE 49

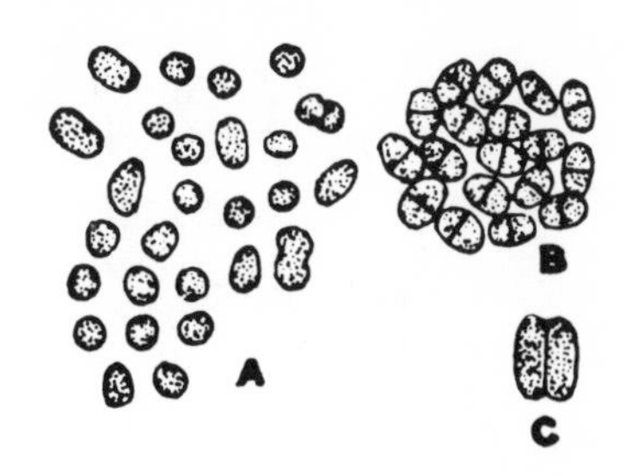

FIGURE 50

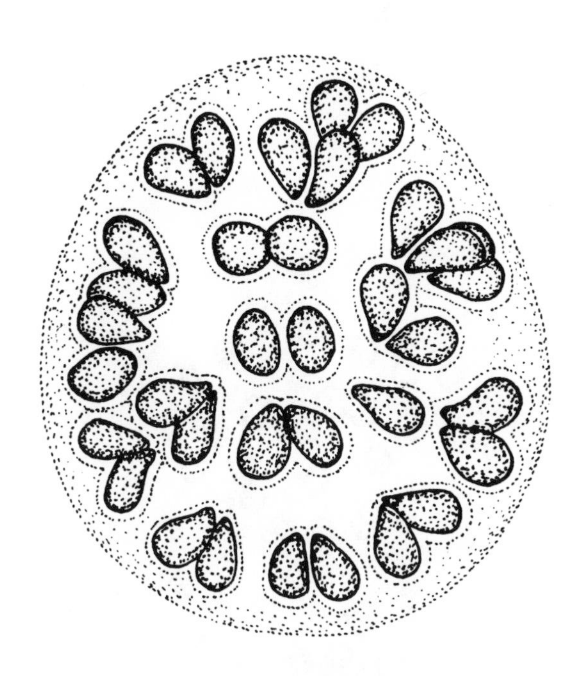

FIGURE 51

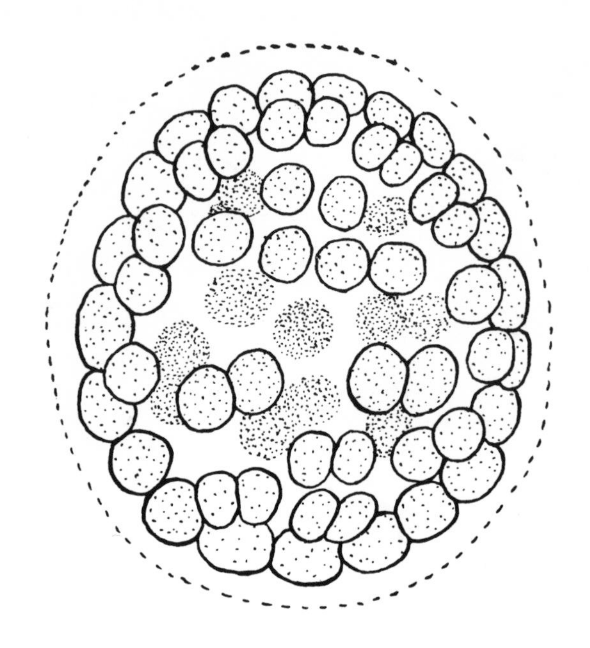

FIGURE 52

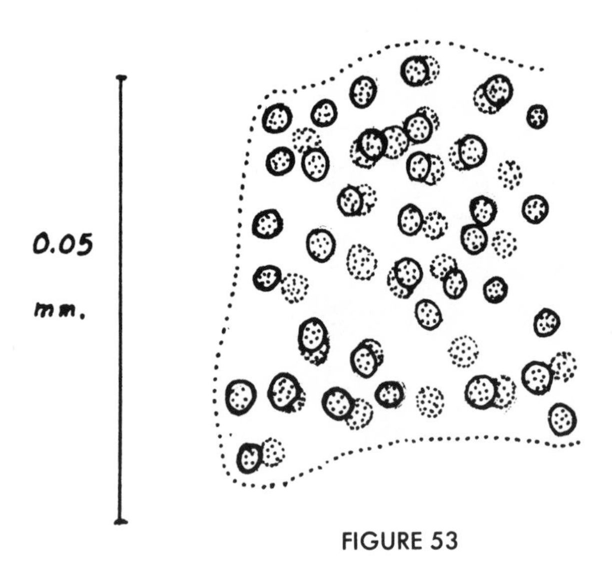

FIGURE 53

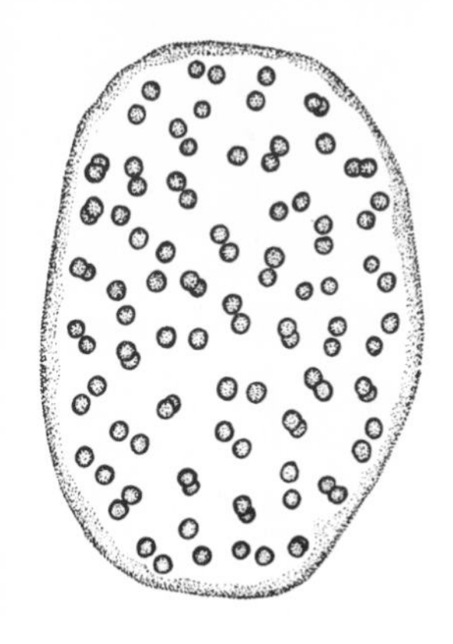

FIGURE 54

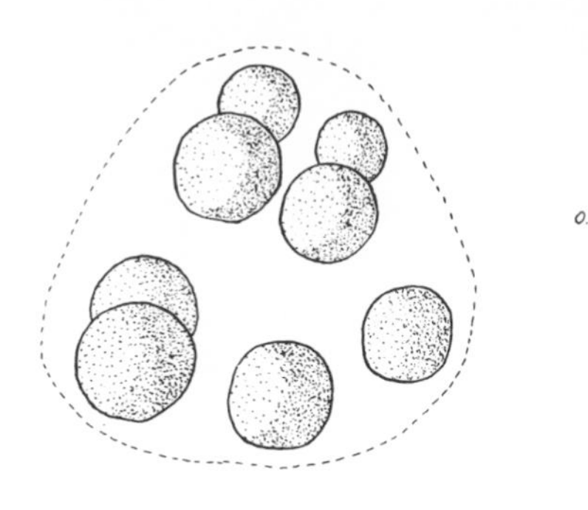

FIGURE 55

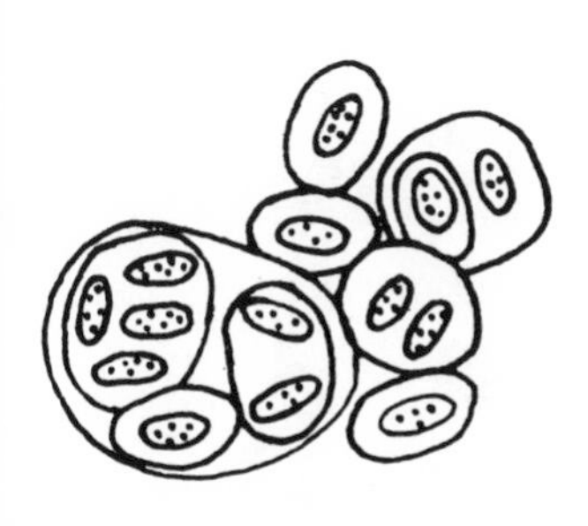

FIGURE 56

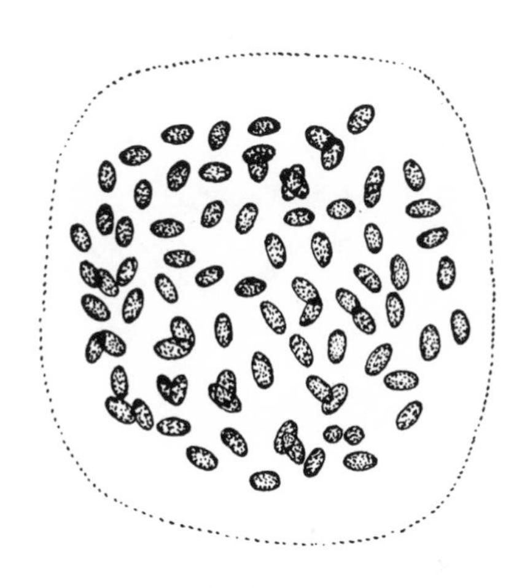

FIGURE 57

FIGURE 58

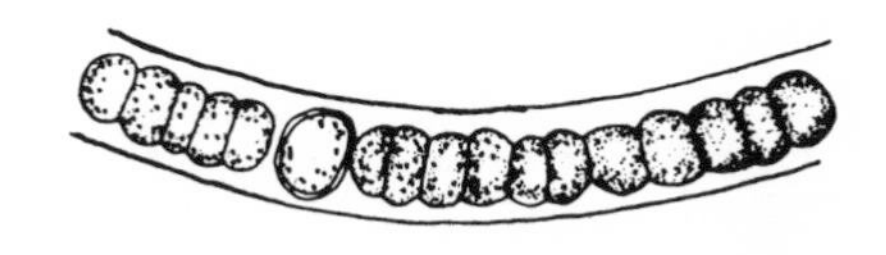

FIGURE 59

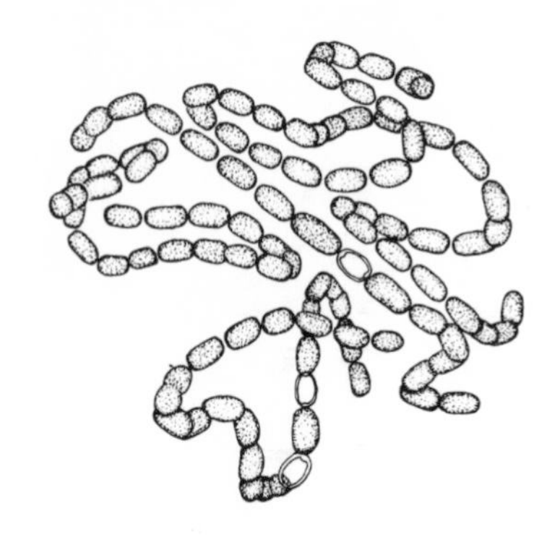

FIGURE 60

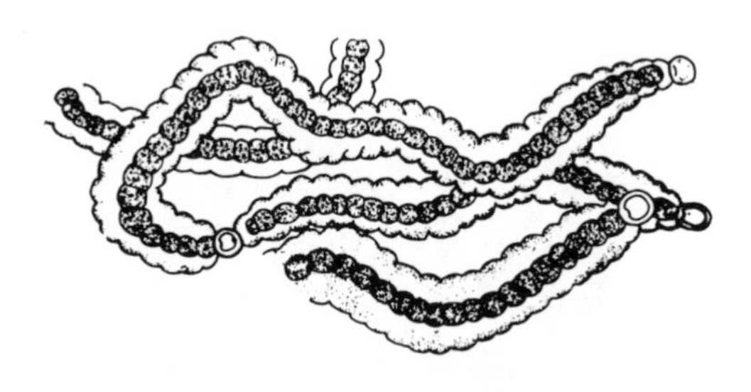

FIGURE 62

FIGURE 61

FIGURE 63

FIGURE 64

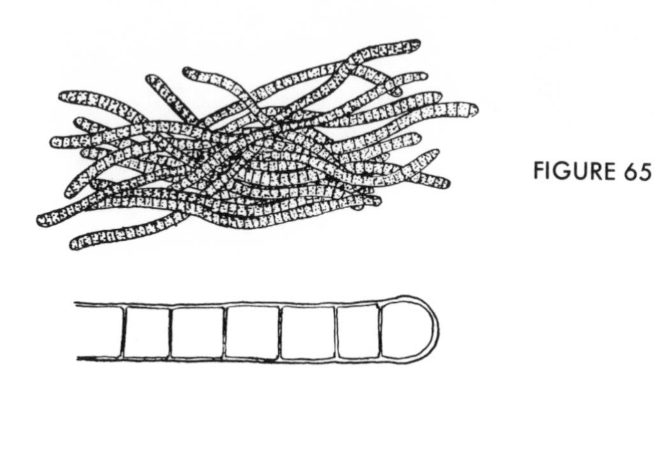

FIGURE 65

0.05 mm.

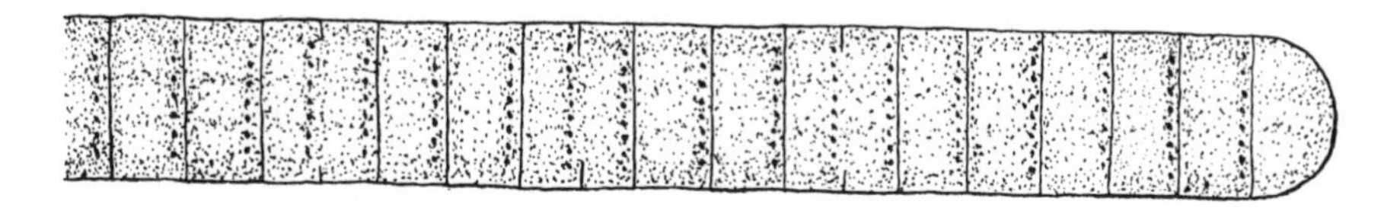

FIGURE 66

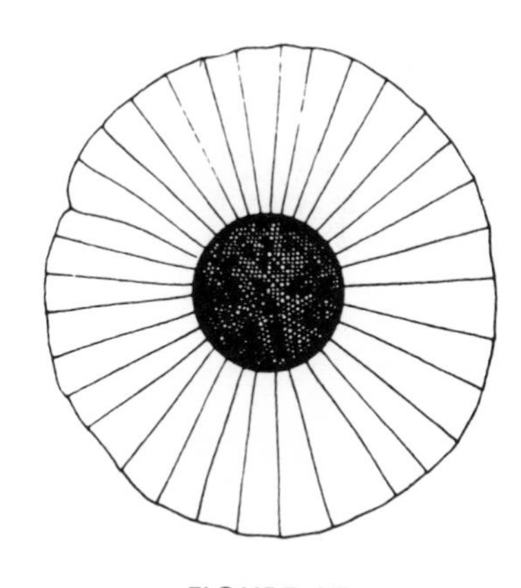

FIGURE 67

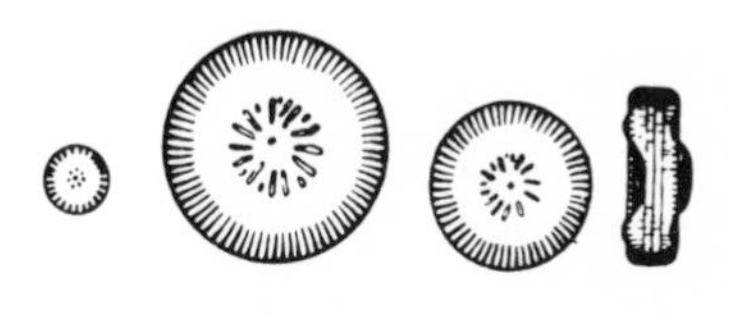

FIGURE 68

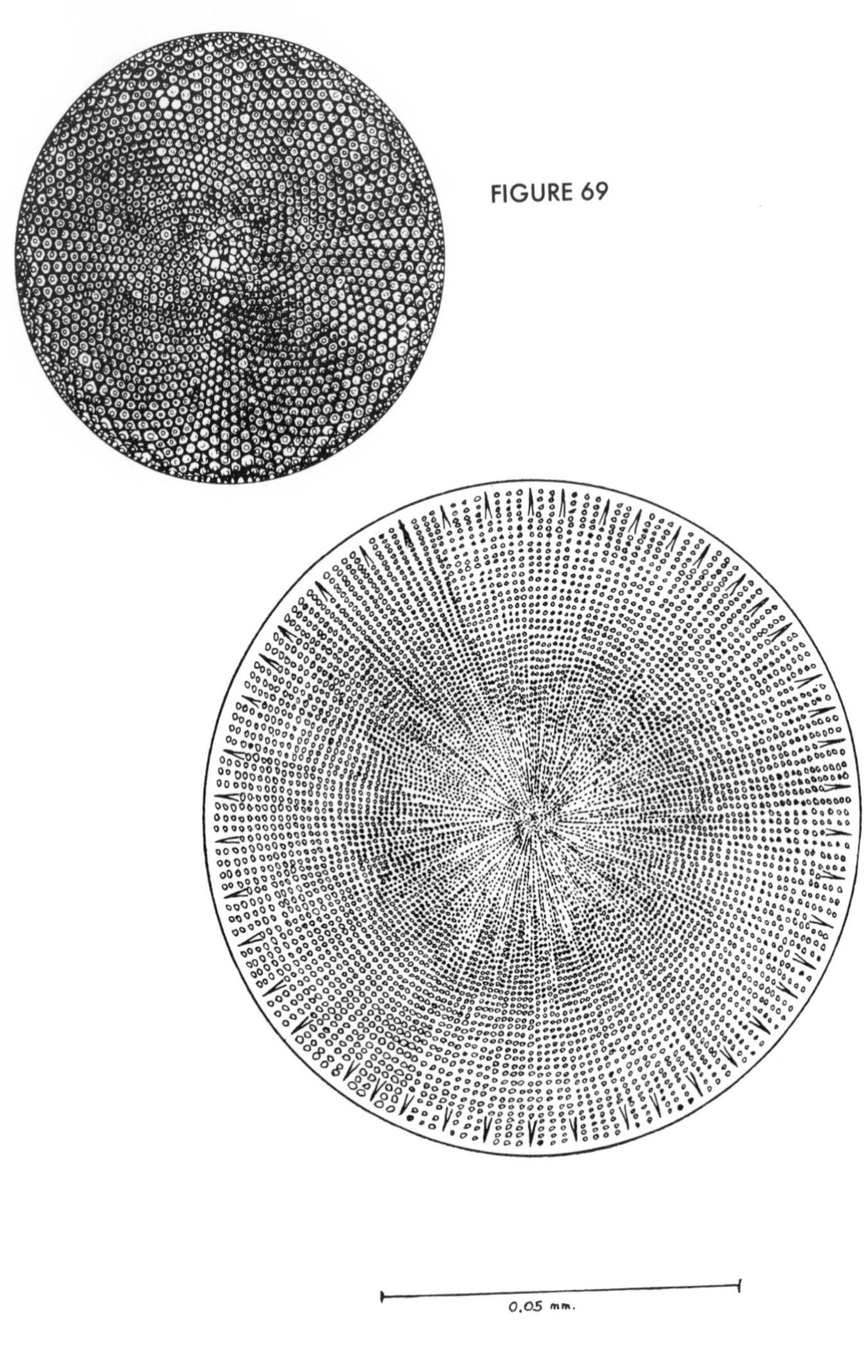

FIGURE 69

FIGURE 70

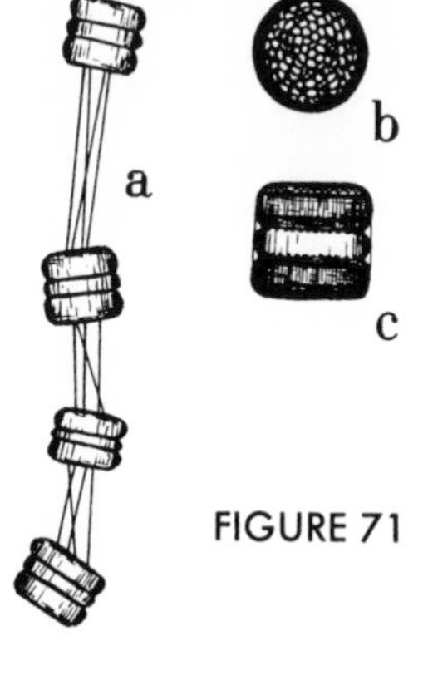
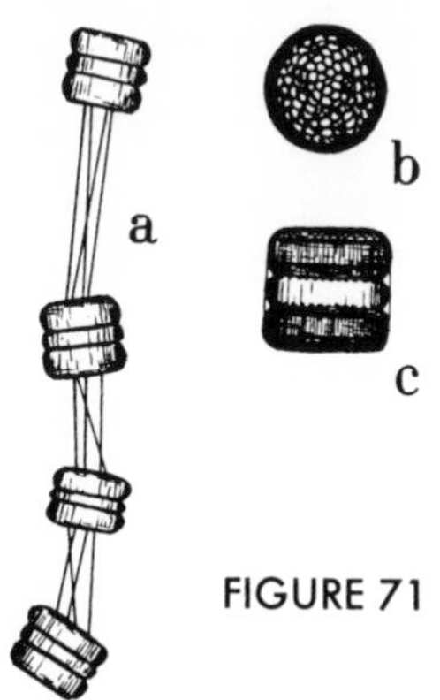

FIGURE 71

FIGURE 72

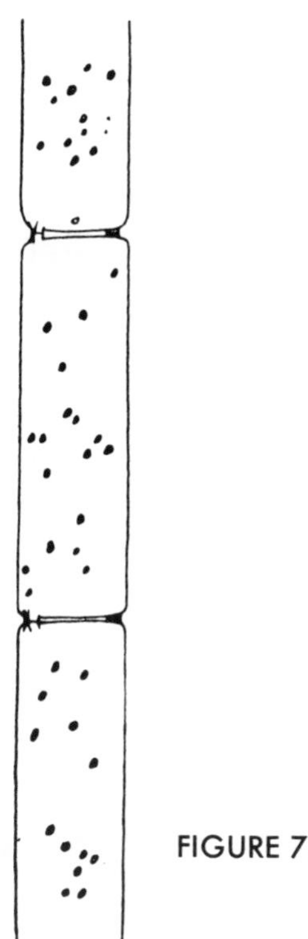

FIGURE 73

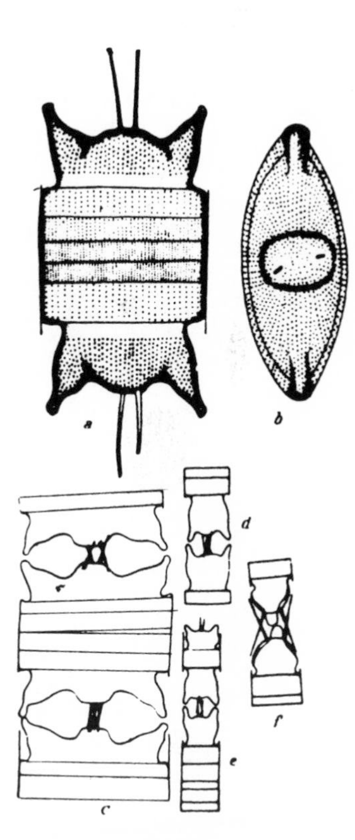

FIGURE 74

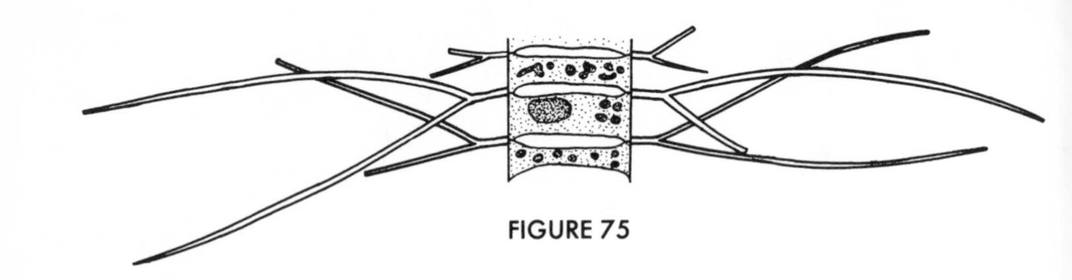

FIGURE 75

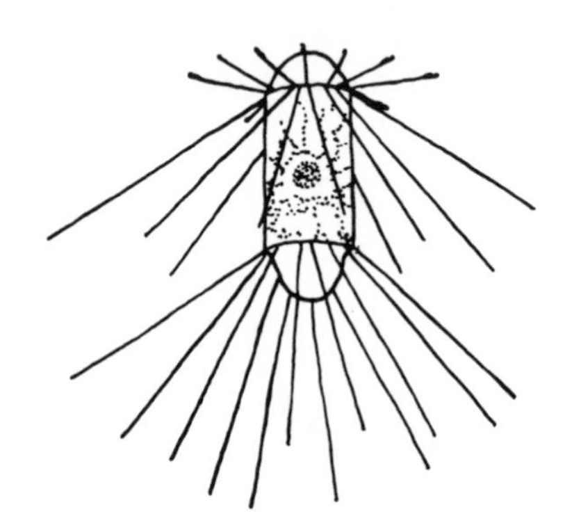

FIGURE 76

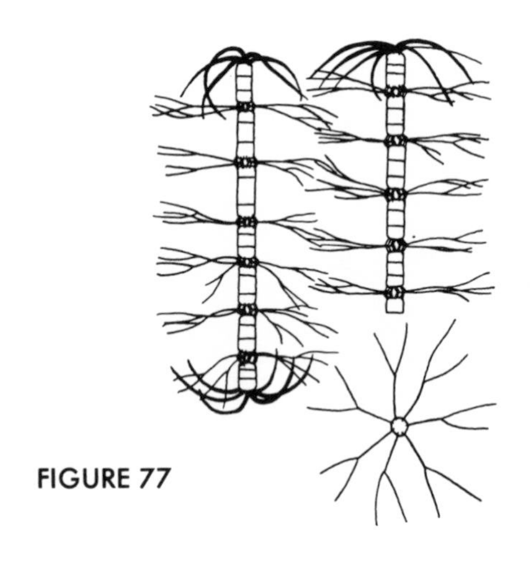

FIGURE 77

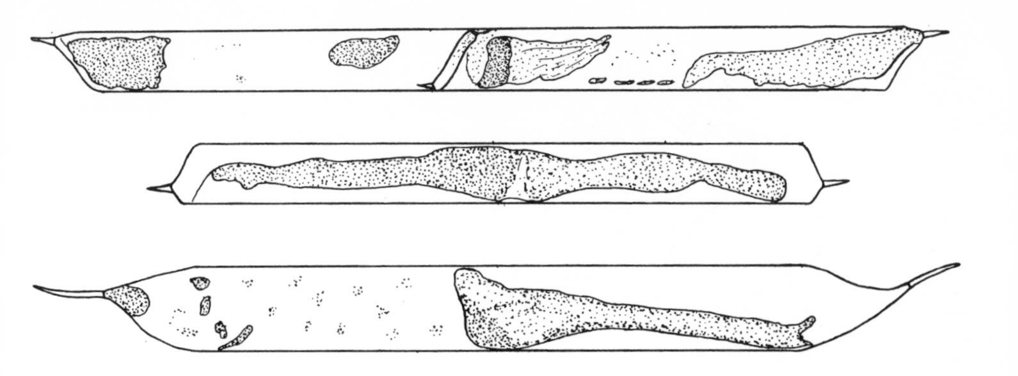

FIGURE 78

FIGURE 79

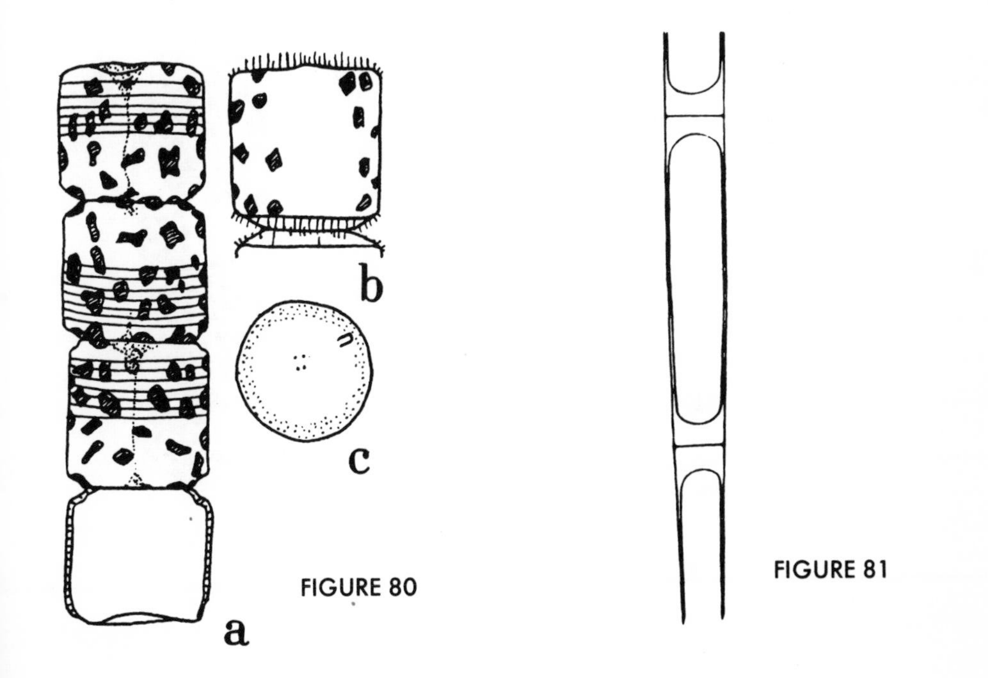

FIGURE 80

FIGURE 81

FIGURE 82

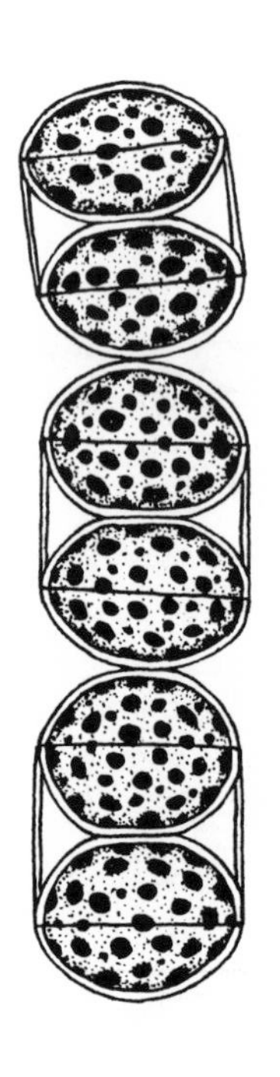
FIGURE 83

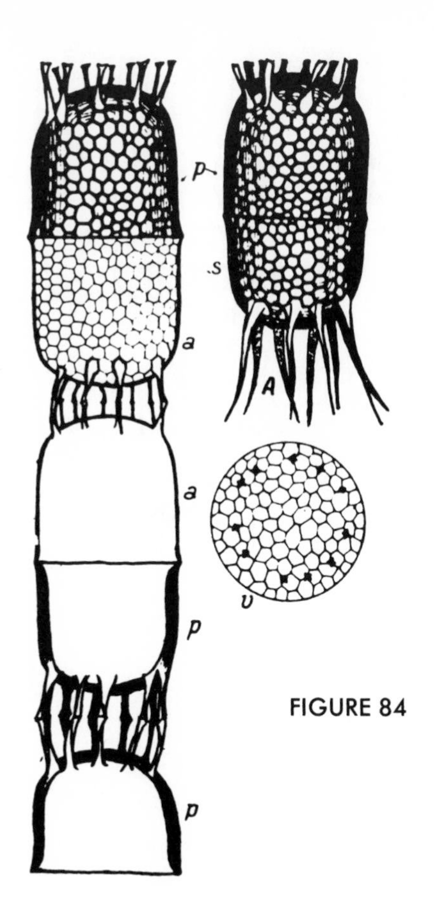

FIGURE 84

0.05 mm.

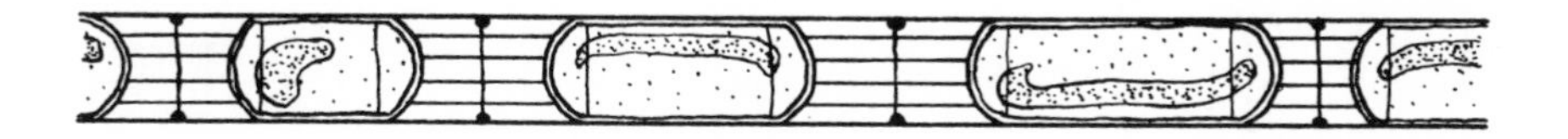

FIGURE 85

FIGURE 86

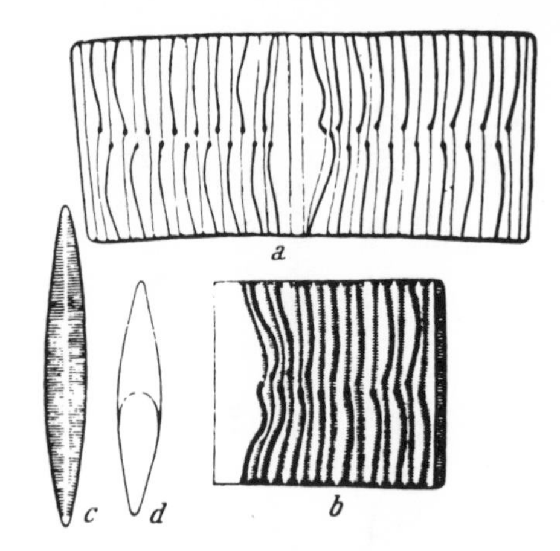

FIGURE 87

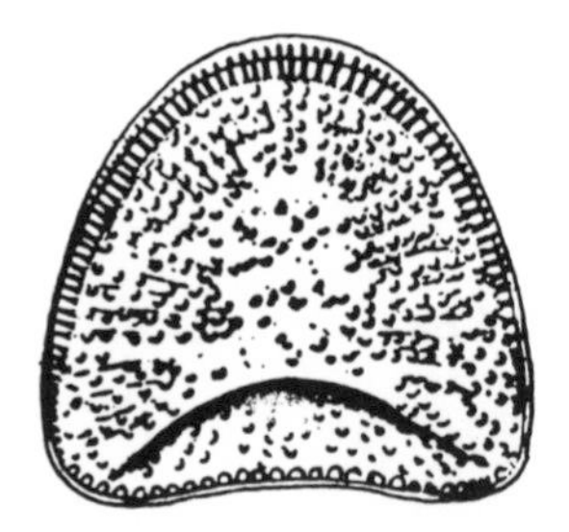

FIGURE 88

FIGURE 89 A

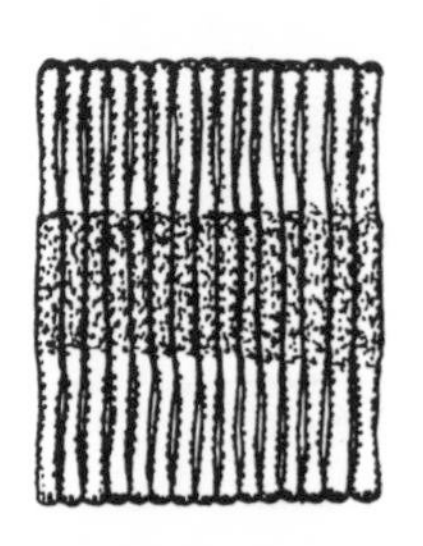

FIGURE 89 B

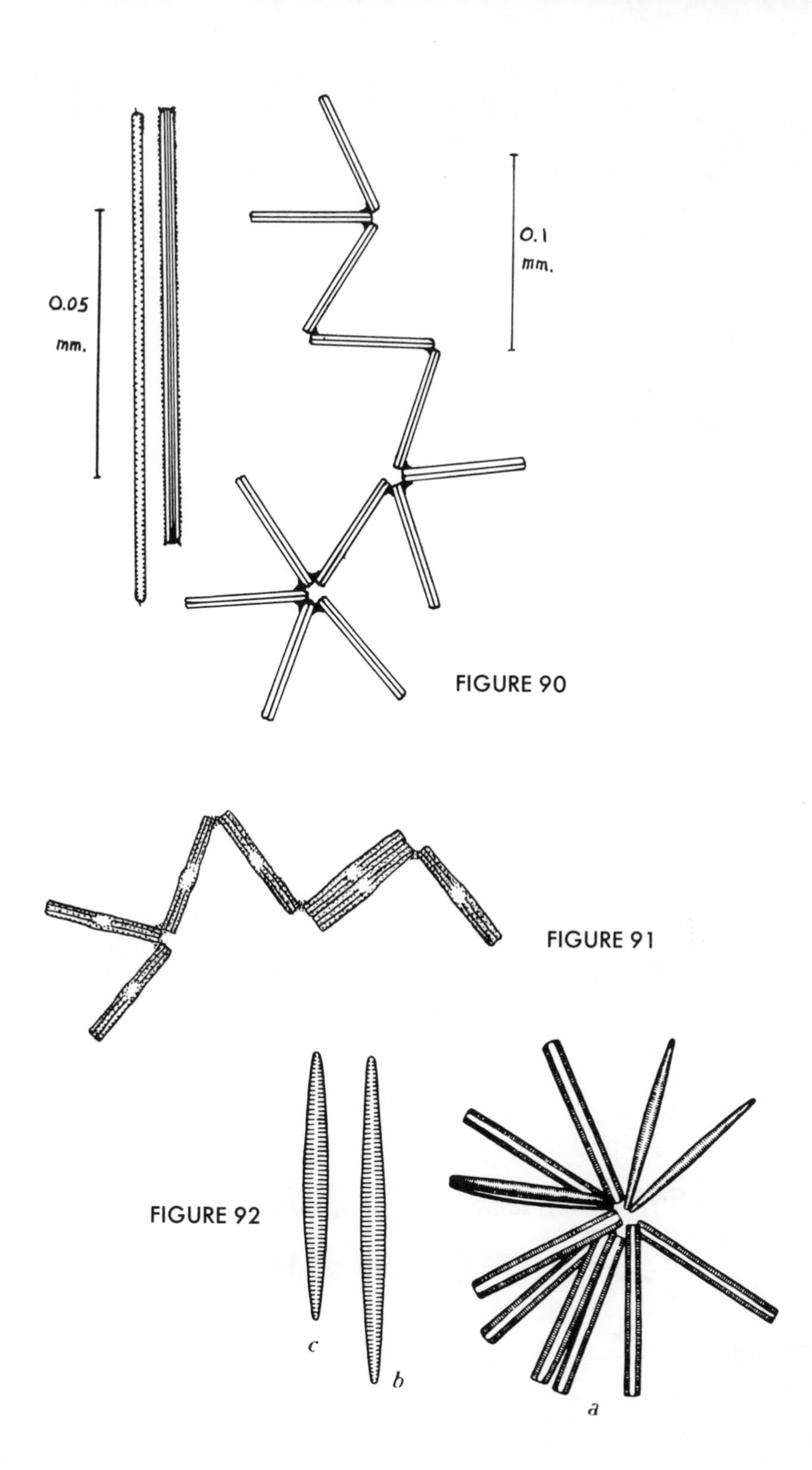

FIGURE 90

FIGURE 91

FIGURE 92

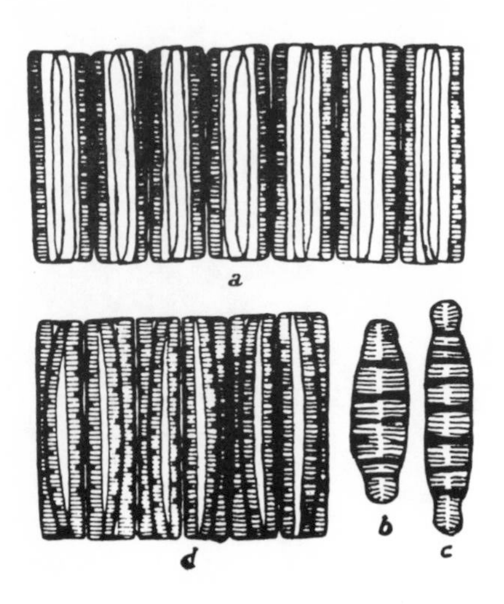

FIGURE 93

FIGURE 94

0.025 mm.

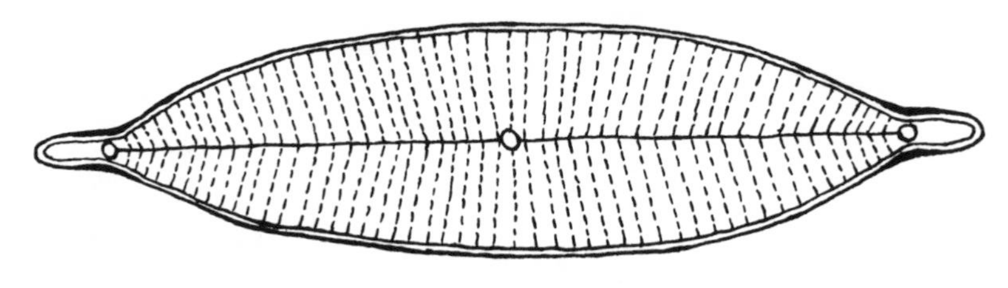

FIGURE 95

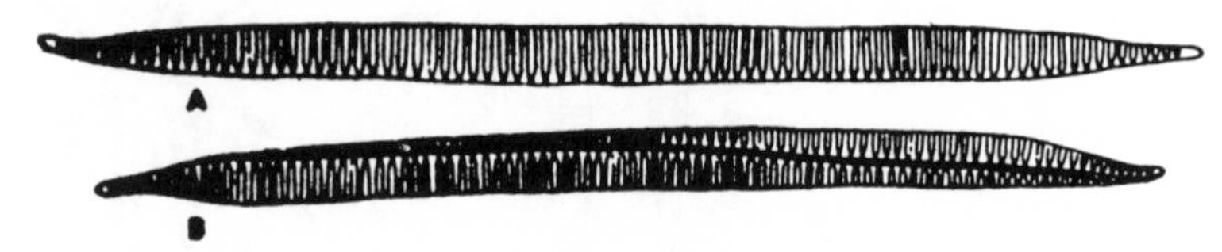

FIGURE 96

0.1 mm.

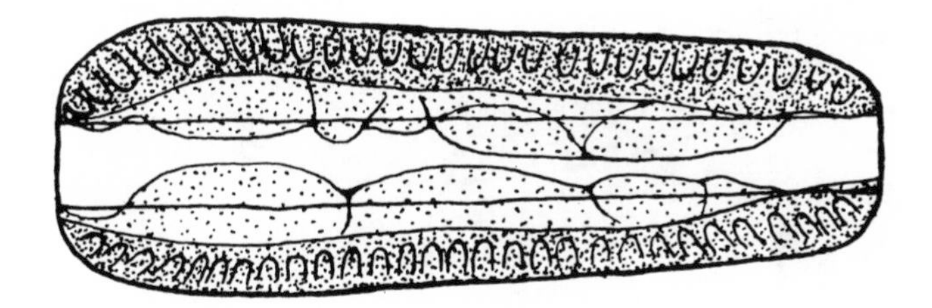

FIGURE 97

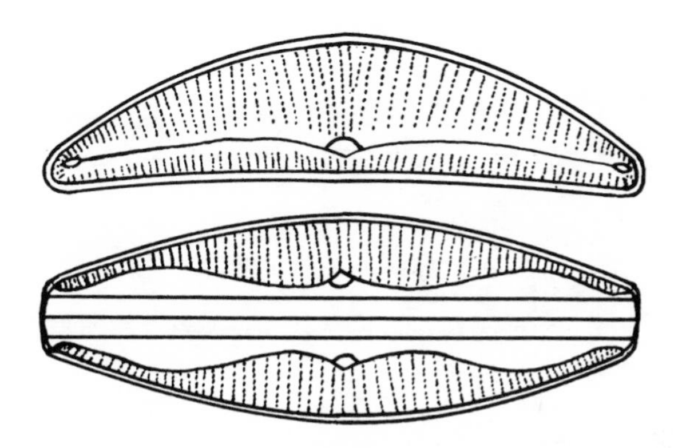

FIGURE 98

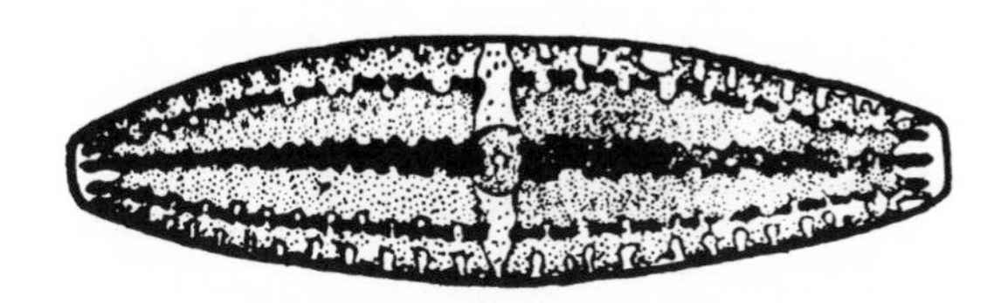

FIGURE 99

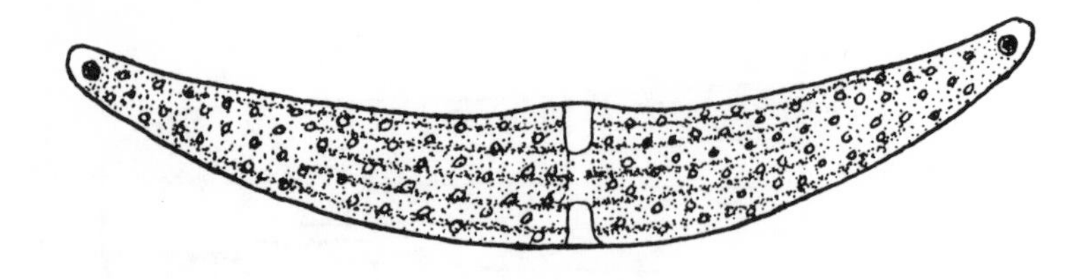

FIGURE 100

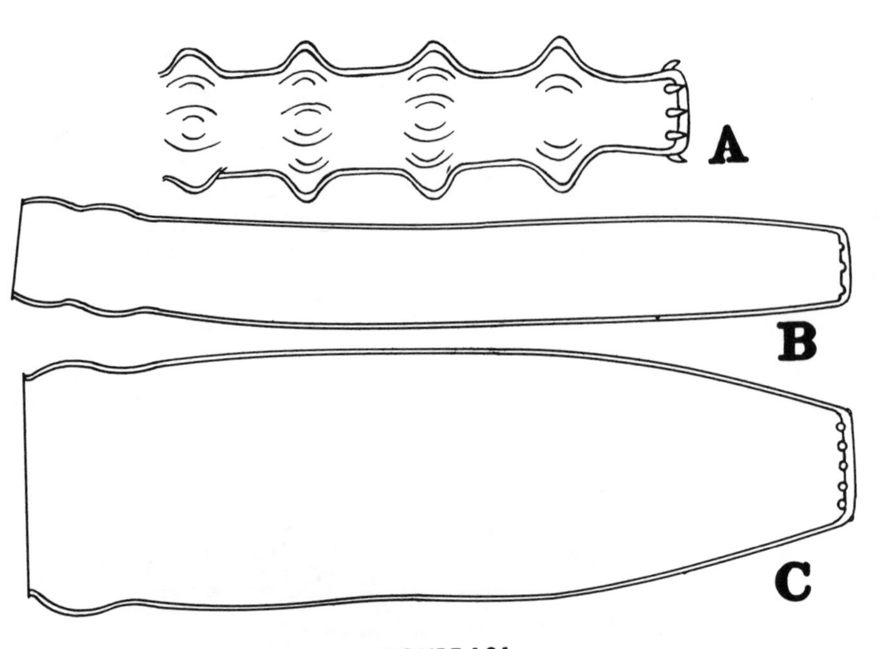

FIGURE 101

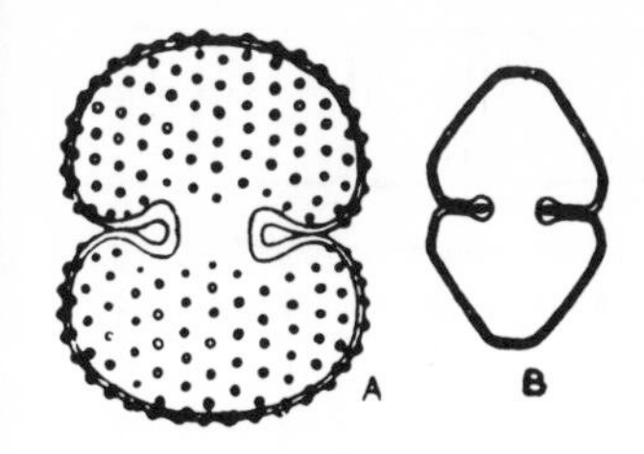

FIGURE 102

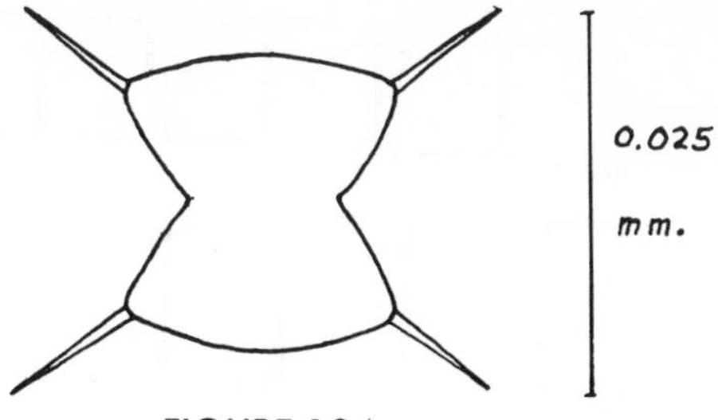

FIGURE 104

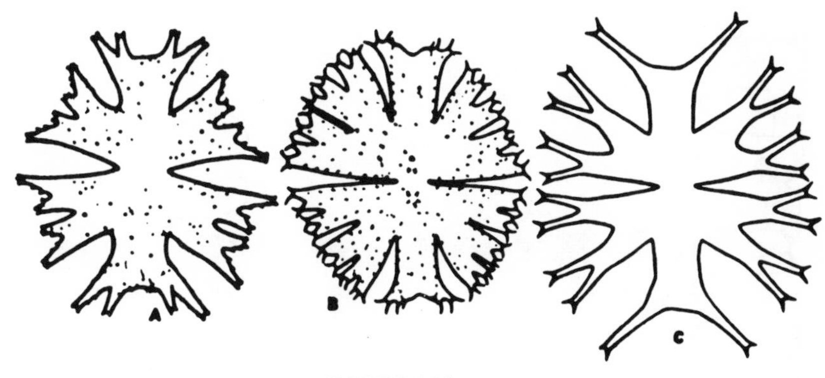

FIGURE 103

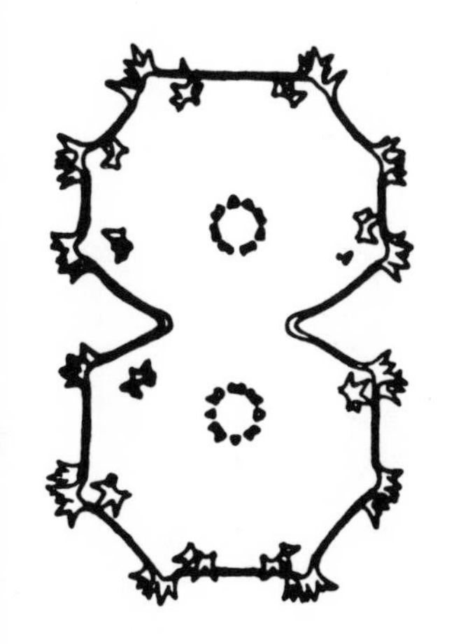

FIGURE 105

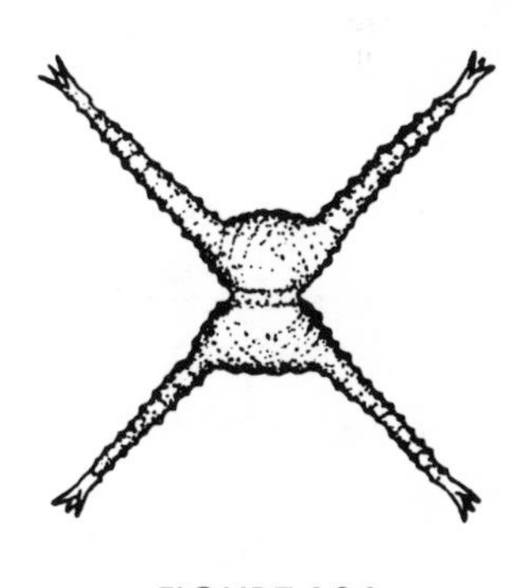

FIGURE 106

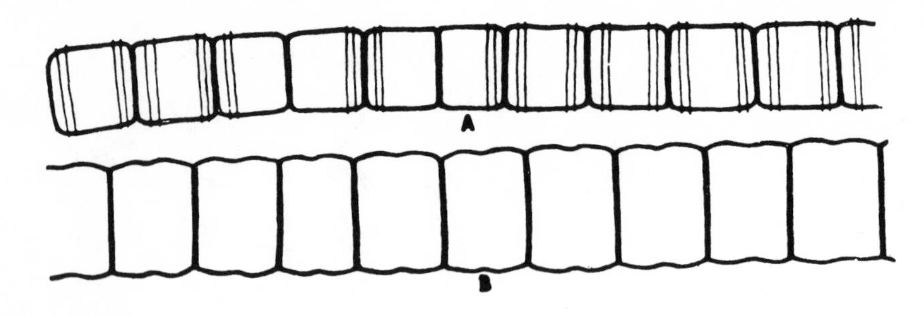

FIGURE 107

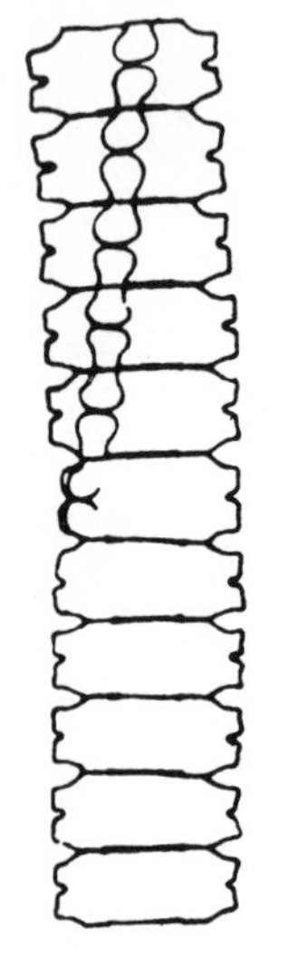

FIGURE 108

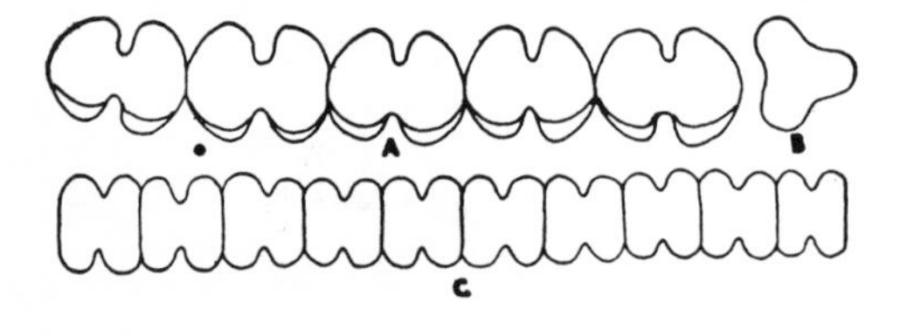

FIGURE 110

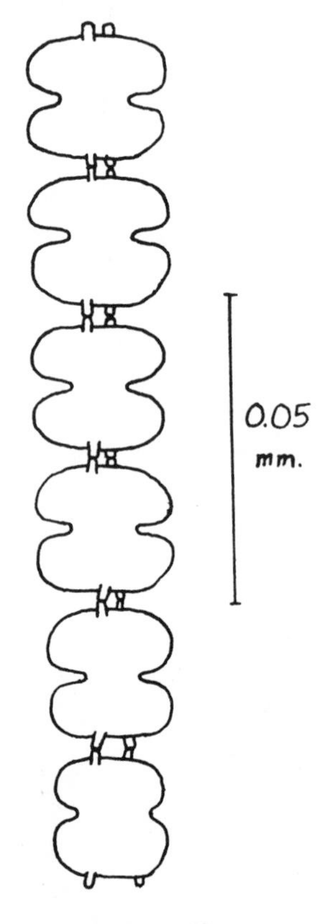

FIGURE 109

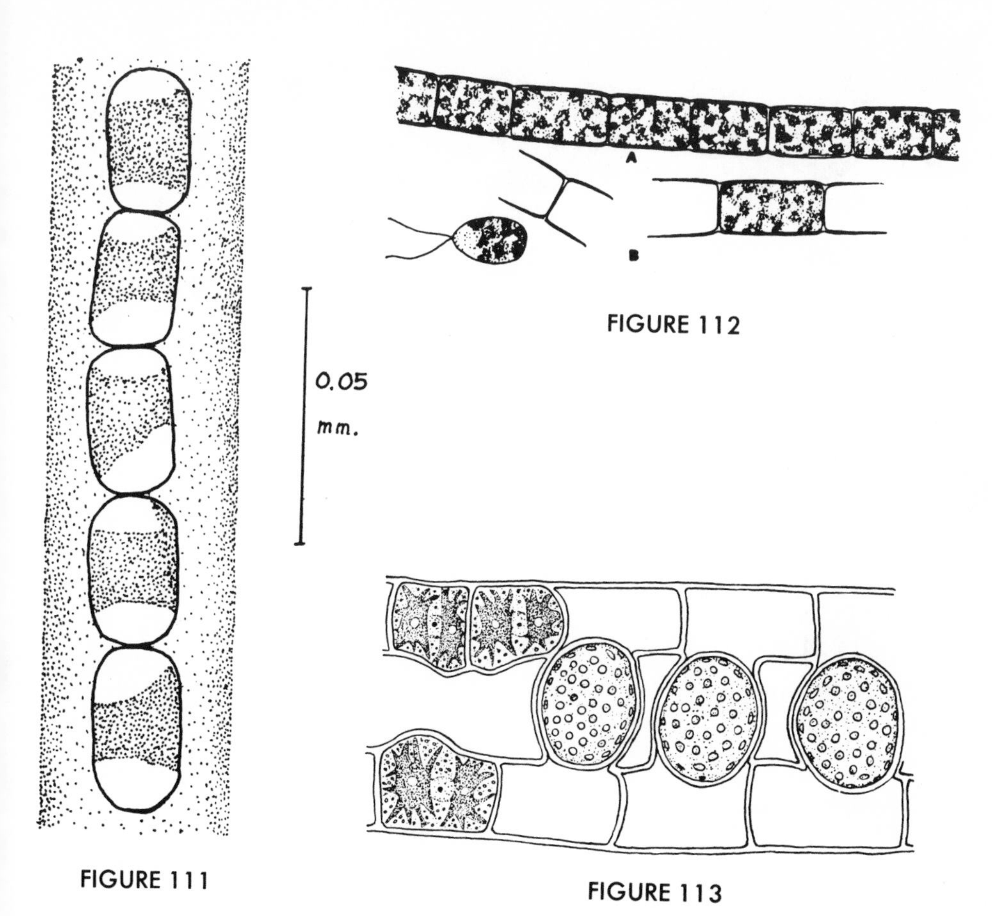

FIGURE 111

FIGURE 112

FIGURE 113

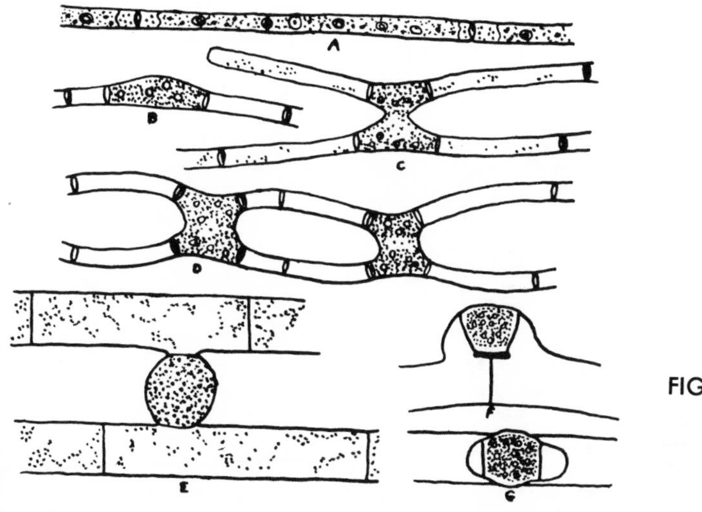

FIGURE 114

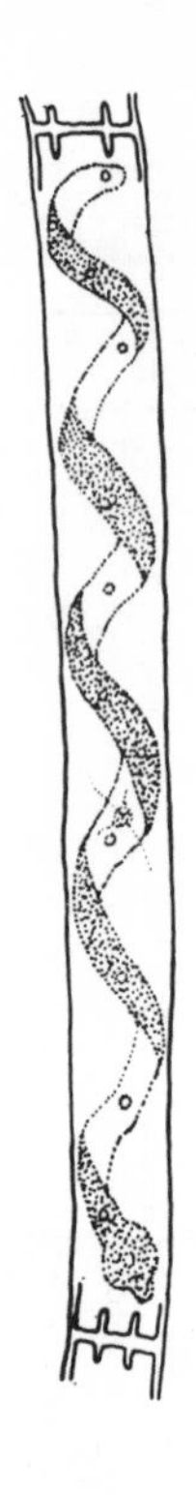

FIGURE 115

FIGURE 116

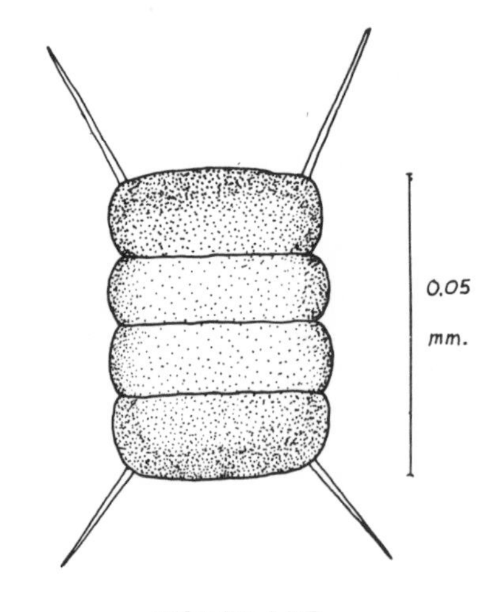

FIGURE 117

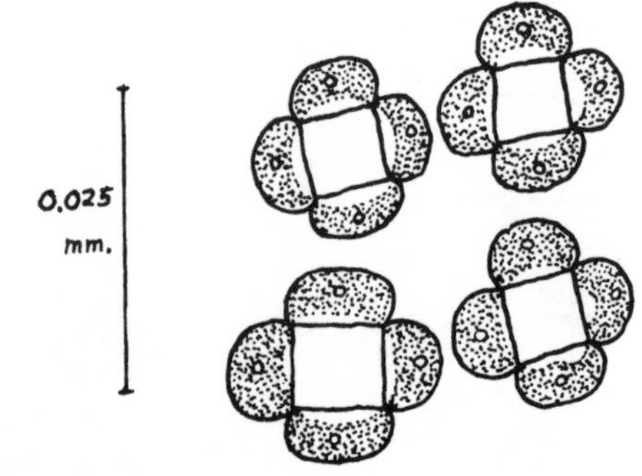

FIGURE 118

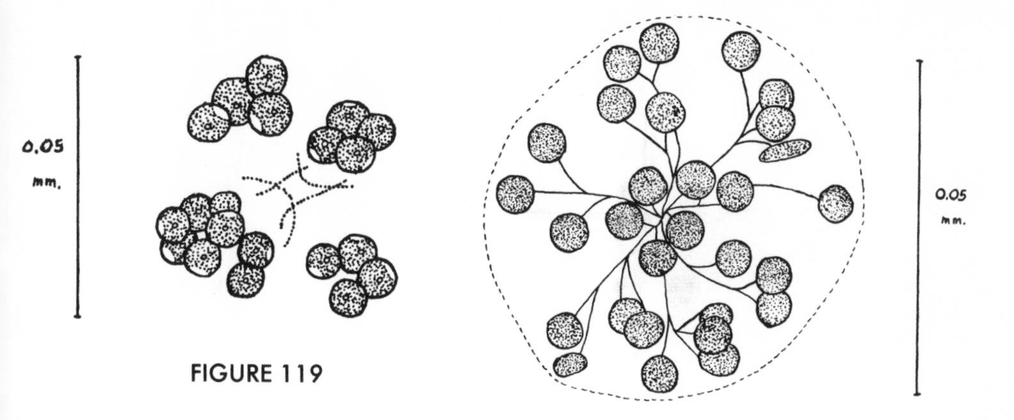

FIGURE 119

FIGURE 120

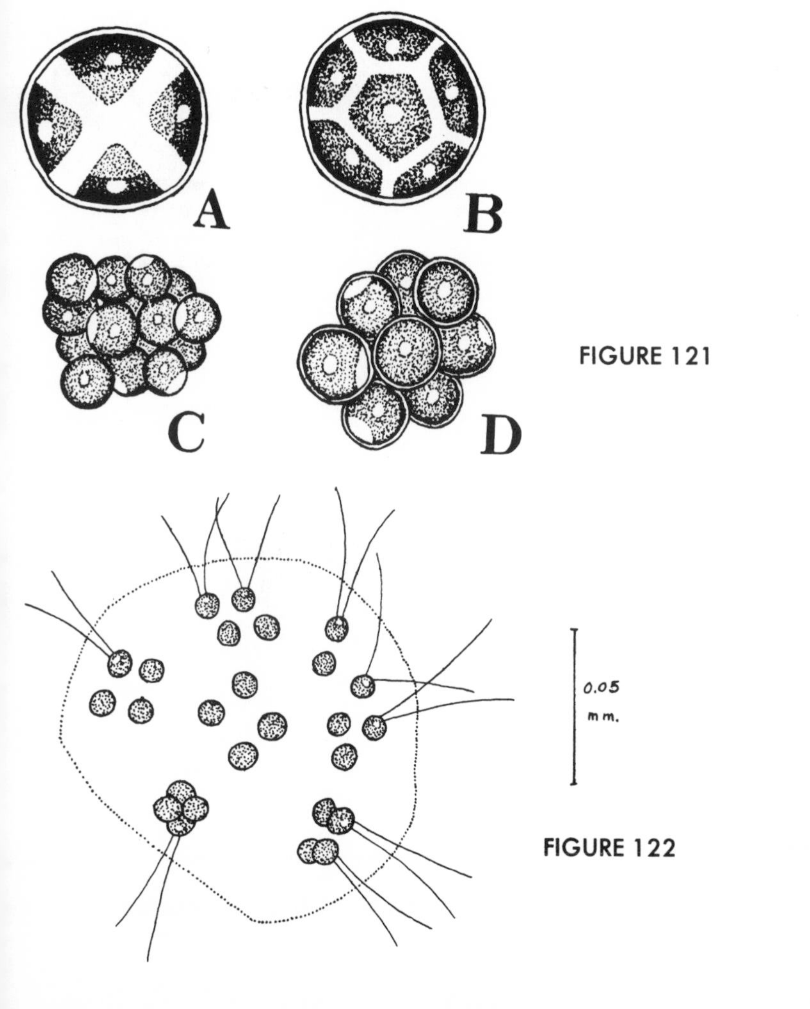

FIGURE 121

FIGURE 122

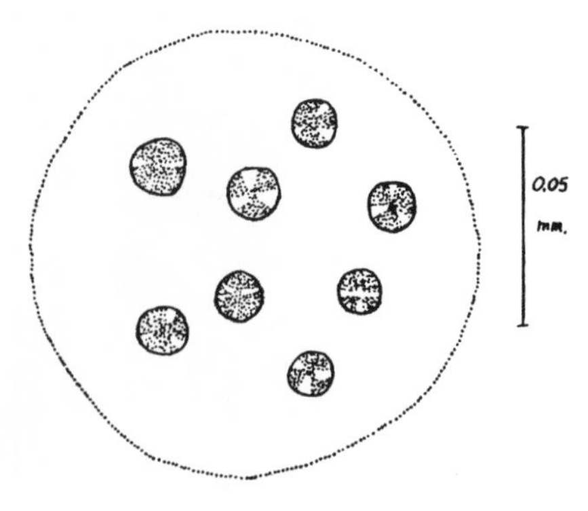

FIGURE 123

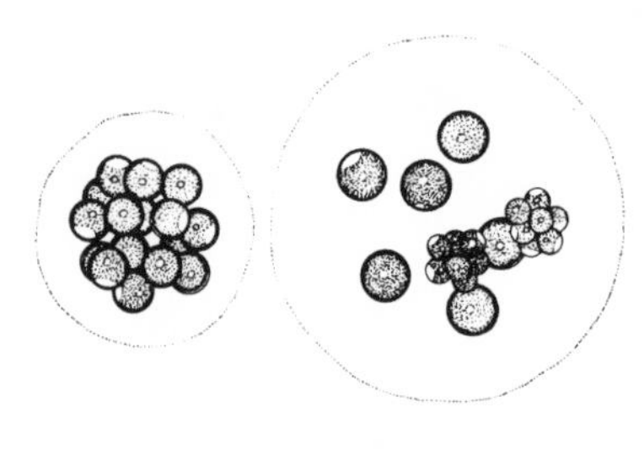

FIGURE 124

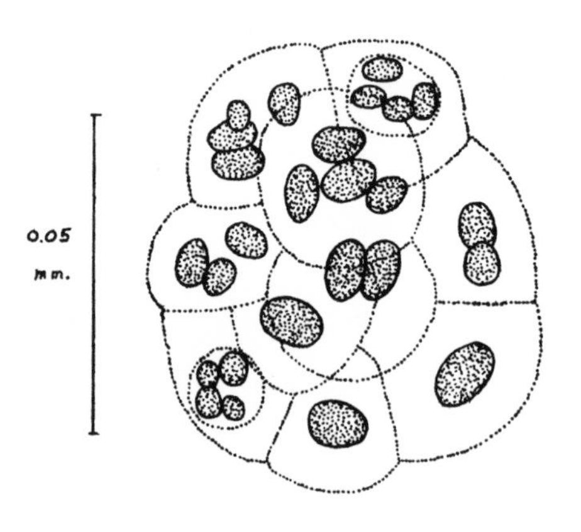

FIGURE 125

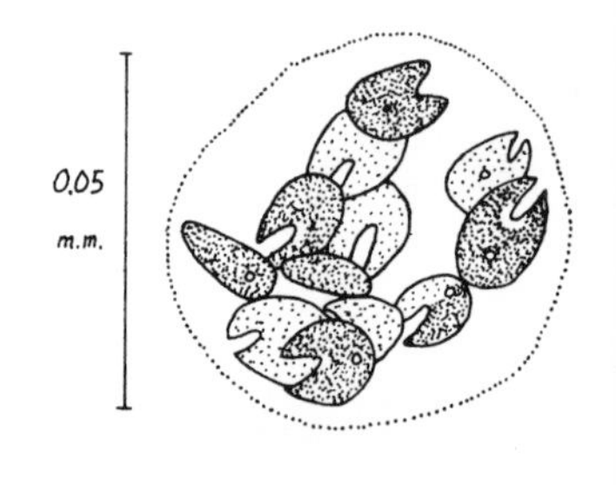

FIGURE 126

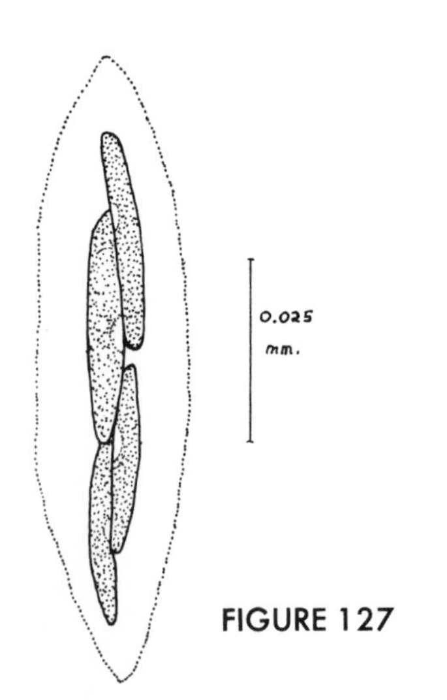

FIGURE 127

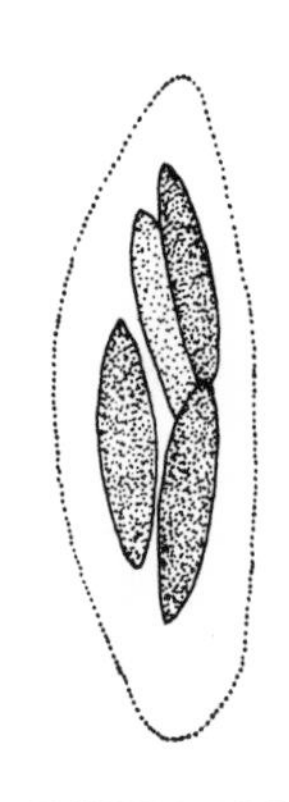

FIGURE 128

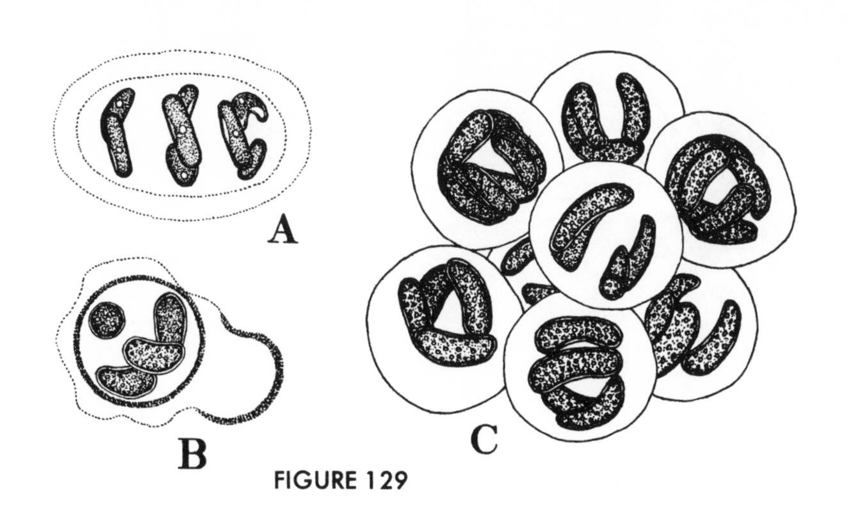

FIGURE 129

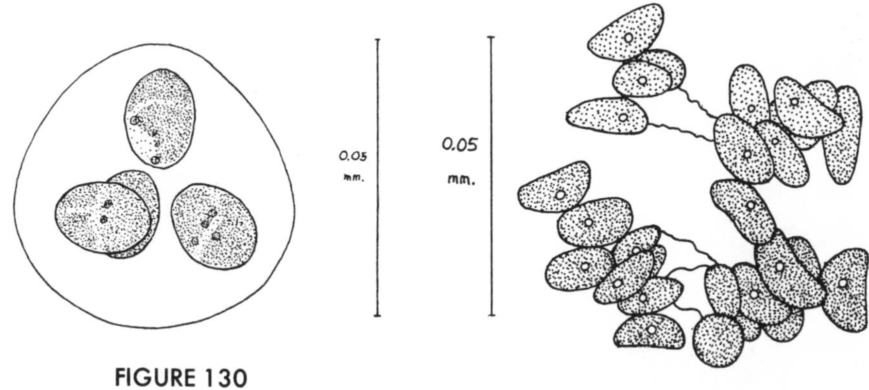

FIGURE 130

FIGURE 131

FIGURE 132

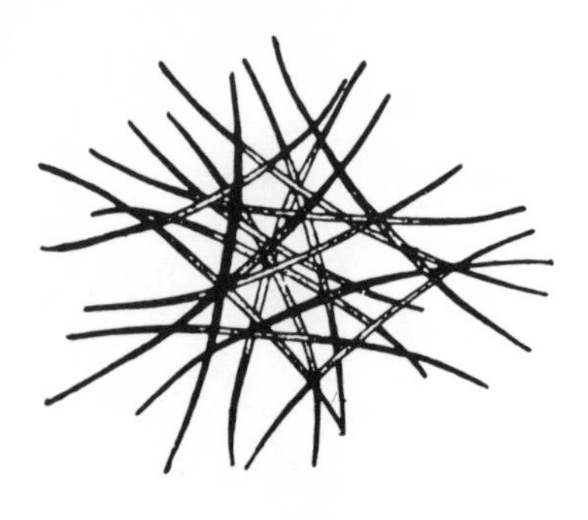

FIGURE 133

FIGURE 134

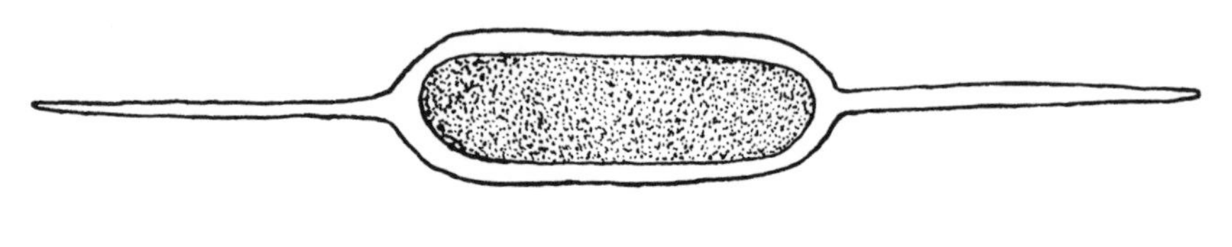

FIGURE 135

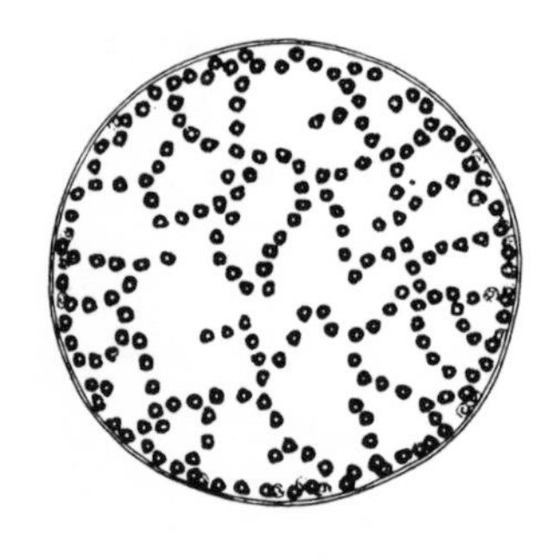

FIGURE 136

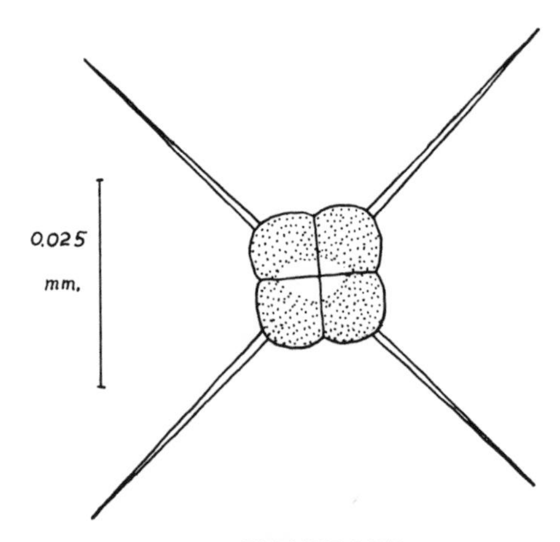

FIGURE 137

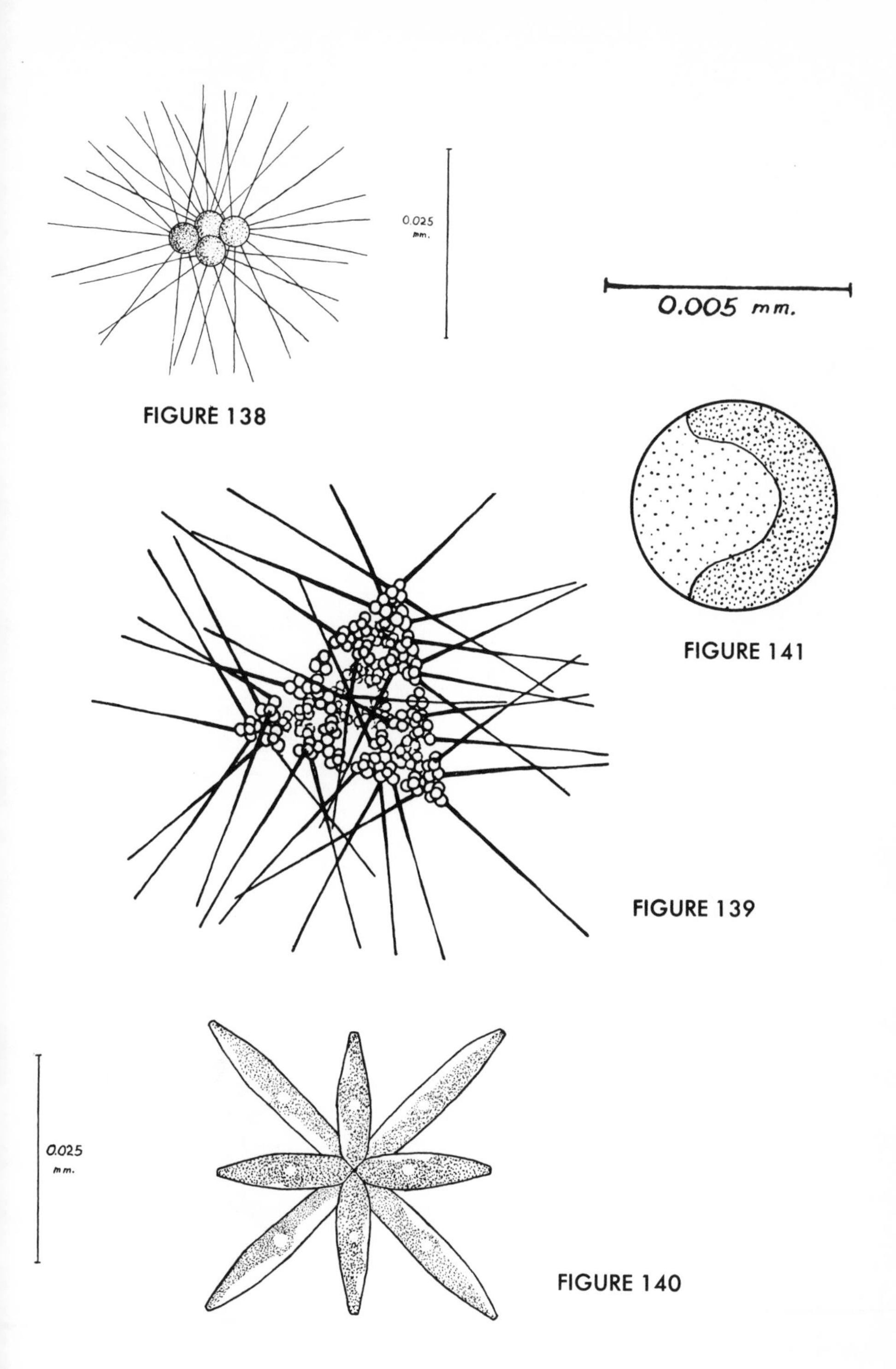

FIGURE 138

FIGURE 141

FIGURE 139

FIGURE 140

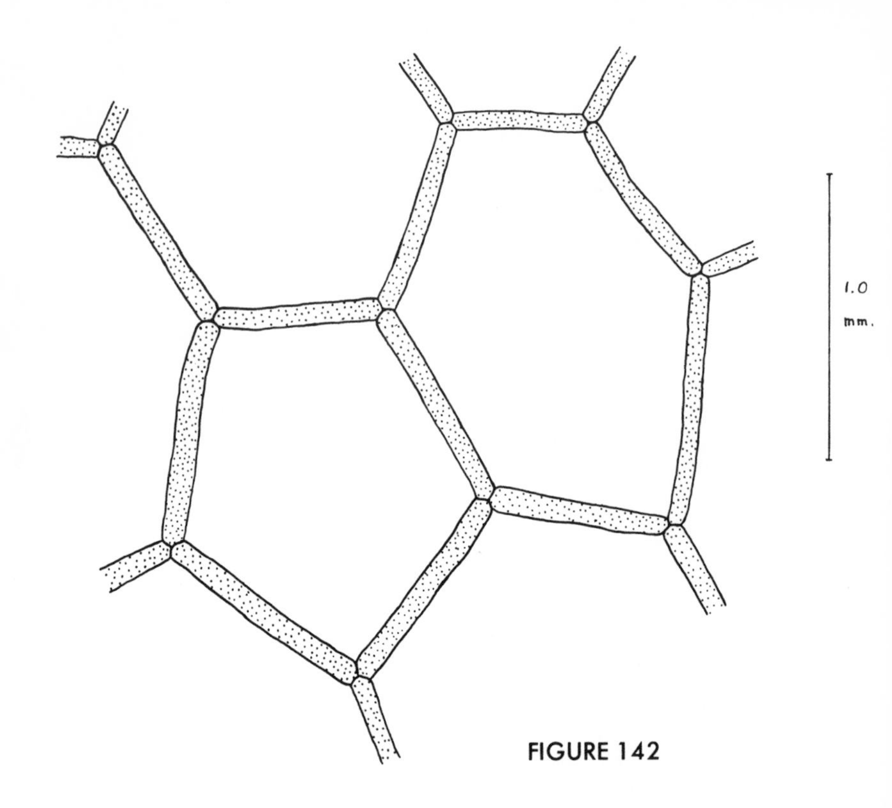

FIGURE 142

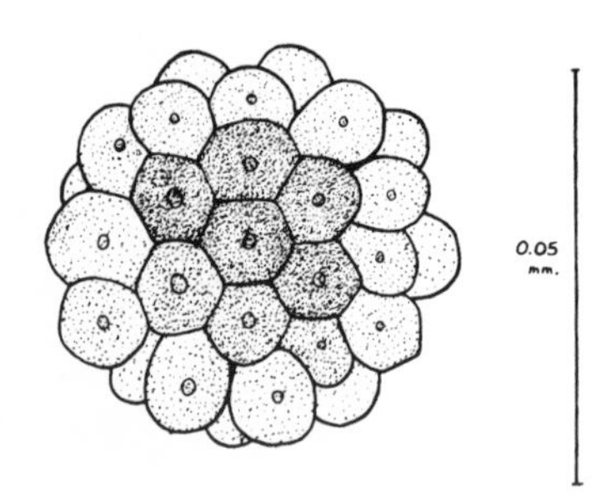

FIGURE 143

FIGURE 144

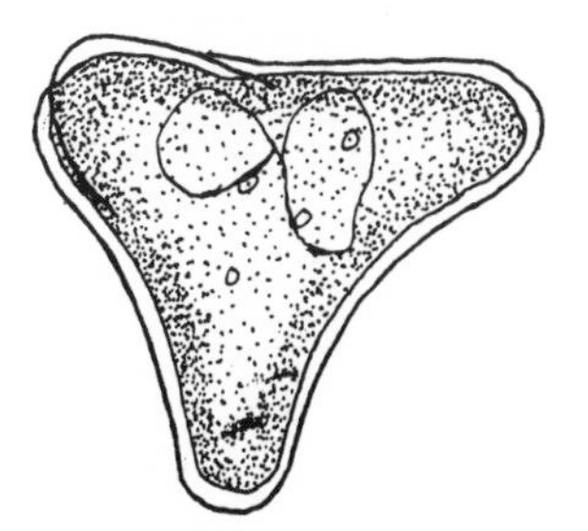

FIGURE 145 A

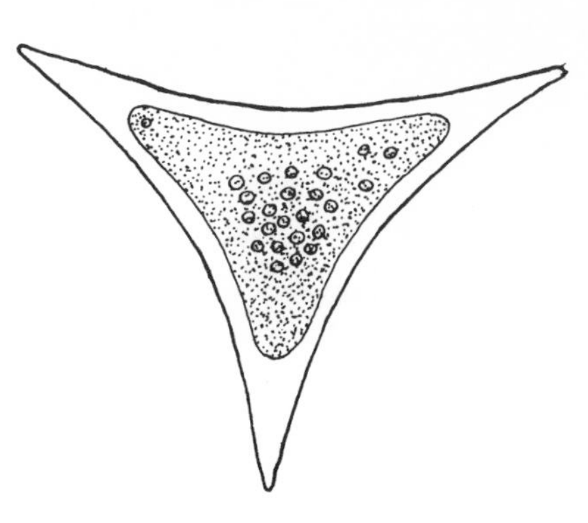
FIGURE 145 B

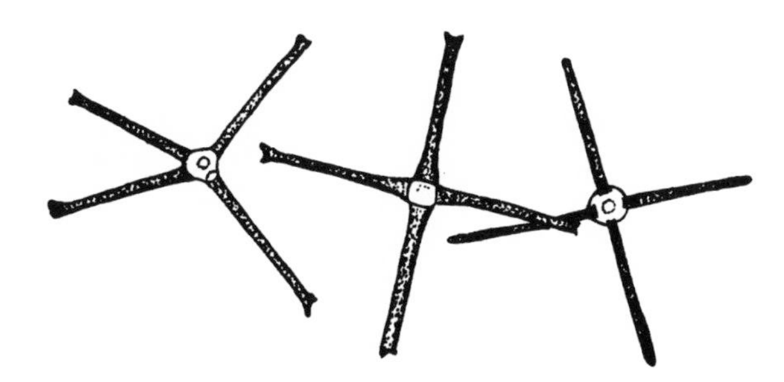
FIGURE 146

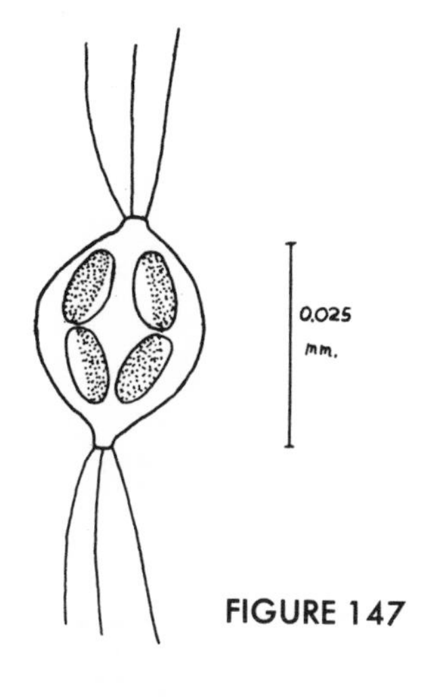

FIGURE 147

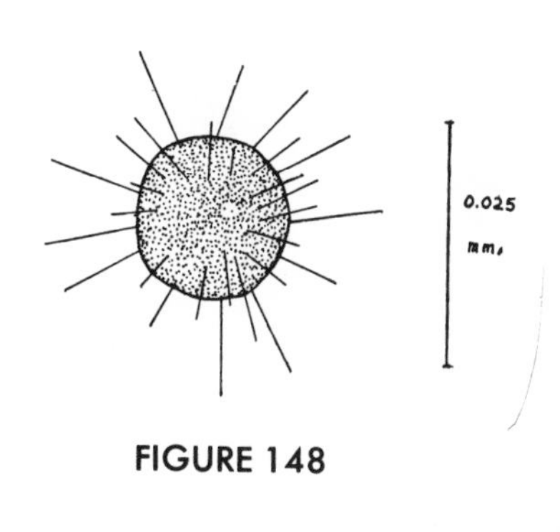

FIGURE 148

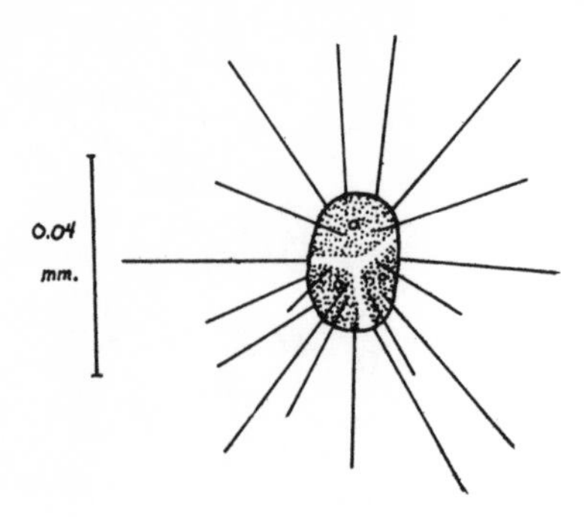

FIGURE 149

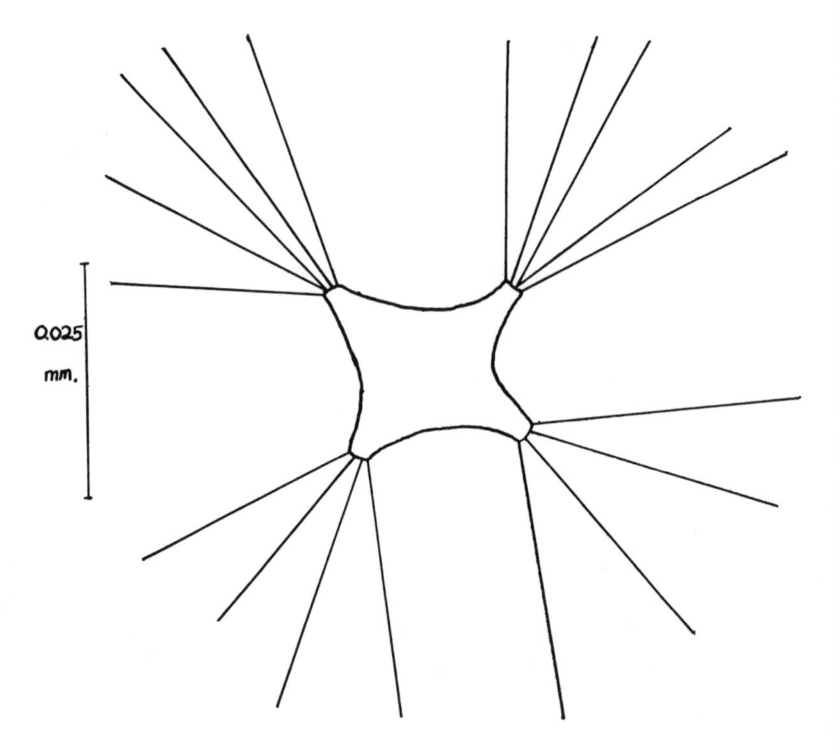

FIGURE 150

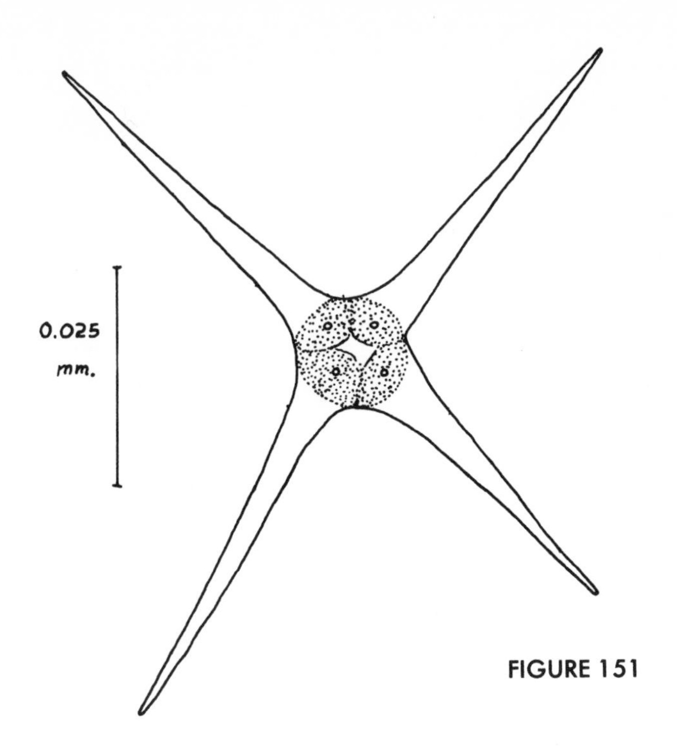

FIGURE 151

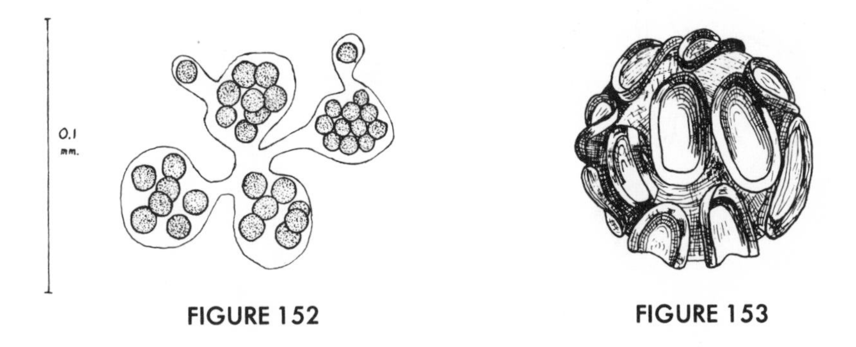

FIGURE 152

FIGURE 153

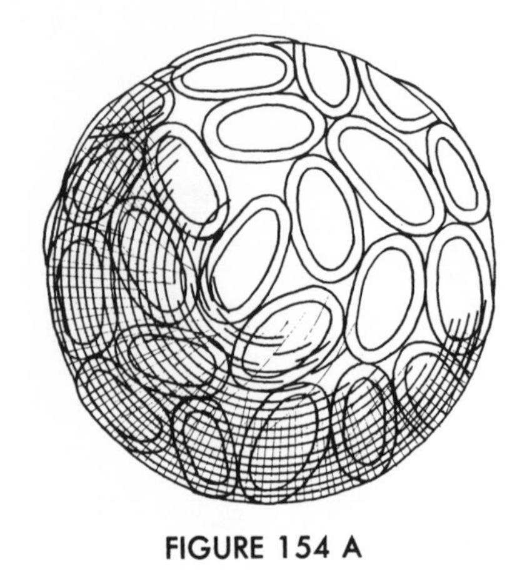

FIGURE 154 A

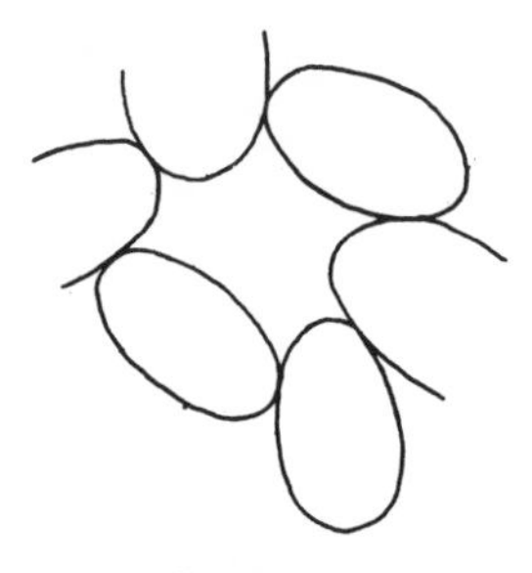

FIGURE 154 B

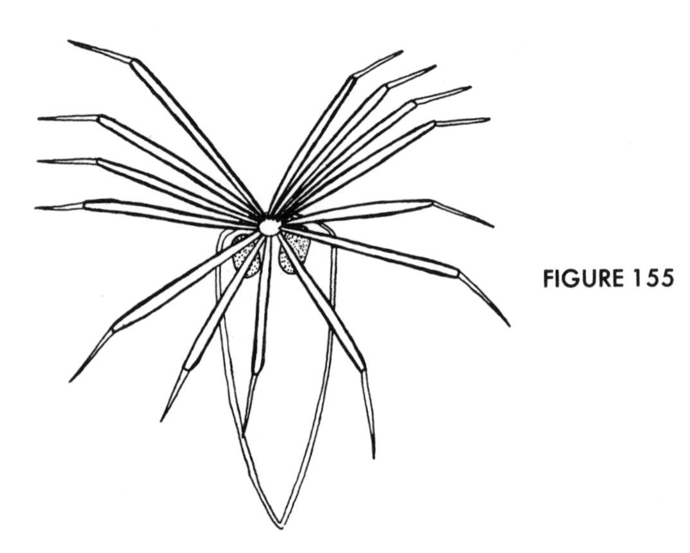

FIGURE 155

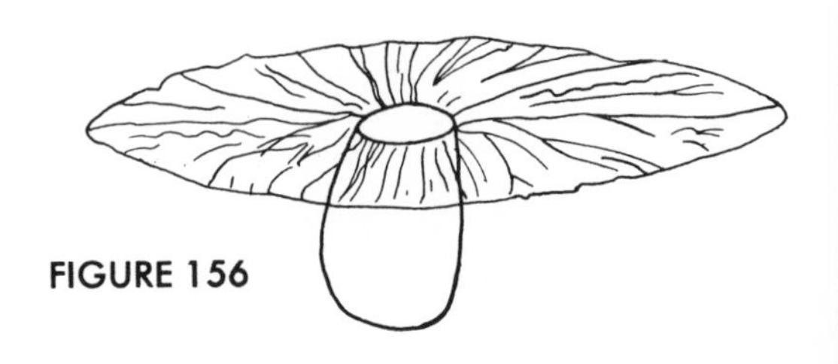

FIGURE 156

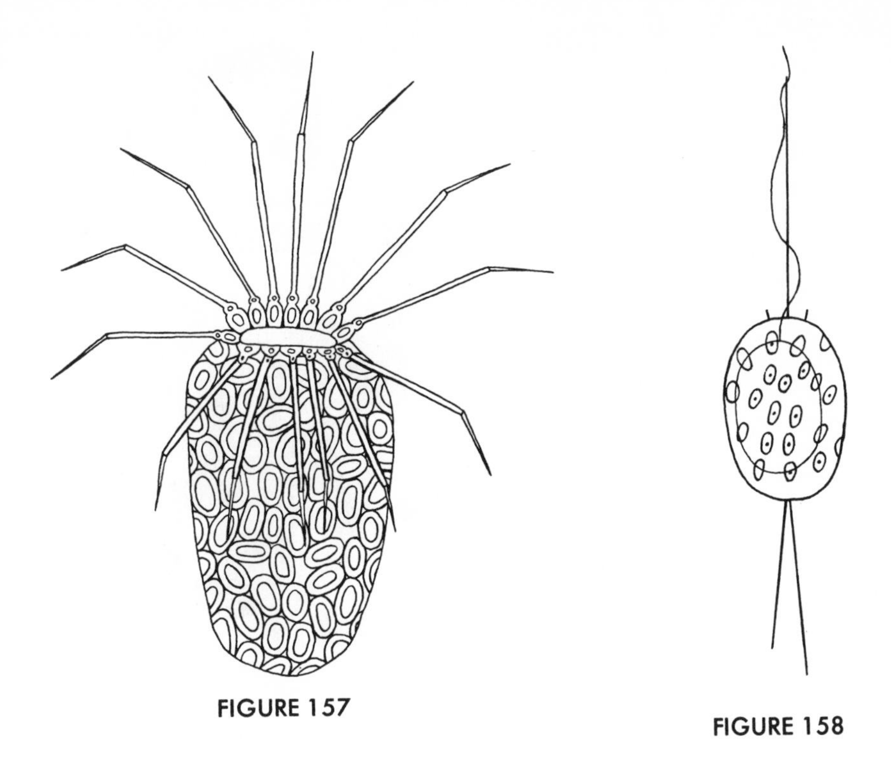

FIGURE 157

FIGURE 158

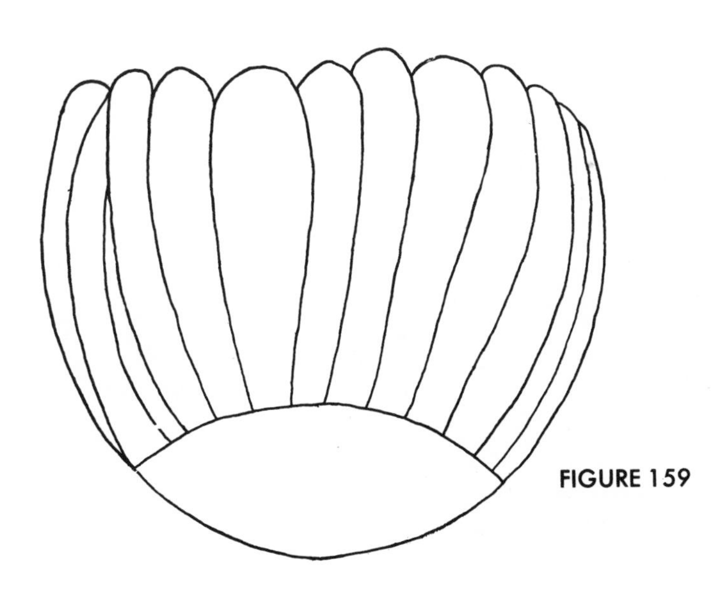

FIGURE 159

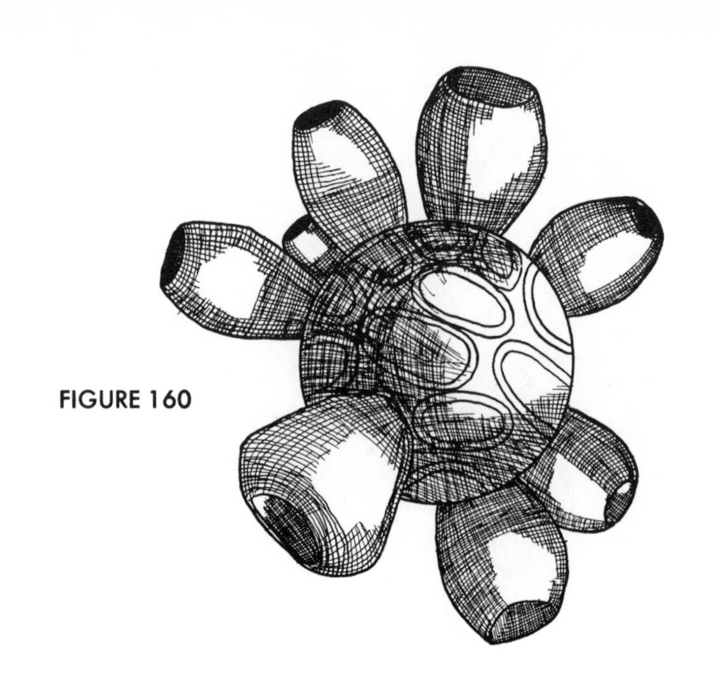

FIGURE 160

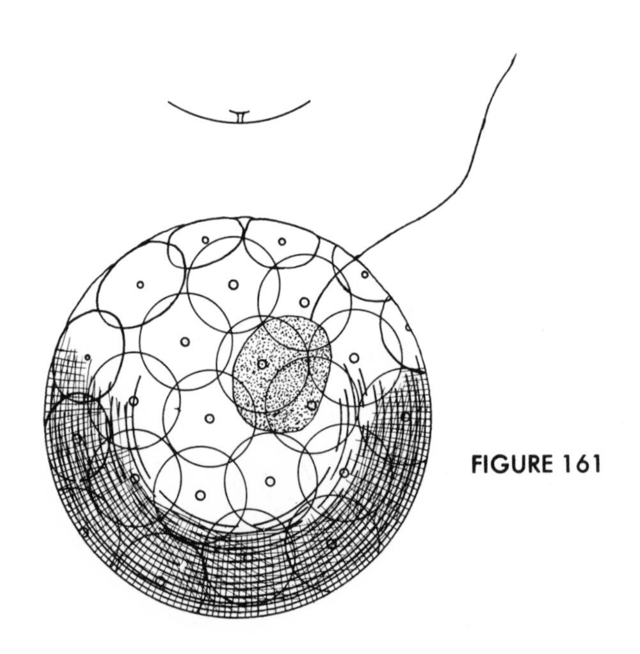

FIGURE 161

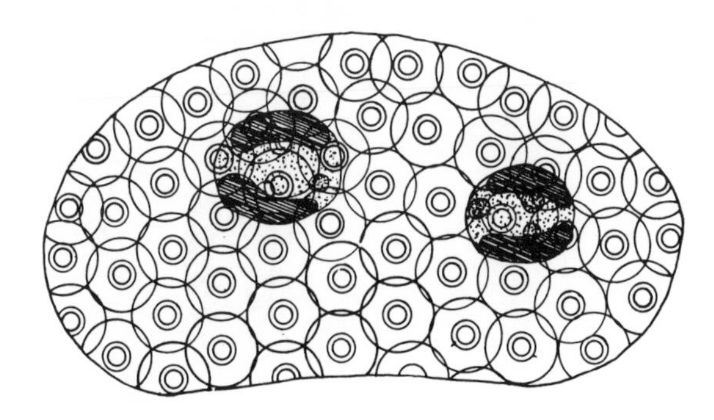

FIGURE 162

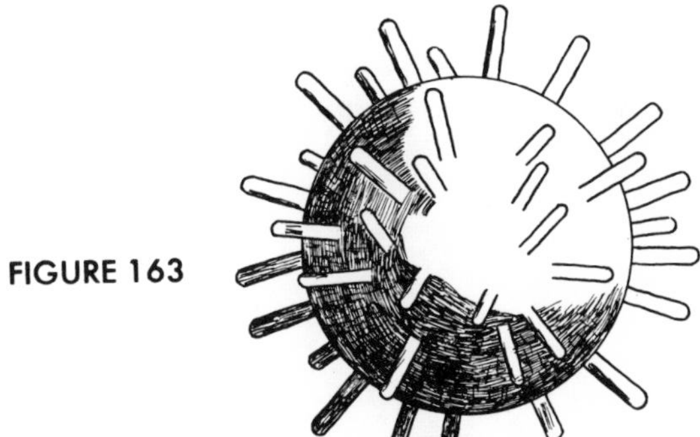

FIGURE 163

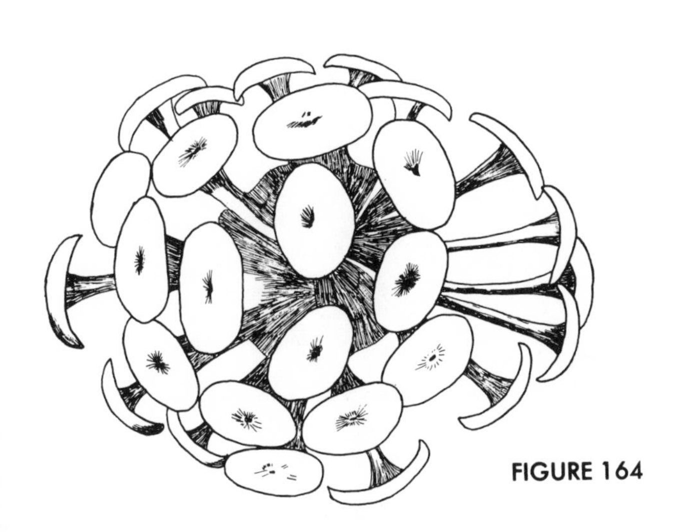

FIGURE 164

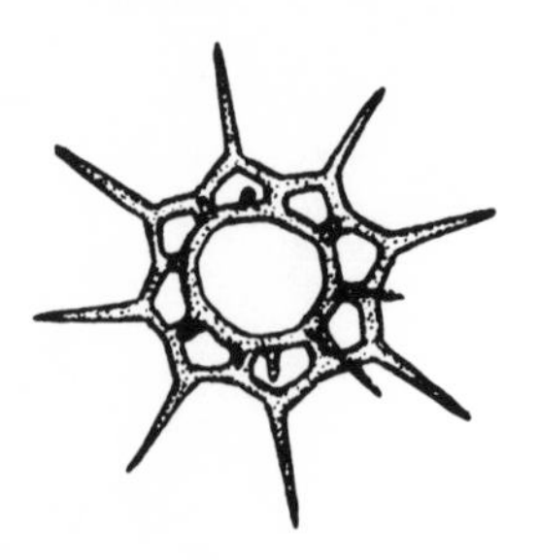

FIGURE 165

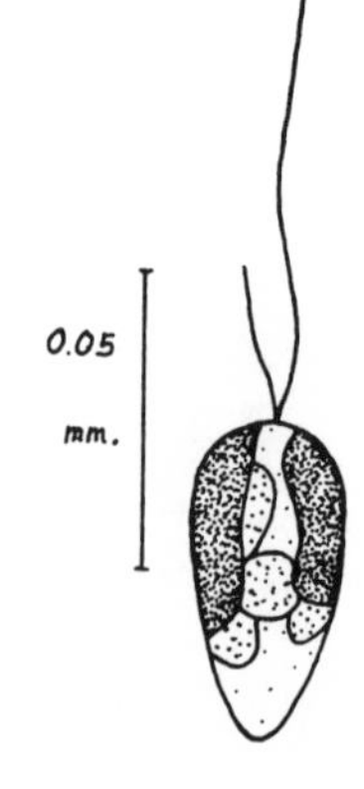

FIGURE 167

FIGURE 166

FIGURE 168

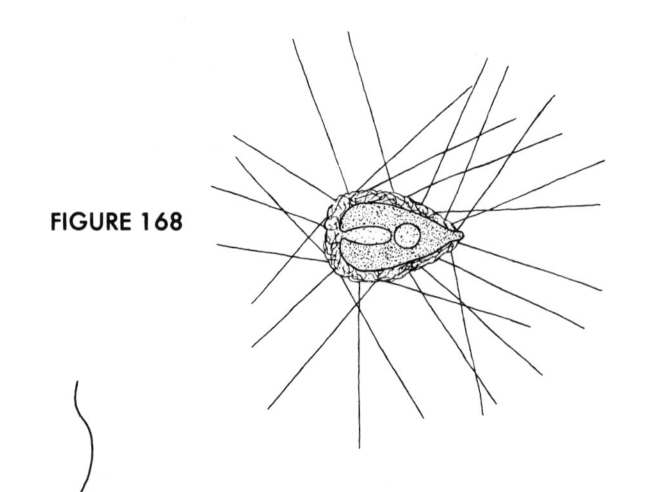

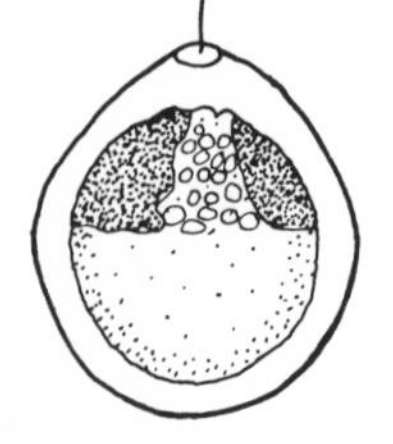

FIGURE 169

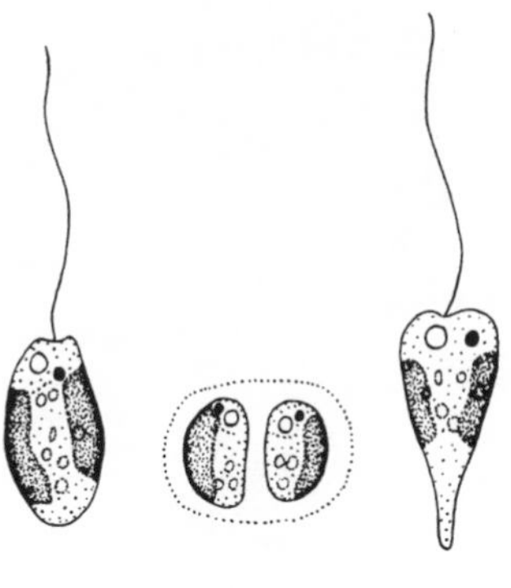

FIGURE 170

FIGURE 171

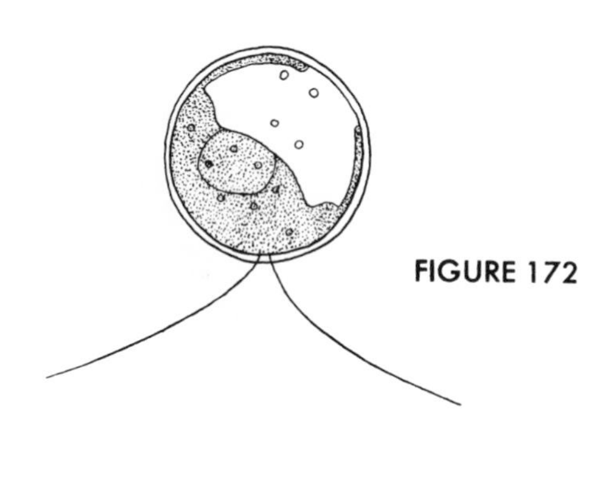

FIGURE 172

0.025 mm.

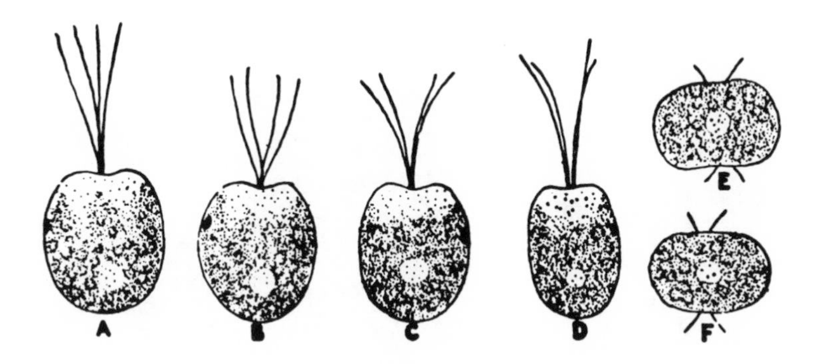

FIGURE 173

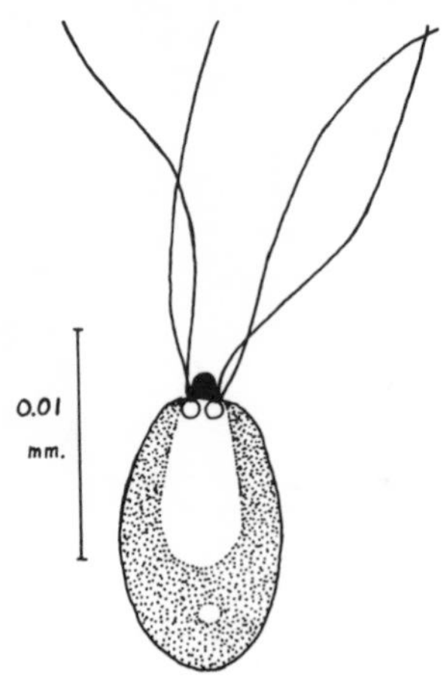

FIGURE 174

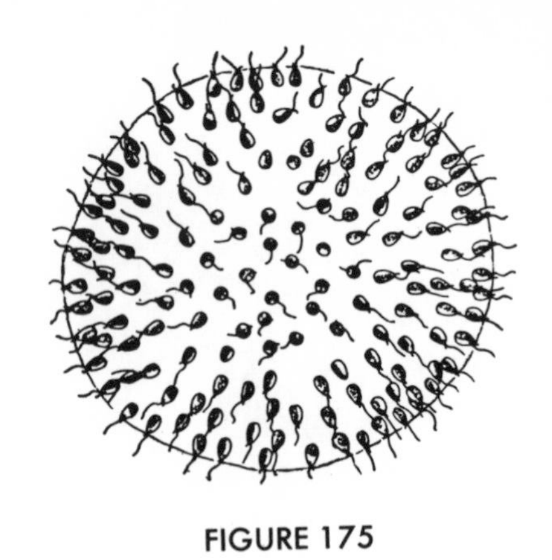

FIGURE 175

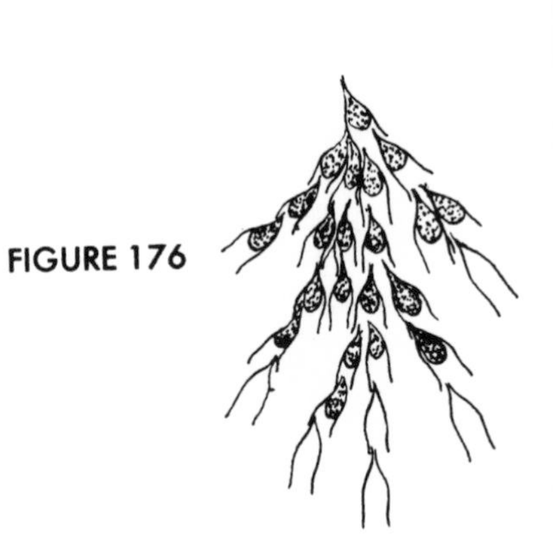

FIGURE 176

FIGURE 177

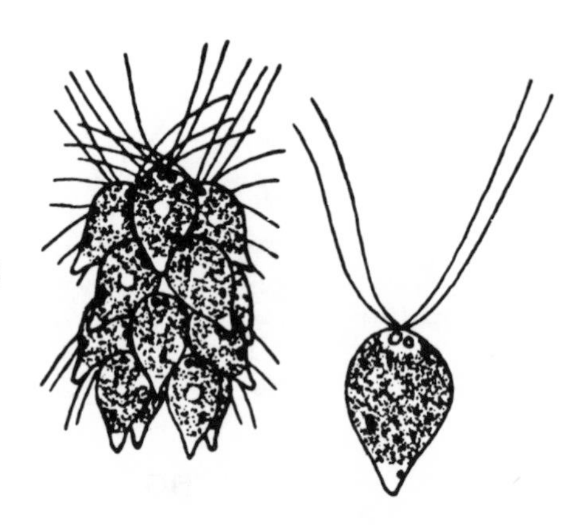

FIGURE 178

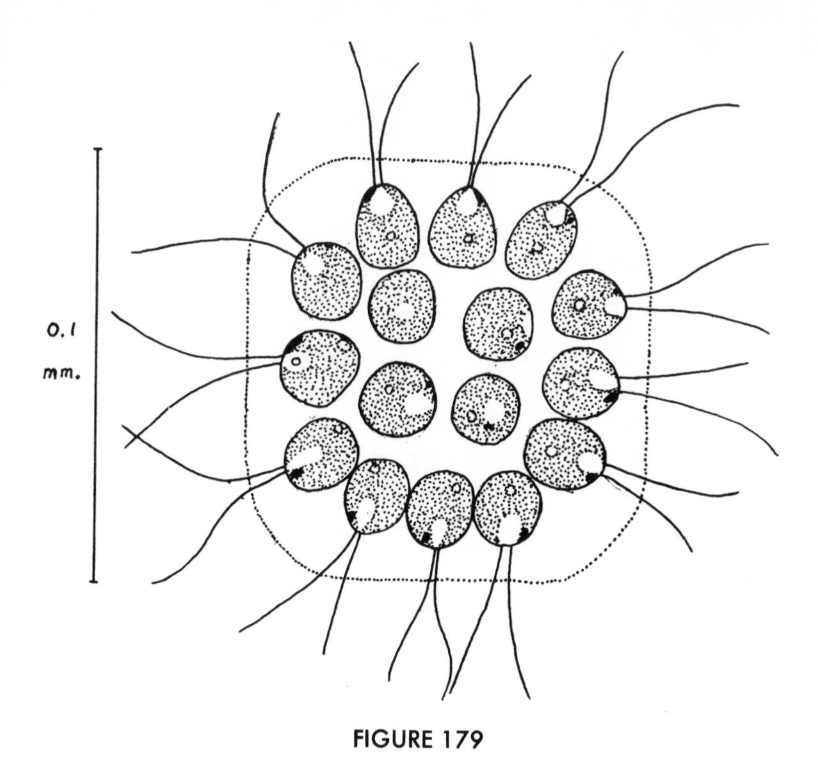

FIGURE 179

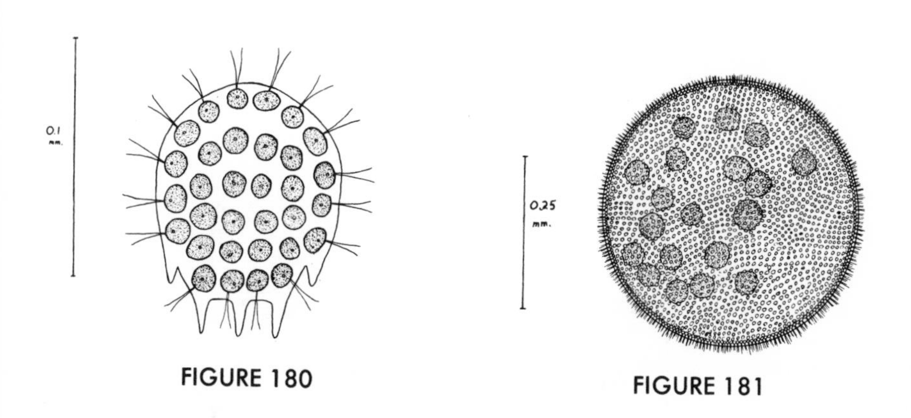

FIGURE 180

FIGURE 181

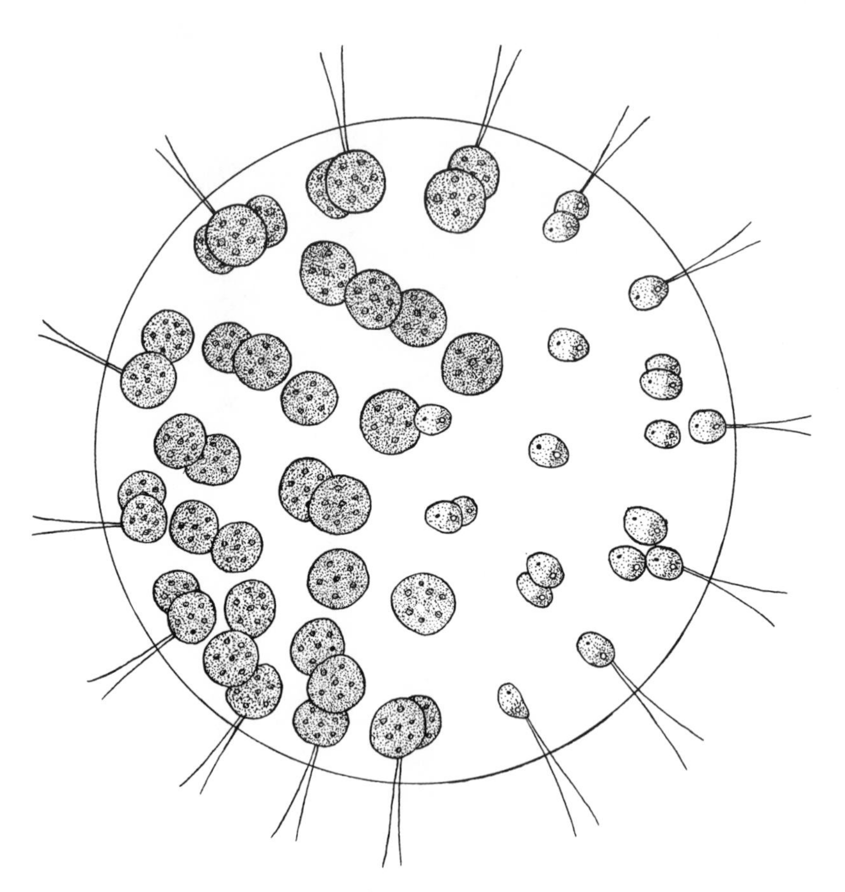

FIGURE 182

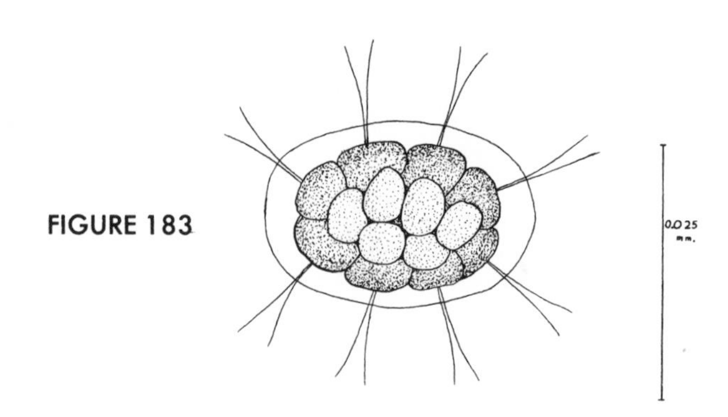

FIGURE 183

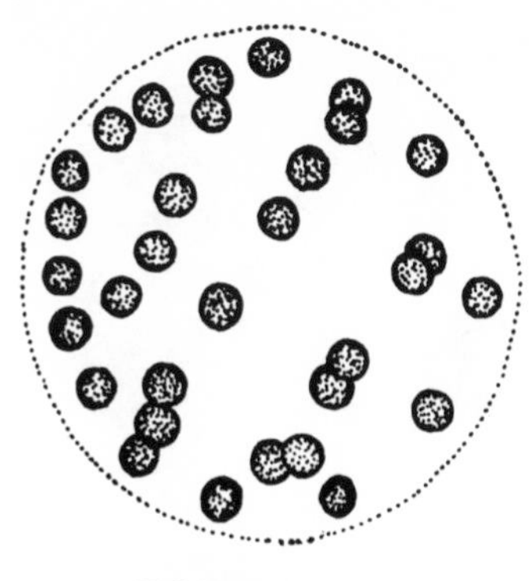

FIGURE 184

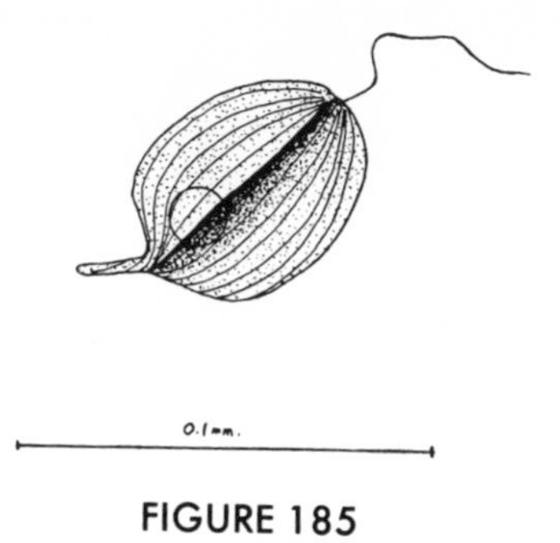

FIGURE 185

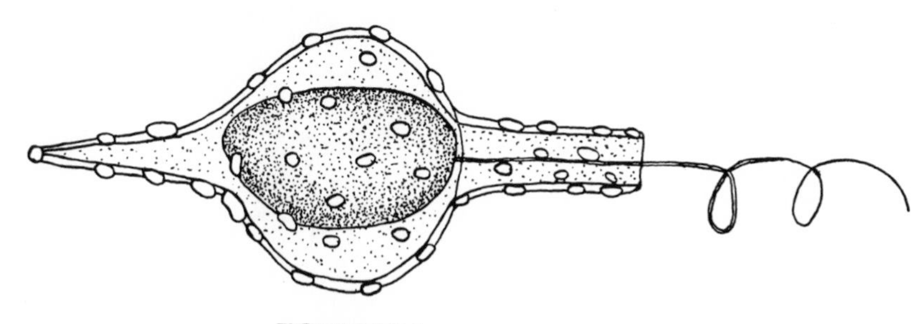

FIGURE 186

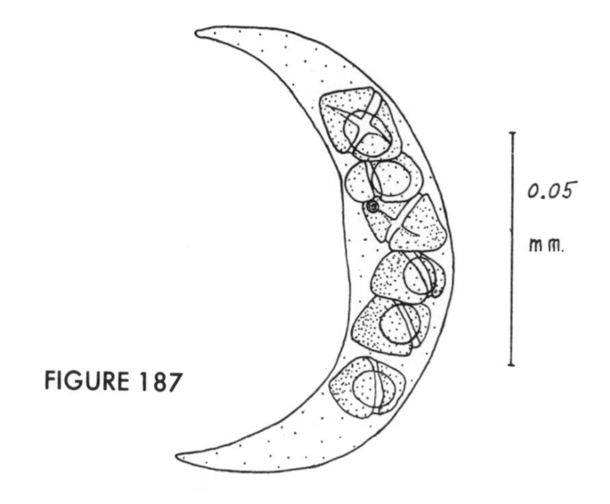

FIGURE 187

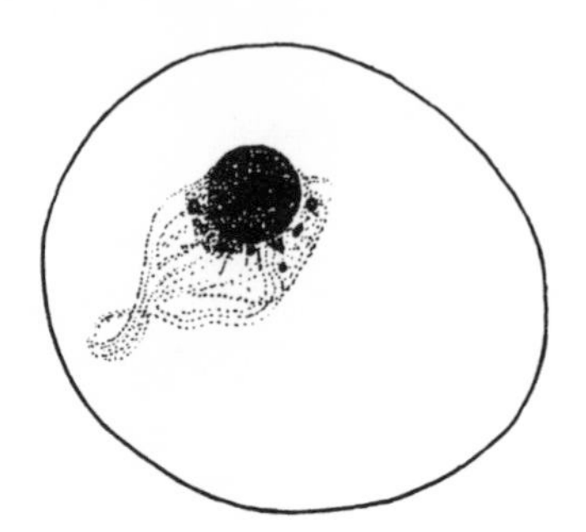

FIGURE 188

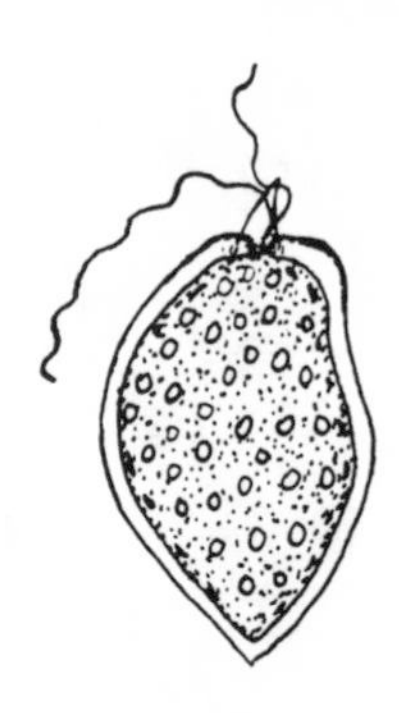

FIGURE 189

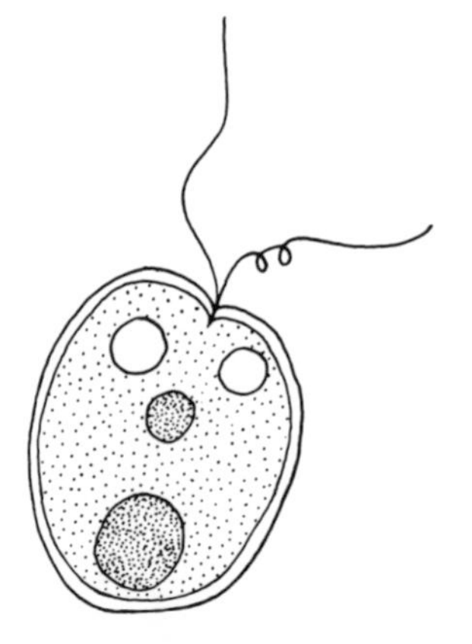

FIGURE 190

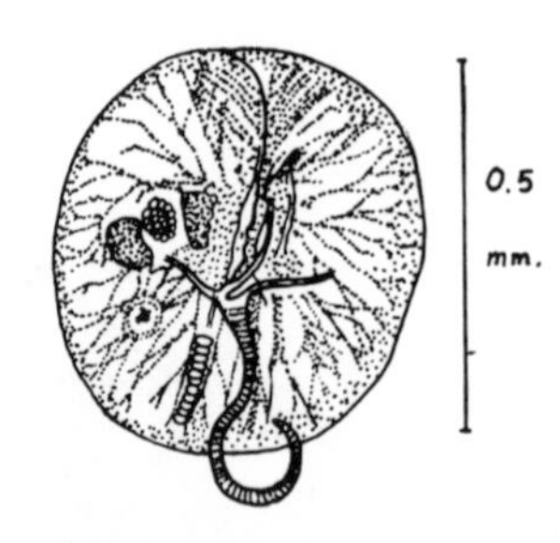

FIGURE 191

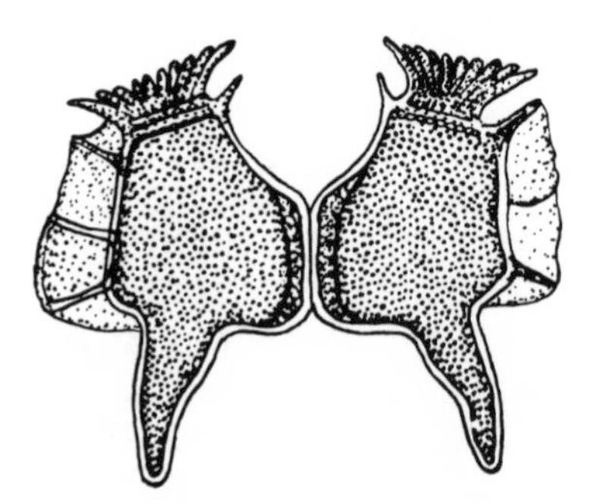

FIGURE 192

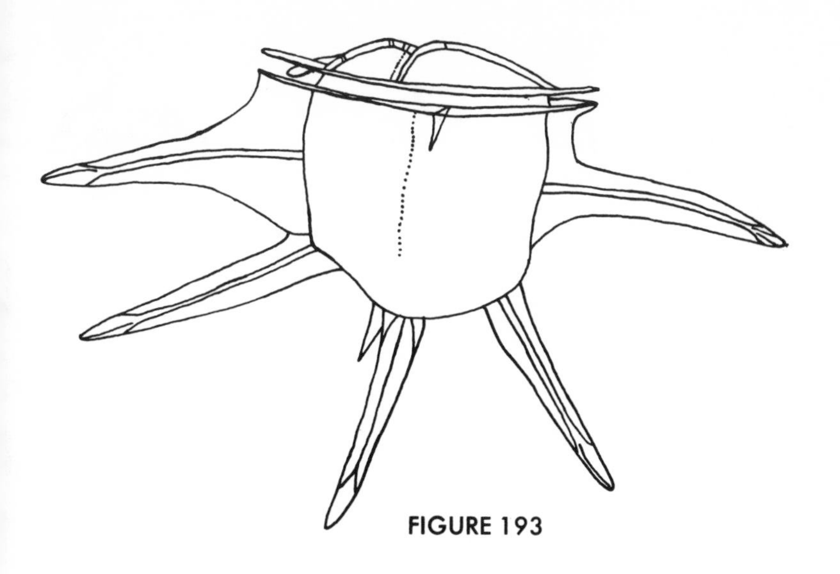

FIGURE 193

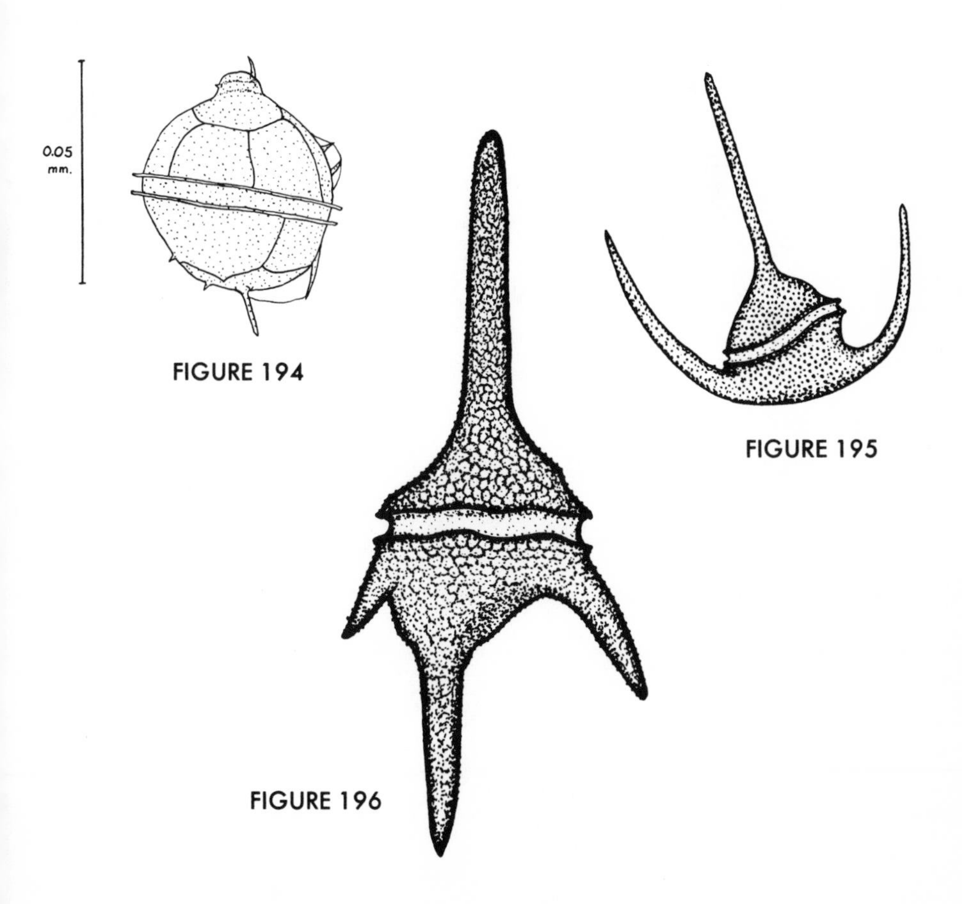

FIGURE 194

FIGURE 195

FIGURE 196

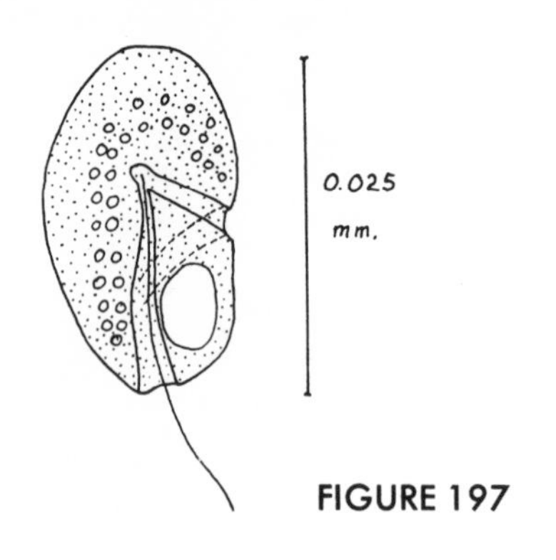

FIGURE 197

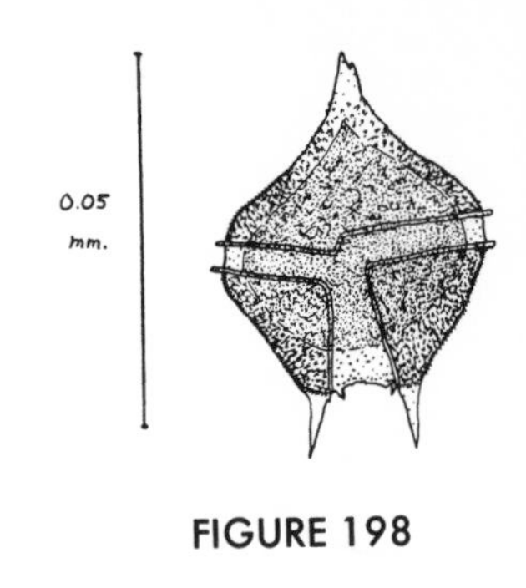

FIGURE 198

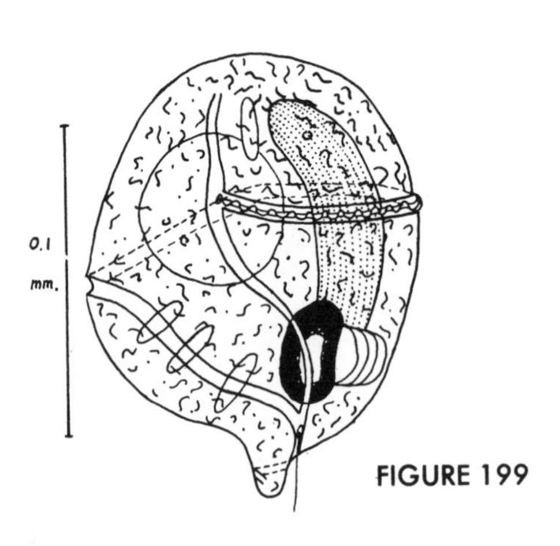

FIGURE 199

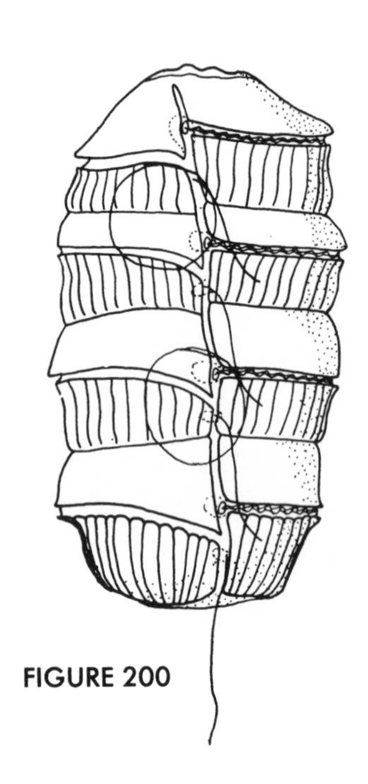

FIGURE 200

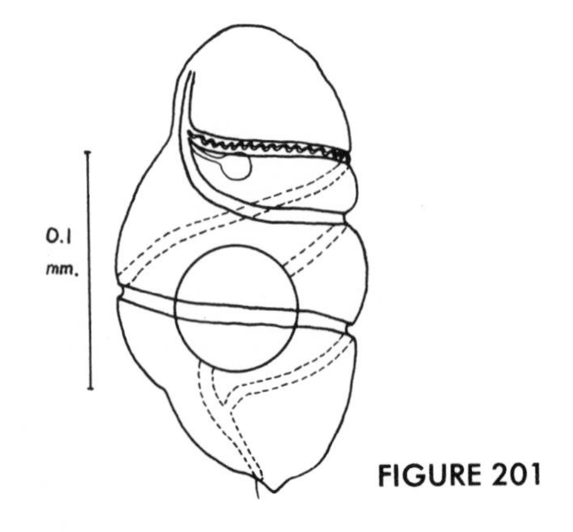

FIGURE 201

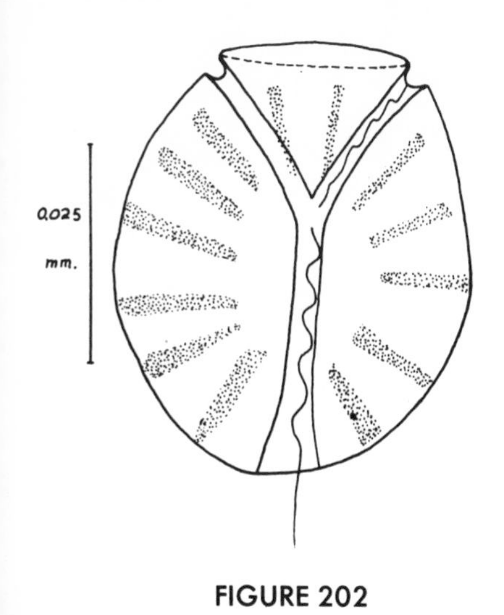

FIGURE 202

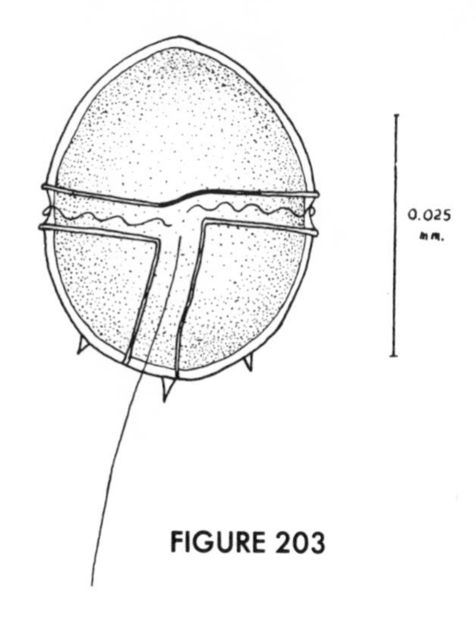

FIGURE 203

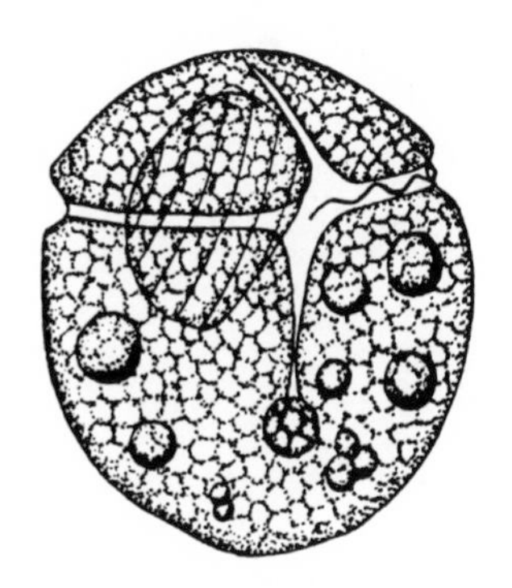

FIGURE 204

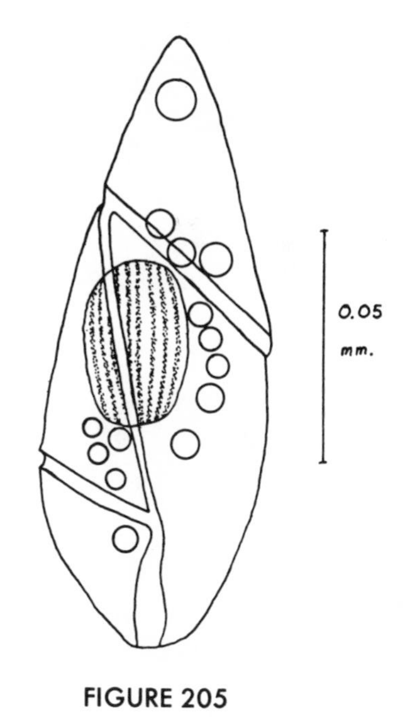

FIGURE 205

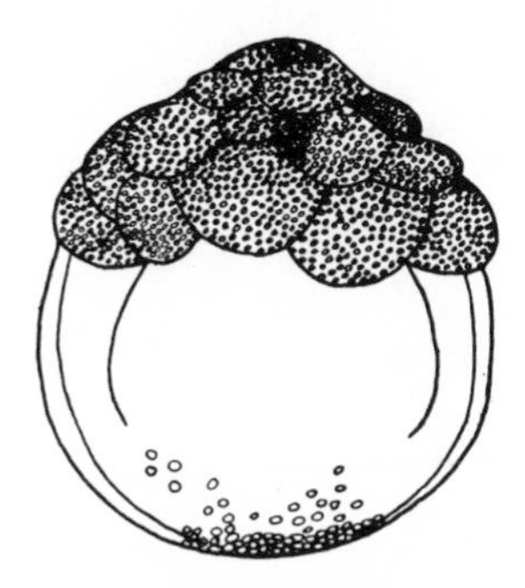

FIGURE 206

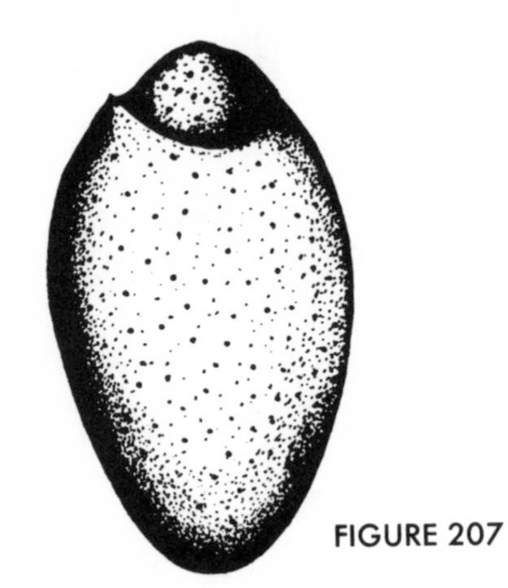
FIGURE 207

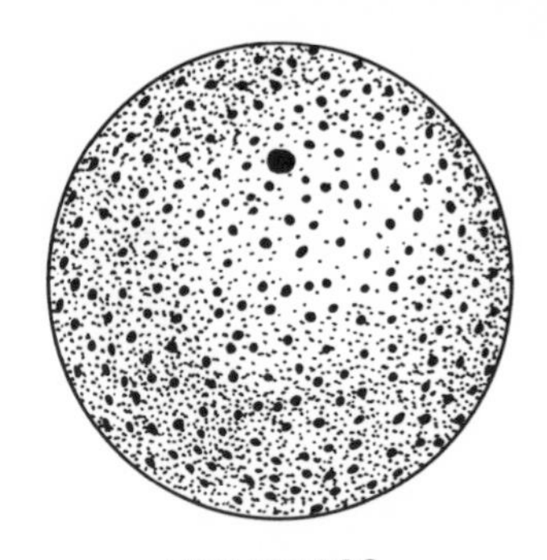
FIGURE 208

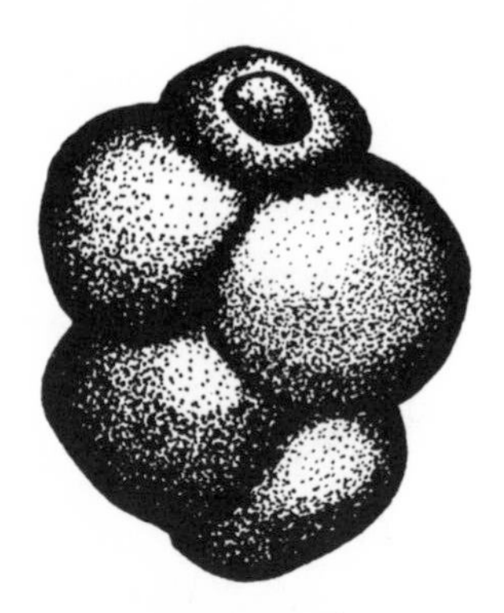
FIGURE 209

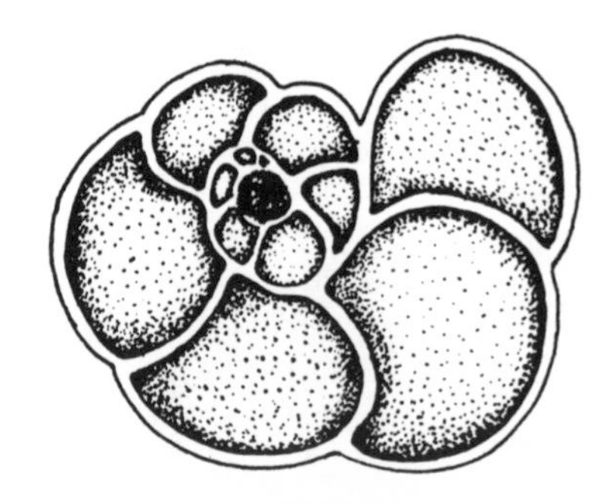
FIGURE 210

FIGURE 211

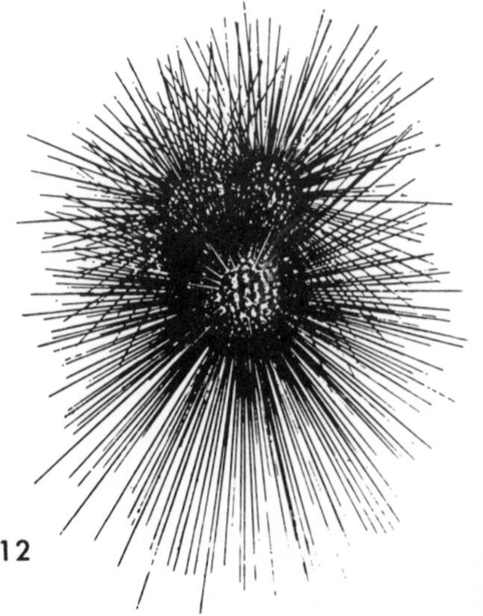
FIGURE 212

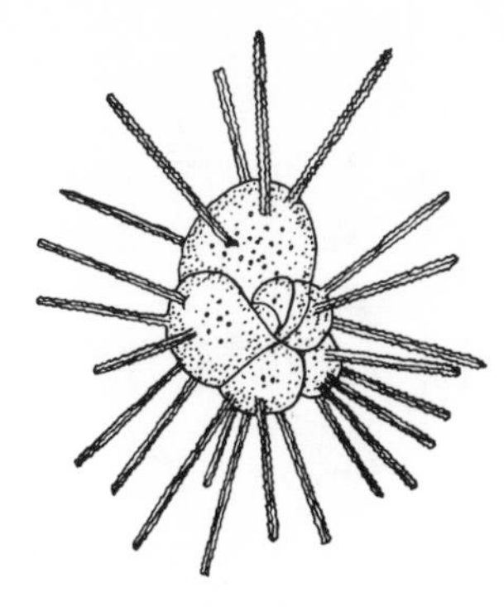

FIGURE 213

FIGURE 214

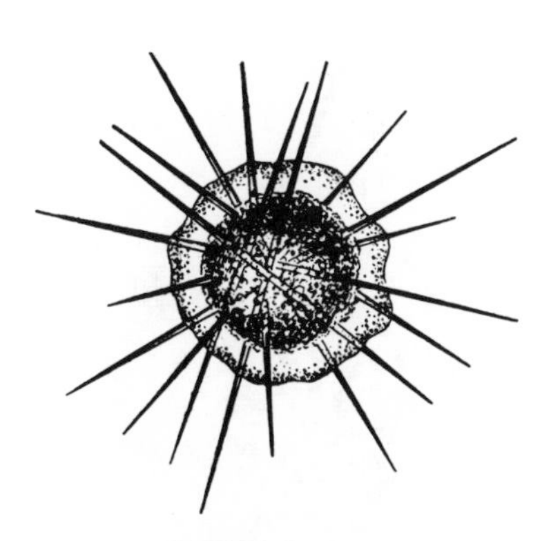

FIGURE 215

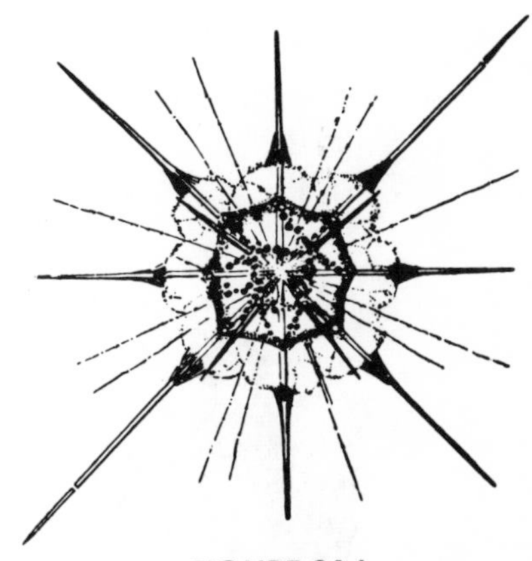

FIGURE 216

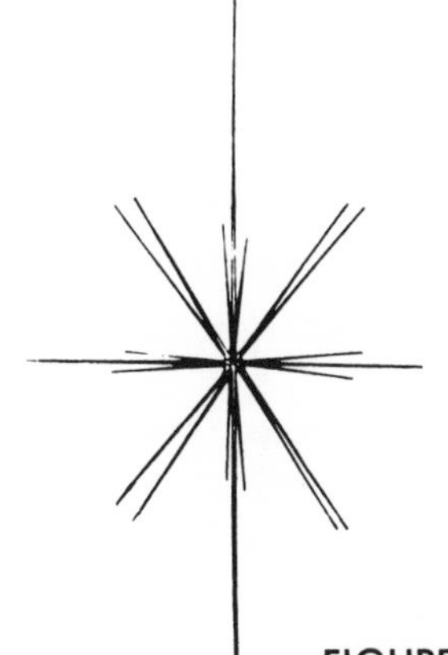

FIGURE 217

FIGURE 218

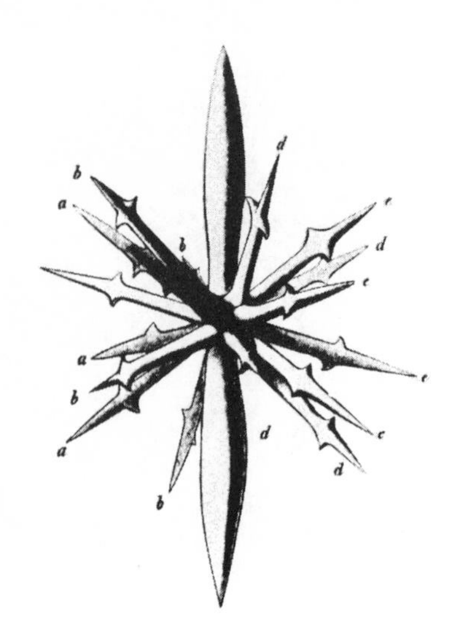

FIGURE 219

FIGURE 220

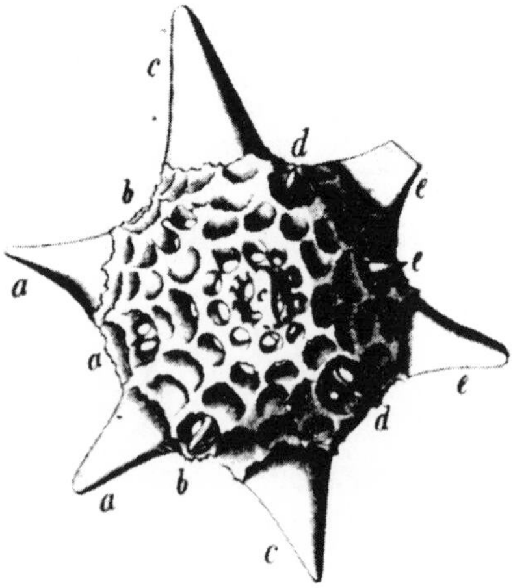

FIGURE 221

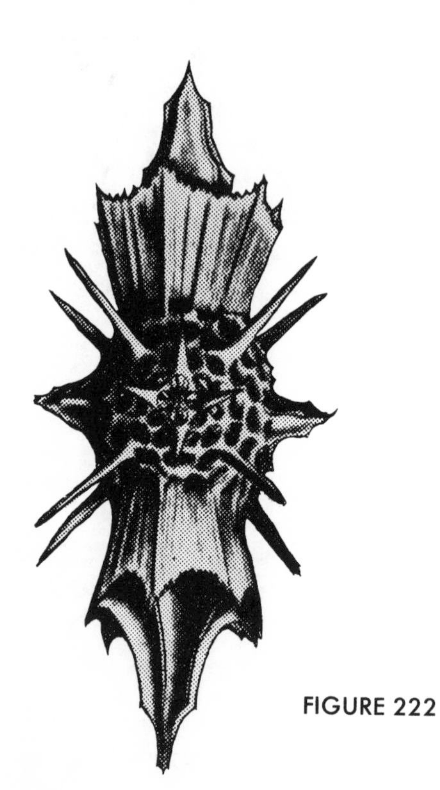

FIGURE 222

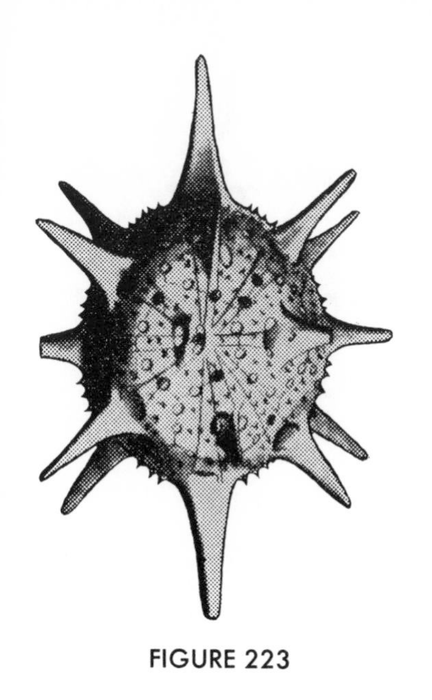

FIGURE 223

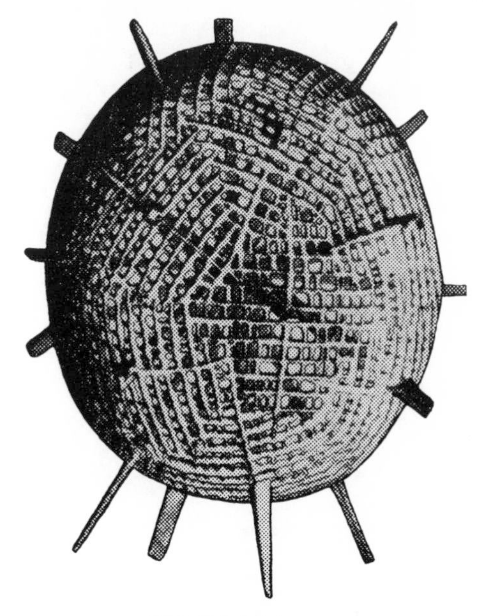

FIGURE 224

FIGURE 225

FIGURE 226

FIGURE 227

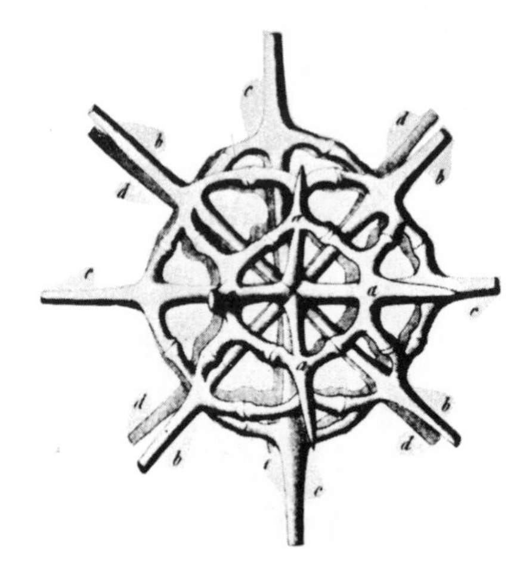

FIGURE 229

FIGURE 228

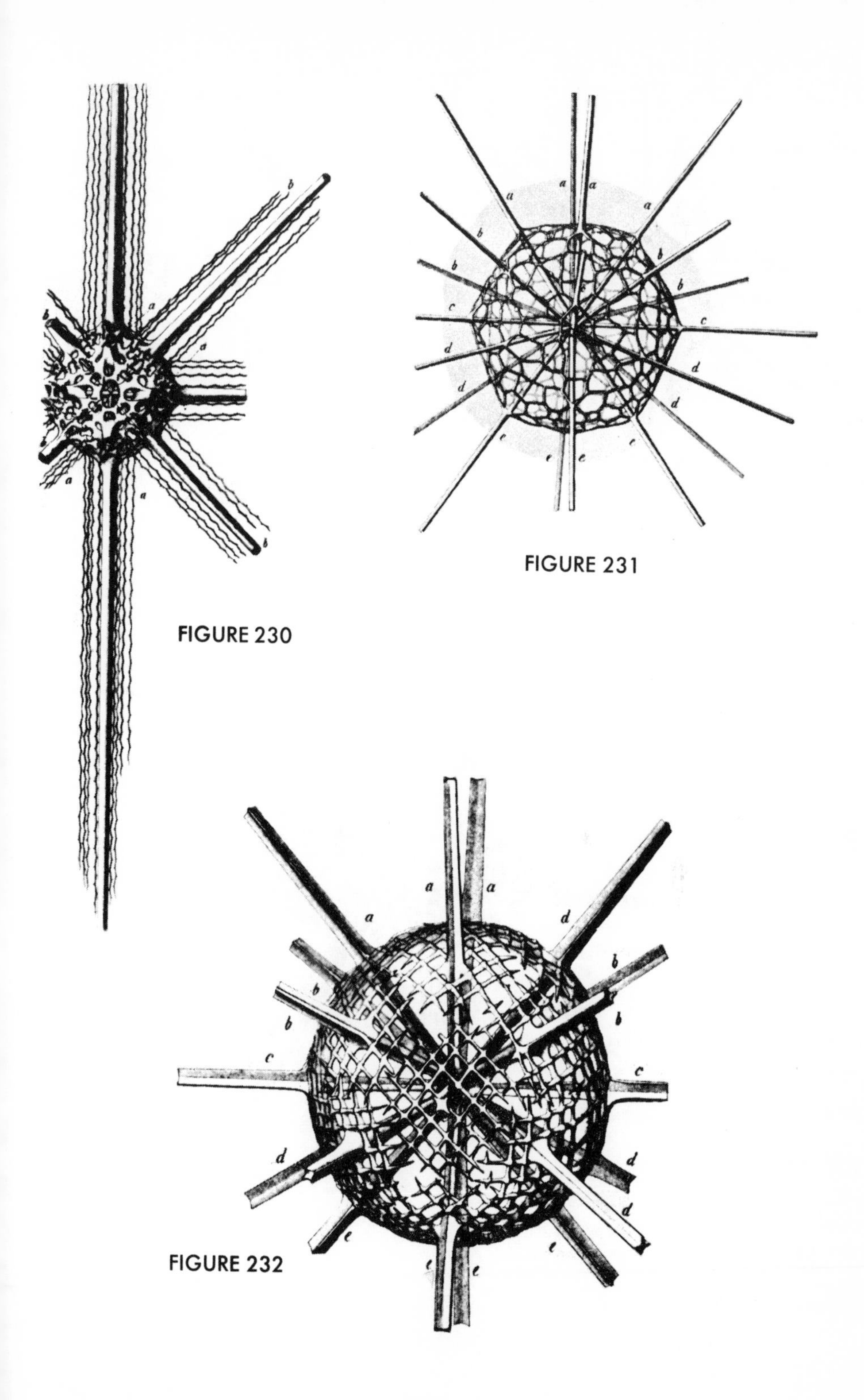

FIGURE 230

FIGURE 231

FIGURE 232

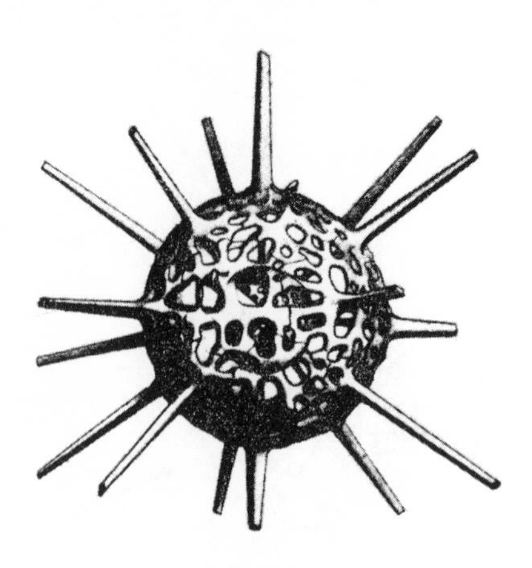

FIGURE 233

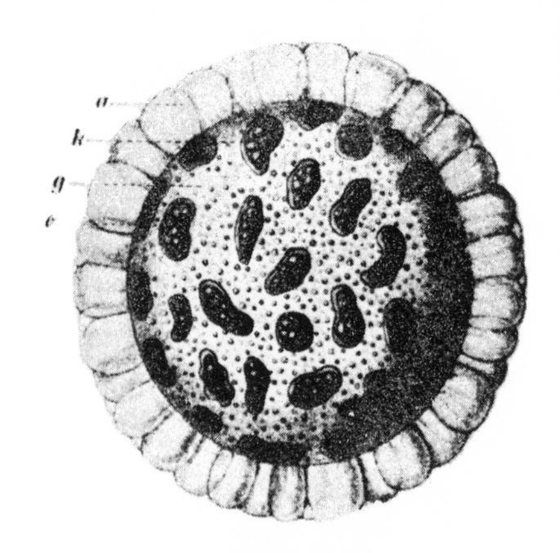

FIGURE 235

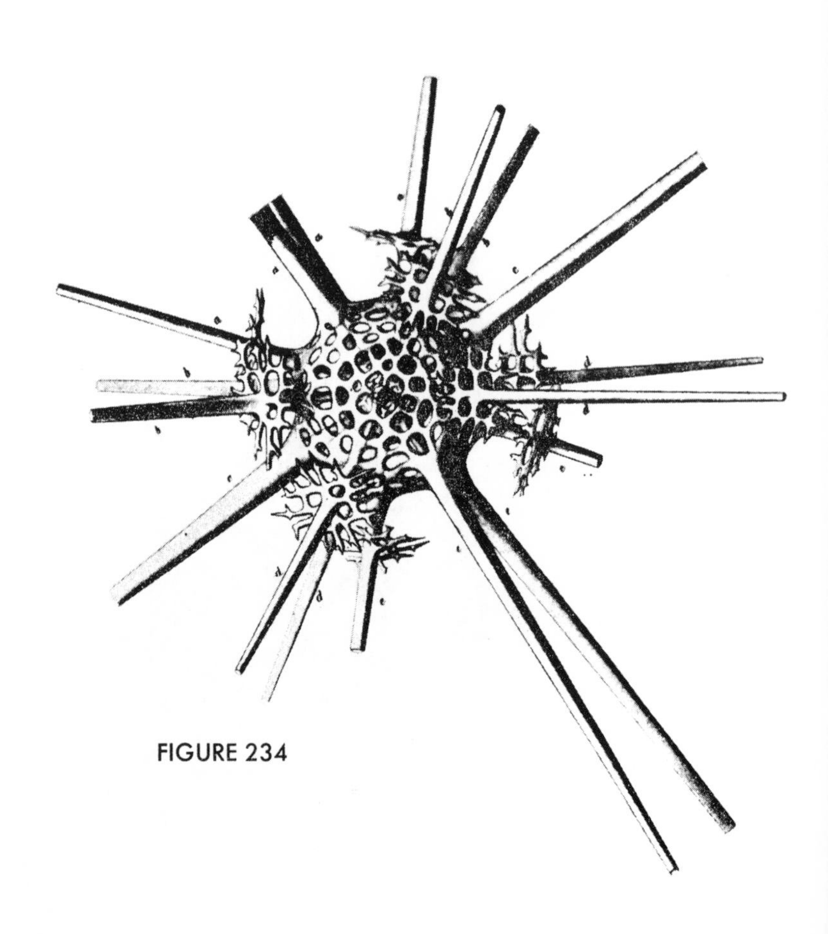

FIGURE 234

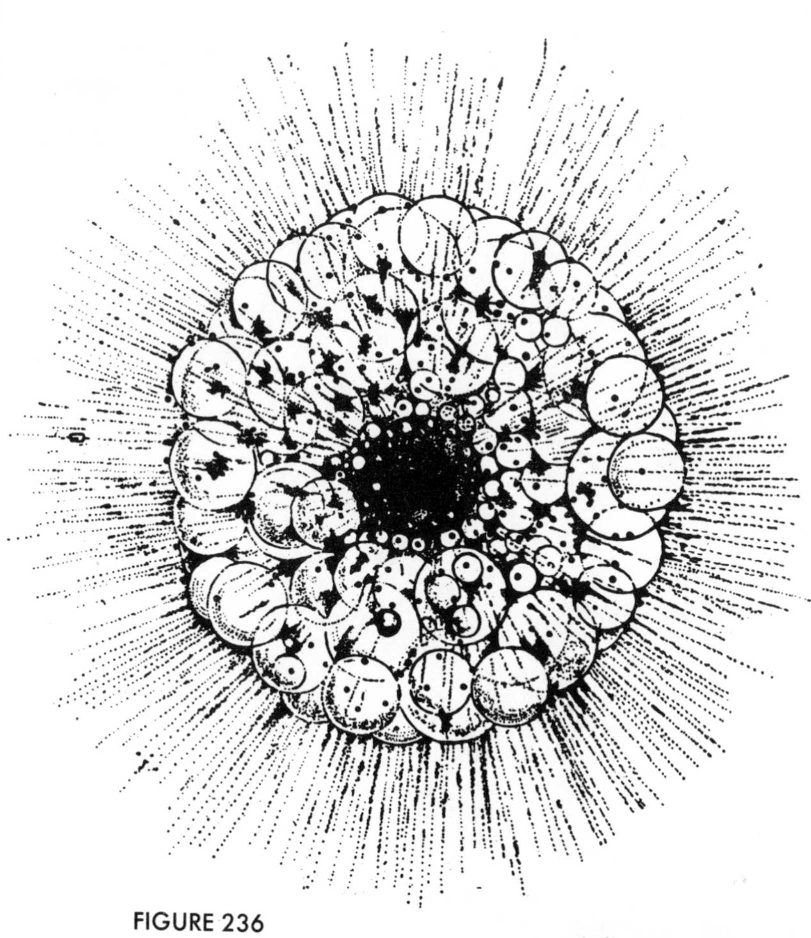

FIGURE 236

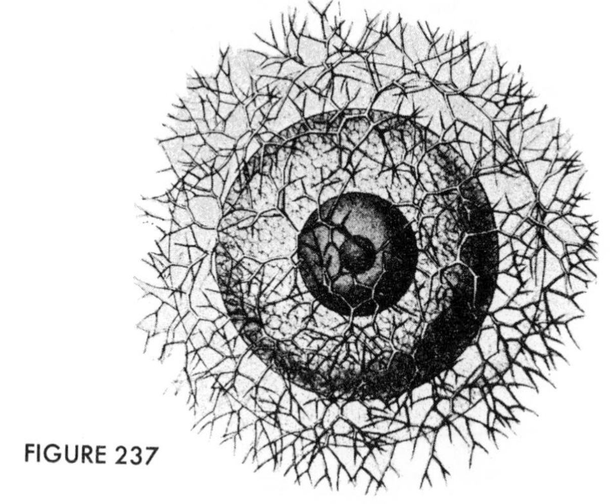

FIGURE 237

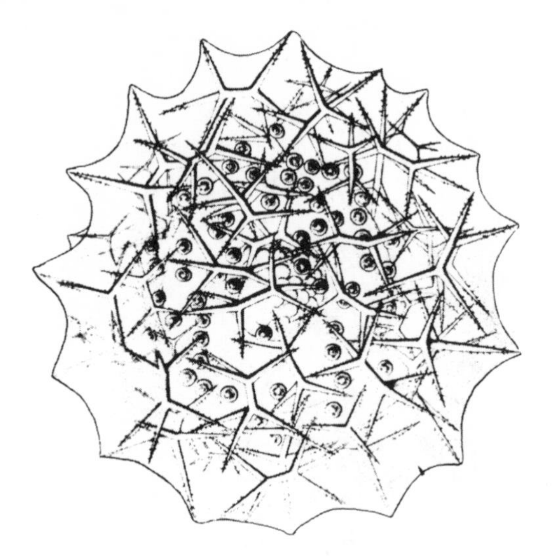

FIGURE 239

FIGURE 238

FIGURE 240

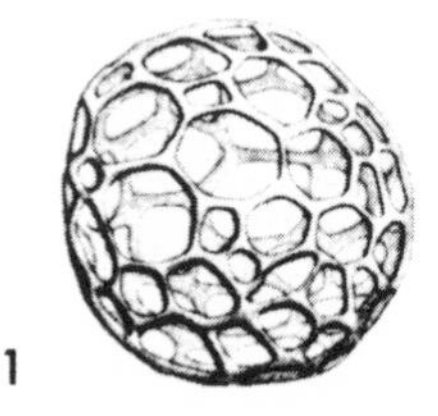

FIGURE 241

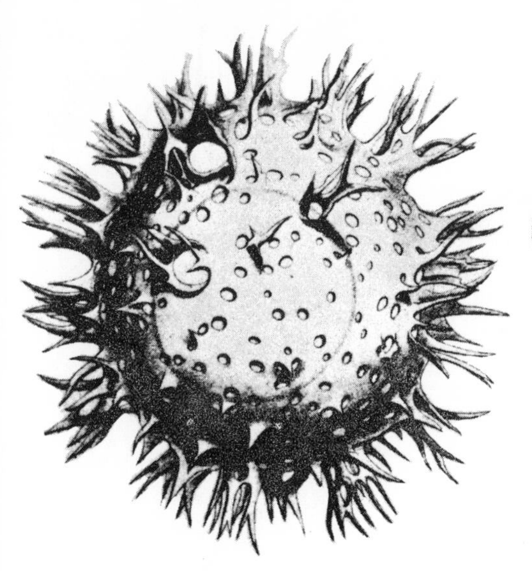

FIGURE 242

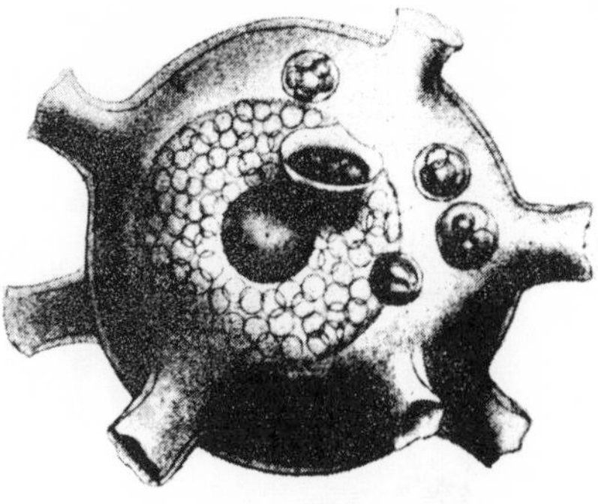

FIGURE 243

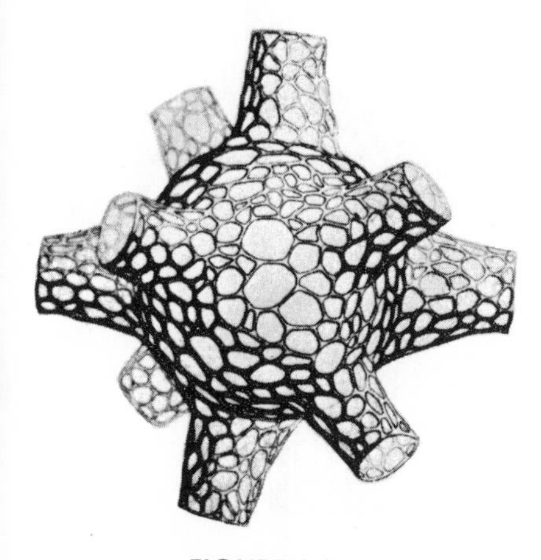

FIGURE 244

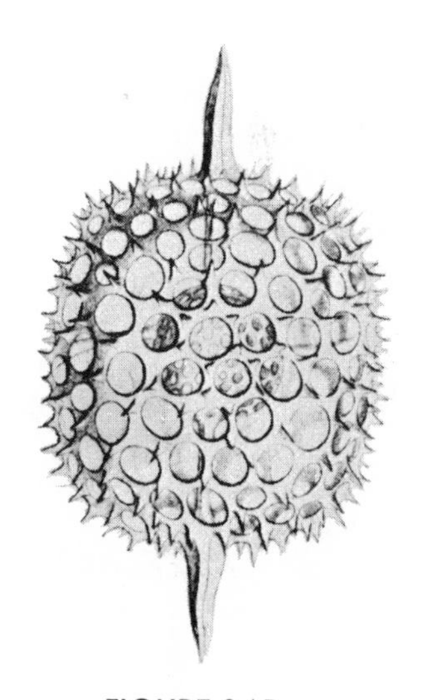

FIGURE 245

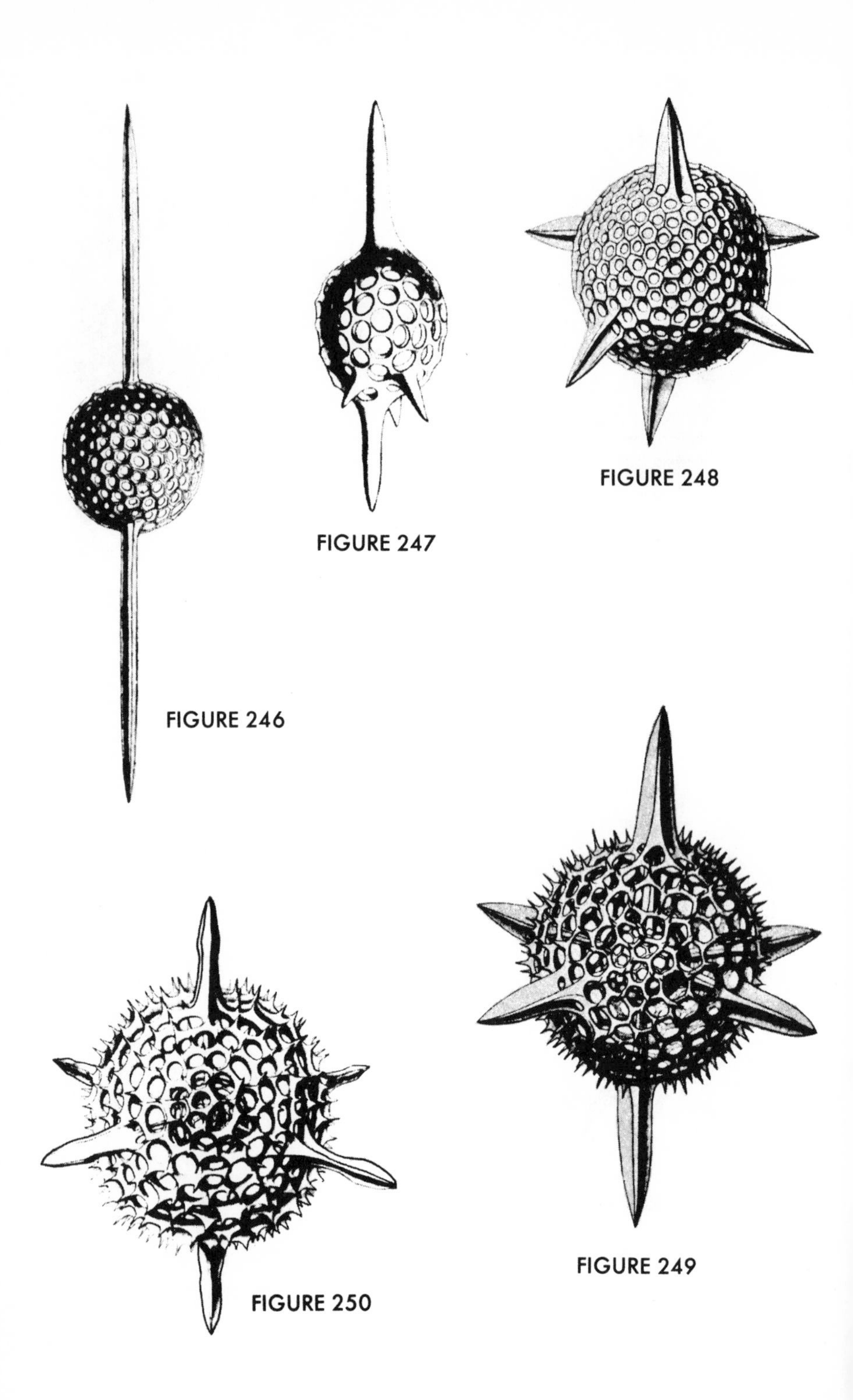
FIGURE 246
FIGURE 247
FIGURE 248
FIGURE 249
FIGURE 250

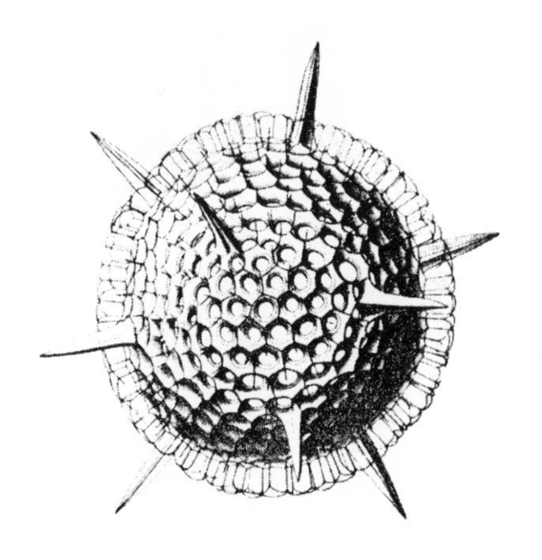

FIGURE 251

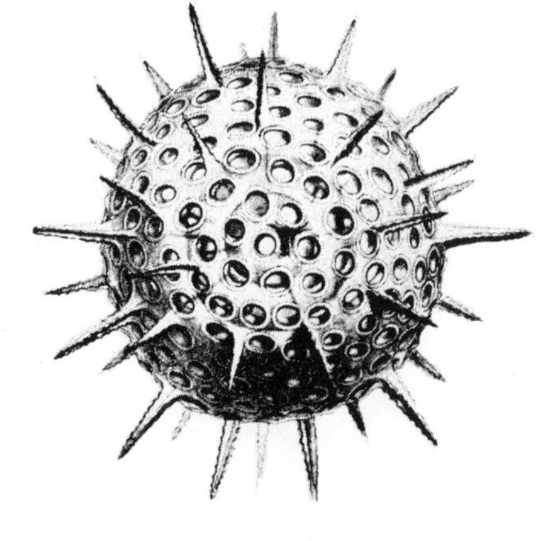

FIGURE 252

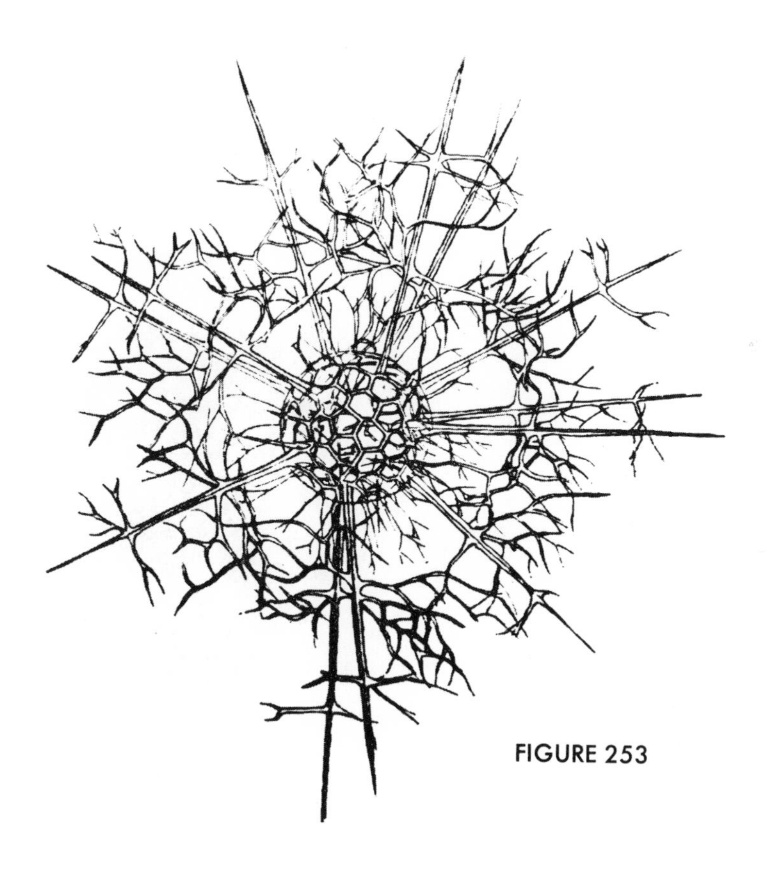

FIGURE 253

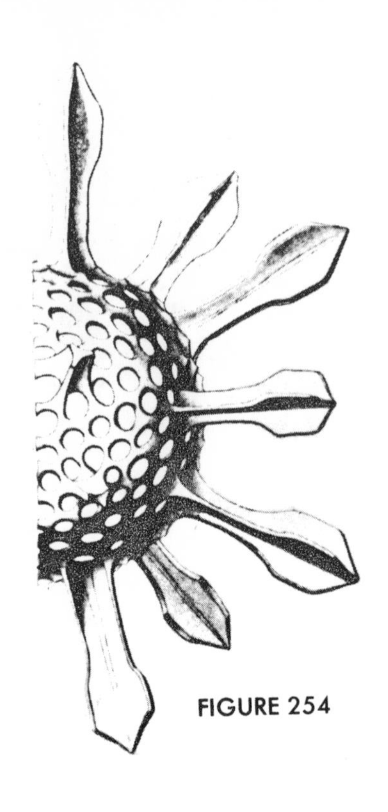
FIGURE 254

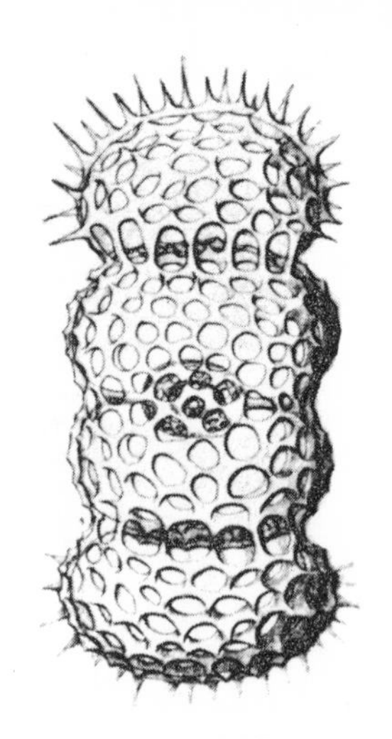
FIGURE 255

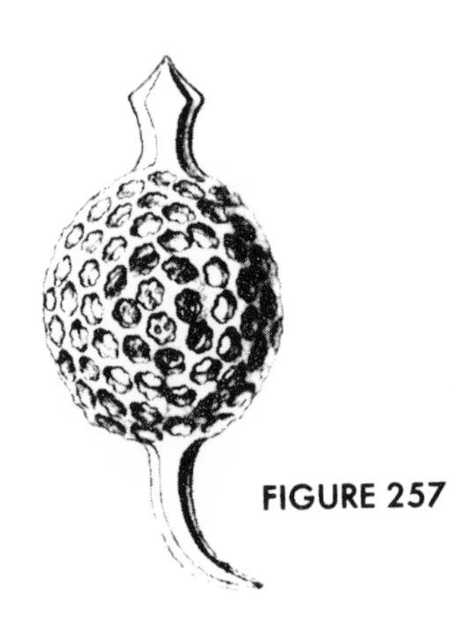
FIGURE 257

FIGURE 256

FIGURE 258

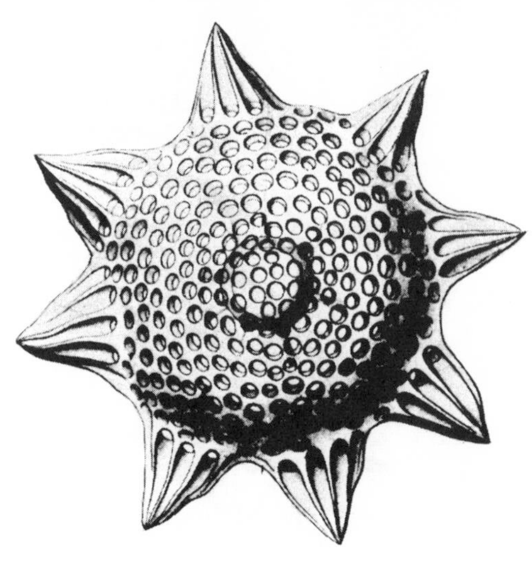

FIGURE 259

FIGURE 260

FIGURE 261

FIGURE 262

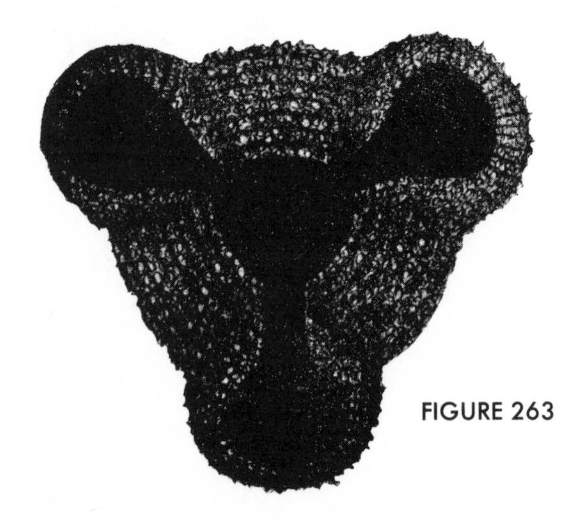

FIGURE 263

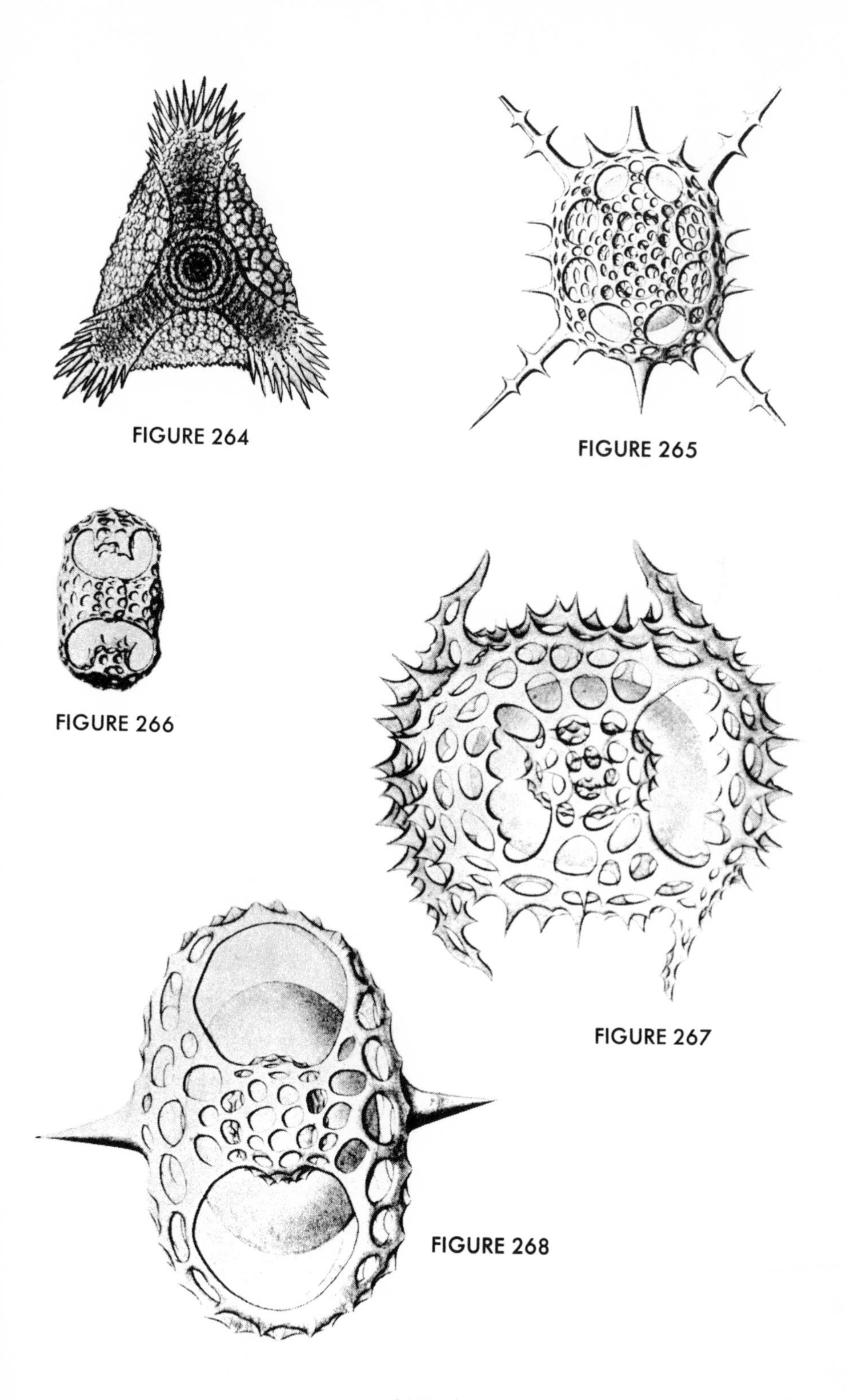
FIGURE 264
FIGURE 265
FIGURE 266
FIGURE 267
FIGURE 268

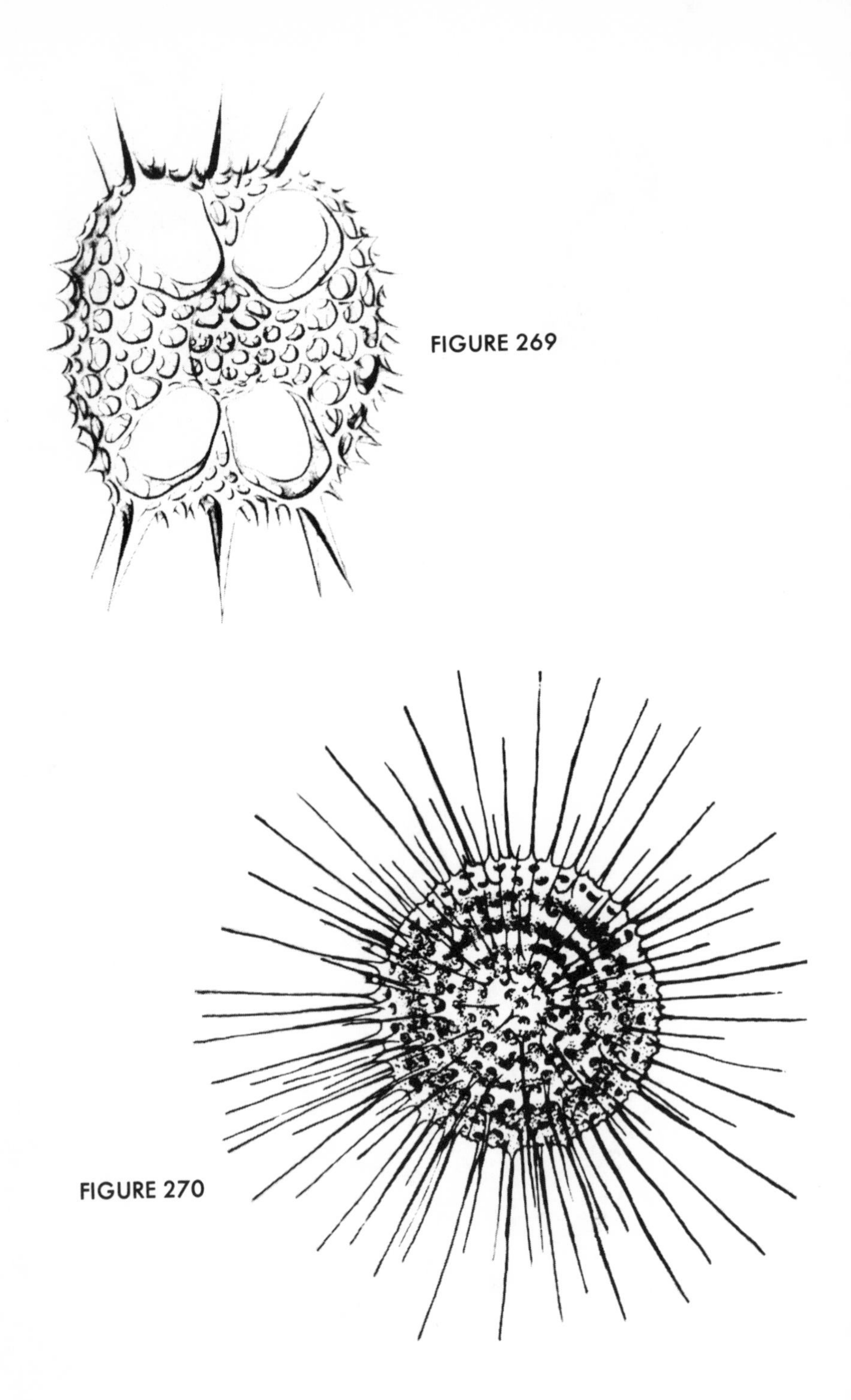

FIGURE 269

FIGURE 270

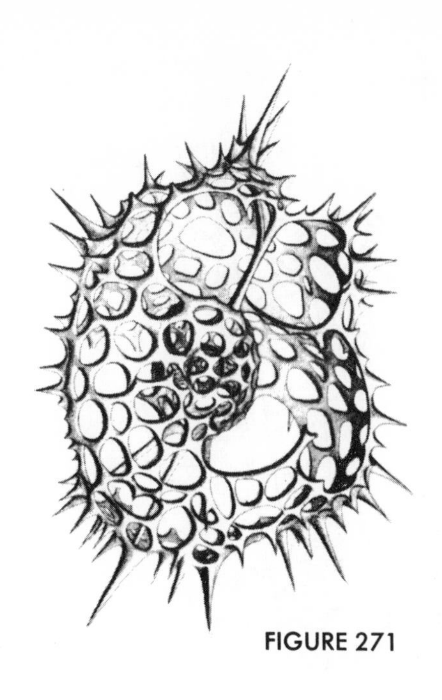

FIGURE 271

FIGURE 272

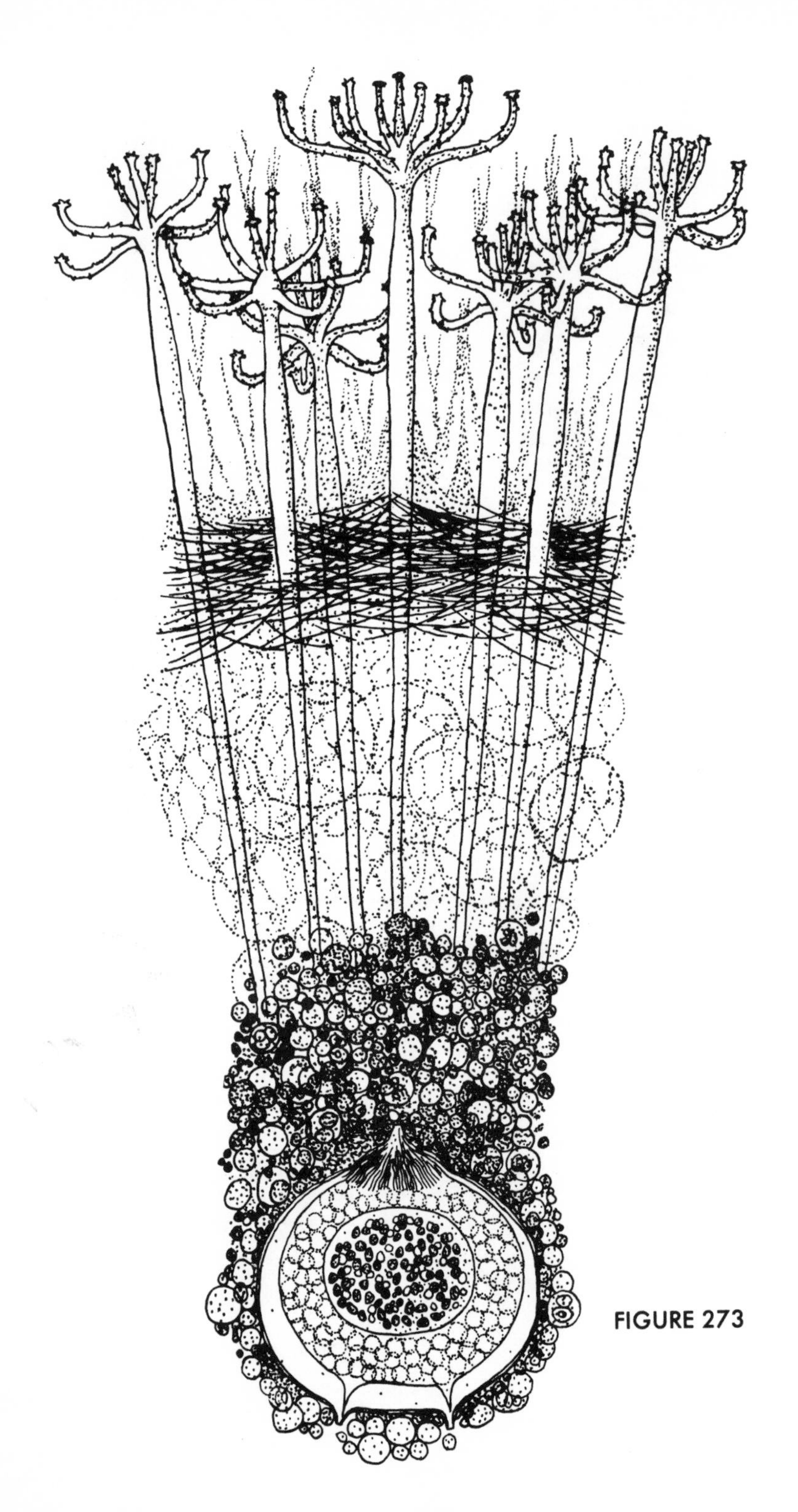

FIGURE 273

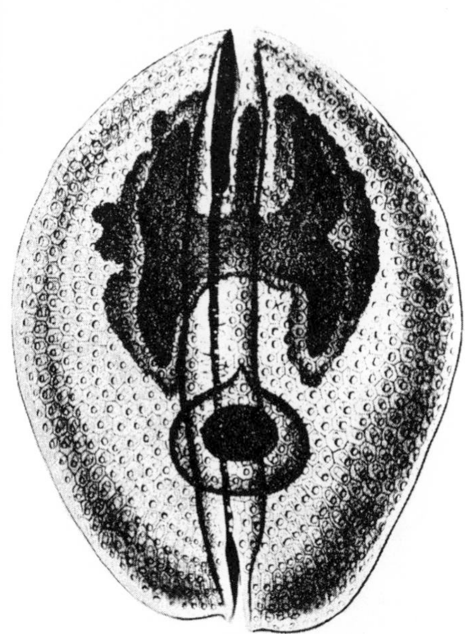

FIGURE 274

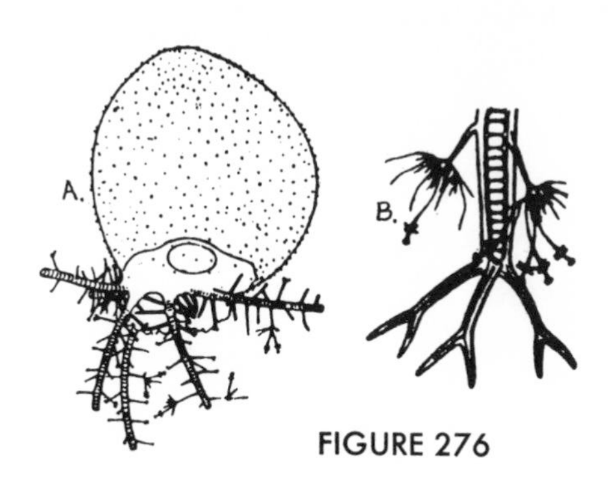

FIGURE 276

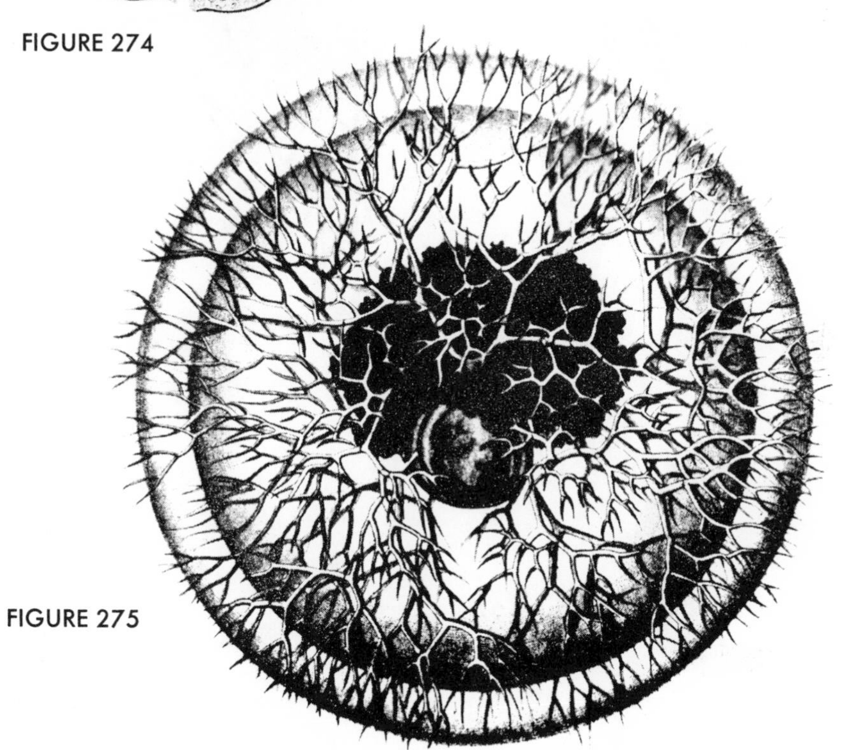

FIGURE 275

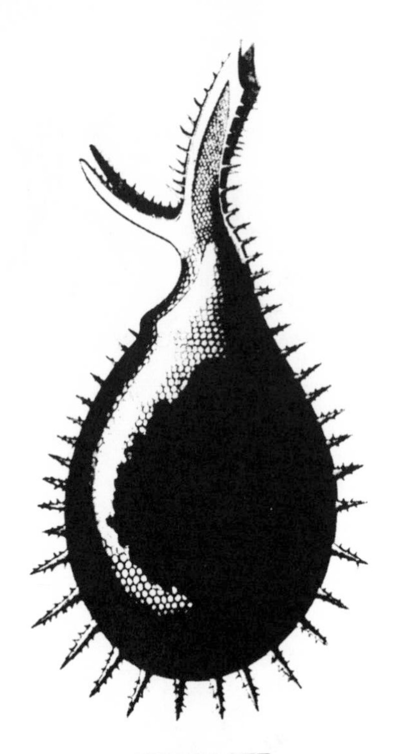
FIGURE 277

FIGURE 279

FIGURE 278

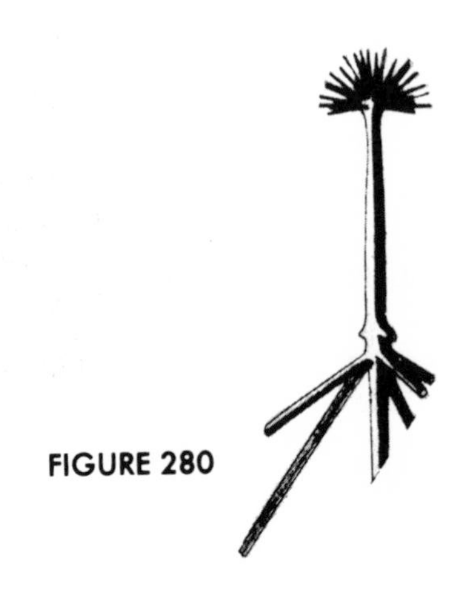
FIGURE 280

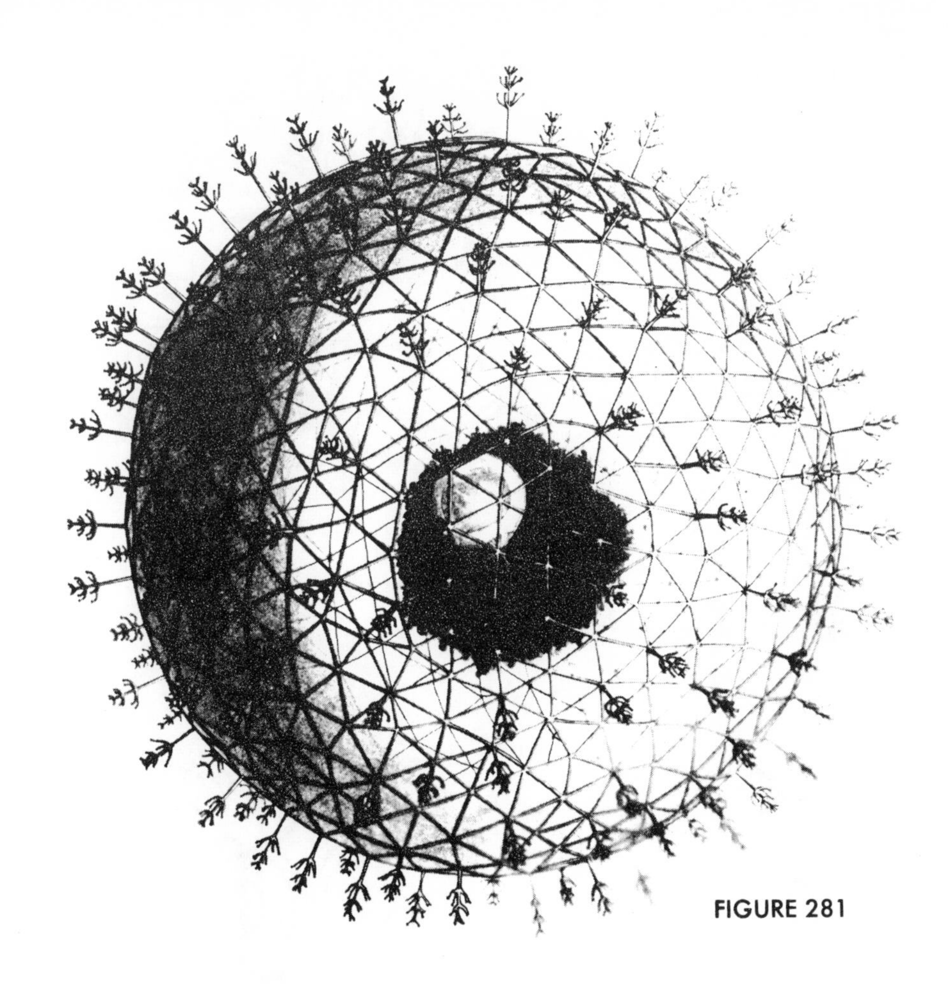

FIGURE 281

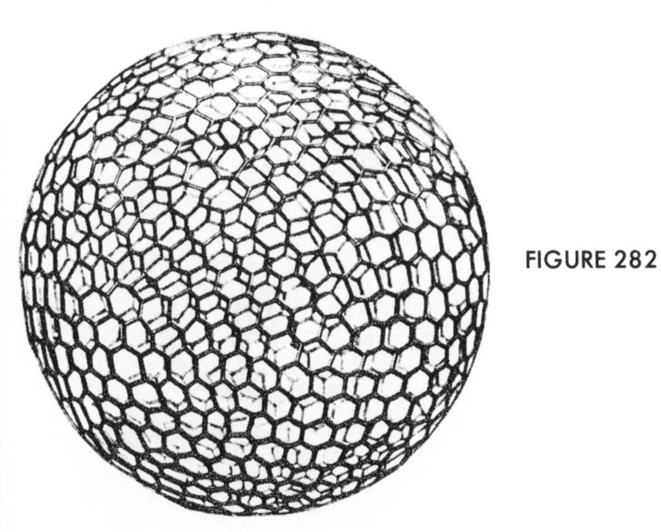

FIGURE 282

FIGURE 283

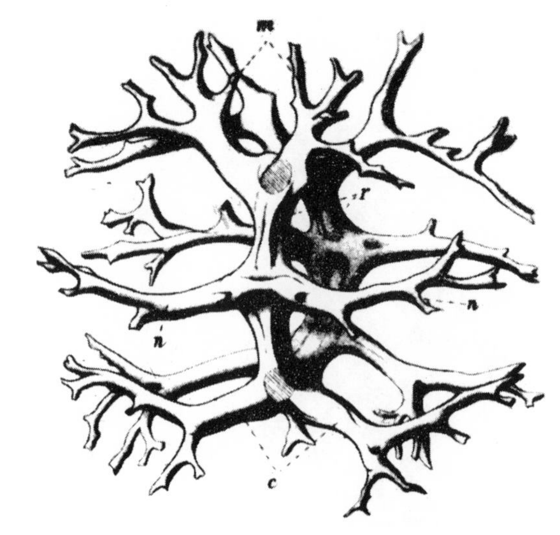

FIGURE 284

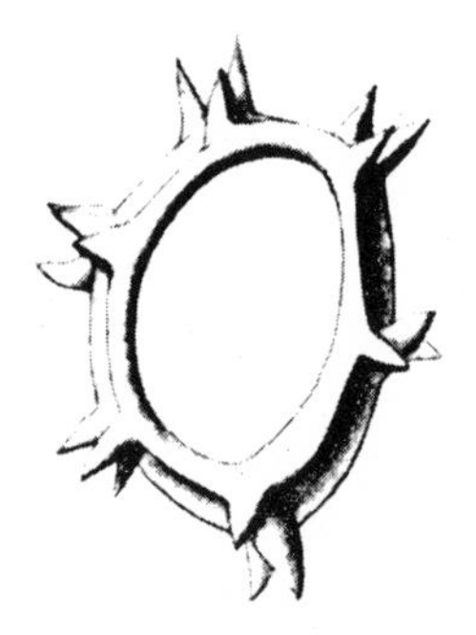

FIGURE 285

FIGURE 286

FIGURE 287

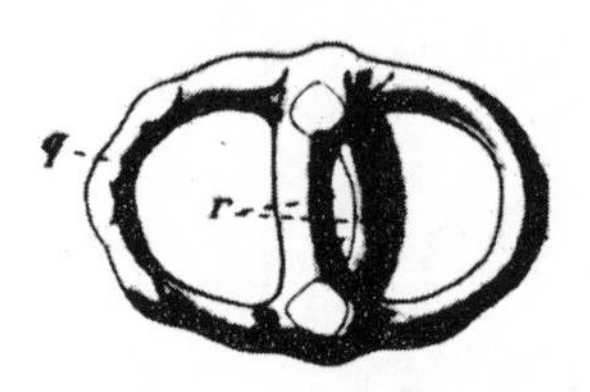

FIGURE 288

FIGURE 289

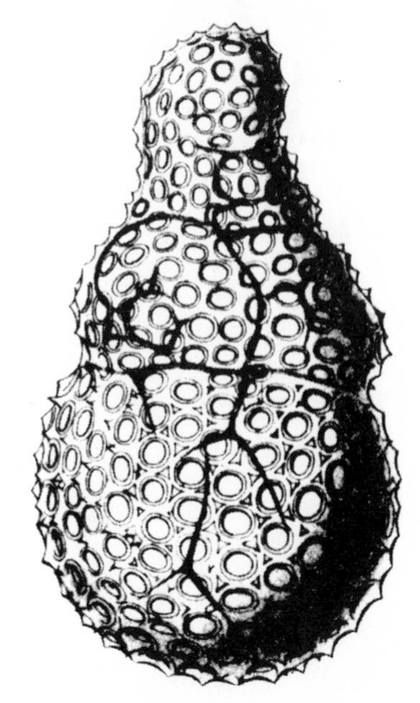

FIGURE 290

FIGURE 291

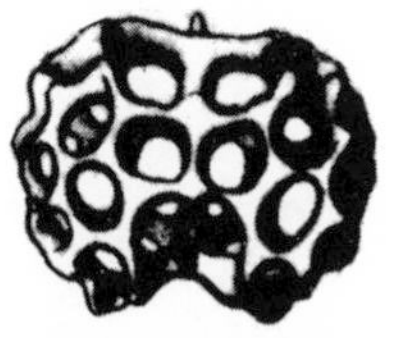

FIGURE 292

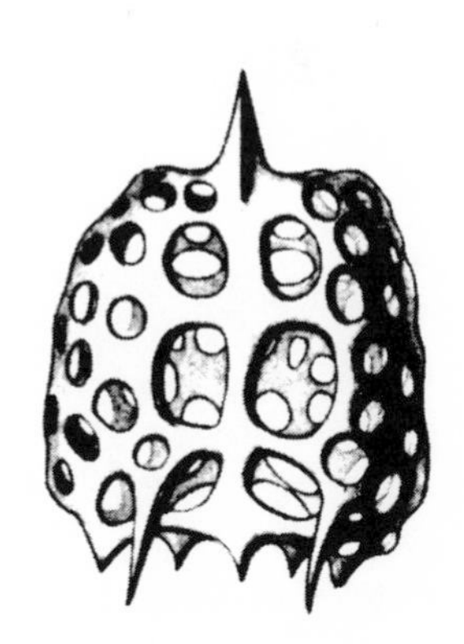

FIGURE 293

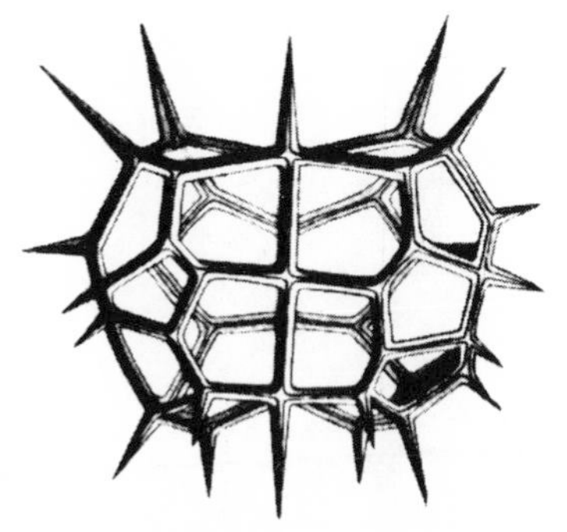

FIGURE 294

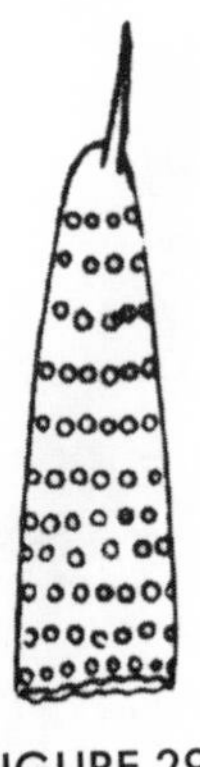
FIGURE 295

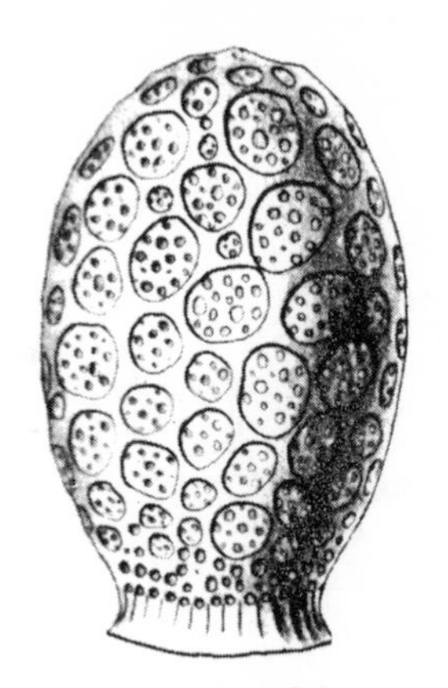
FIGURE 296

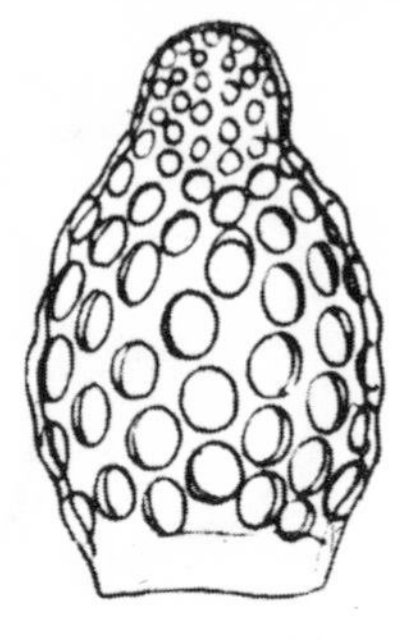
FIGURE 298

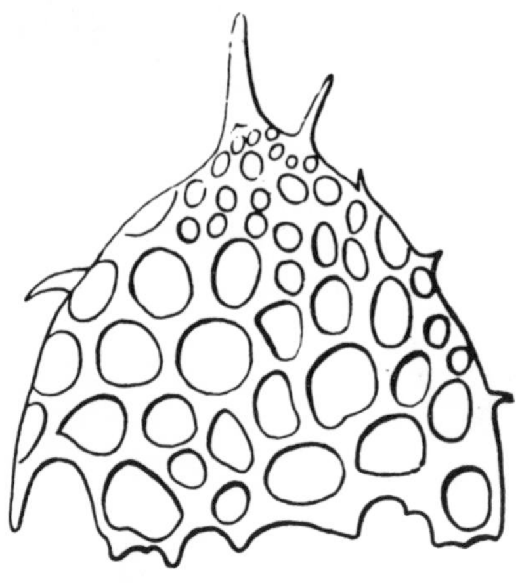
FIGURE 297

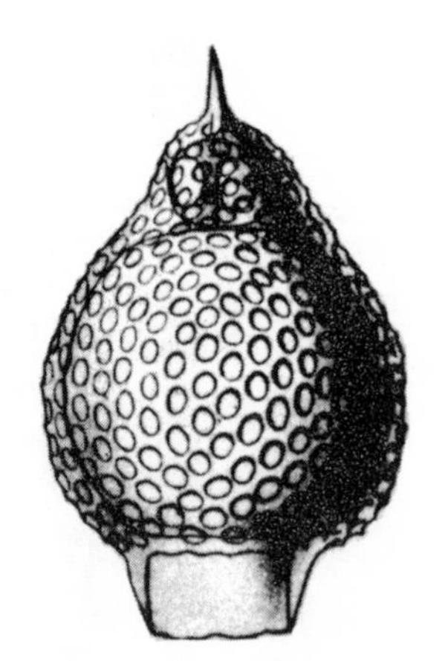
FIGURE 299

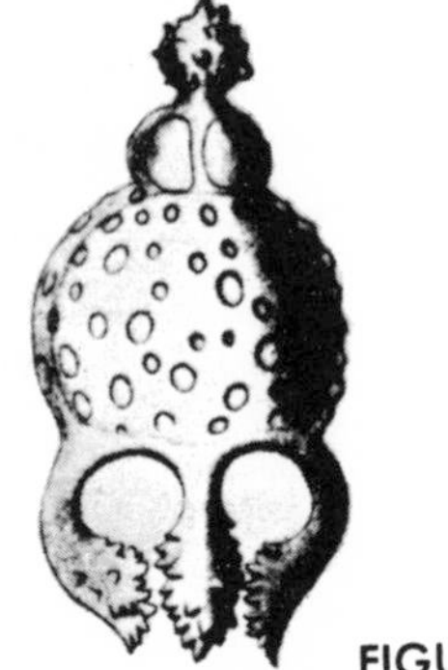
FIGURE 300

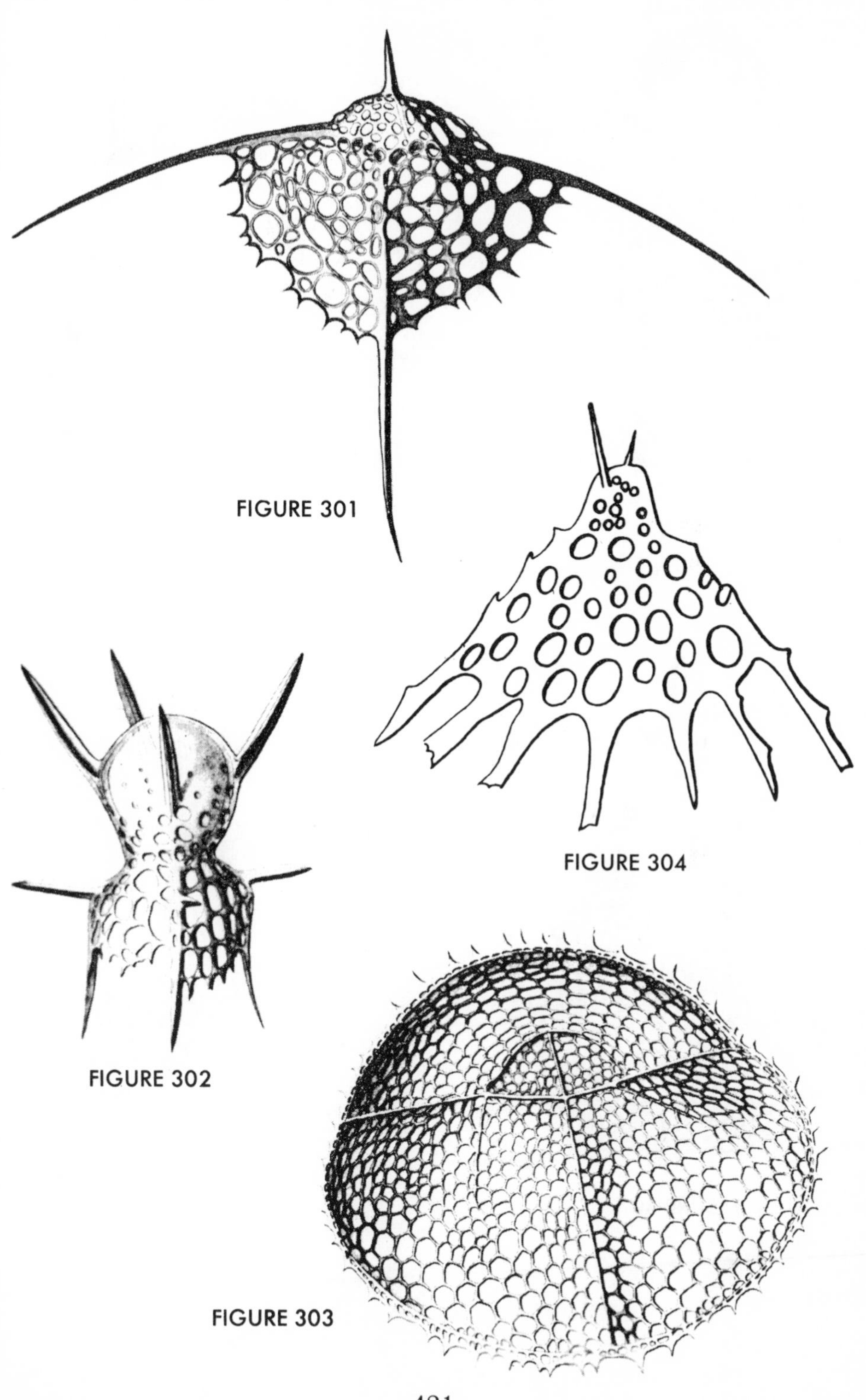
FIGURE 301
FIGURE 304
FIGURE 302
FIGURE 303

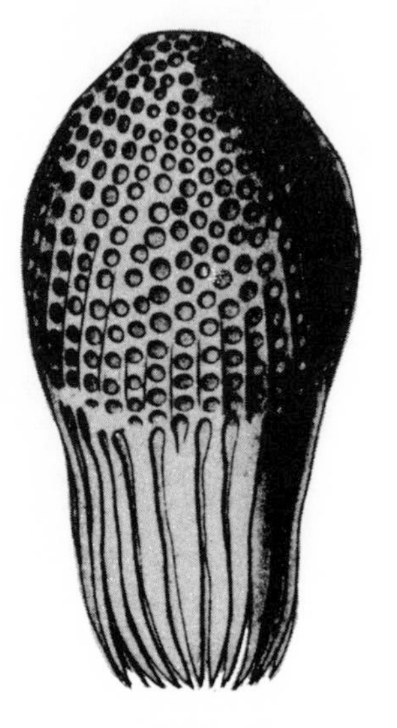

FIGURE 305

FIGURE 308

FIGURE 306

FIGURE 307

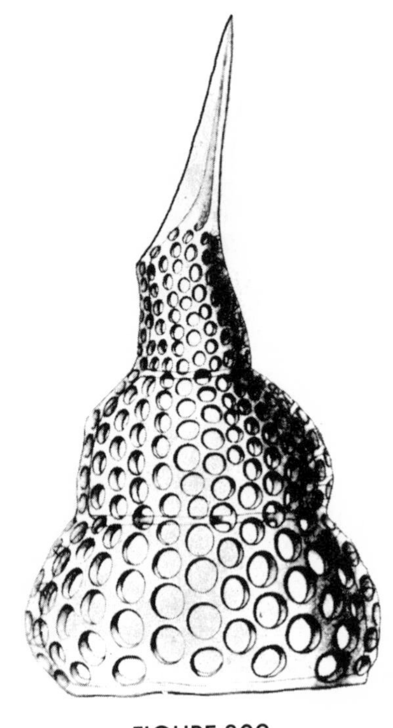

FIGURE 309

FIGURE 310

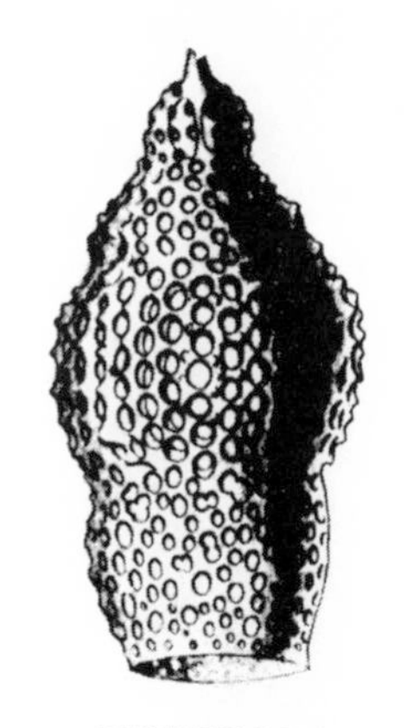
FIGURE 314

FIGURE 311

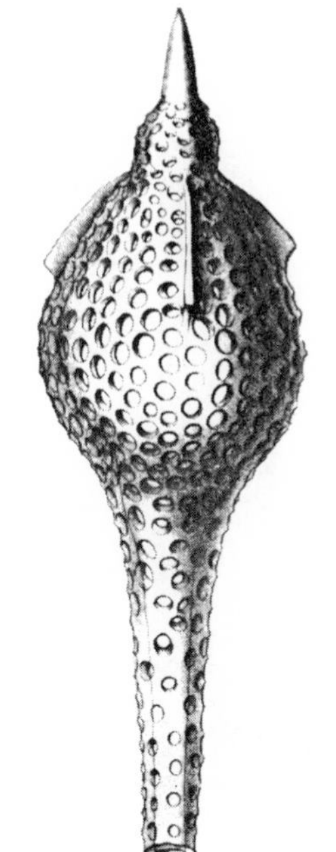
FIGURE 312

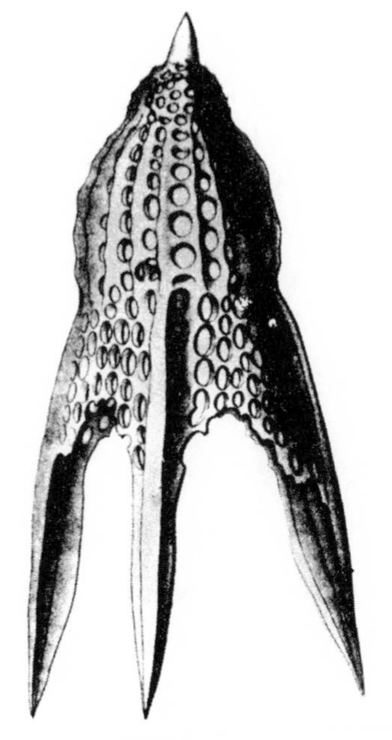
FIGURE 313

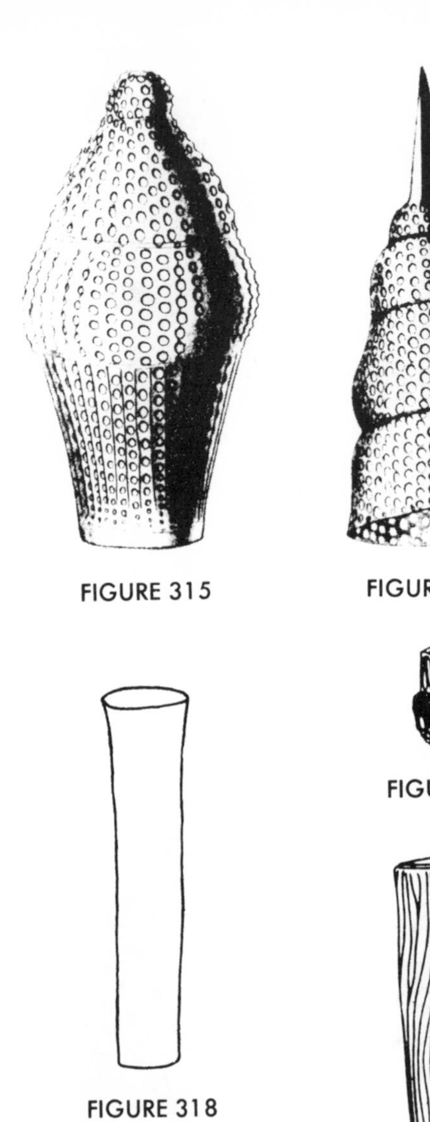

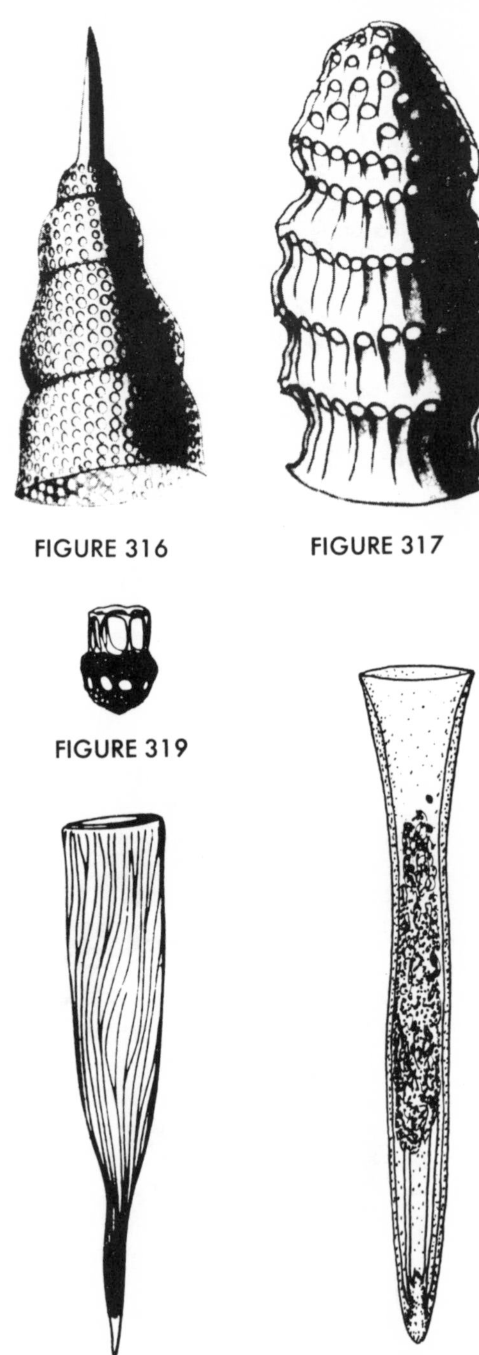

FIGURE 315

FIGURE 316

FIGURE 317

FIGURE 318

FIGURE 319

FIGURE 322

FIGURE 320

FIGURE 321

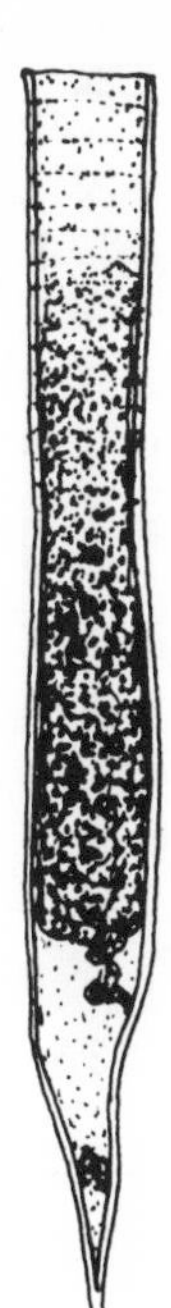

FIGURE 323

FIGURE 324

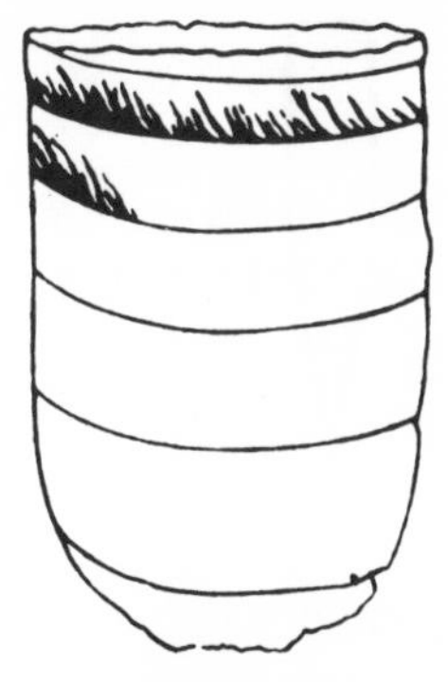

FIGURE 325

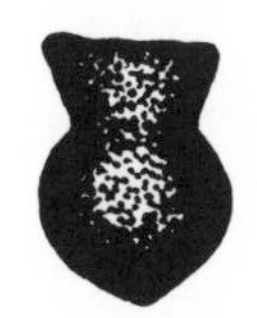

FIGURE 326

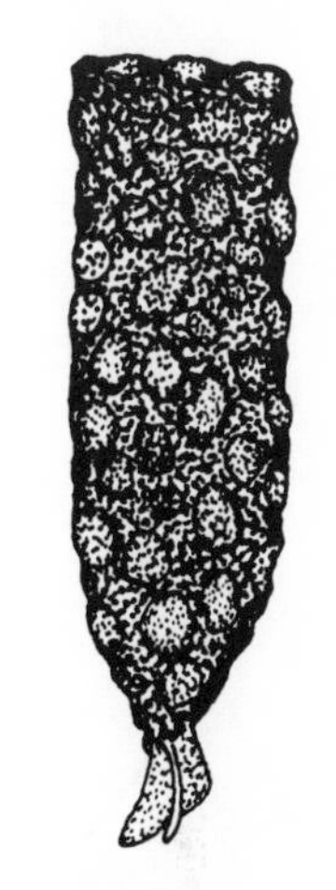

FIGURE 327

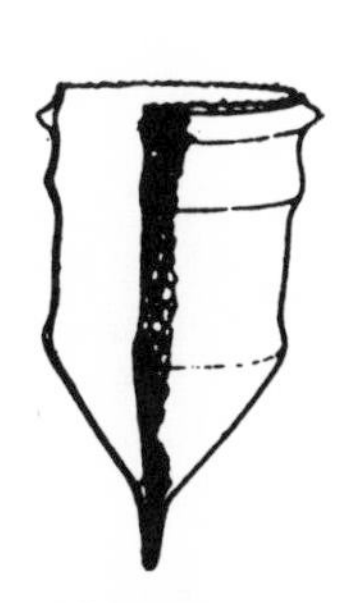

FIGURE 328

FIGURE 329

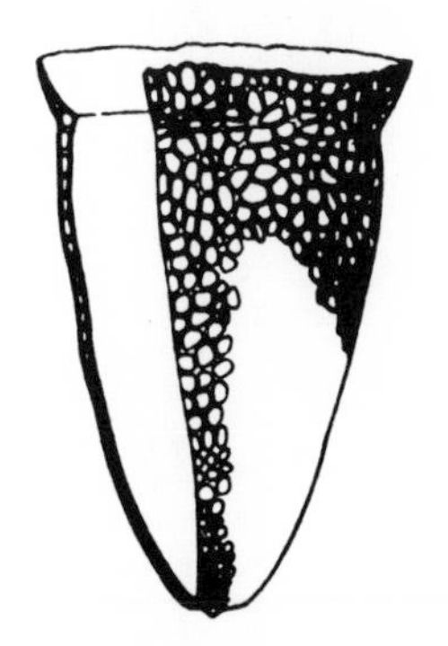

FIGURE 330

FIGURE 331

FIGURE 332

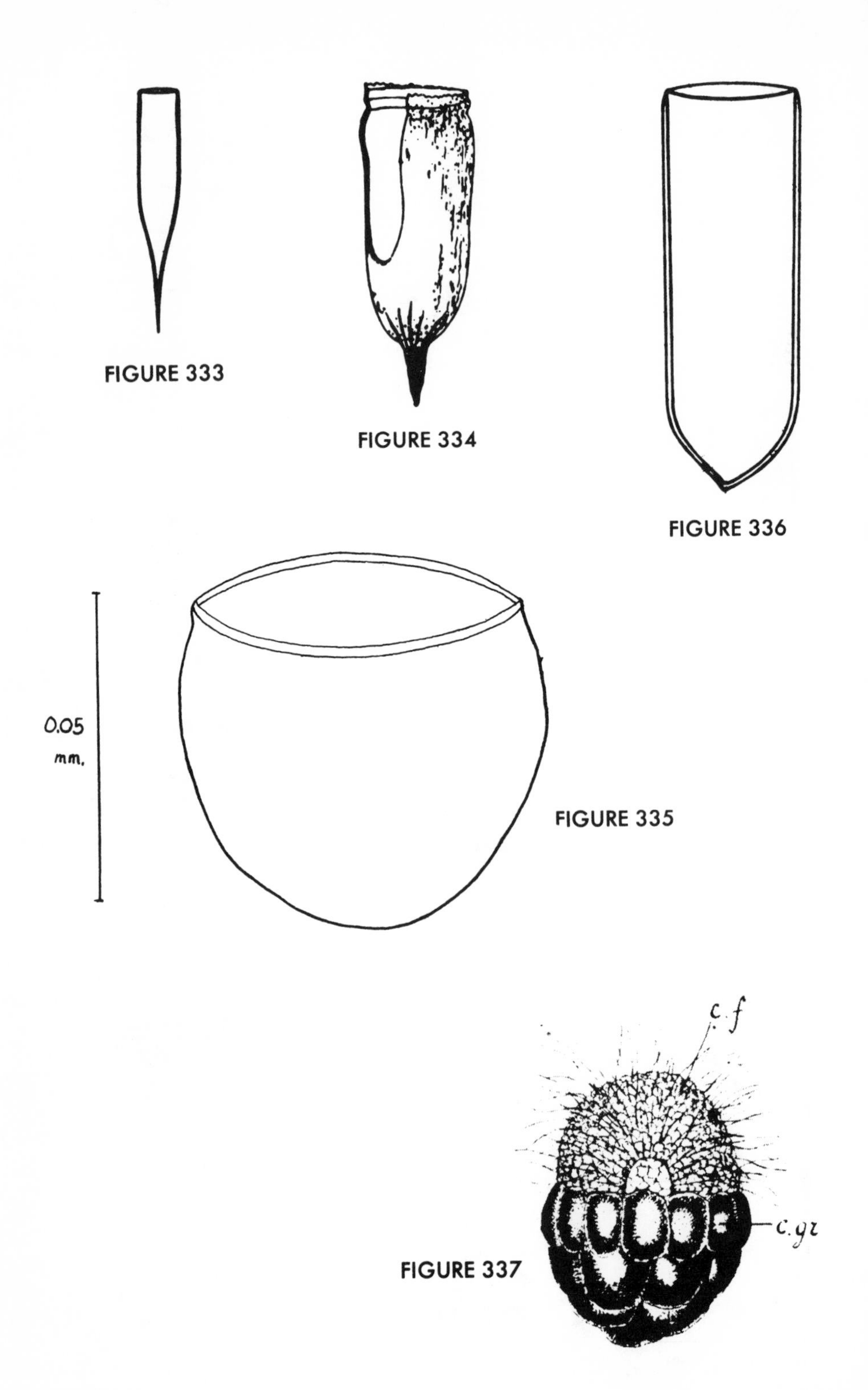

FIGURE 333

FIGURE 334

FIGURE 336

FIGURE 335

FIGURE 337

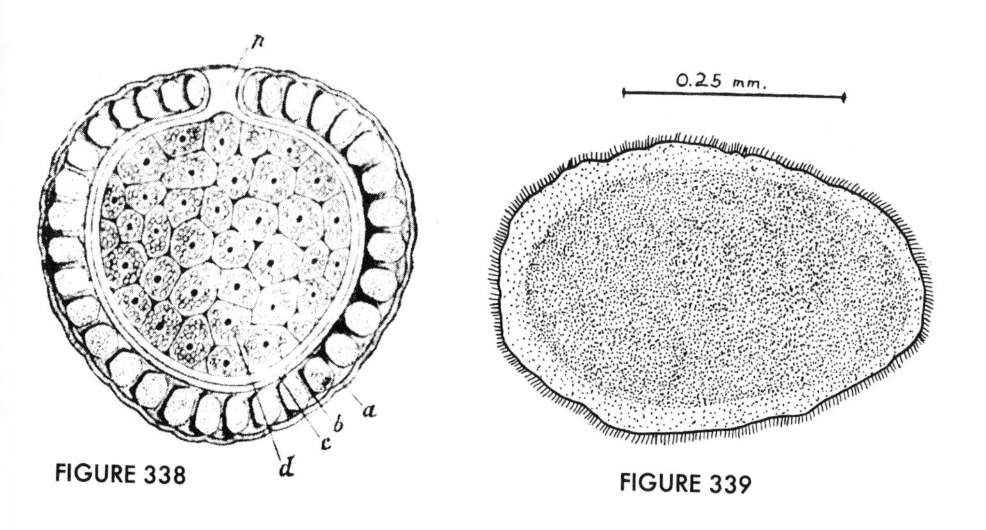

FIGURE 338

FIGURE 339

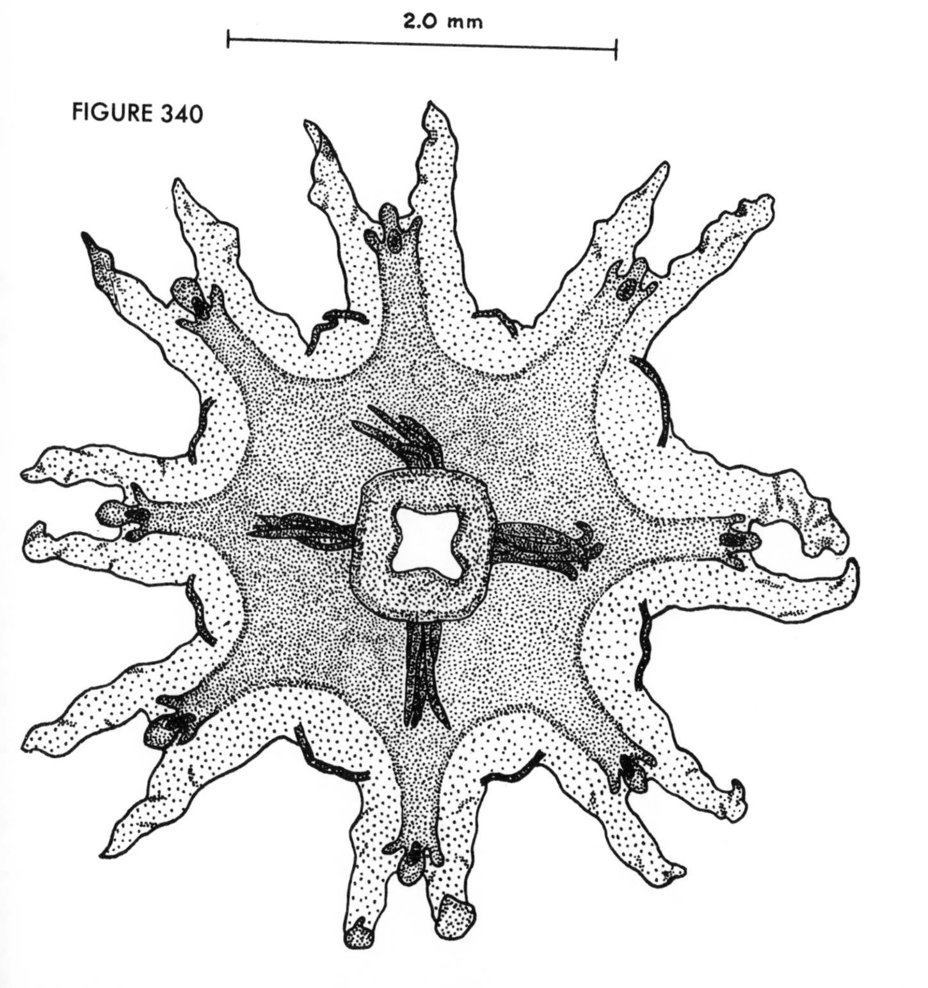

FIGURE 340

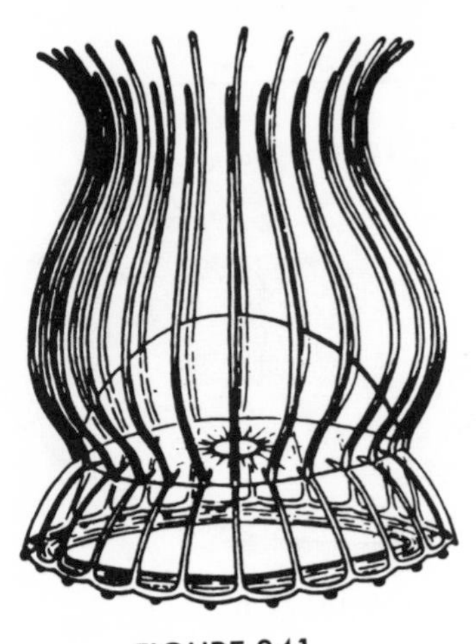

FIGURE 341

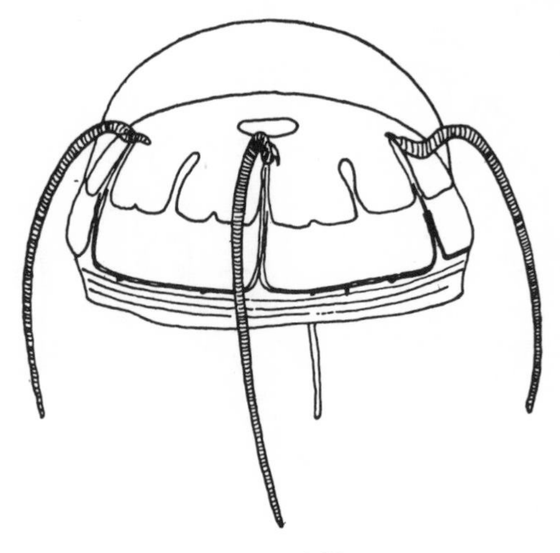

FIGURE 343

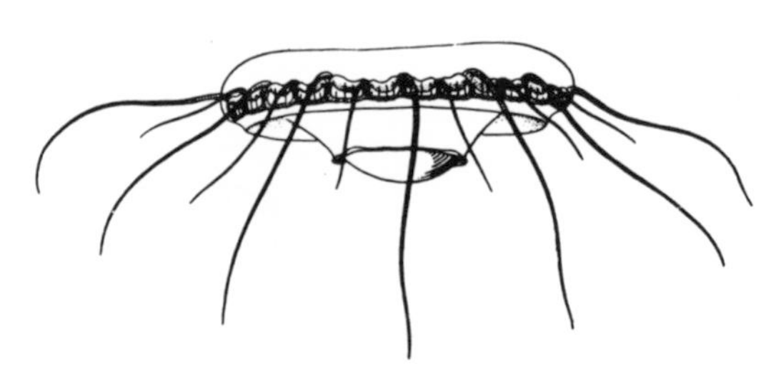

FIGURE 342

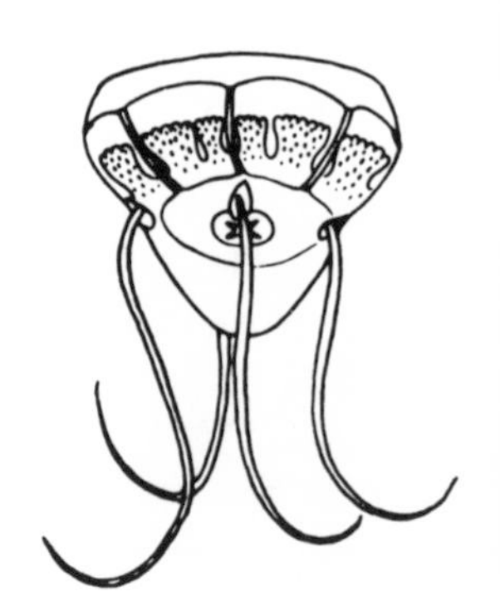

FIGURE 344

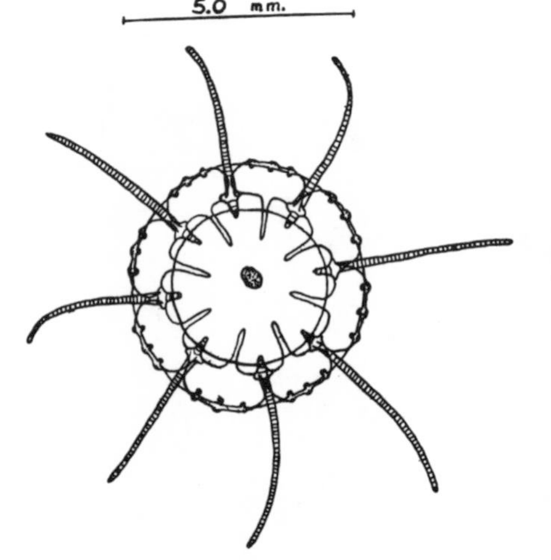

FIGURE 345

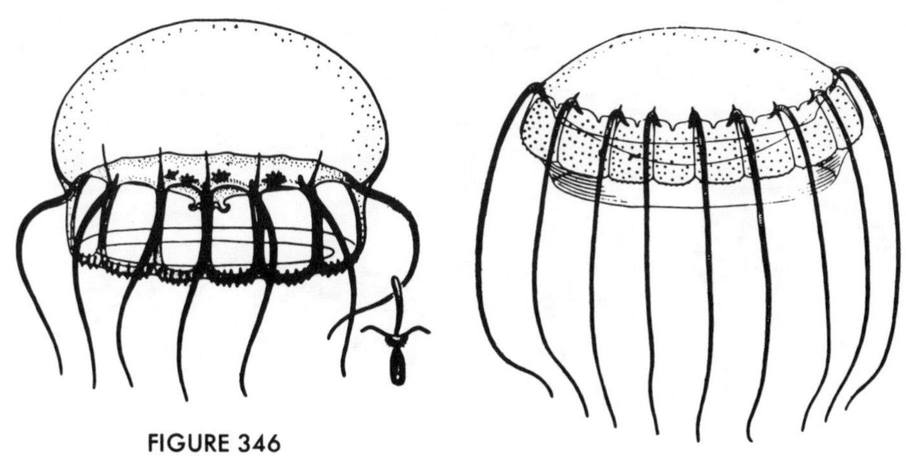

FIGURE 346

FIGURE 347

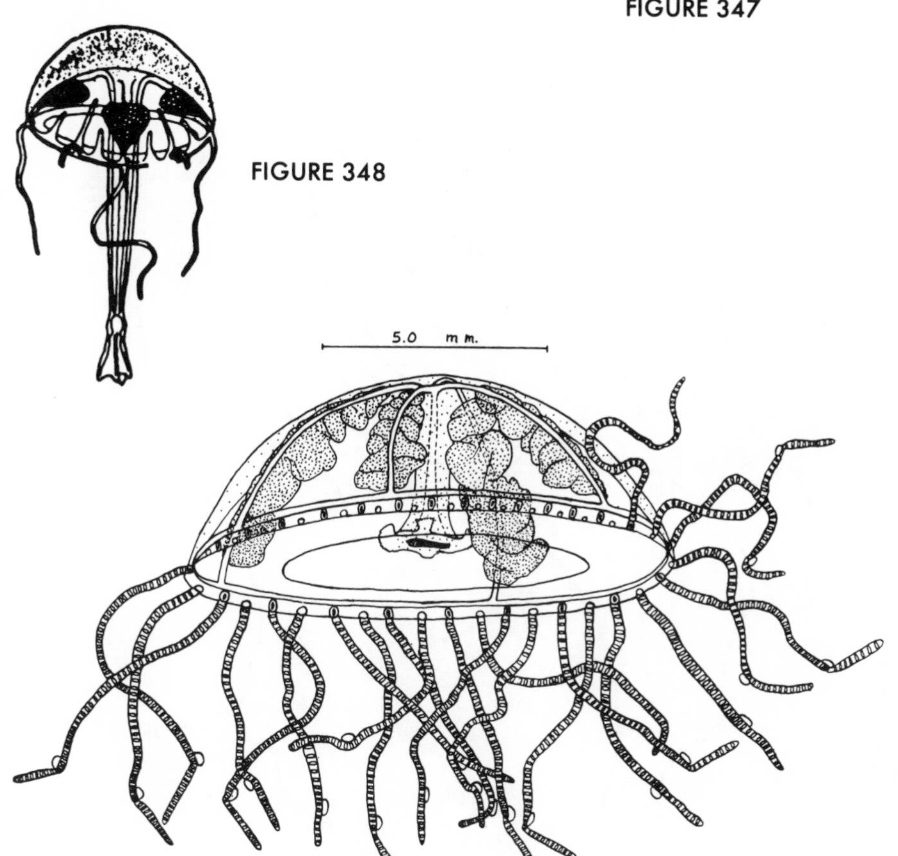

FIGURE 348

FIGURE 349

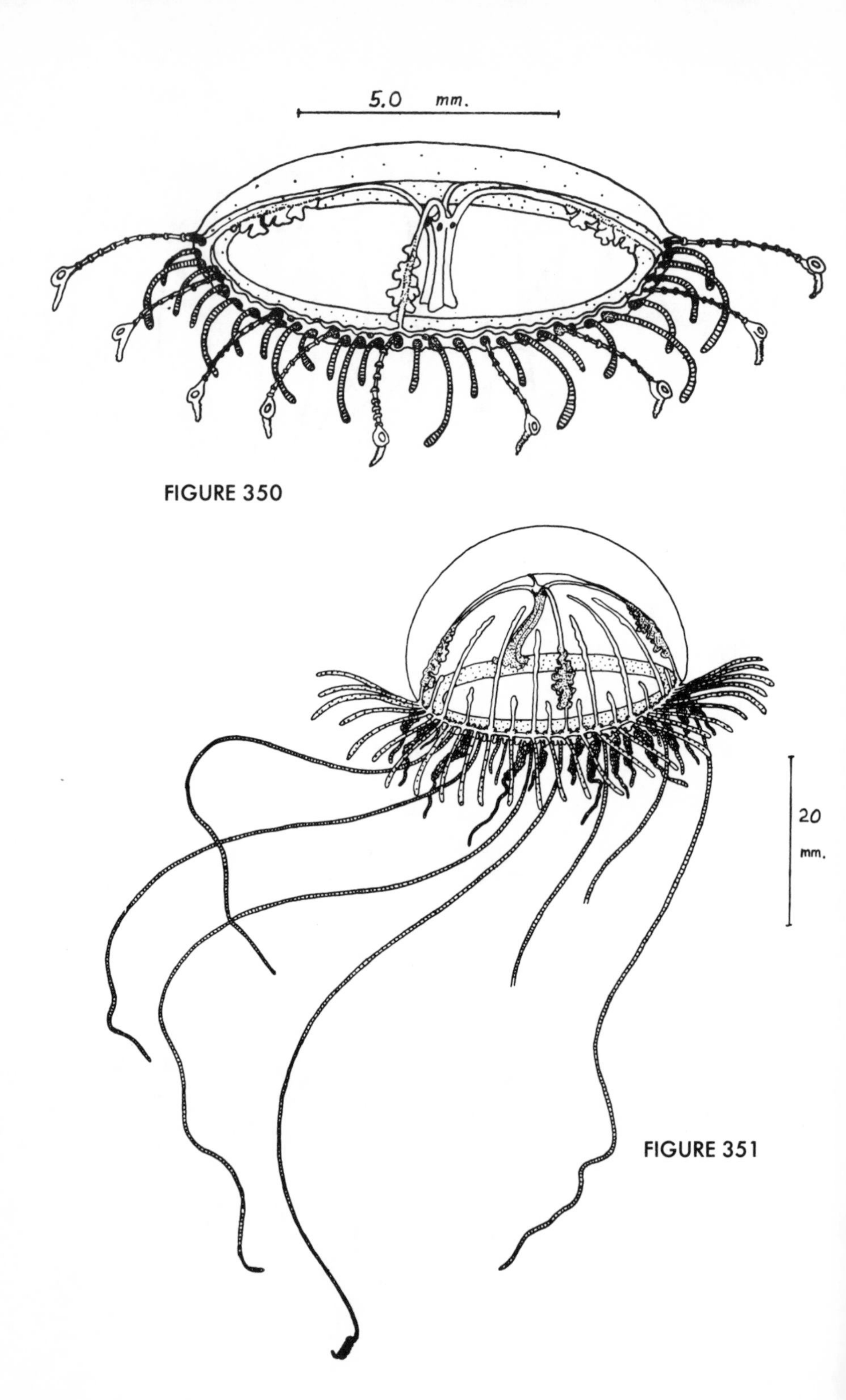

FIGURE 350

FIGURE 351

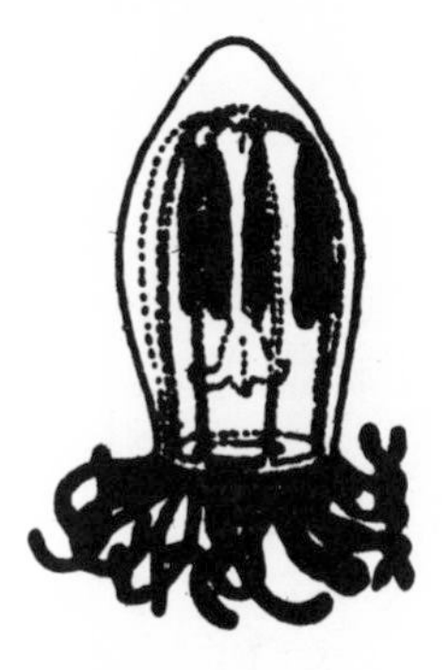
FIGURE 352

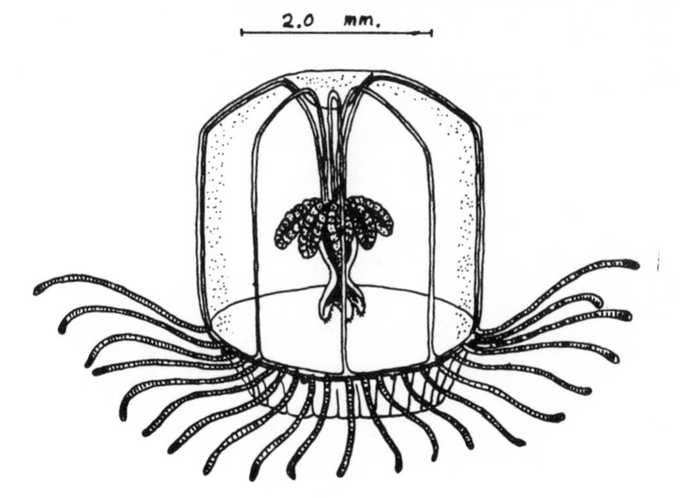

FIGURE 353

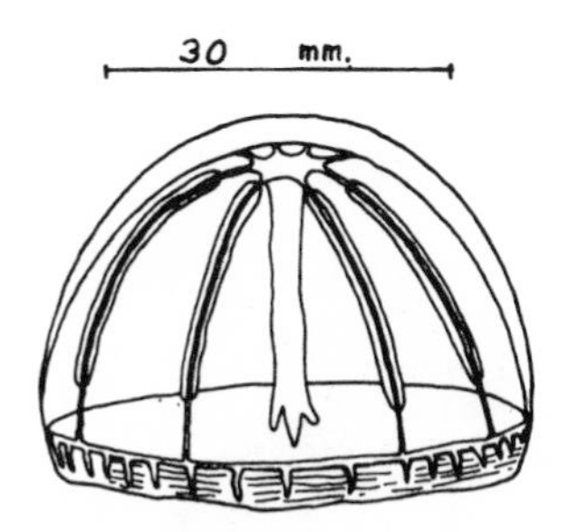

FIGURE 354

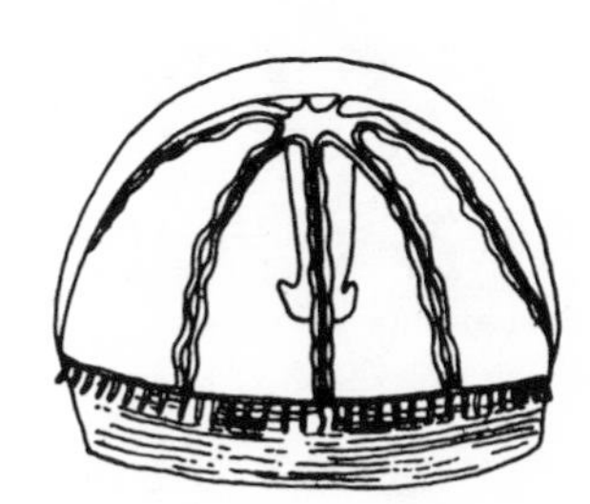
FIGURE 355

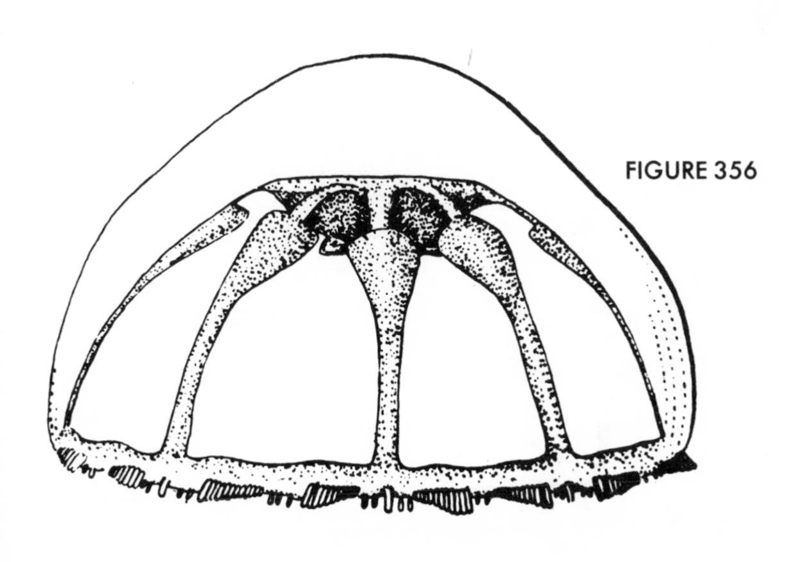
FIGURE 356

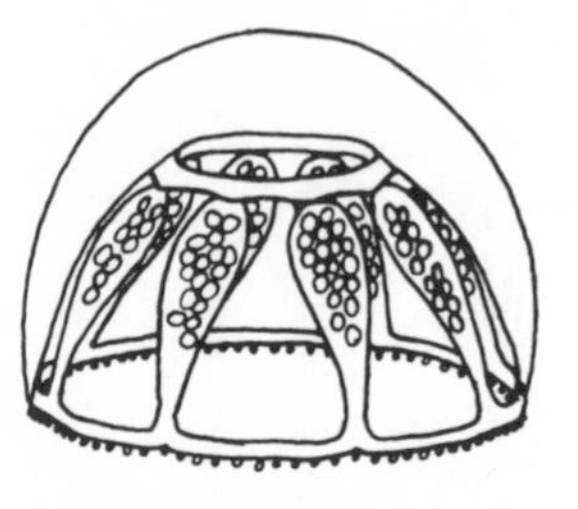

FIGURE 357

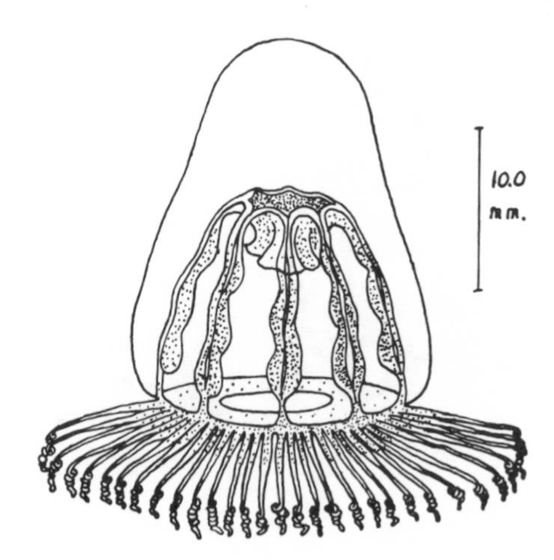

FIGURE 359

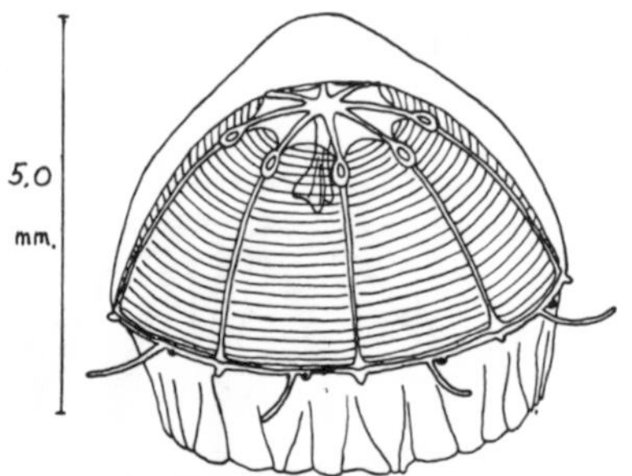

FIGURE 358

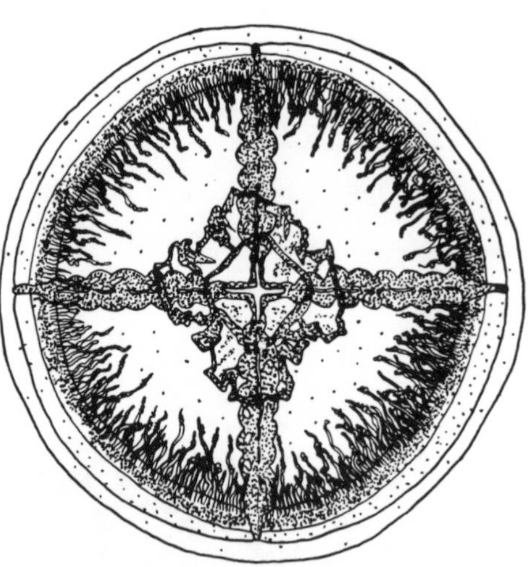

FIGURE 360 A

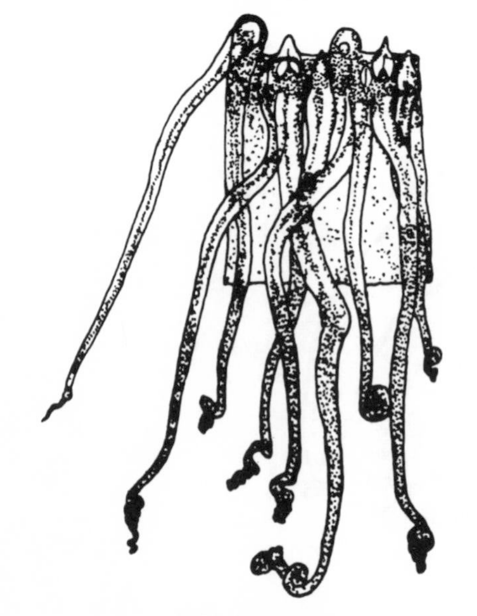

FIGURE 360 B

FIGURE 361

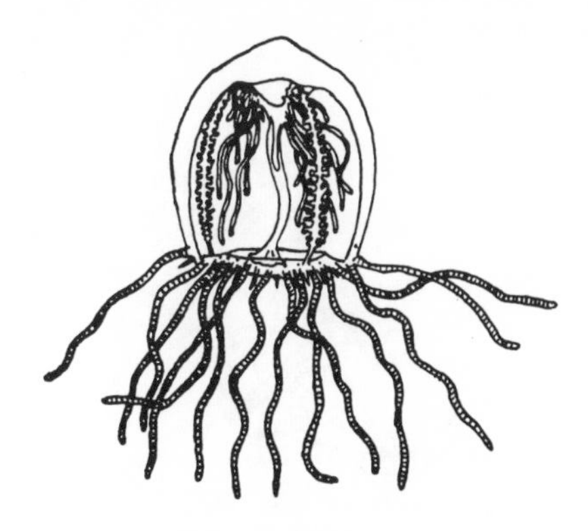

FIGURE 362

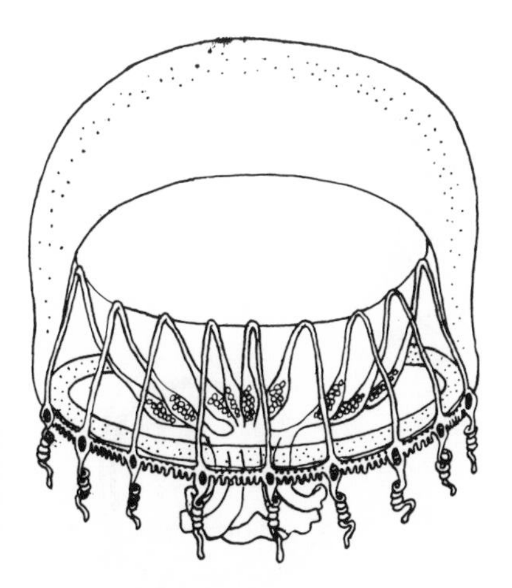

FIGURE 363

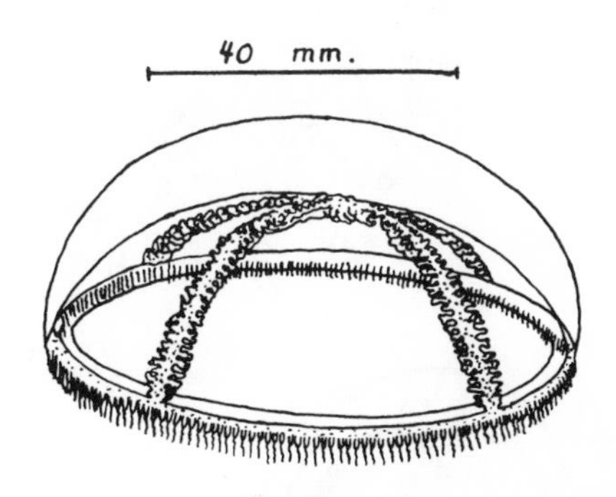

FIGURE 364

FIGURE 365

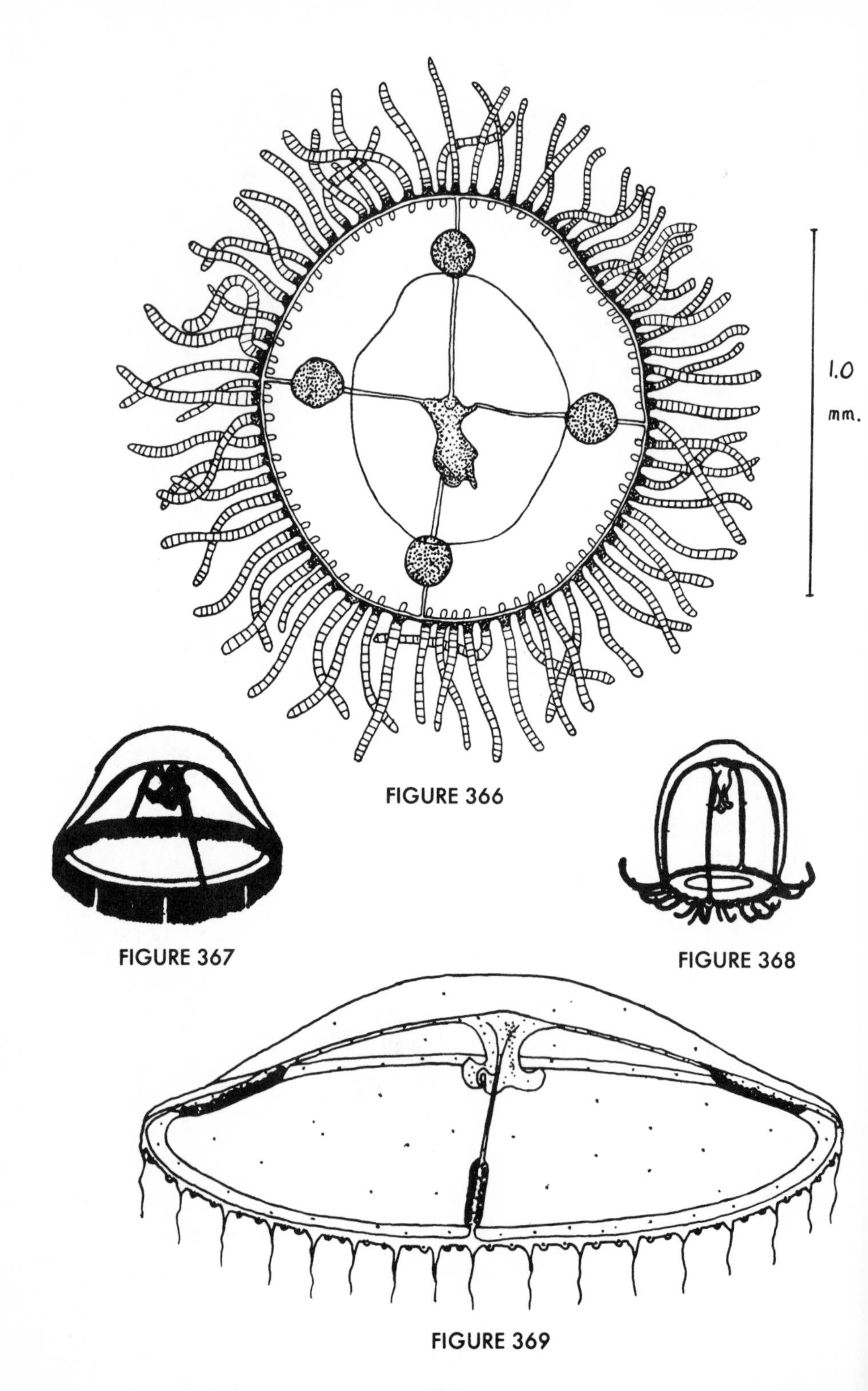

FIGURE 366

FIGURE 367

FIGURE 368

FIGURE 369

FIGURE 370

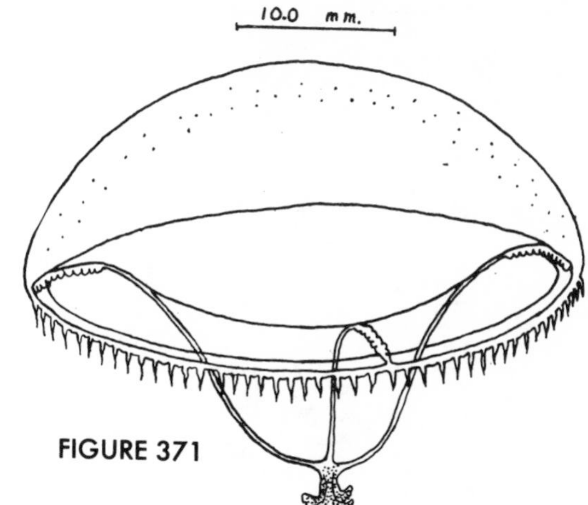

FIGURE 371

FIGURE 372

FIGURE 373

FIGURE 374

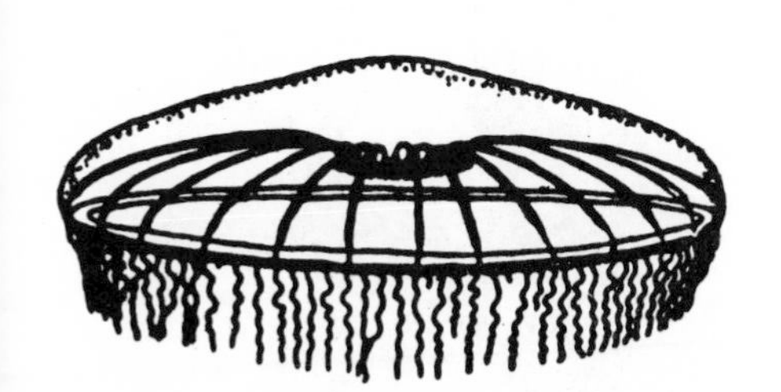
FIGURE 375

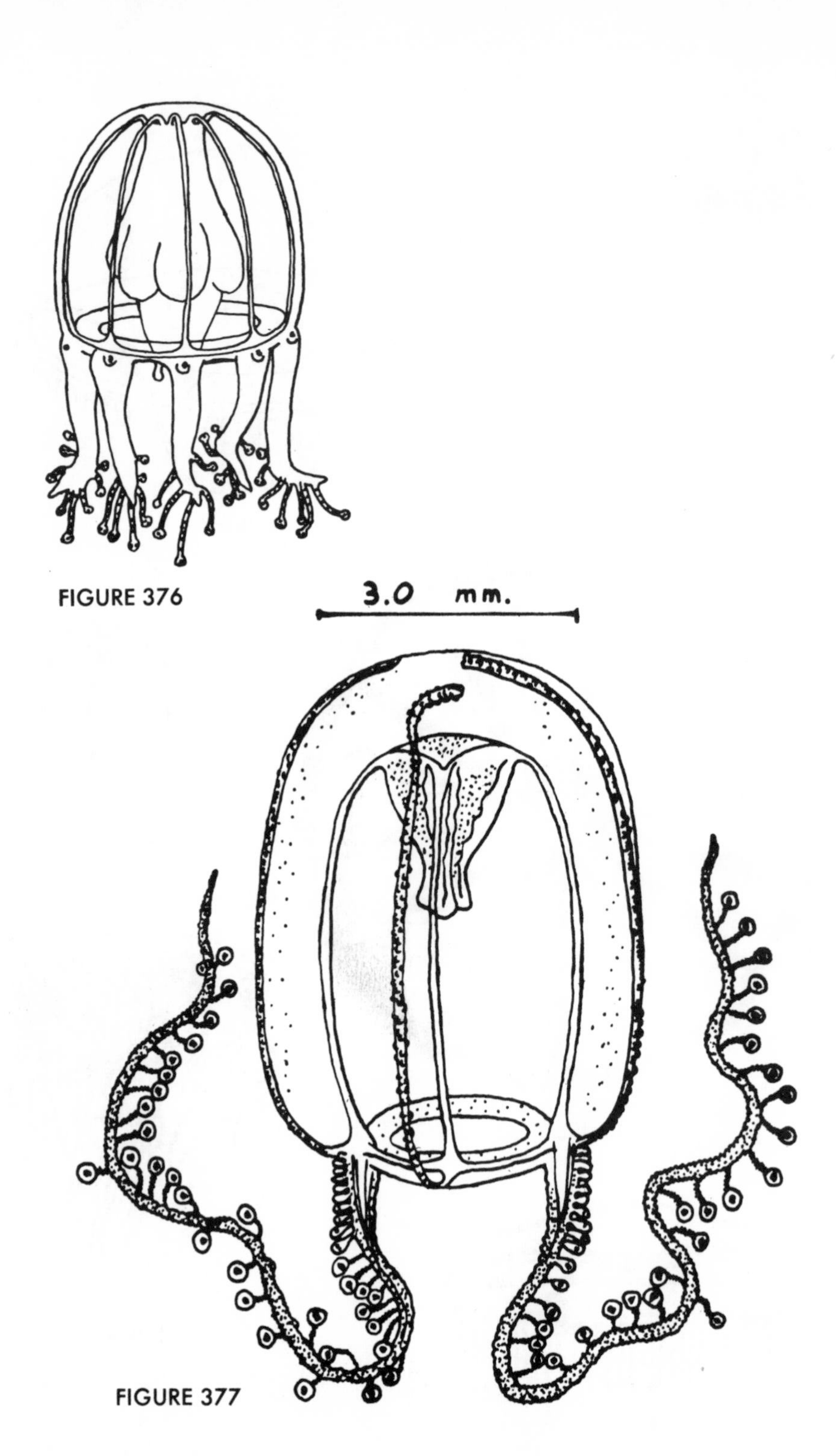

FIGURE 376

FIGURE 377

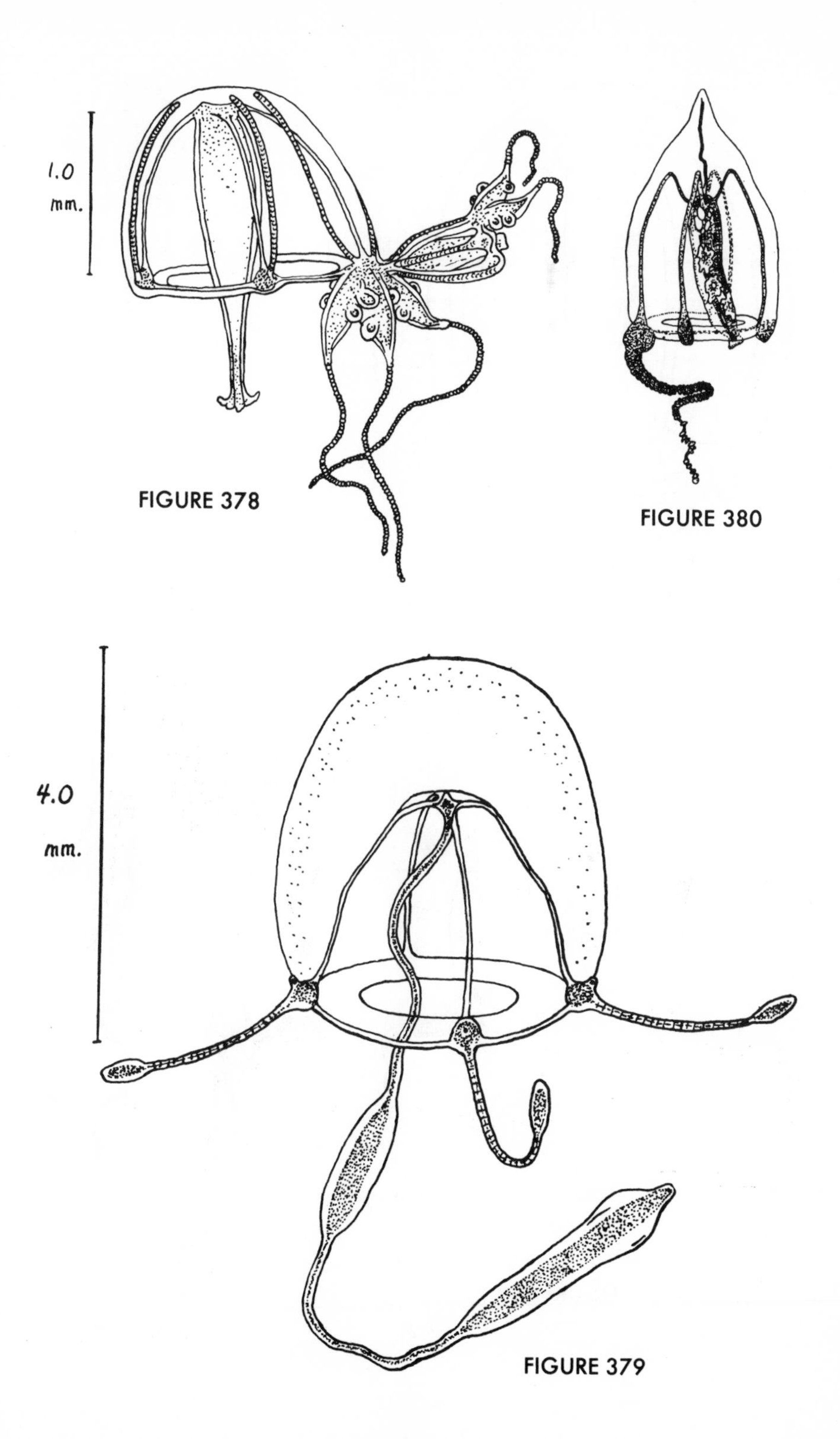

FIGURE 378

FIGURE 380

FIGURE 379

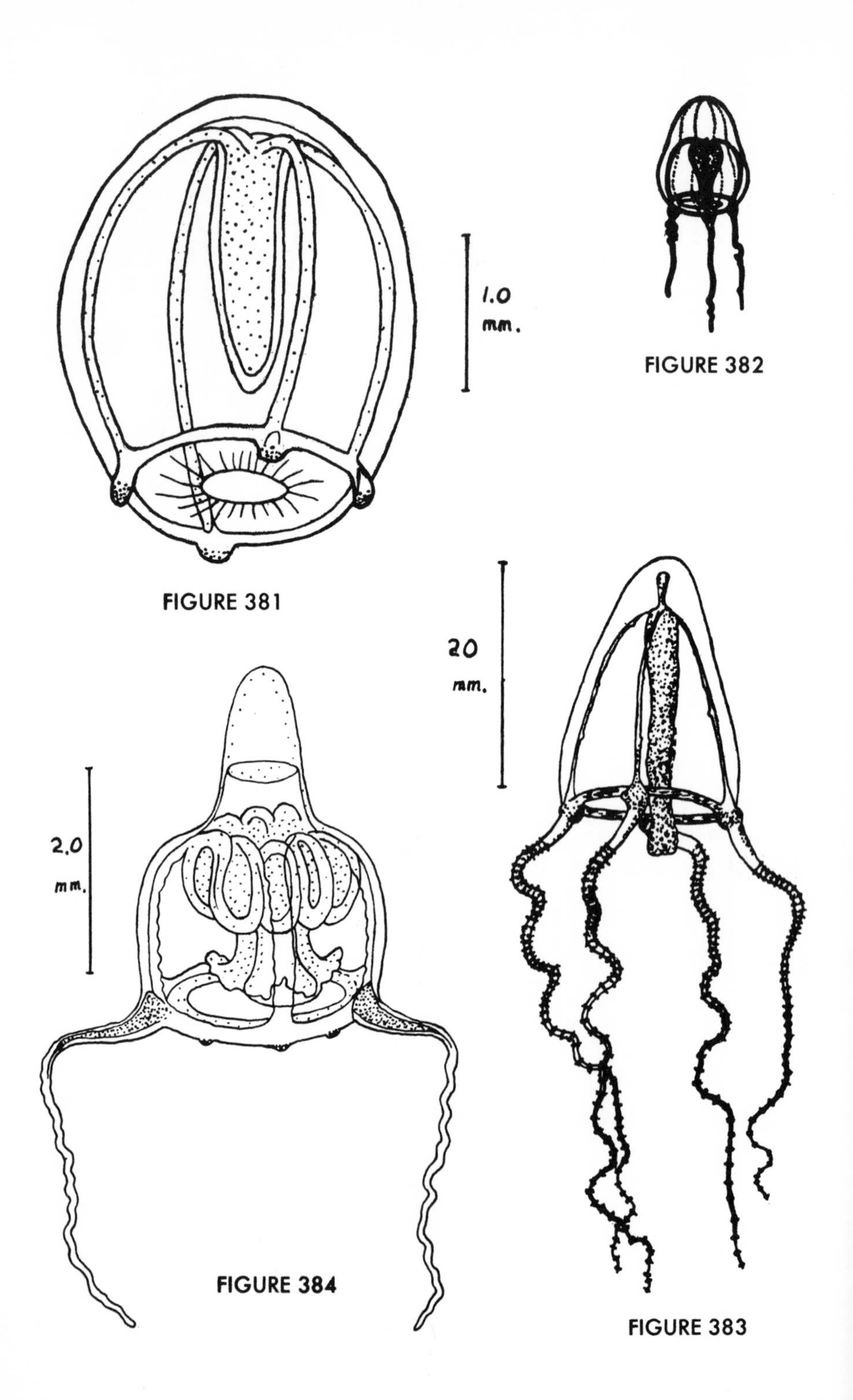

FIGURE 381

FIGURE 382

FIGURE 383

FIGURE 384

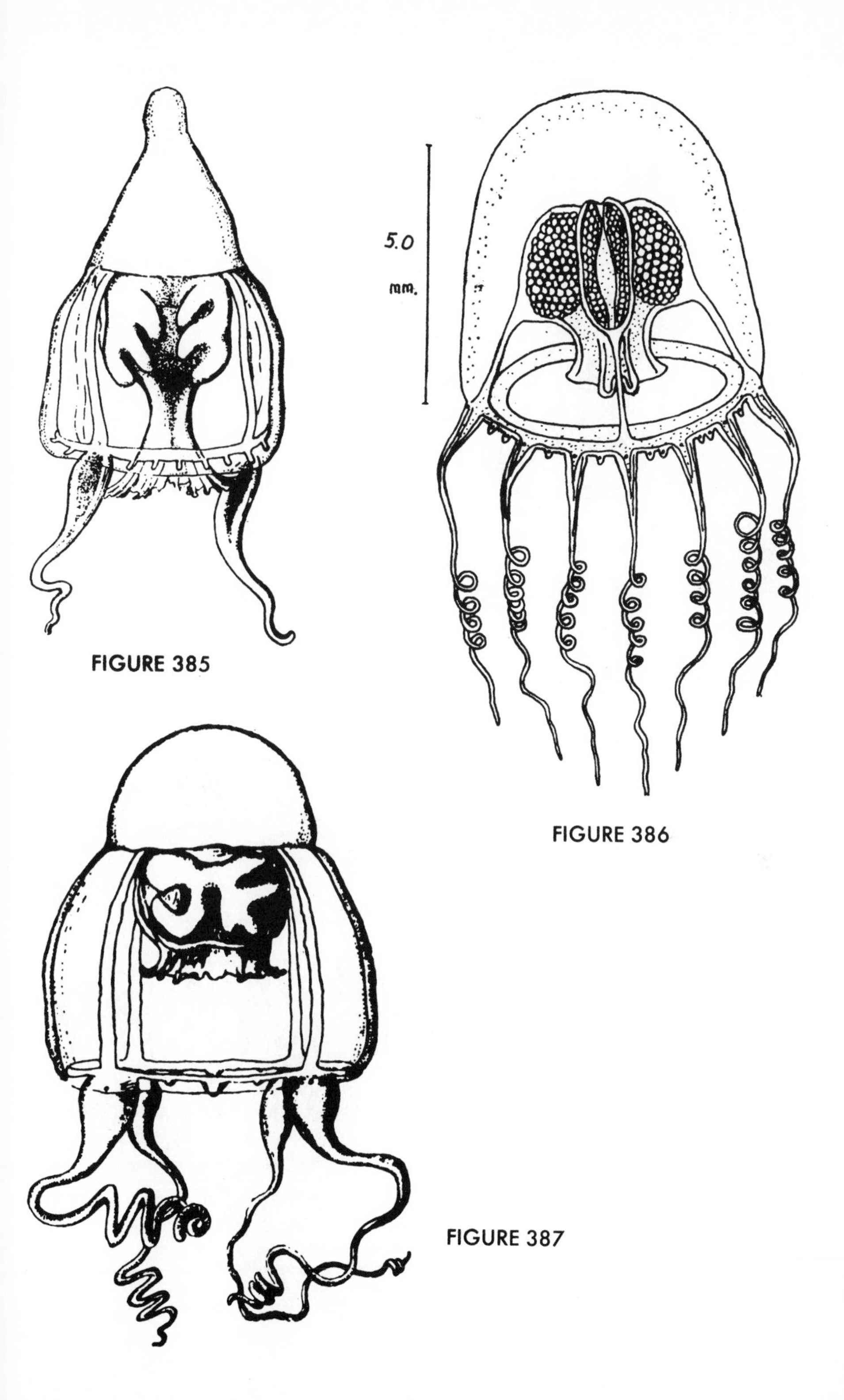

FIGURE 385

FIGURE 386

FIGURE 387

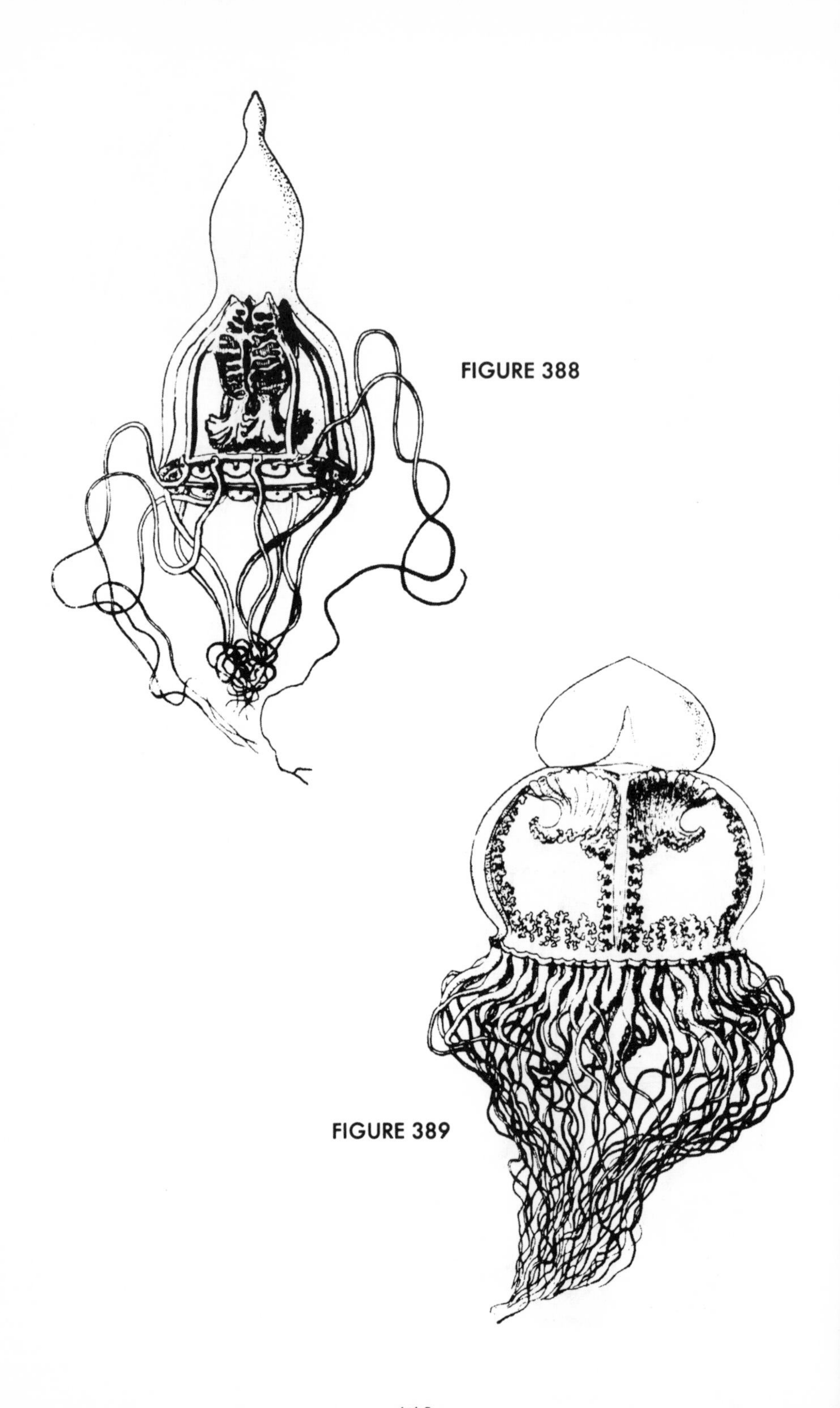

FIGURE 388

FIGURE 389

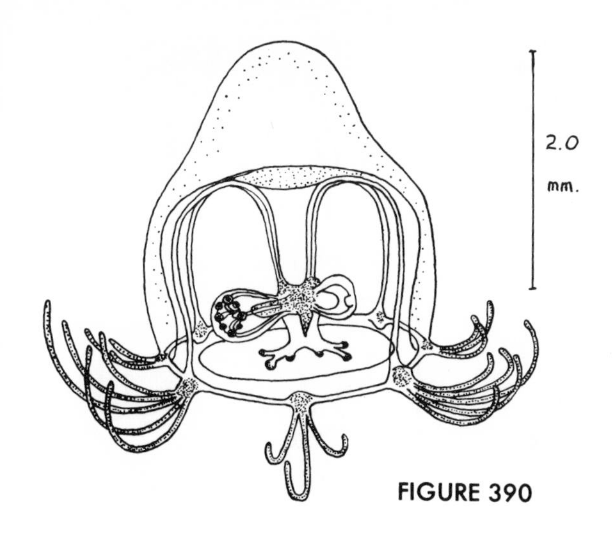

FIGURE 390

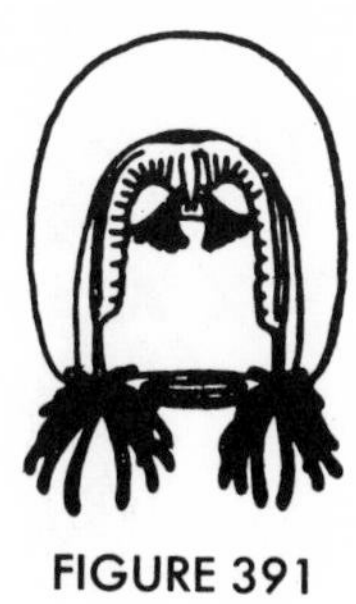

FIGURE 391

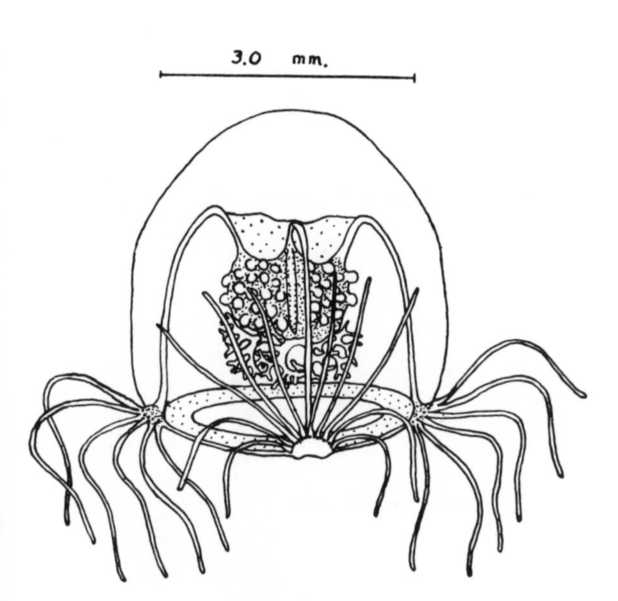

FIGURE 392

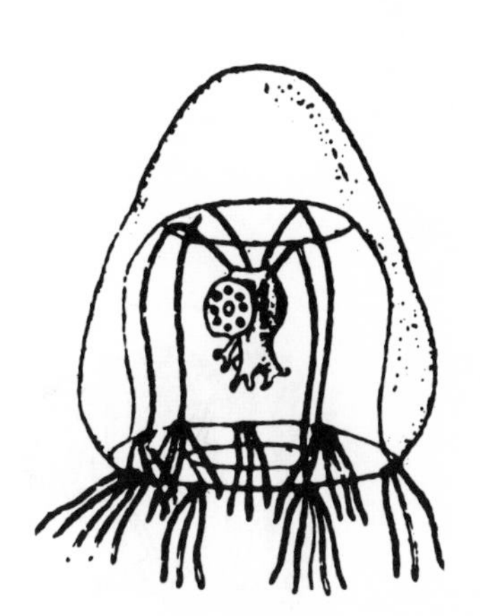

FIGURE 393

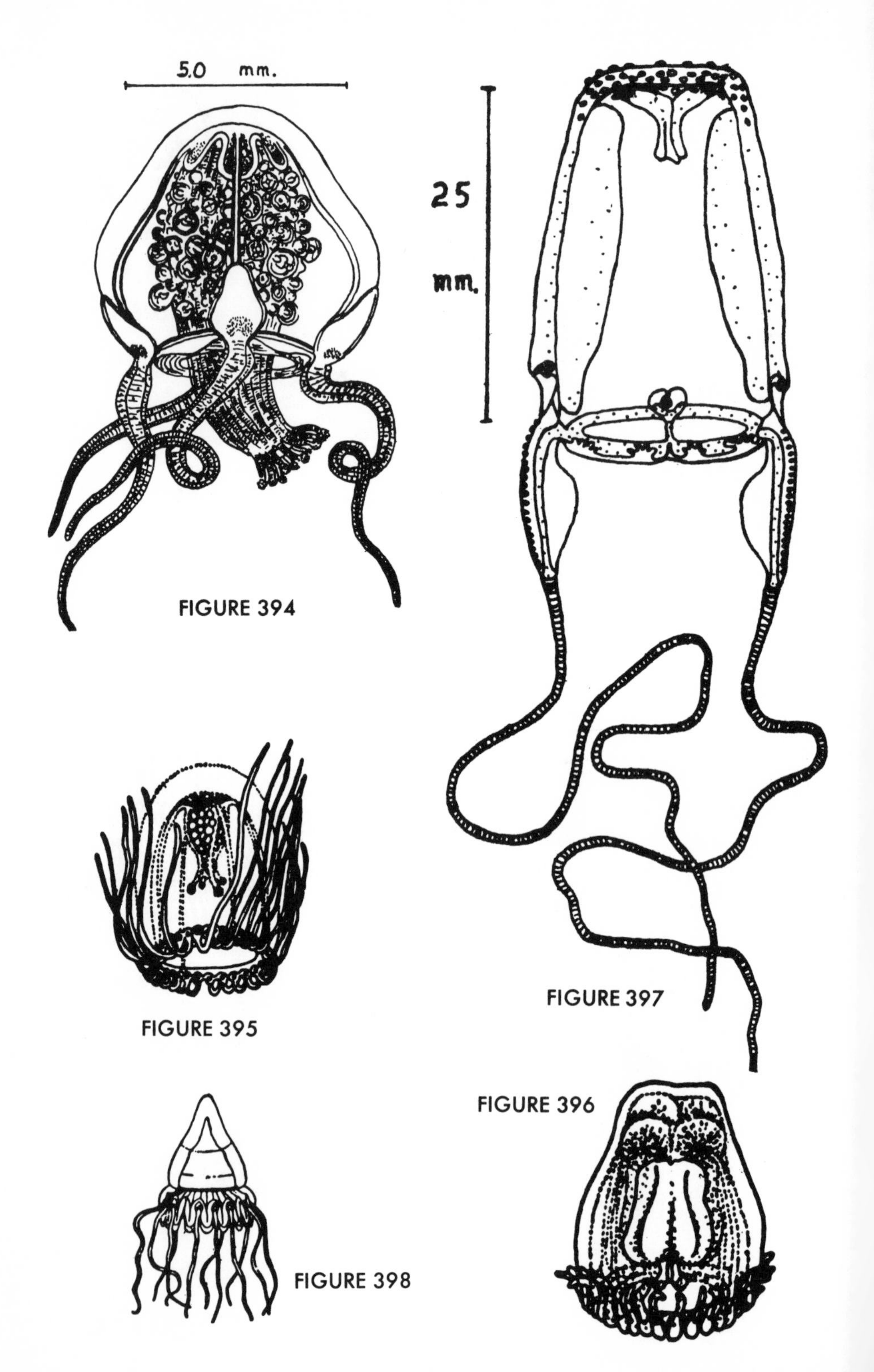

FIGURE 394

FIGURE 395

FIGURE 397

FIGURE 396

FIGURE 398

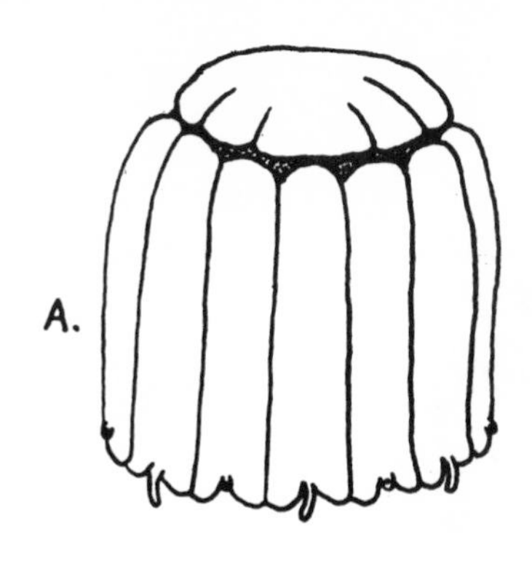

FIGURE 400

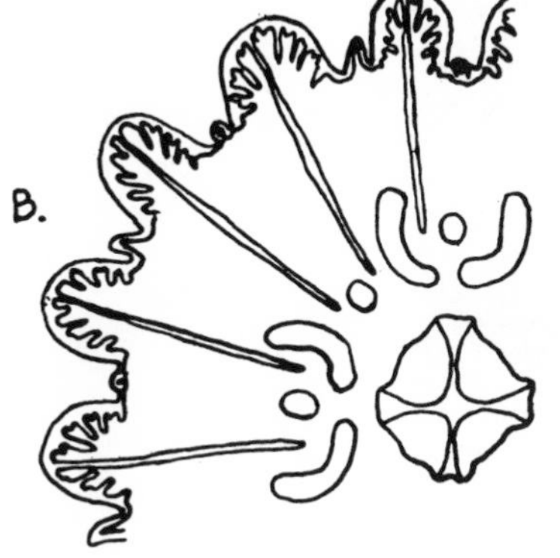

FIGURE 399

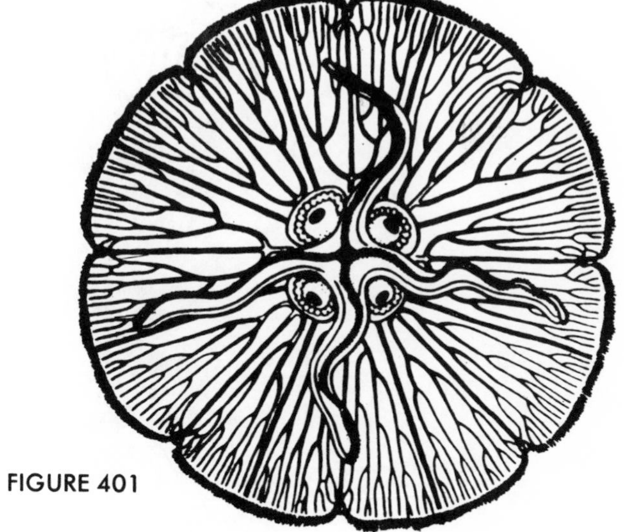

FIGURE 401

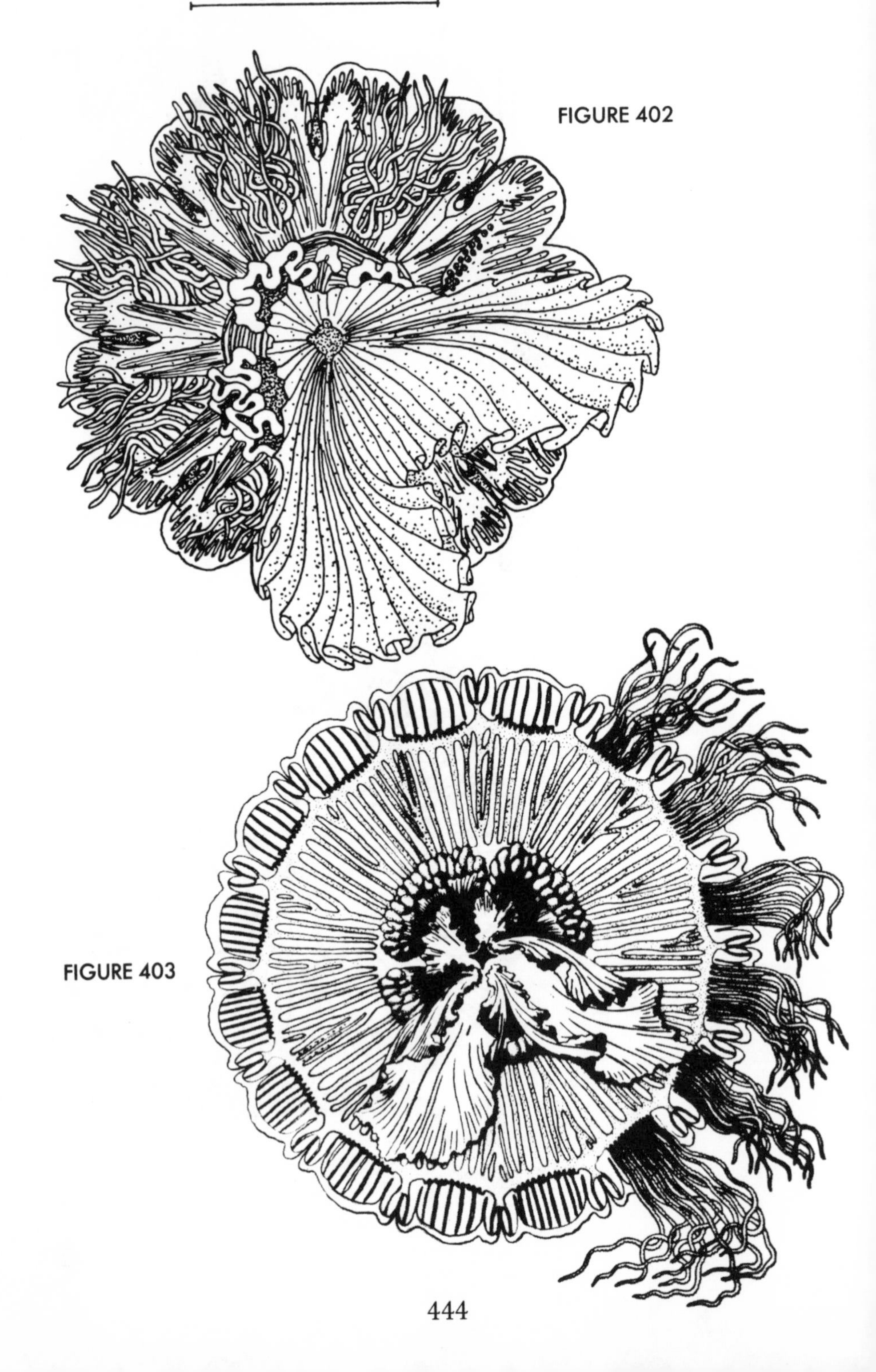
50 mm.
FIGURE 402
FIGURE 403

FIGURE 405

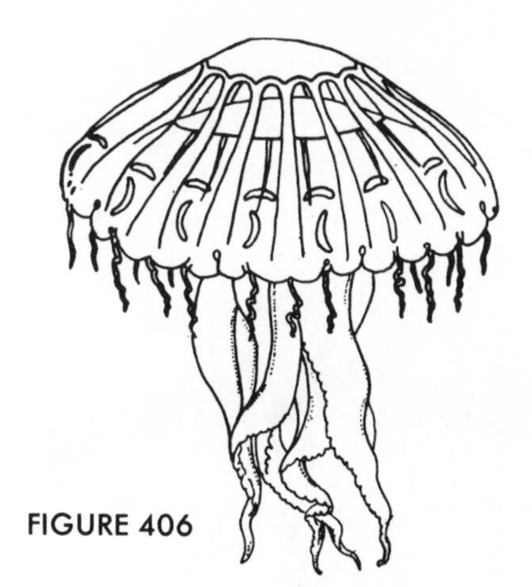

FIGURE 406

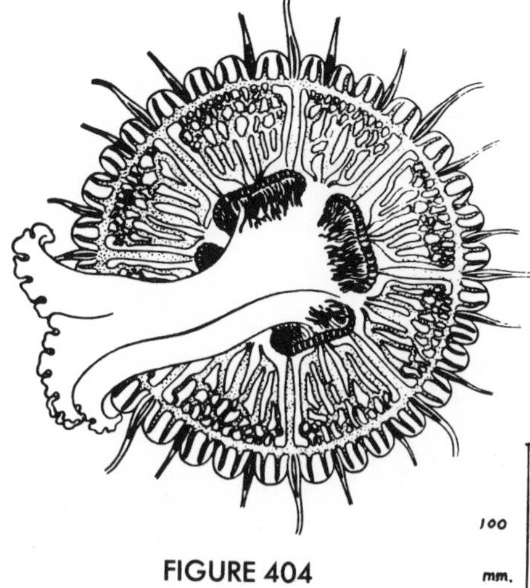

FIGURE 404

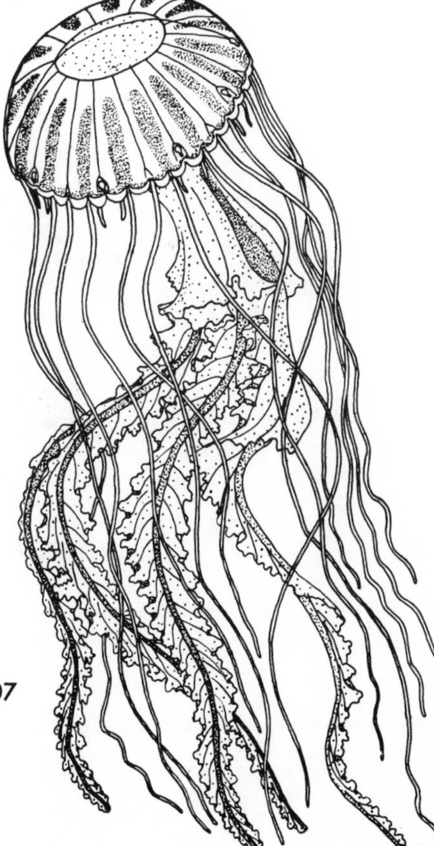

FIGURE 407

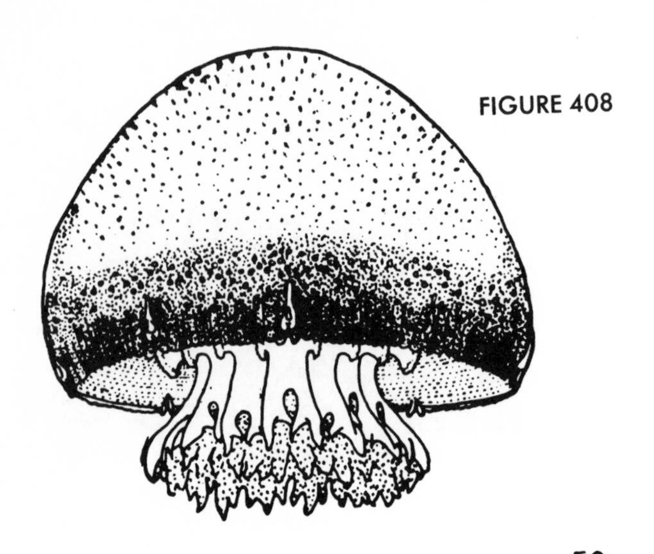

FIGURE 408

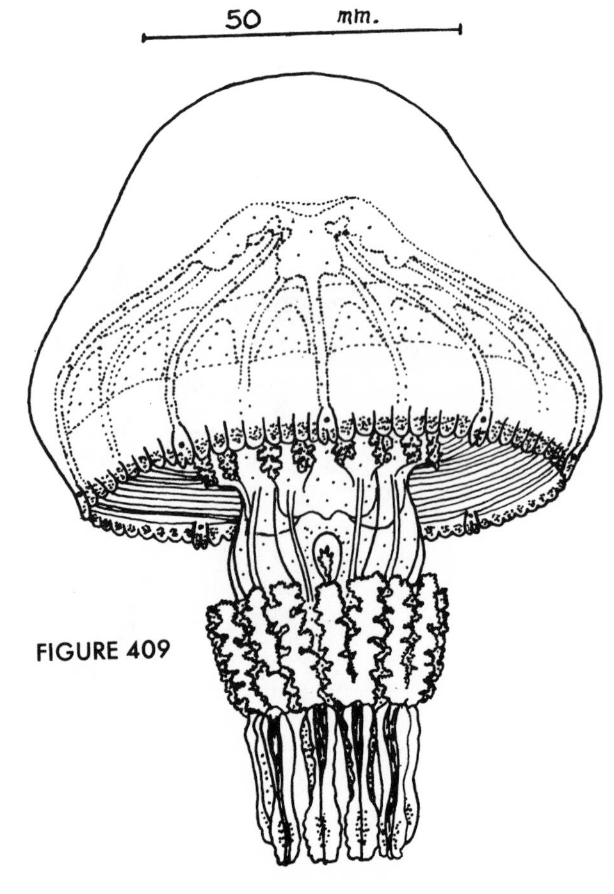

FIGURE 409

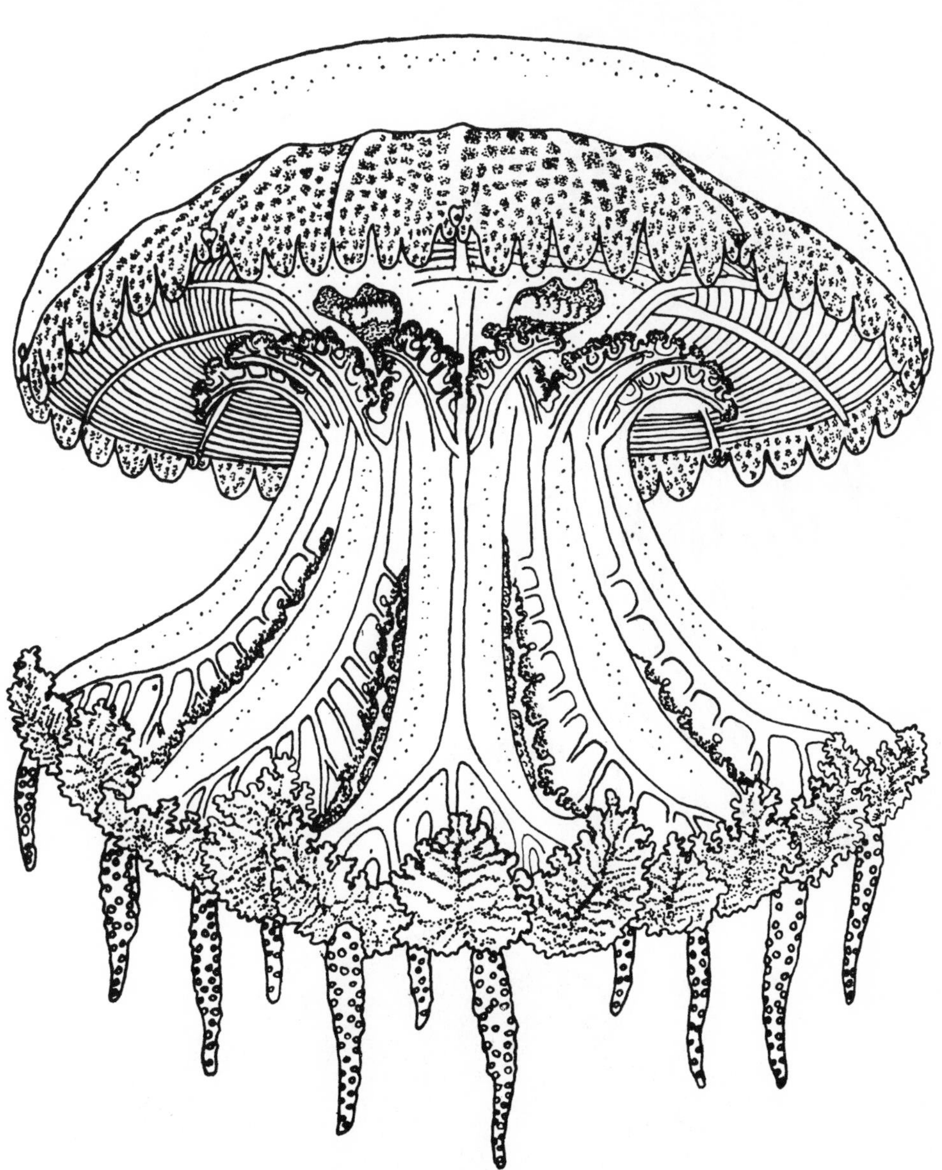

FIGURE 410

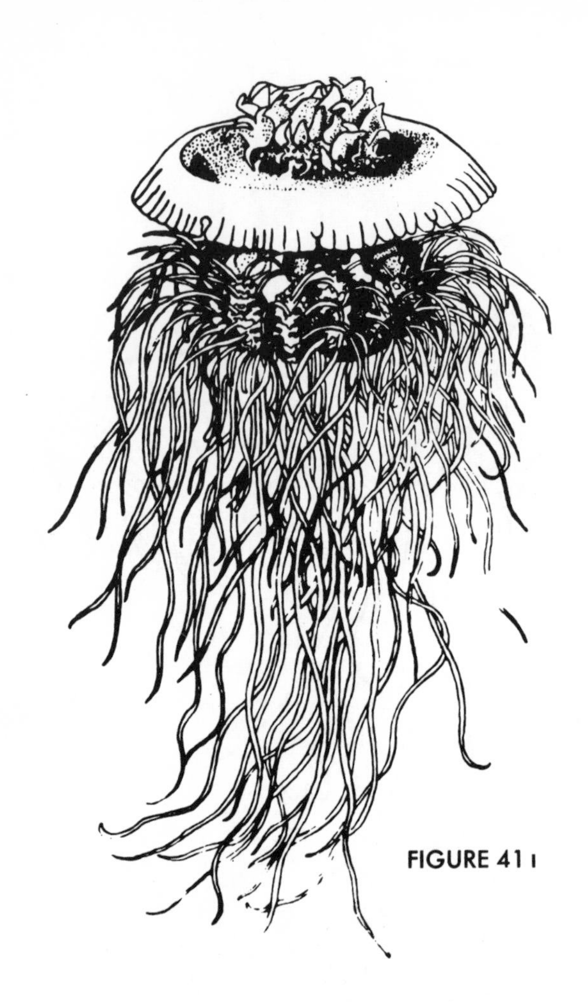

FIGURE 411

FIGURE 412

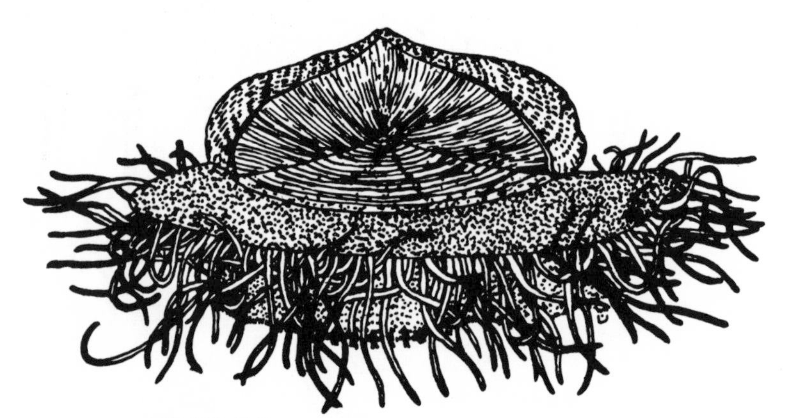

FIGURE 413

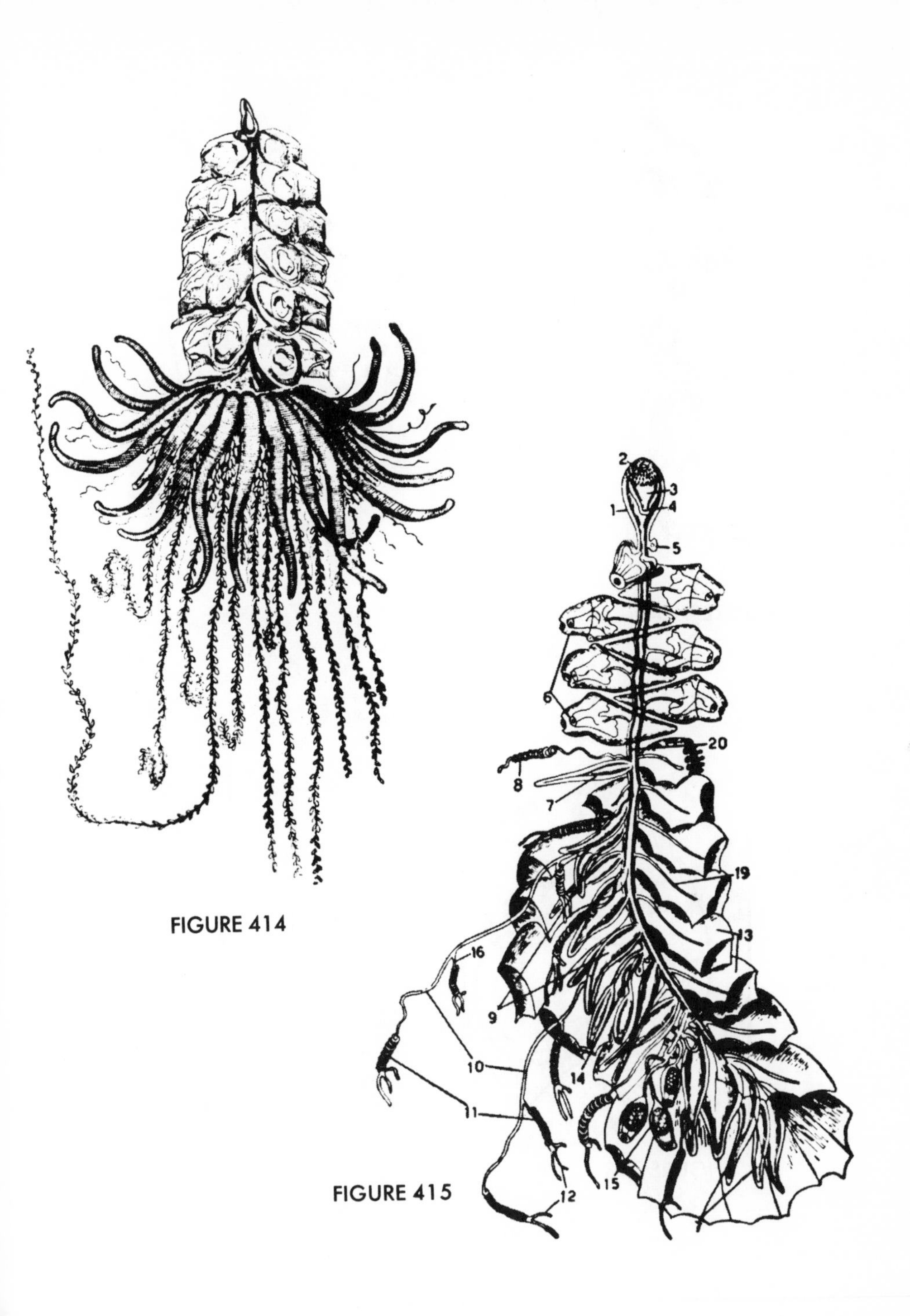

FIGURE 414

FIGURE 415

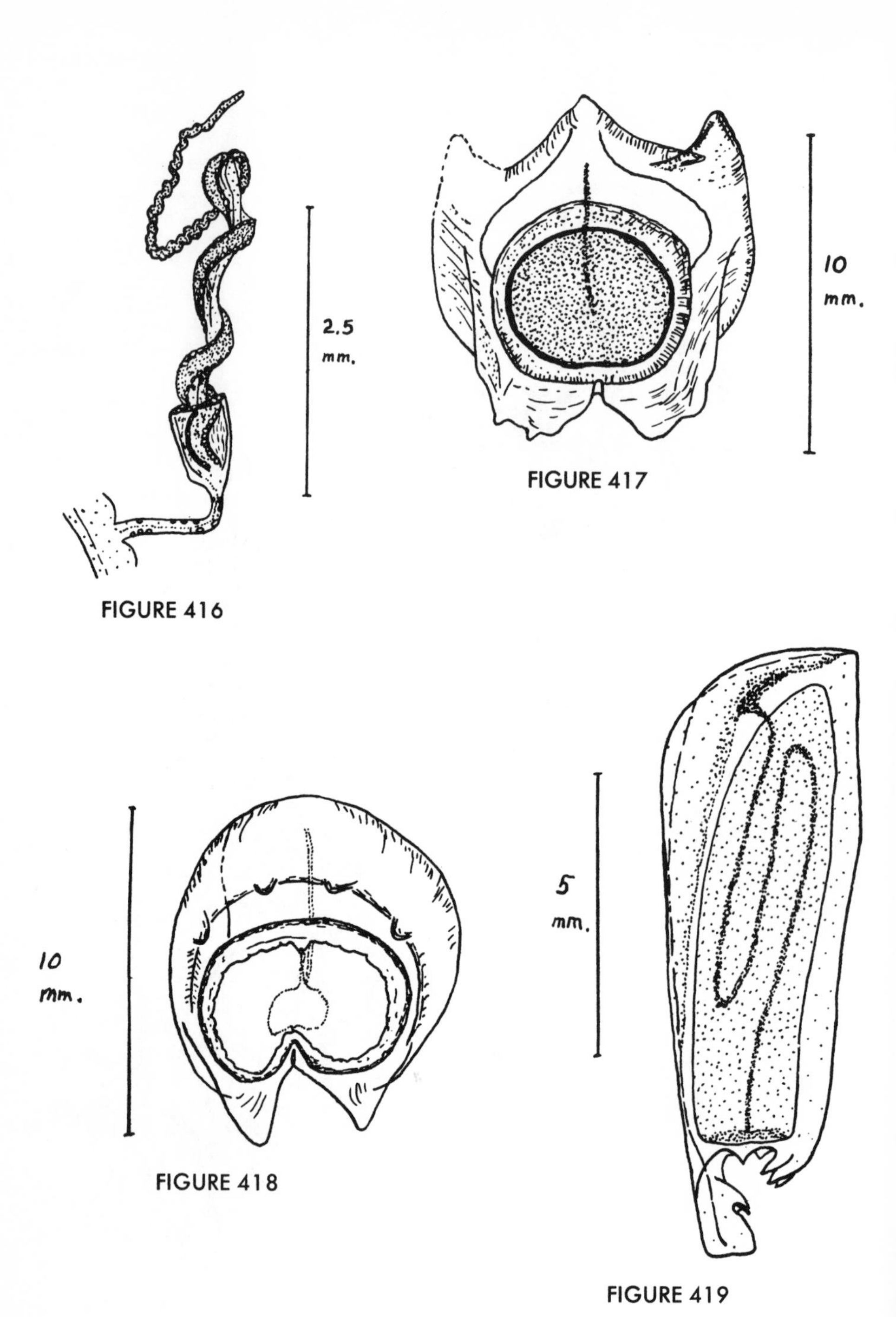

FIGURE 416

FIGURE 417

FIGURE 418

FIGURE 419

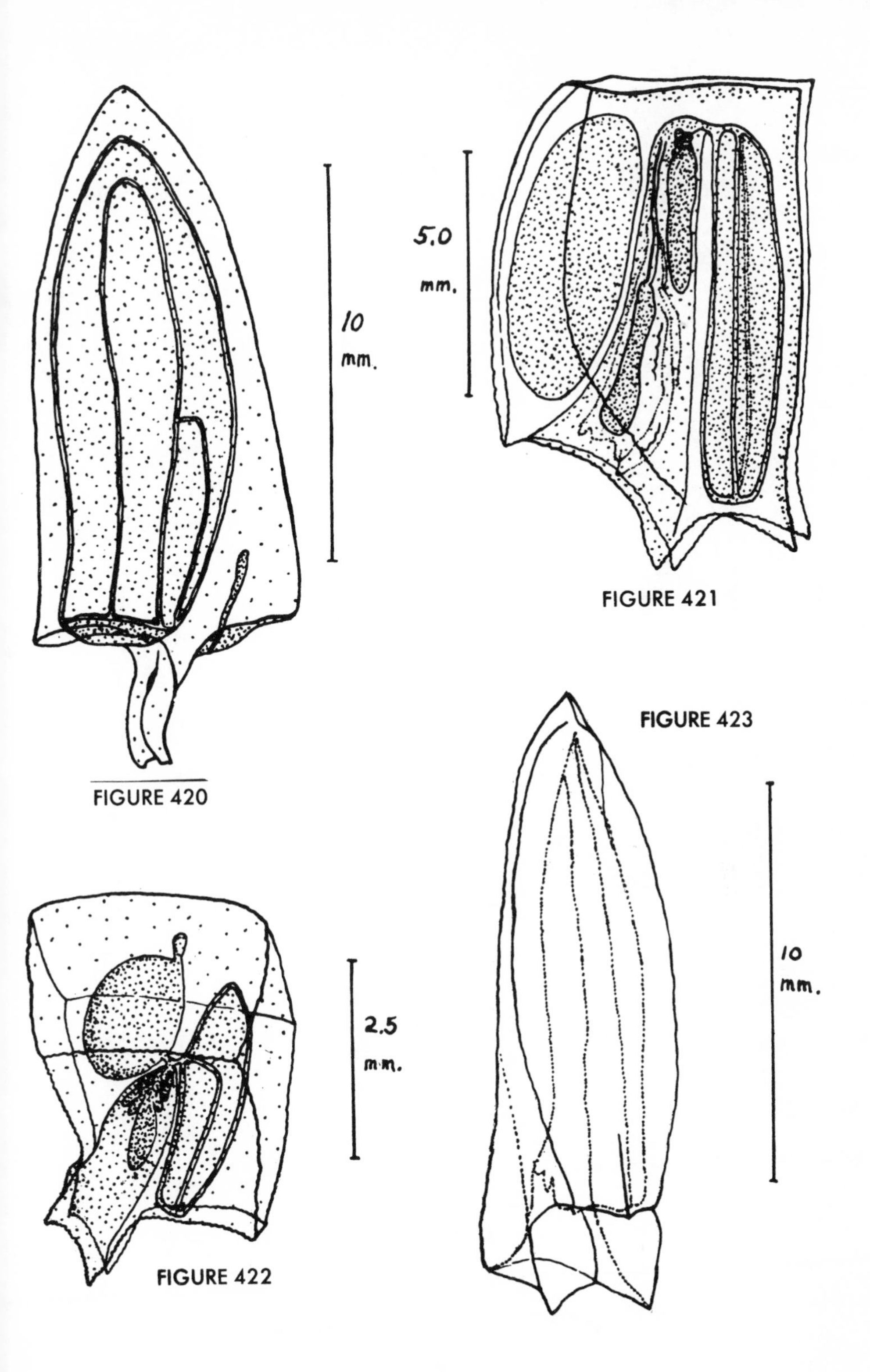

FIGURE 420

FIGURE 421

FIGURE 422

FIGURE 423

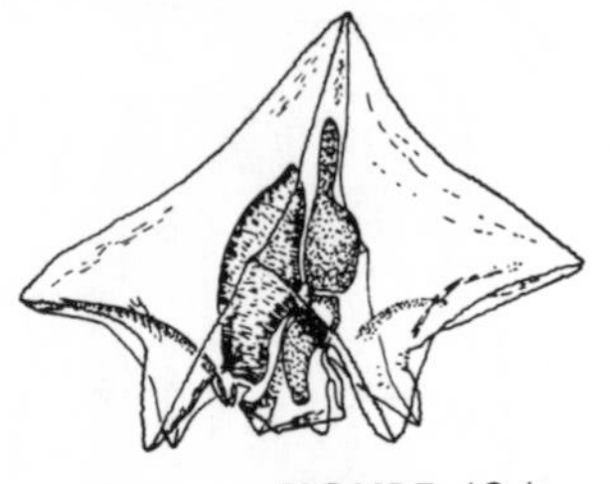

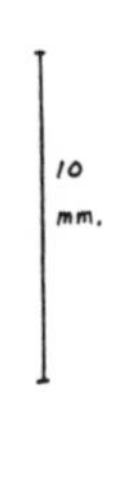

FIGURE 424

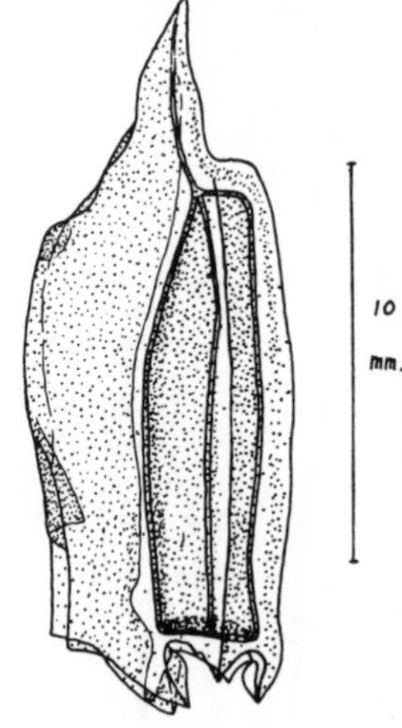

FIGURE 425

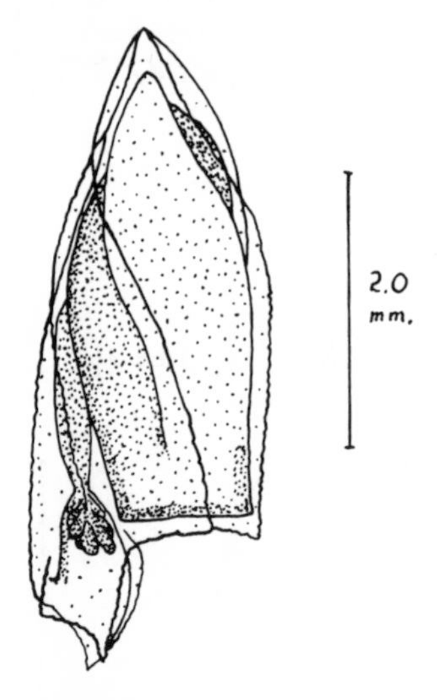

FIGURE 427

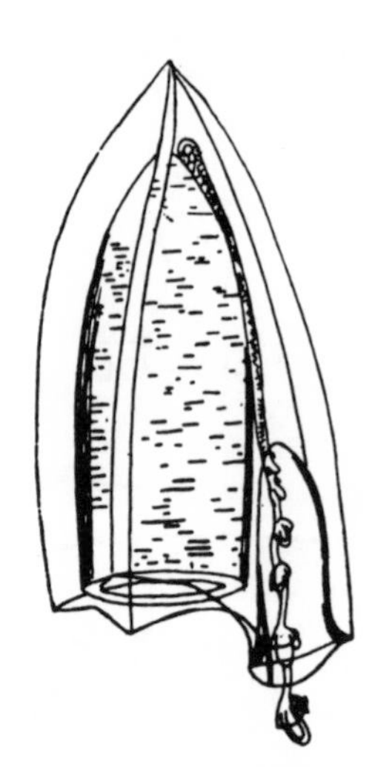

FIGURE 426

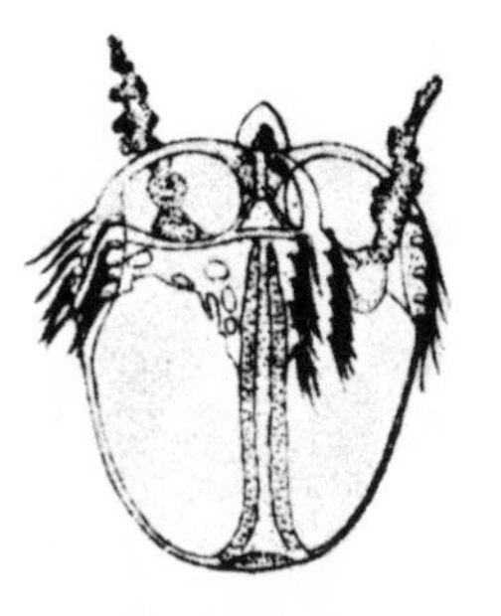

FIGURE 429

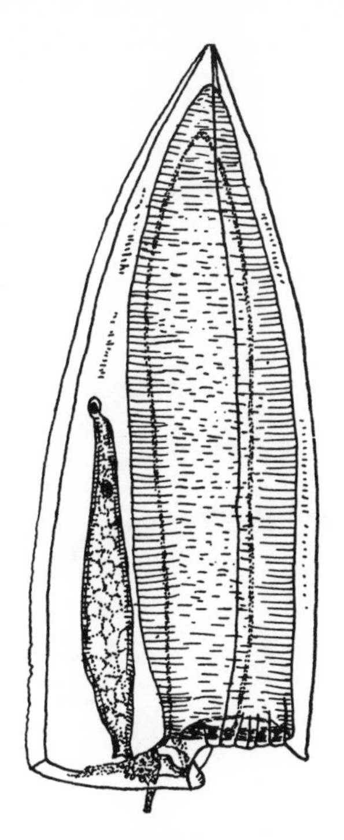

5mm.

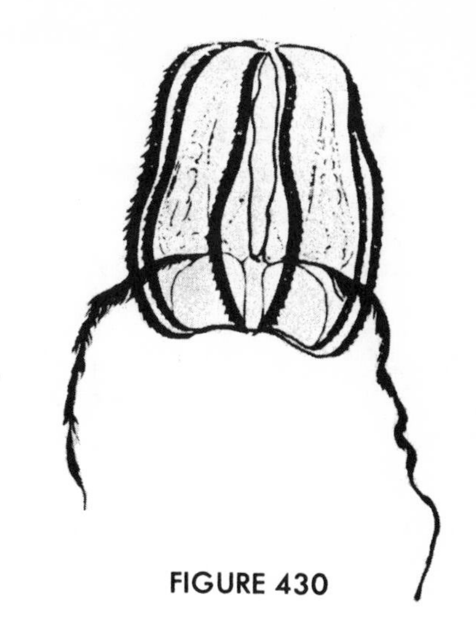

FIGURE 430

FIGURE 428

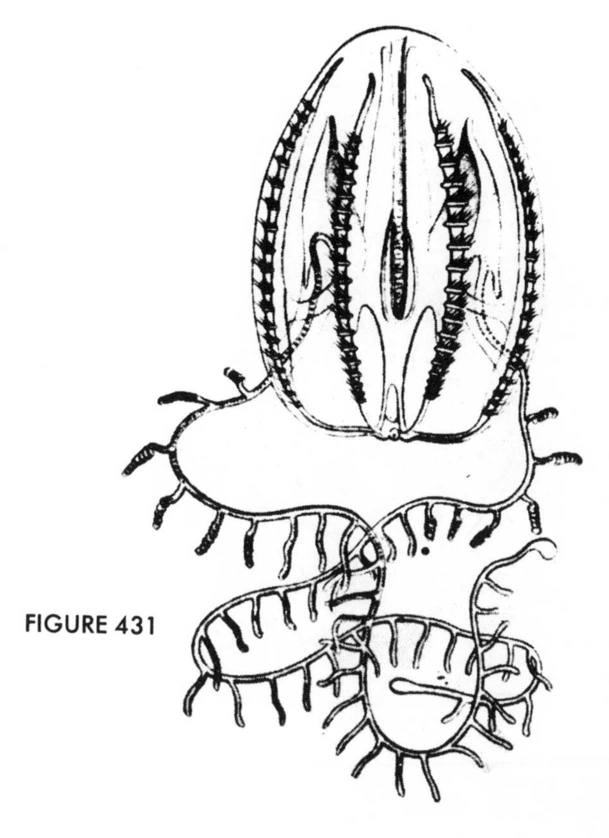

FIGURE 431

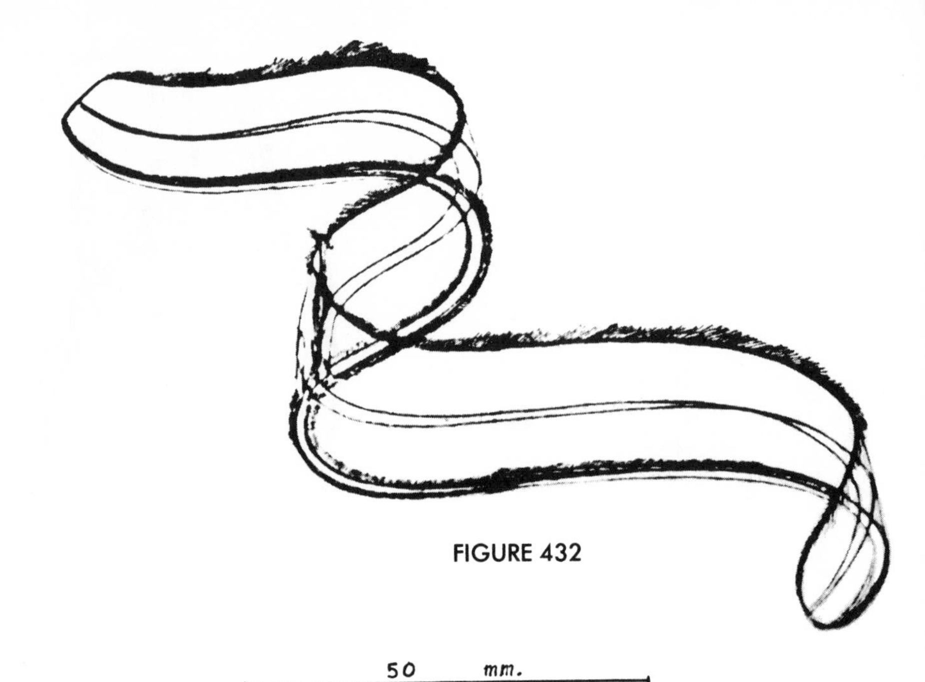

FIGURE 432

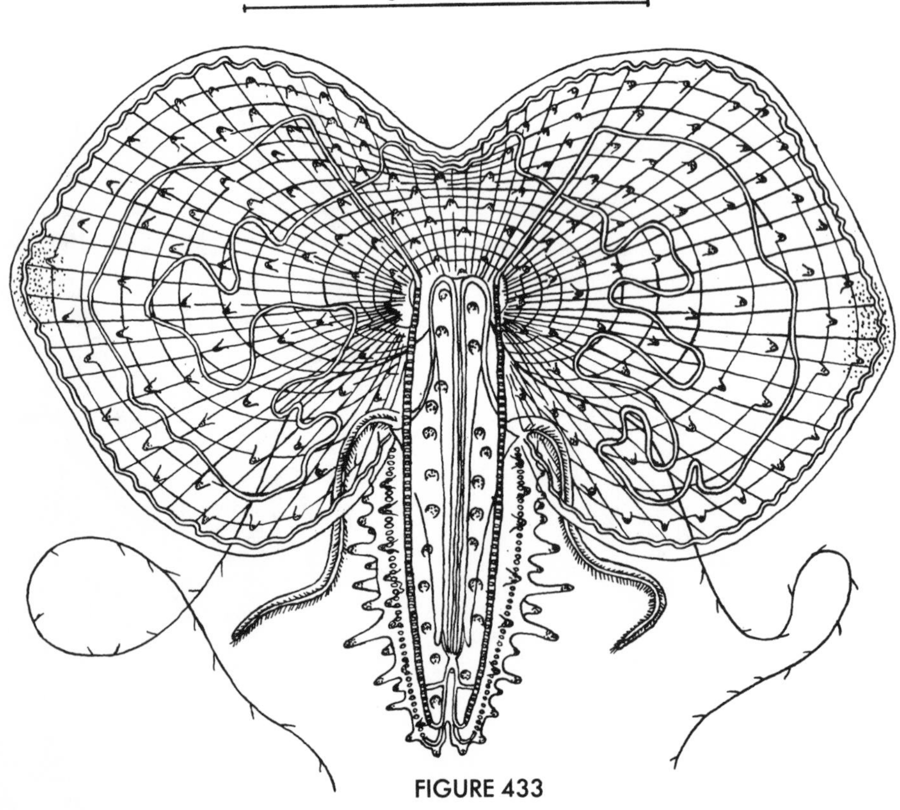

FIGURE 433

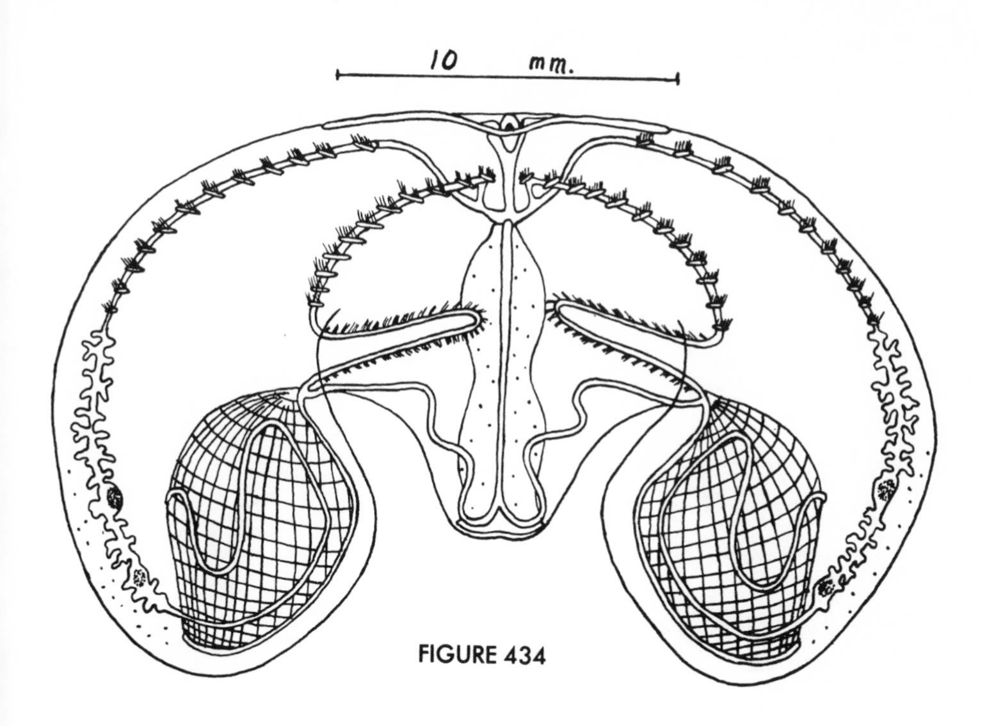

FIGURE 434

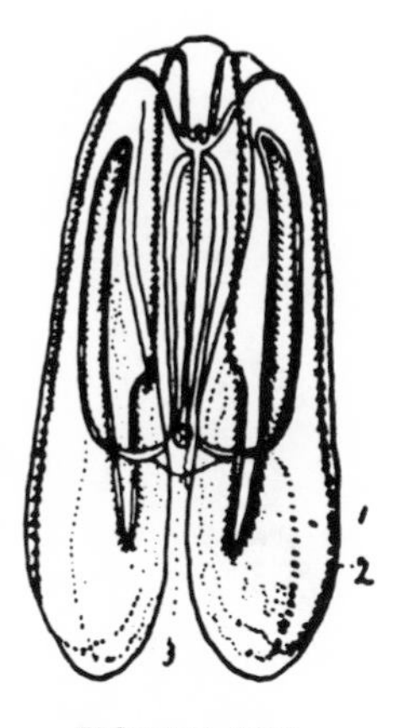

FIGURE 435

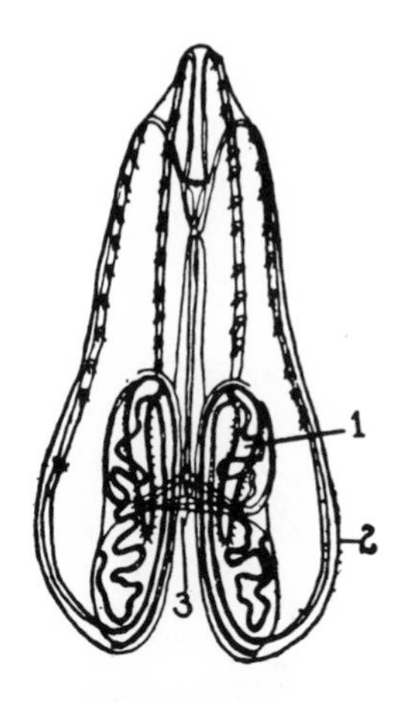

FIGURE 436

FIGURE 437

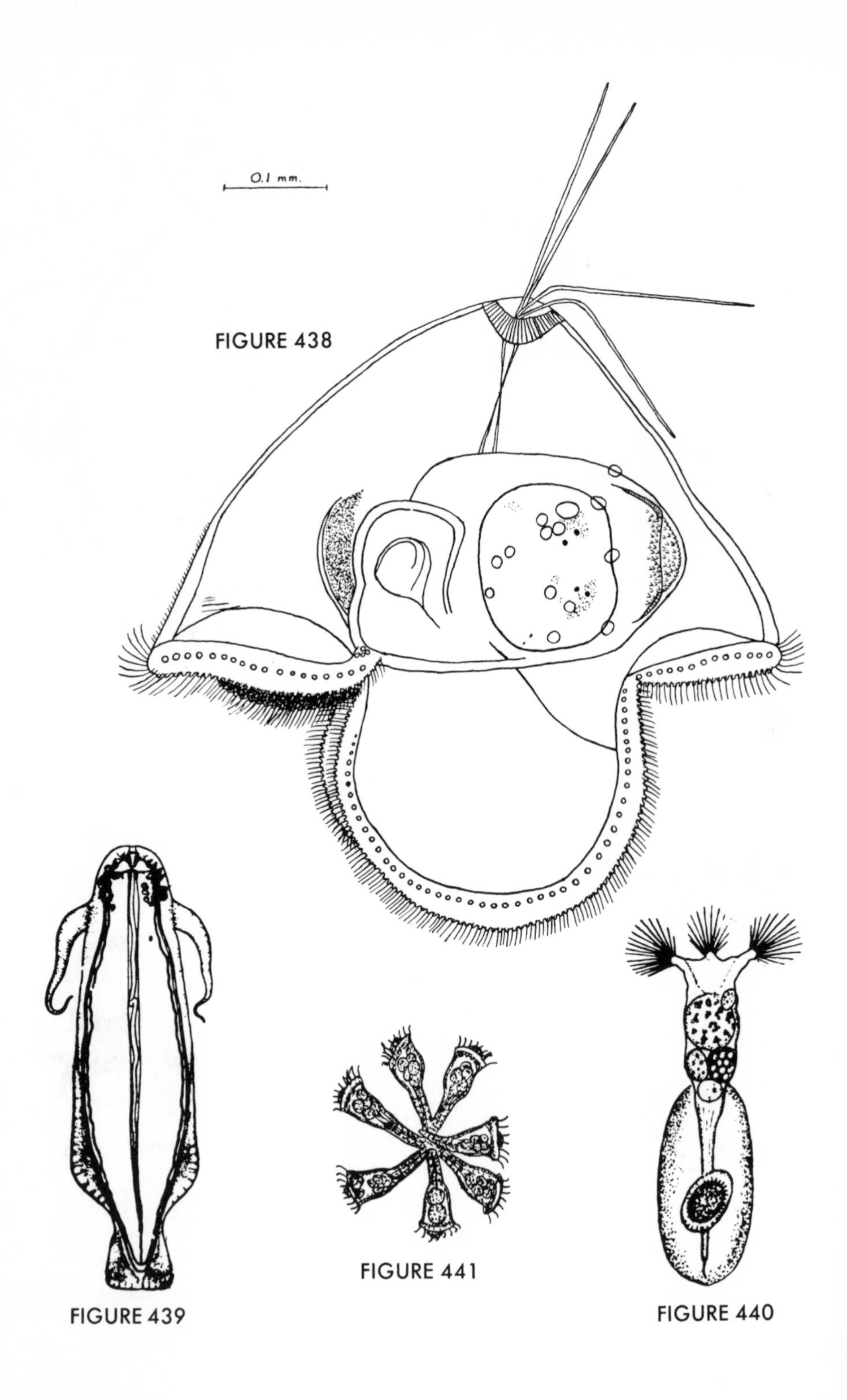

FIGURE 438

FIGURE 439

FIGURE 441

FIGURE 440

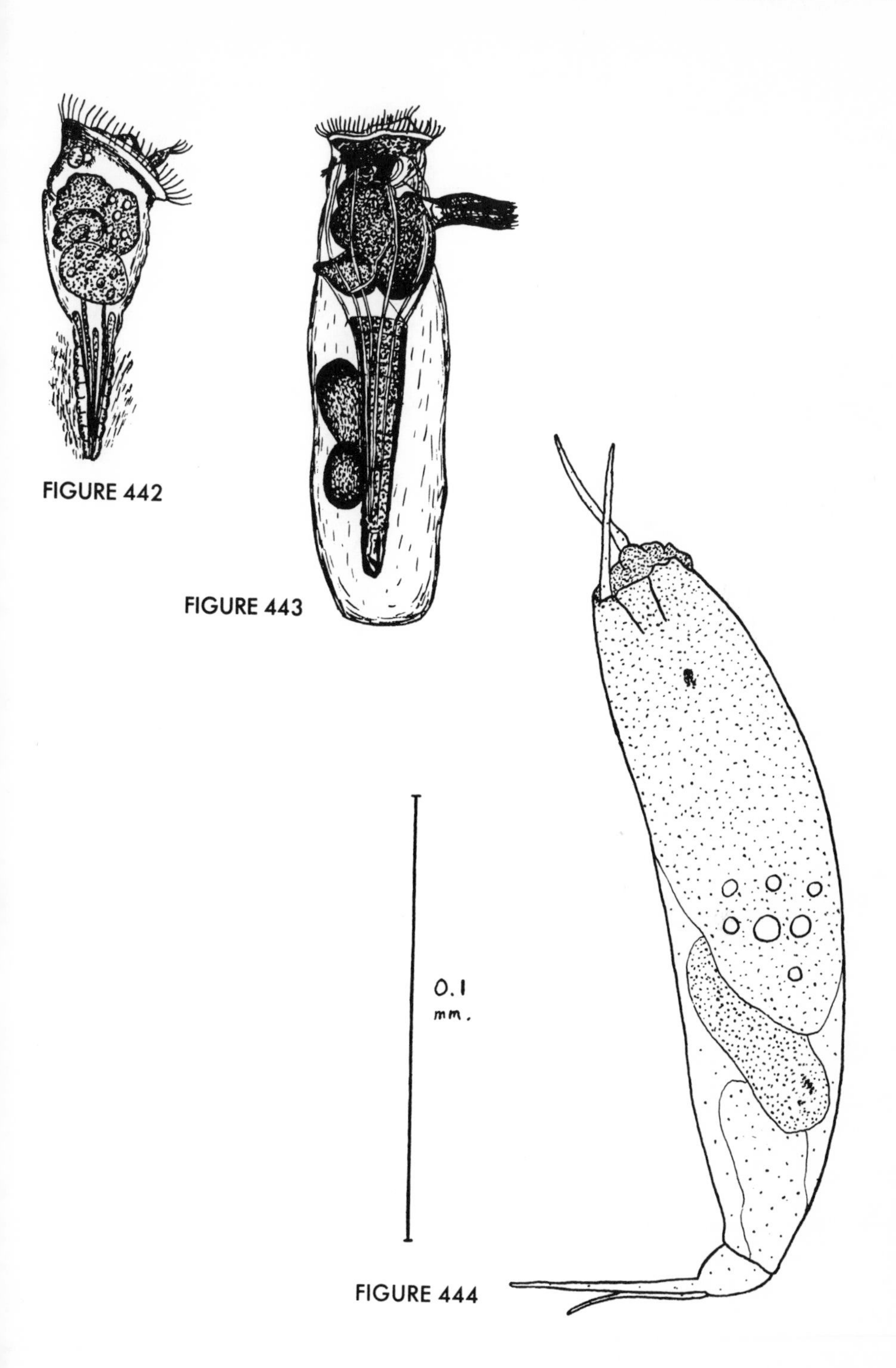

FIGURE 442

FIGURE 443

FIGURE 444

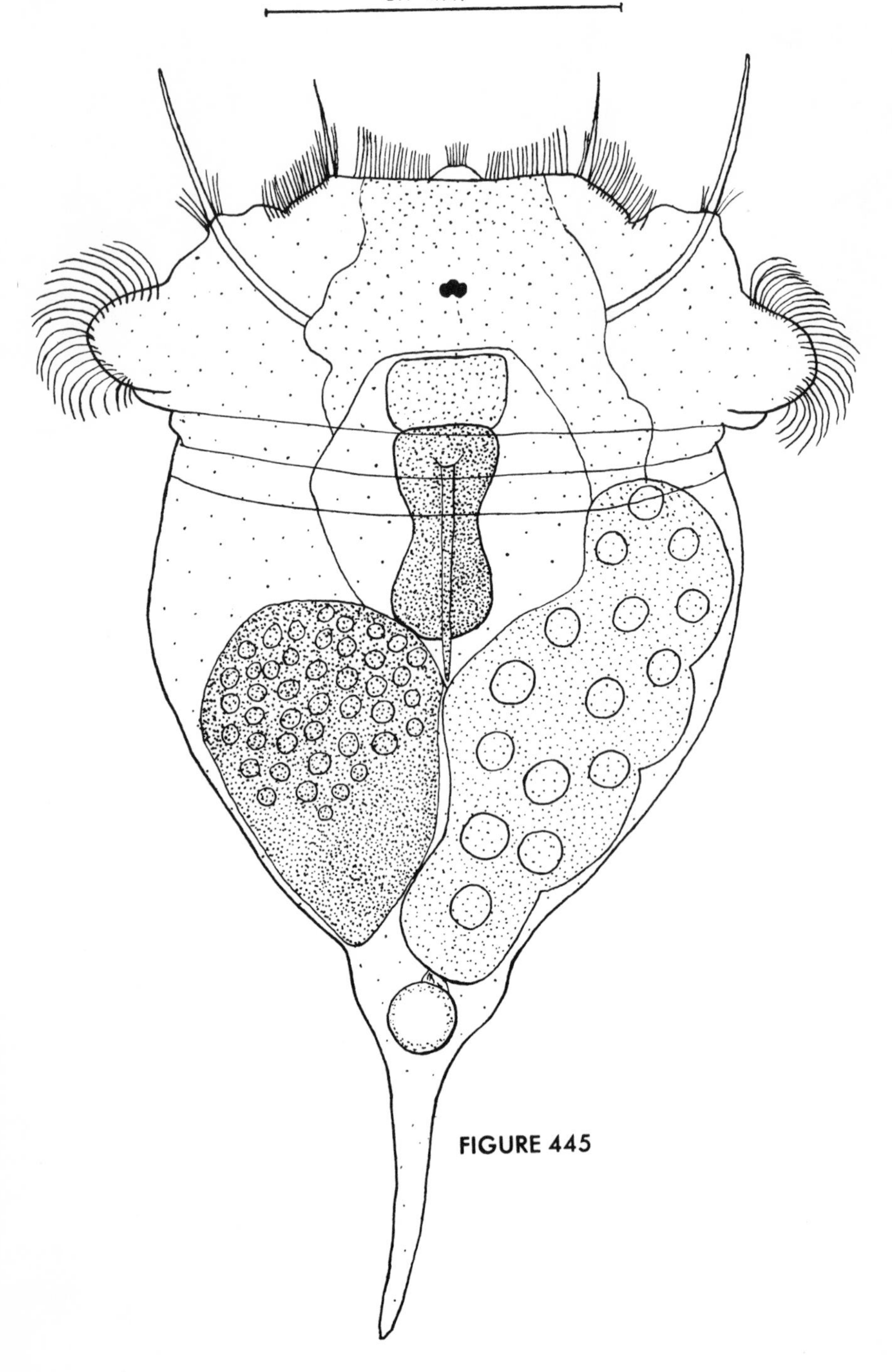

FIGURE 445

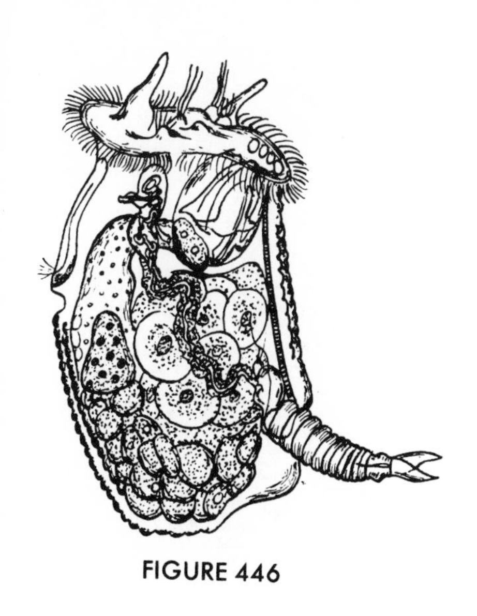

FIGURE 446

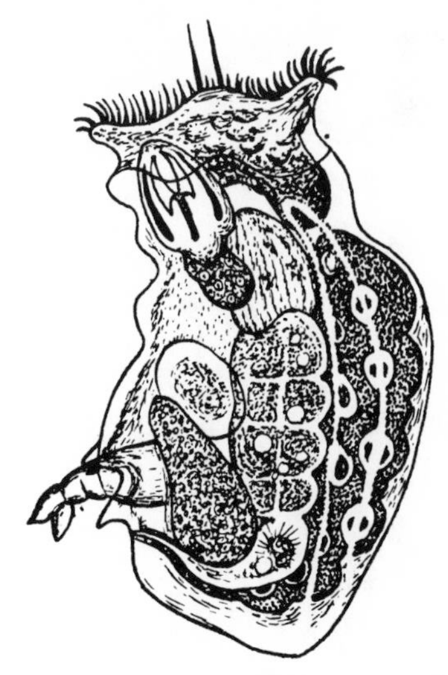

FIGURE 447

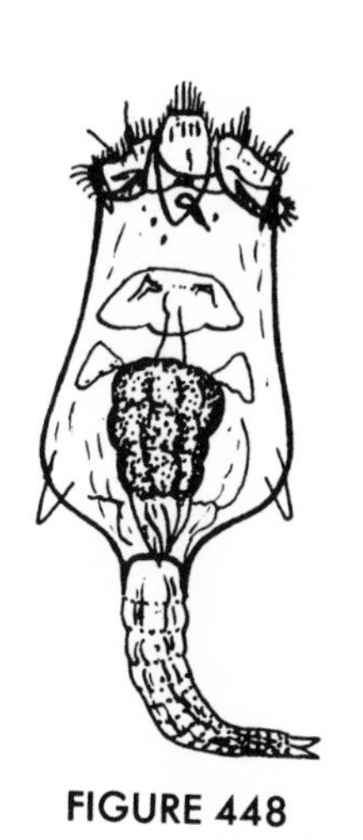

FIGURE 448

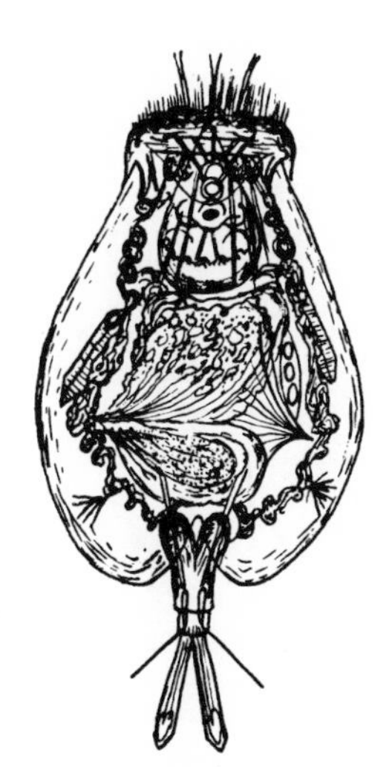

FIGURE 449

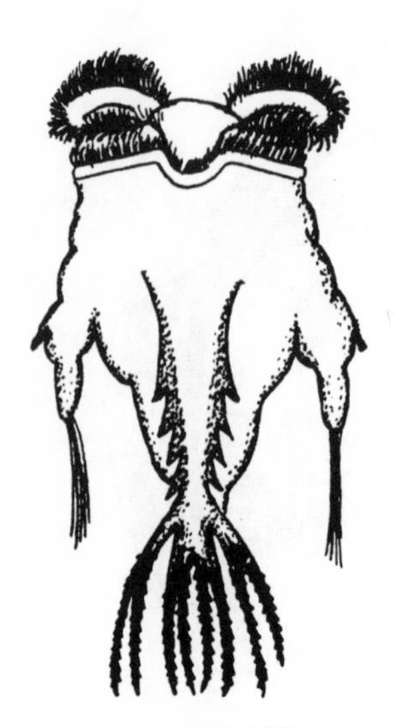

FIGURE 450

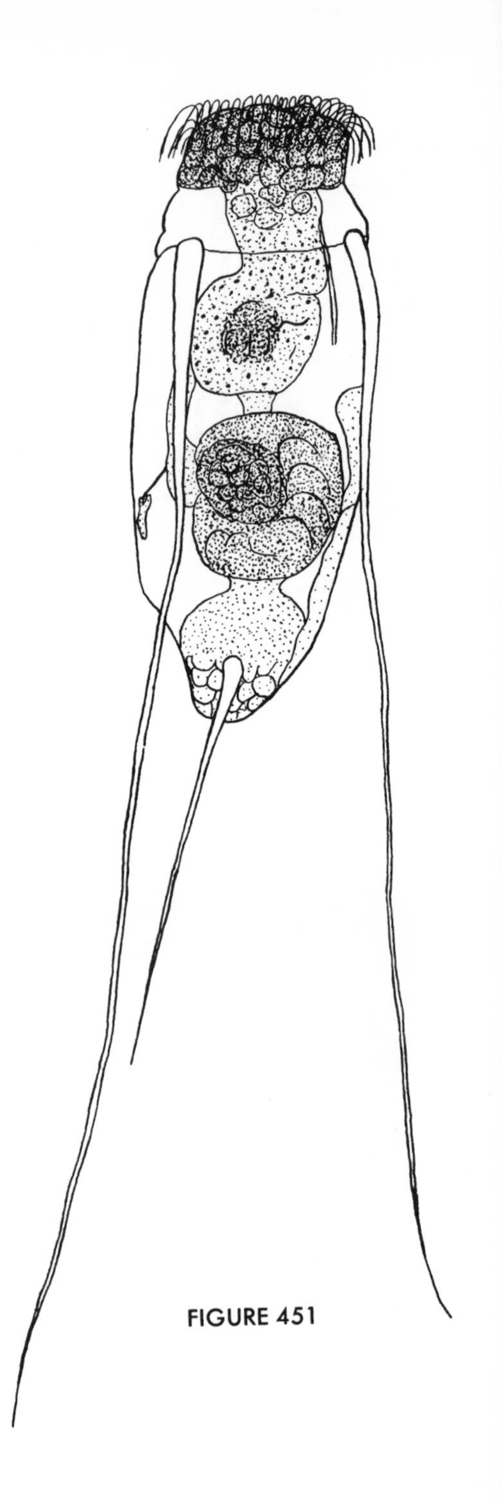

FIGURE 451

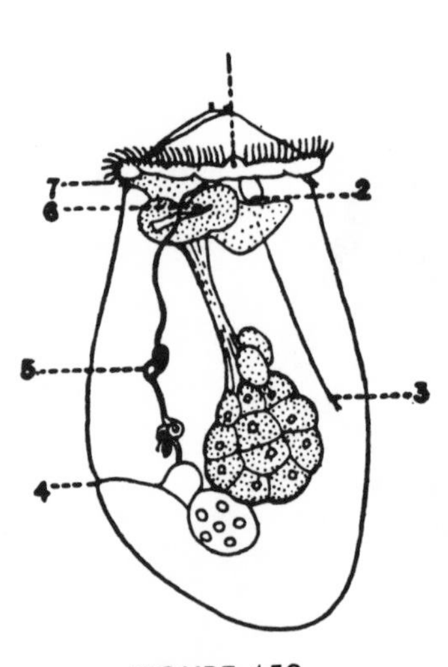

FIGURE 452

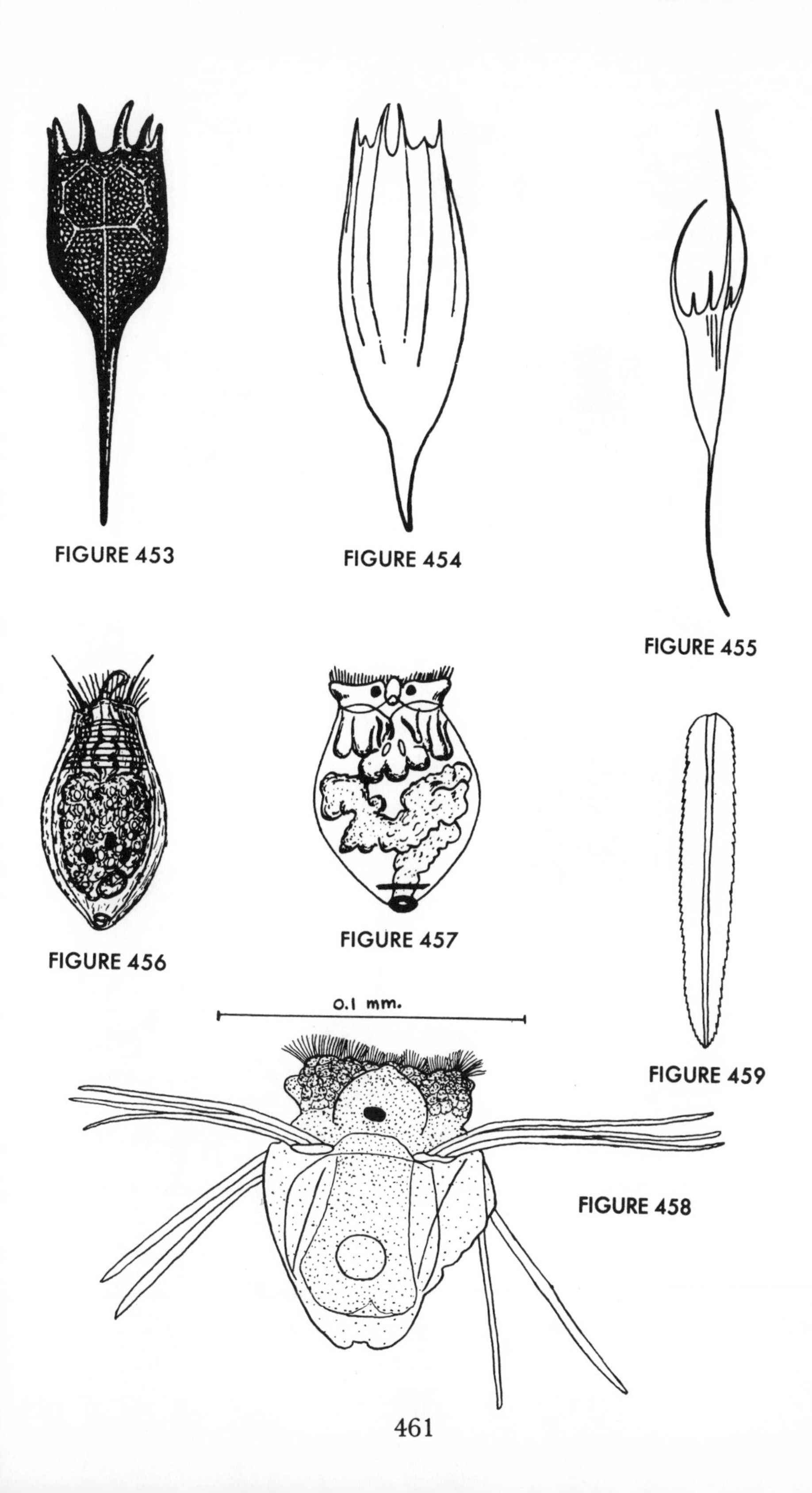

FIGURE 453

FIGURE 454

FIGURE 455

FIGURE 456

FIGURE 457

FIGURE 458

FIGURE 459

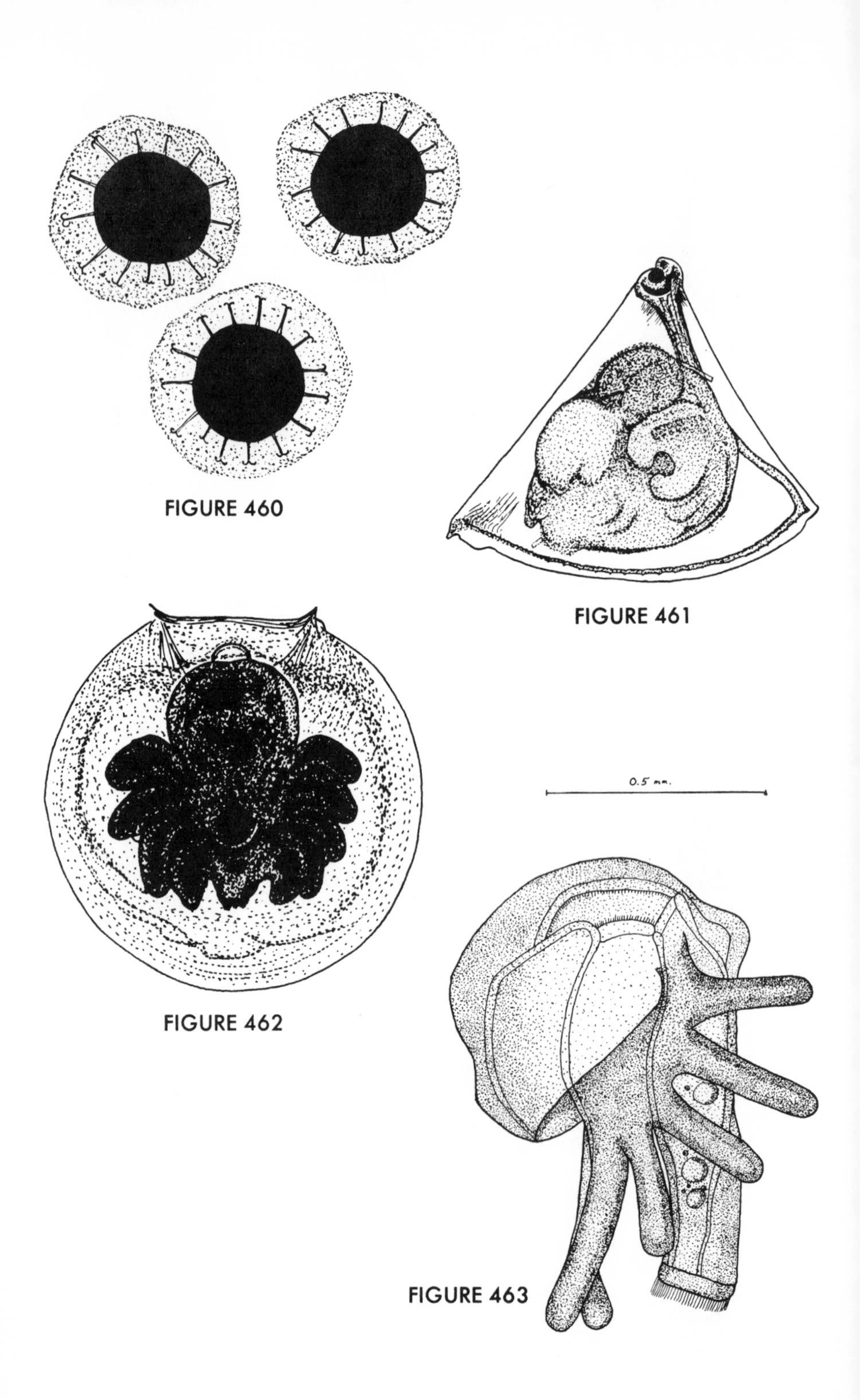

FIGURE 460

FIGURE 461

FIGURE 462

FIGURE 463

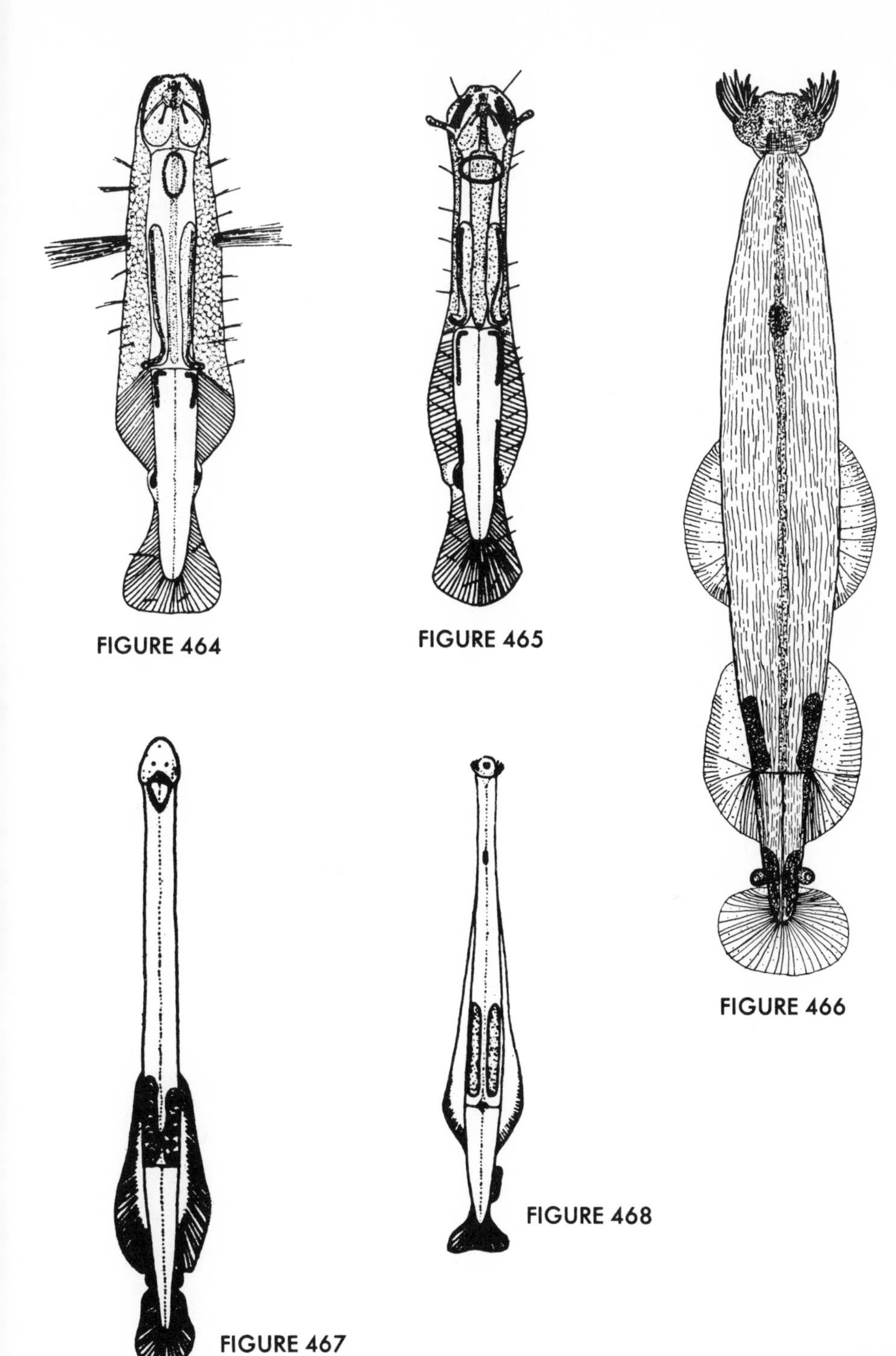
FIGURE 464
FIGURE 465
FIGURE 466
FIGURE 467
FIGURE 468

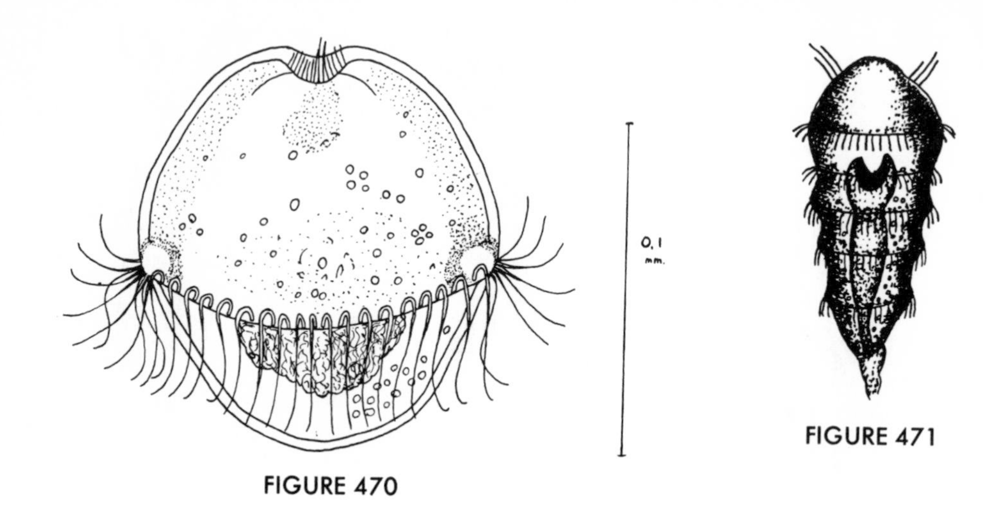

FIGURE 470

FIGURE 471

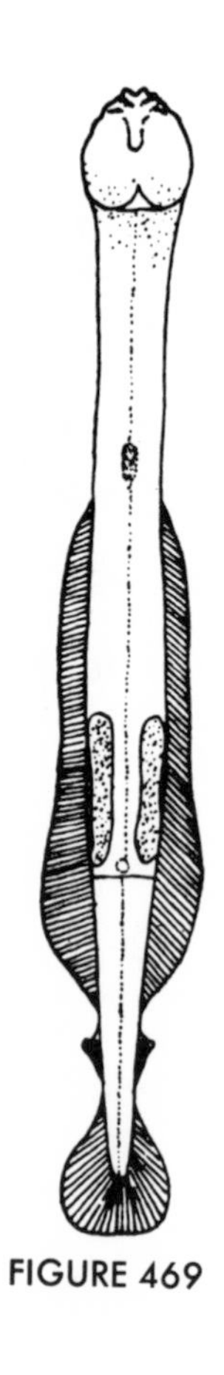

FIGURE 469

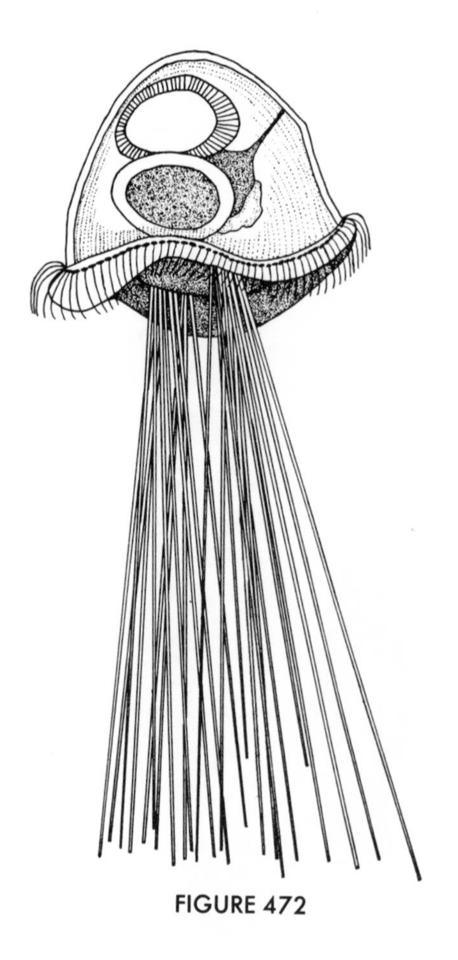

FIGURE 472

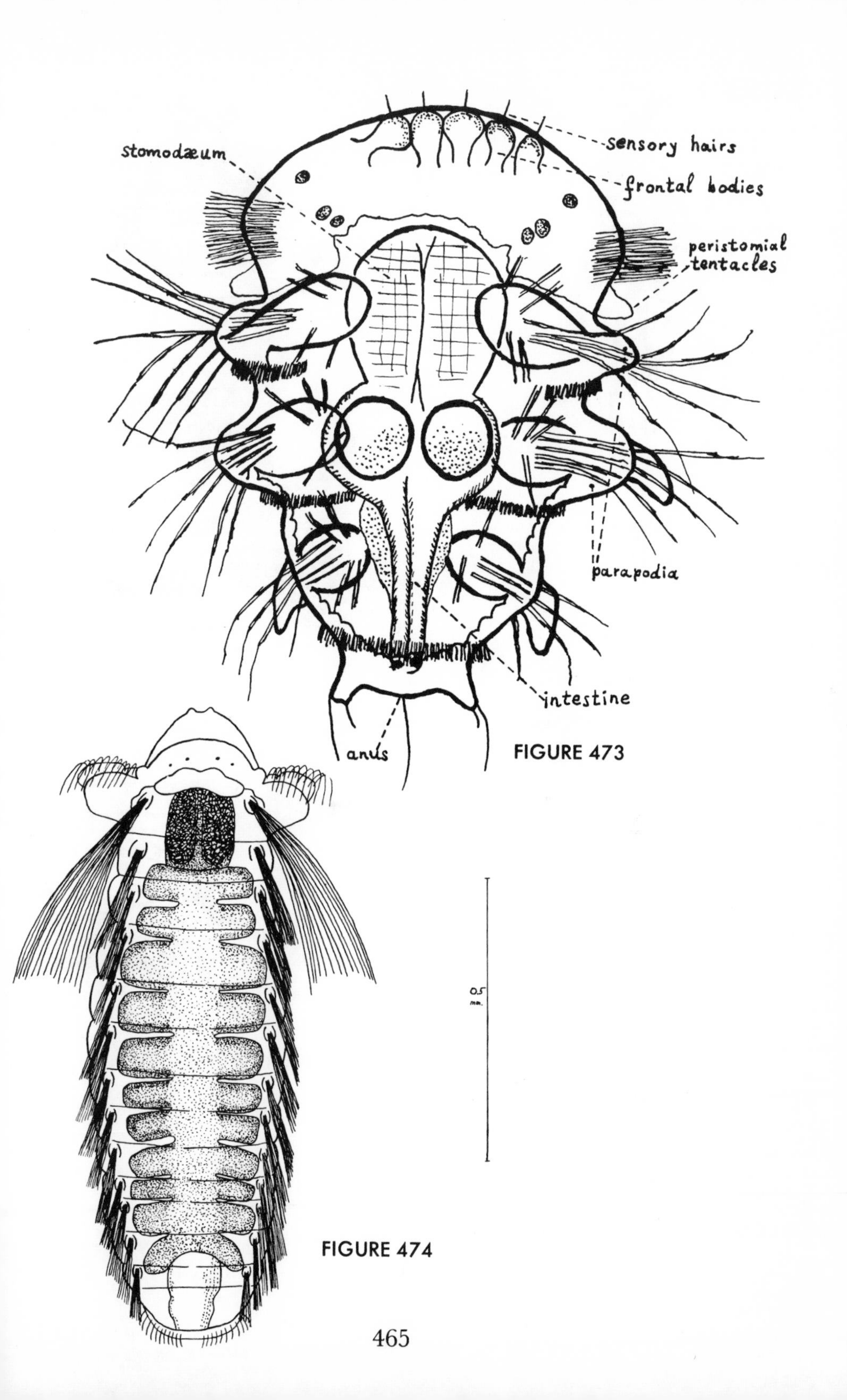

FIGURE 473

FIGURE 474

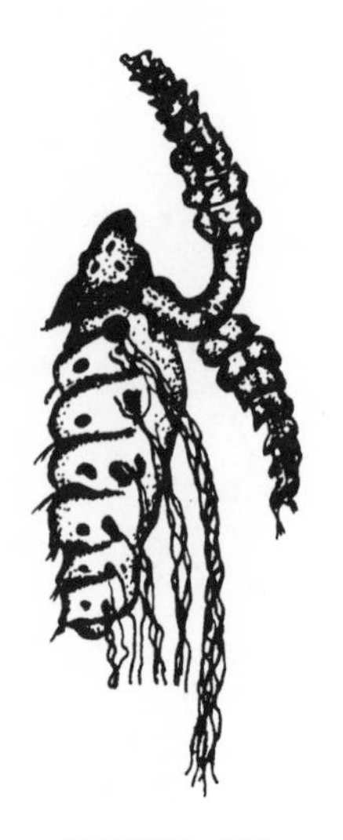

FIGURE 475

FIGURE 476

FIGURE 477

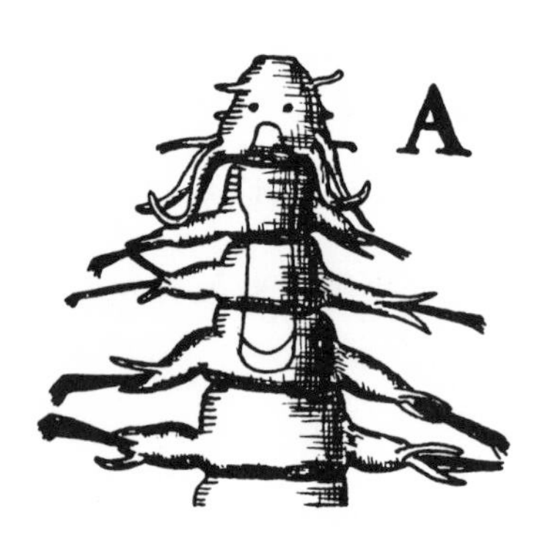

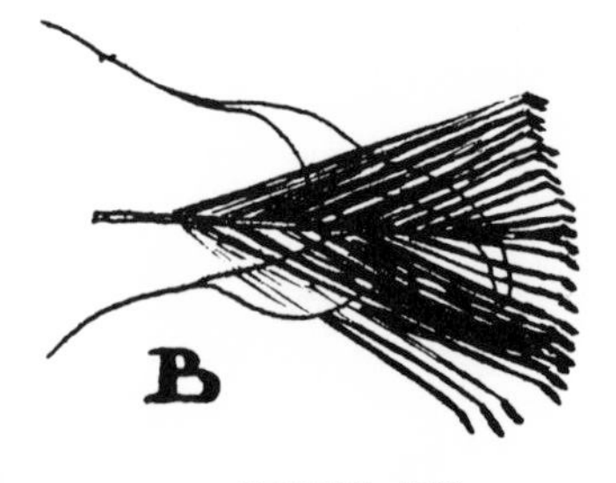

FIGURE 478

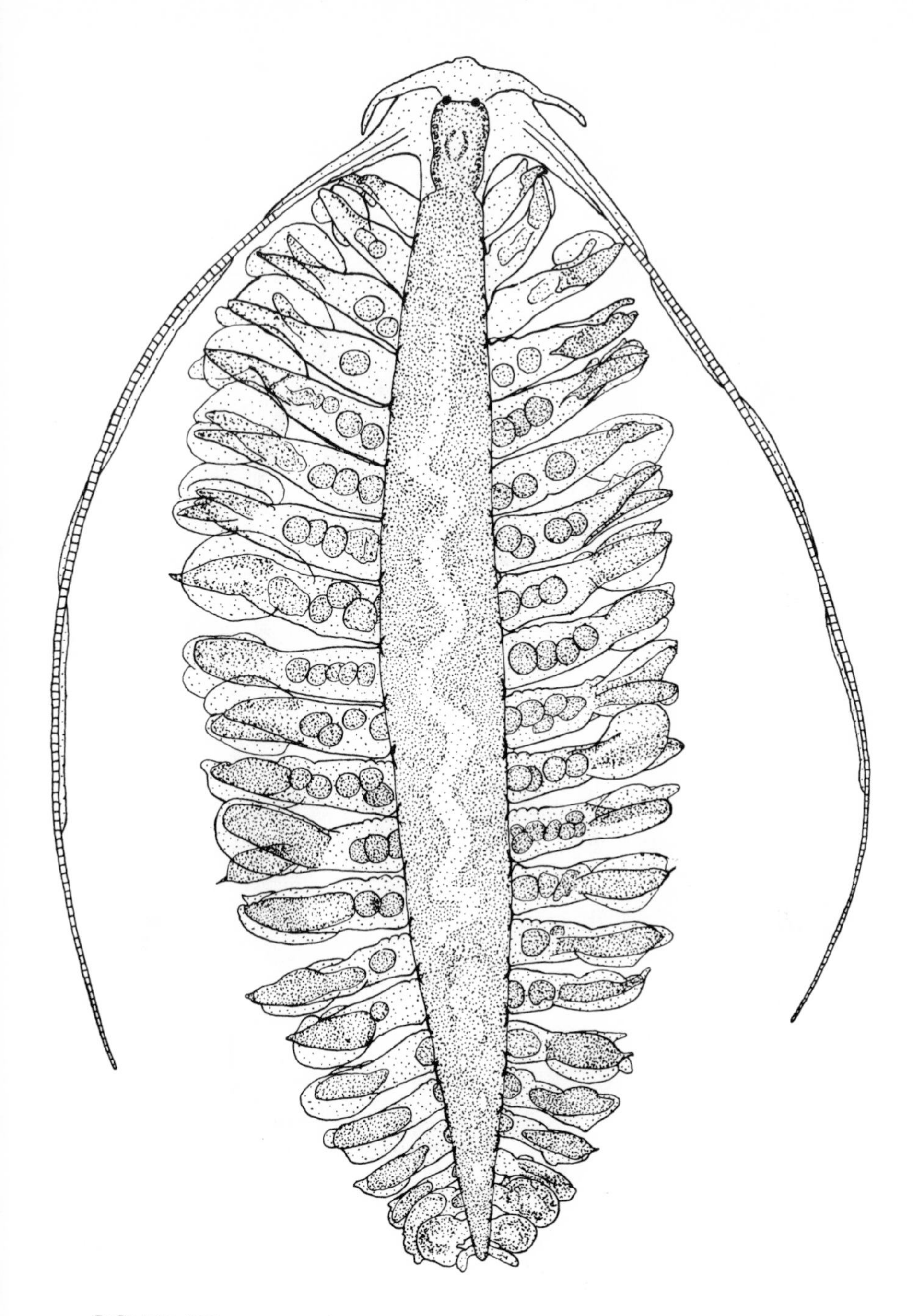

FIGURE 479

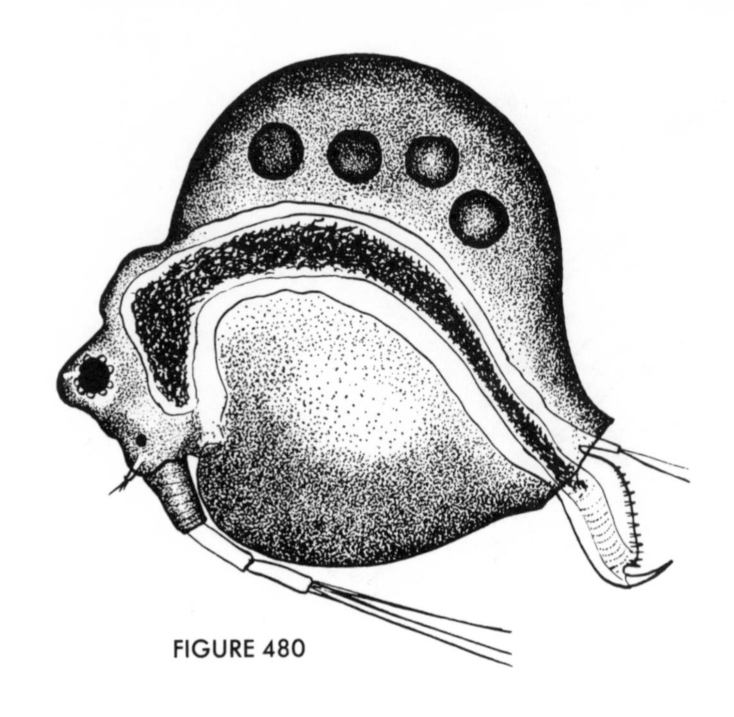

FIGURE 480

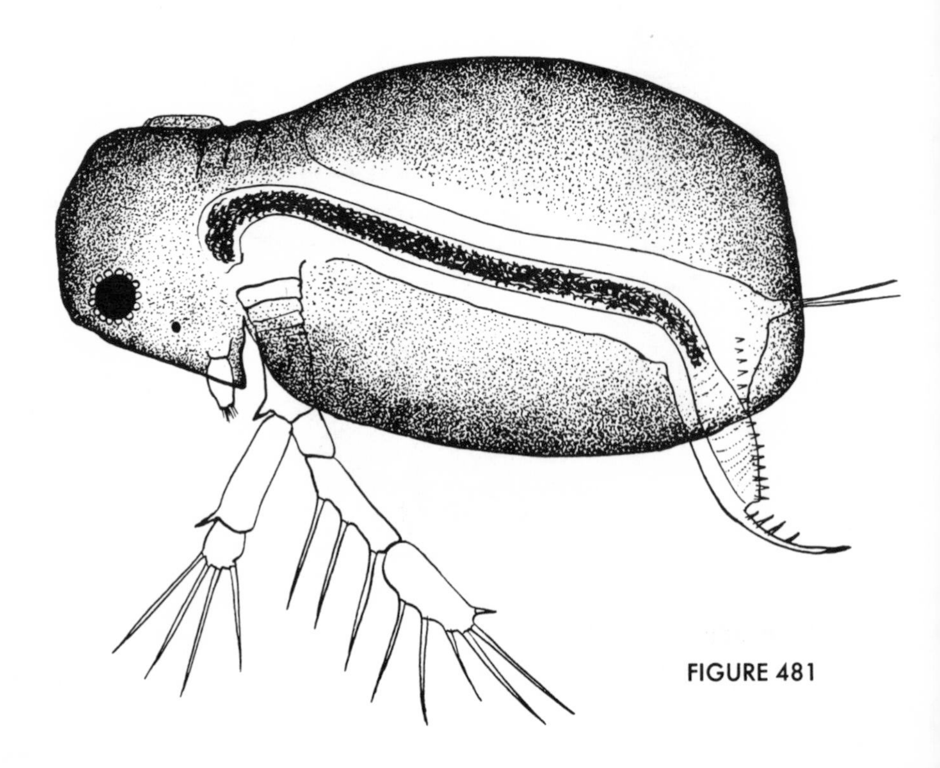

FIGURE 481

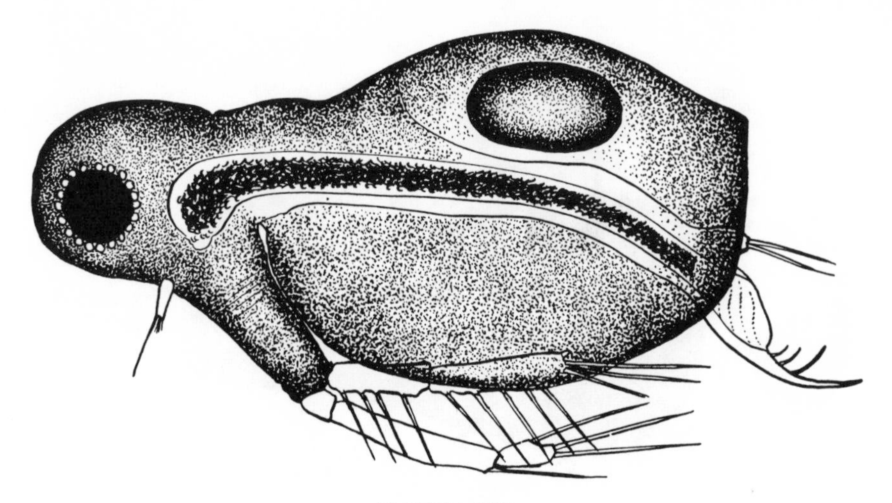

FIGURE 482

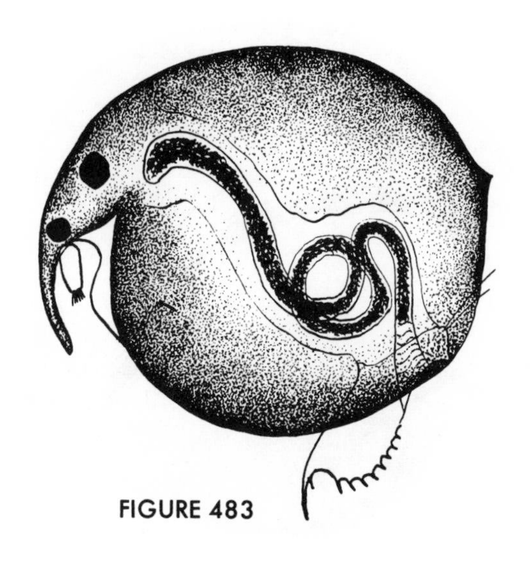

FIGURE 483

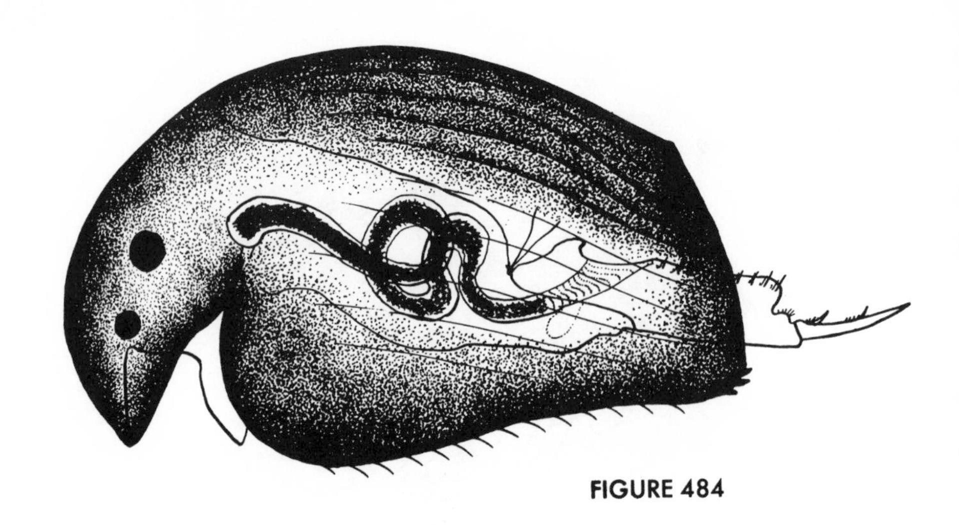

FIGURE 484

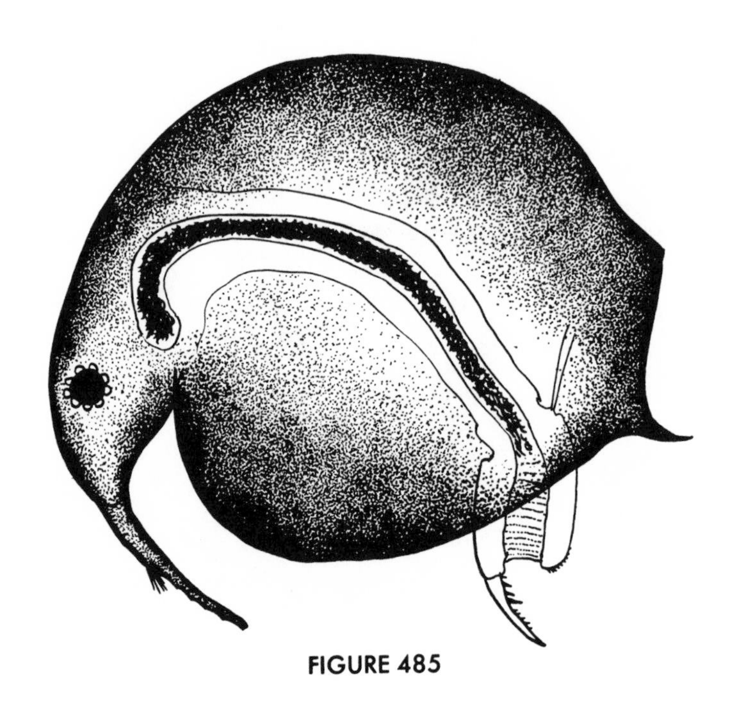

FIGURE 485

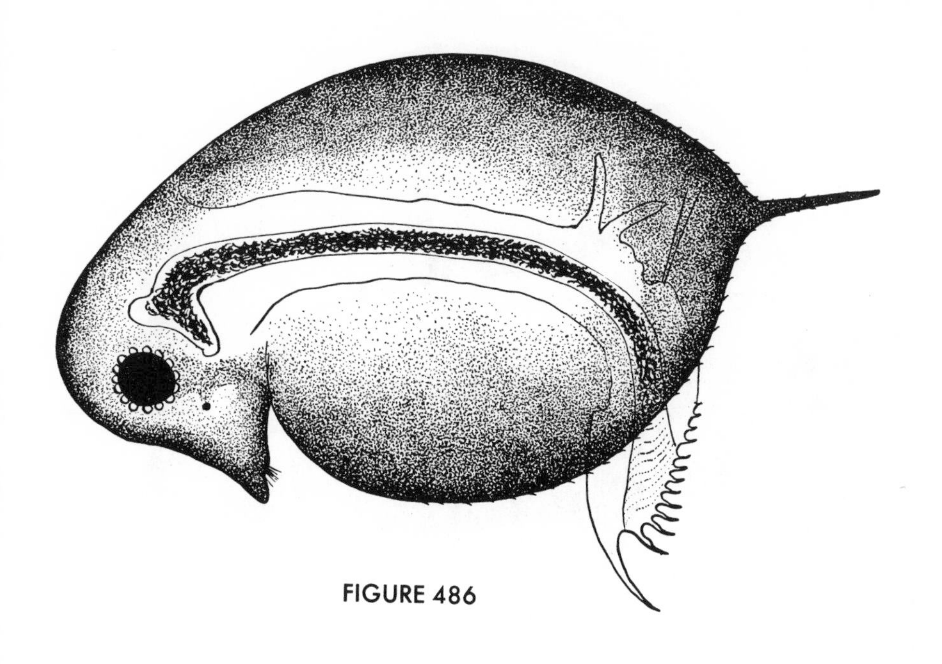

FIGURE 486

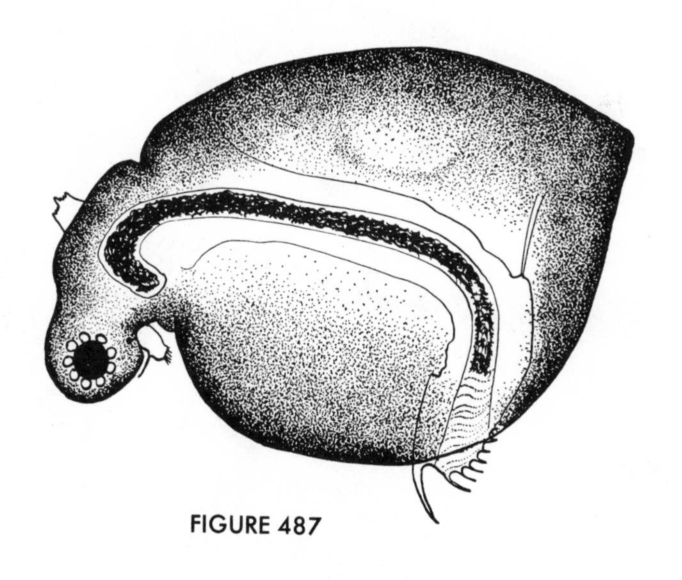

FIGURE 487

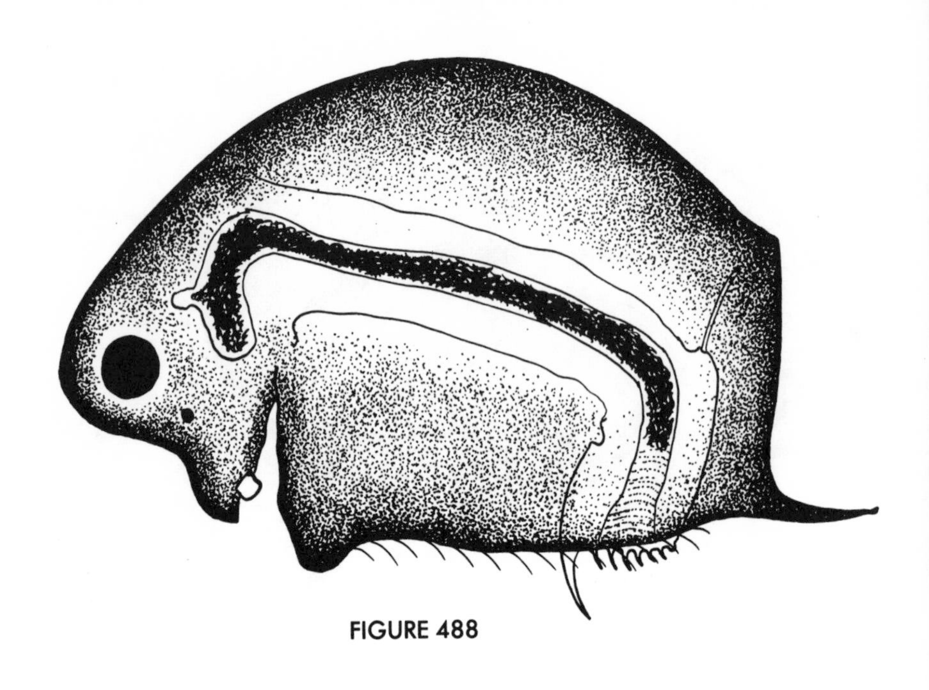

FIGURE 488

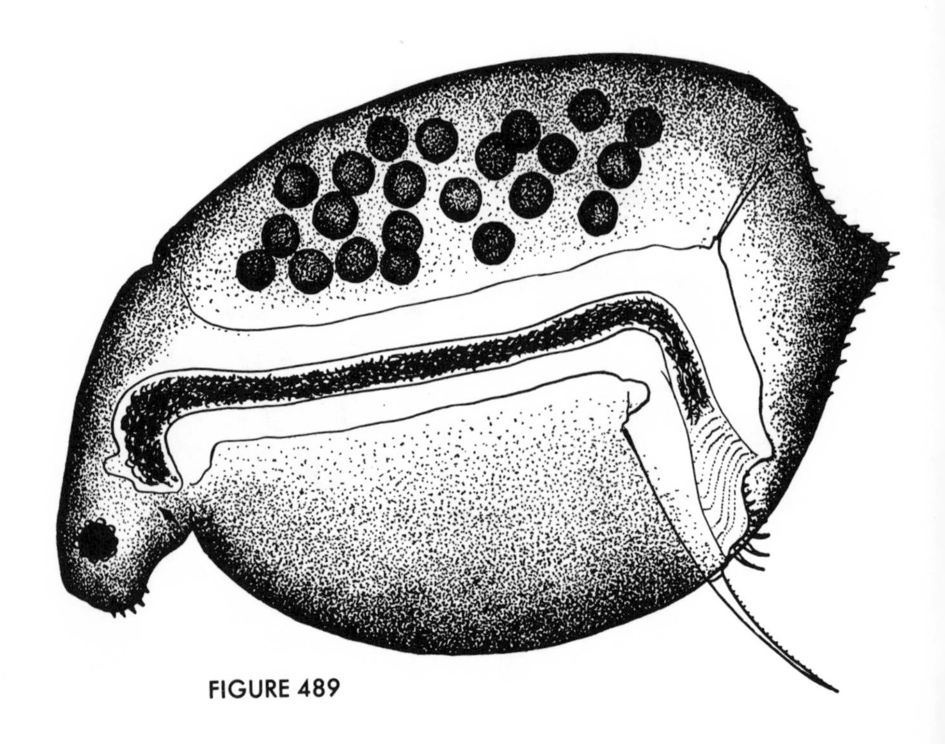

FIGURE 489

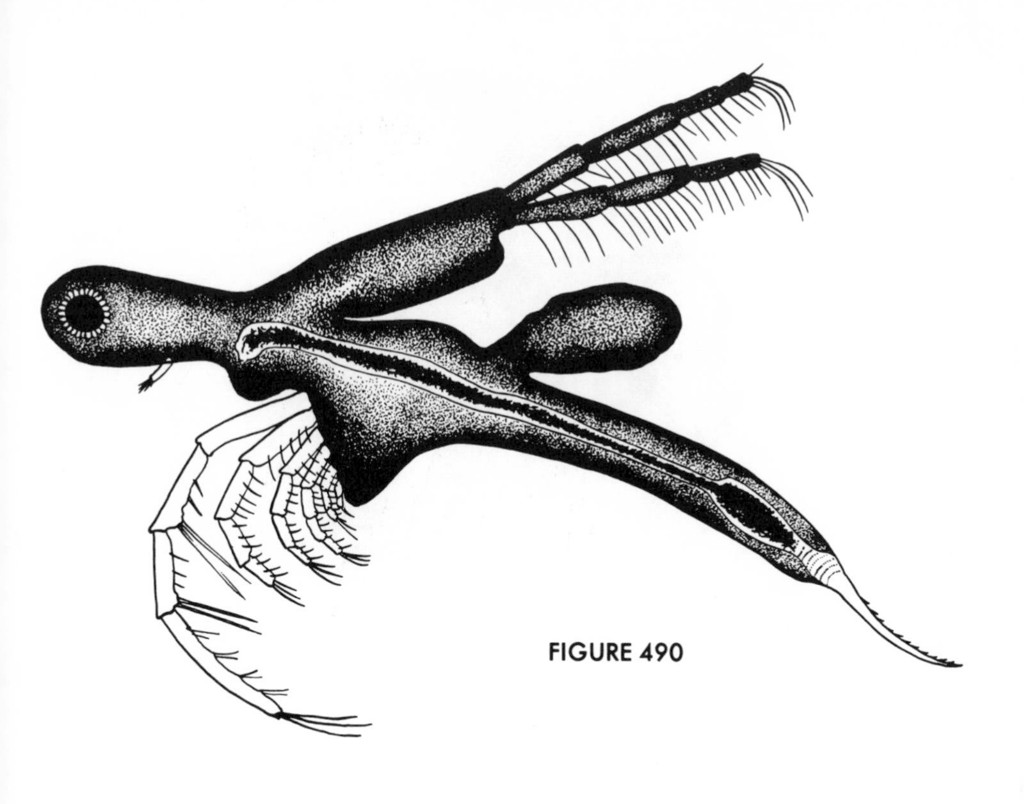

FIGURE 490

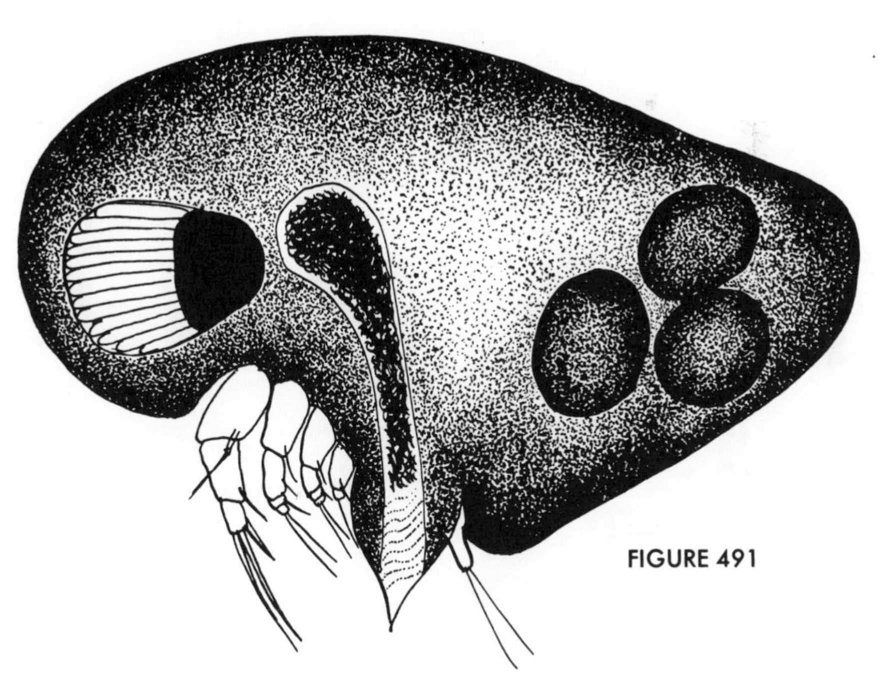

FIGURE 491

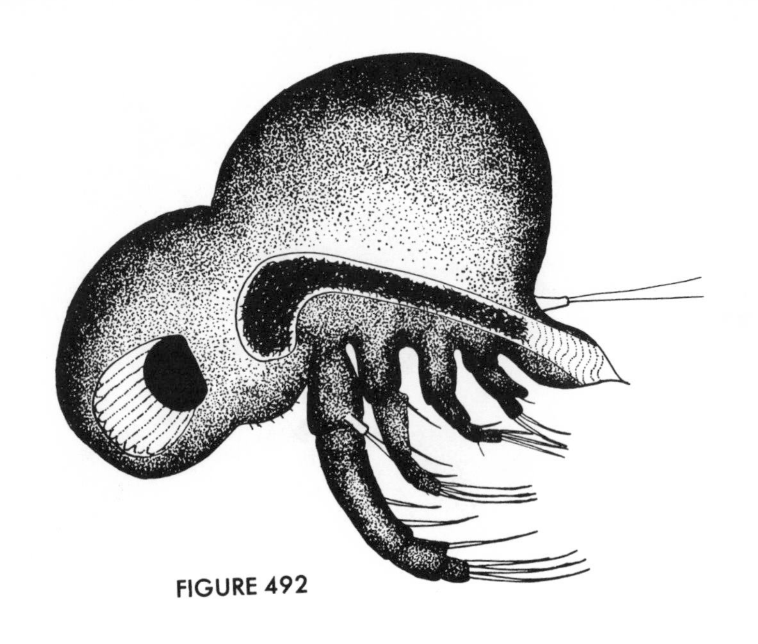
FIGURE 492

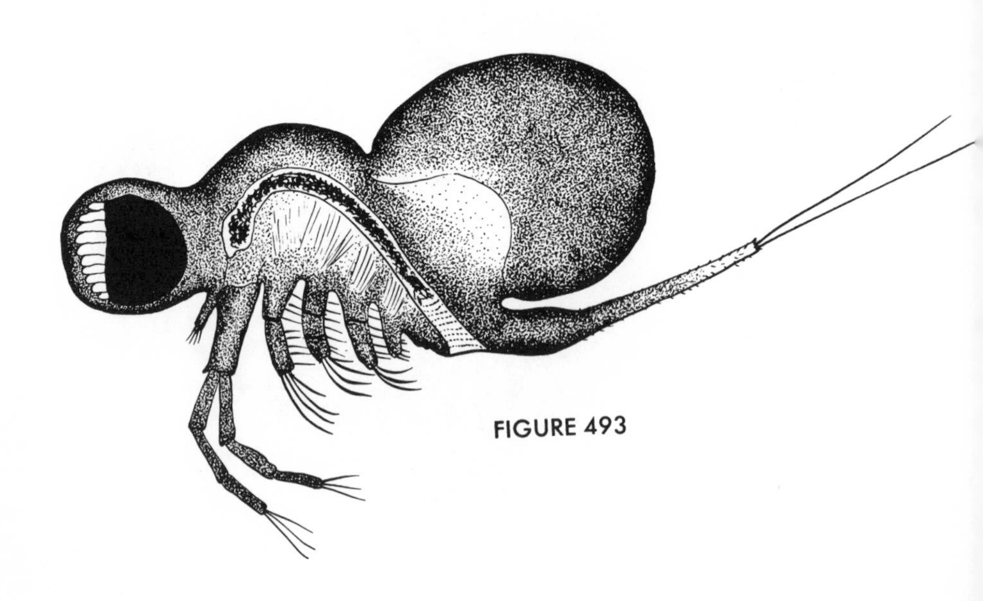
FIGURE 493

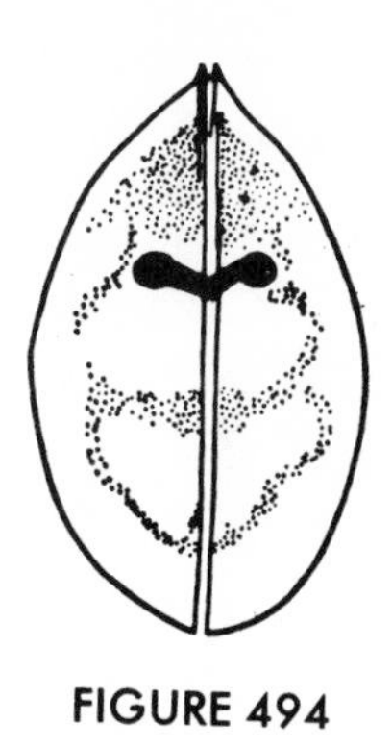

FIGURE 494

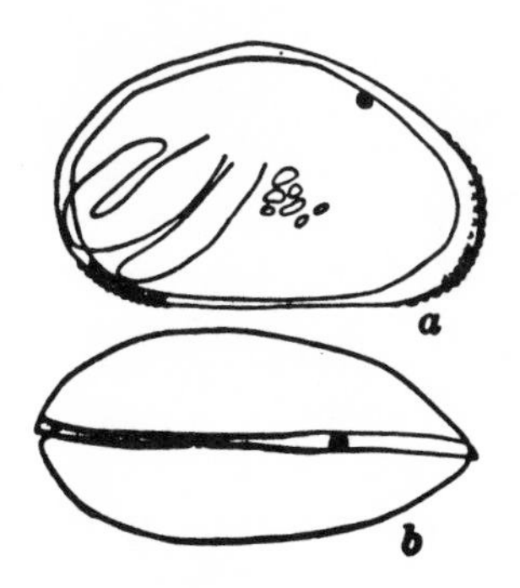

FIGURE 495

1.0 mm.

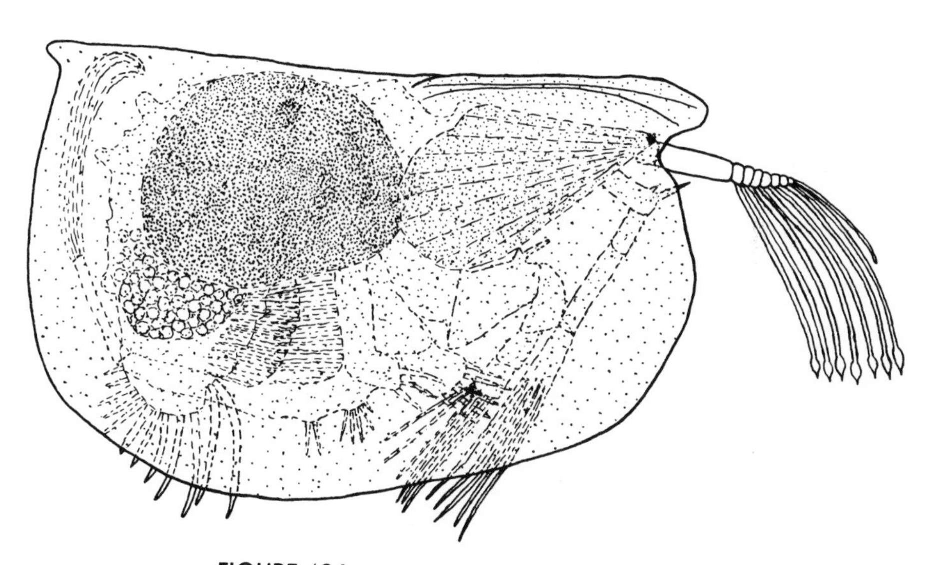

FIGURE 496

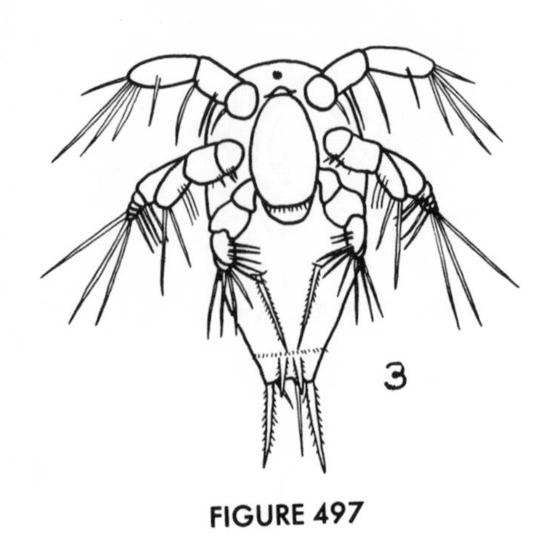

FIGURE 497

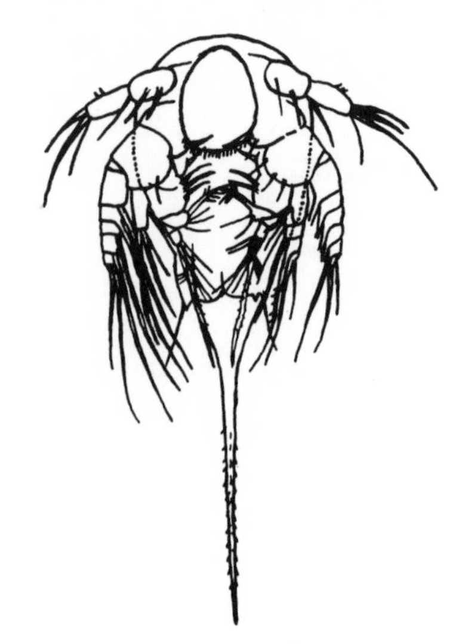

FIGURE 498

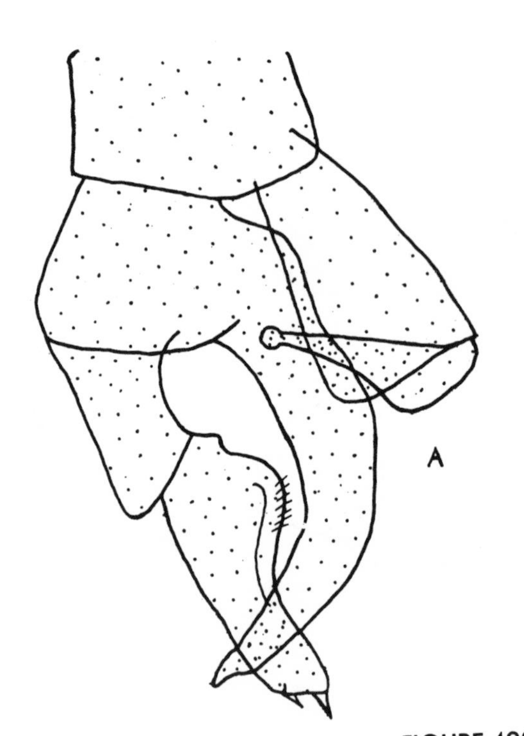

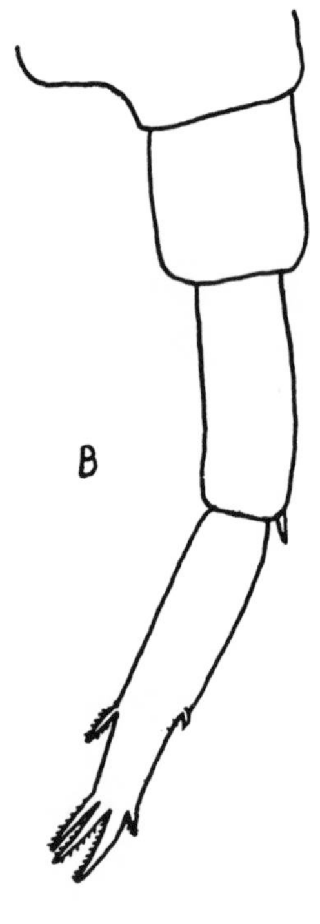

FIGURE 499

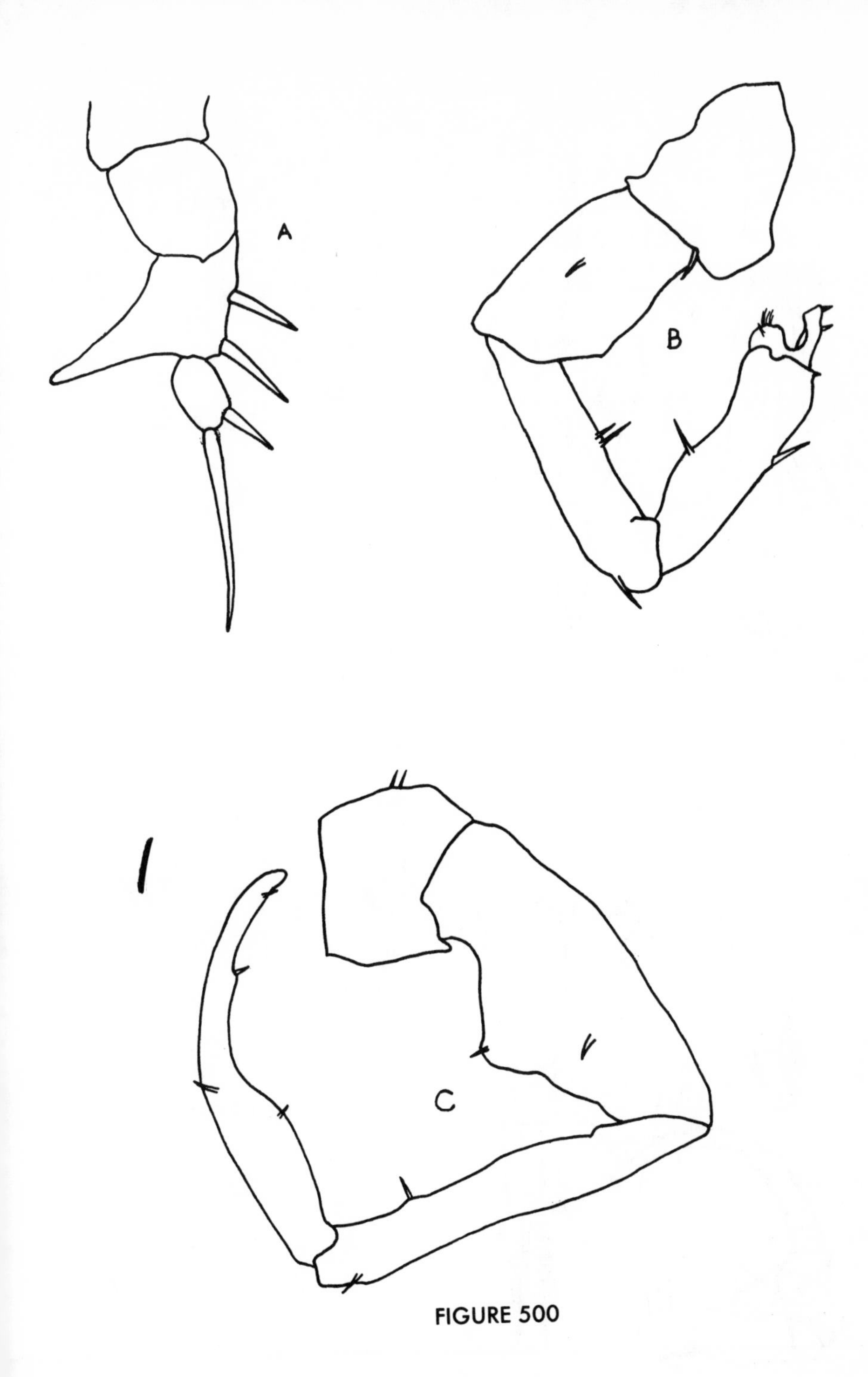

FIGURE 500

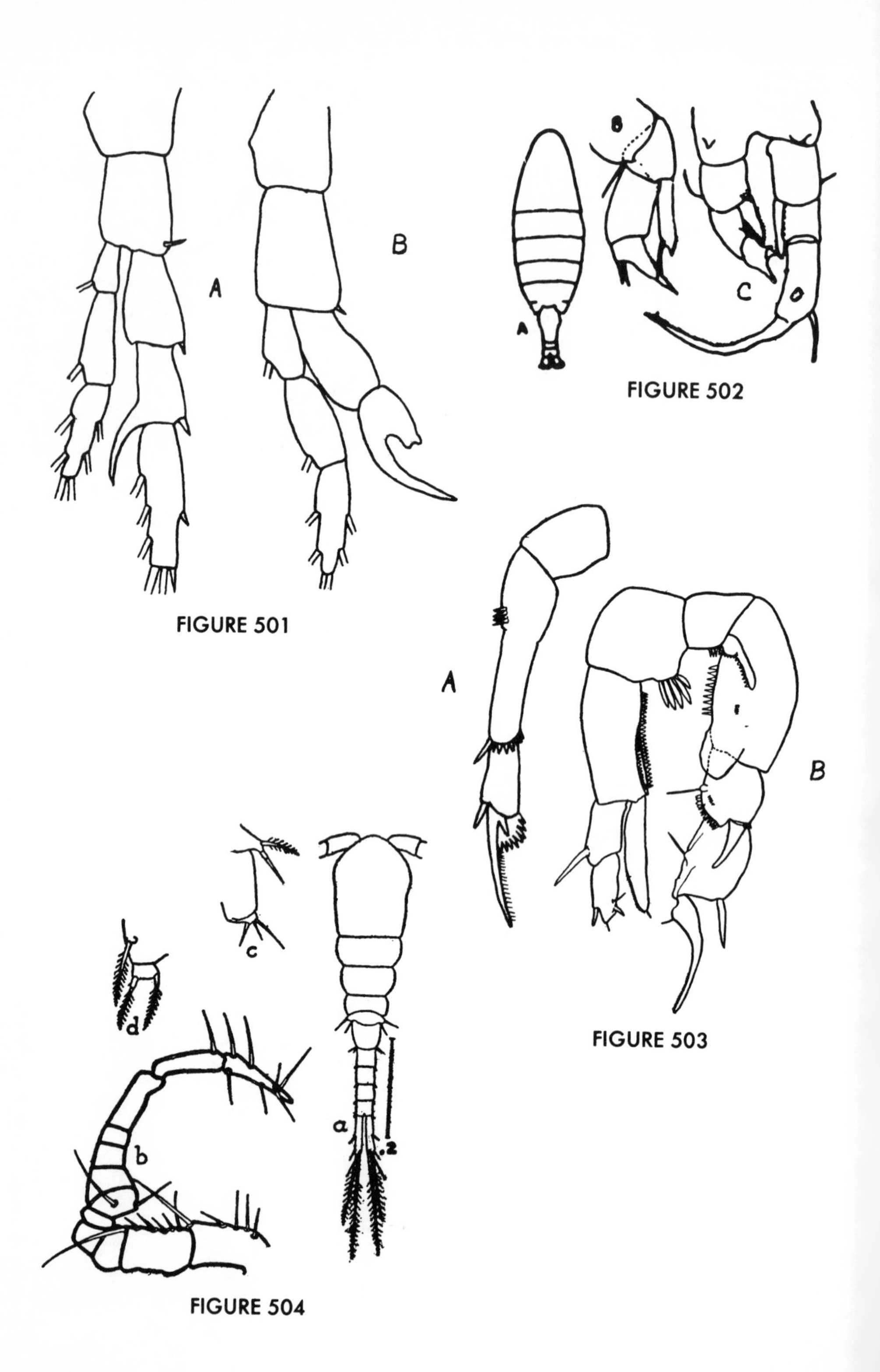
A
B
FIGURE 501
A
B
C
FIGURE 502
A
B
FIGURE 503
a
b
c
d
FIGURE 504

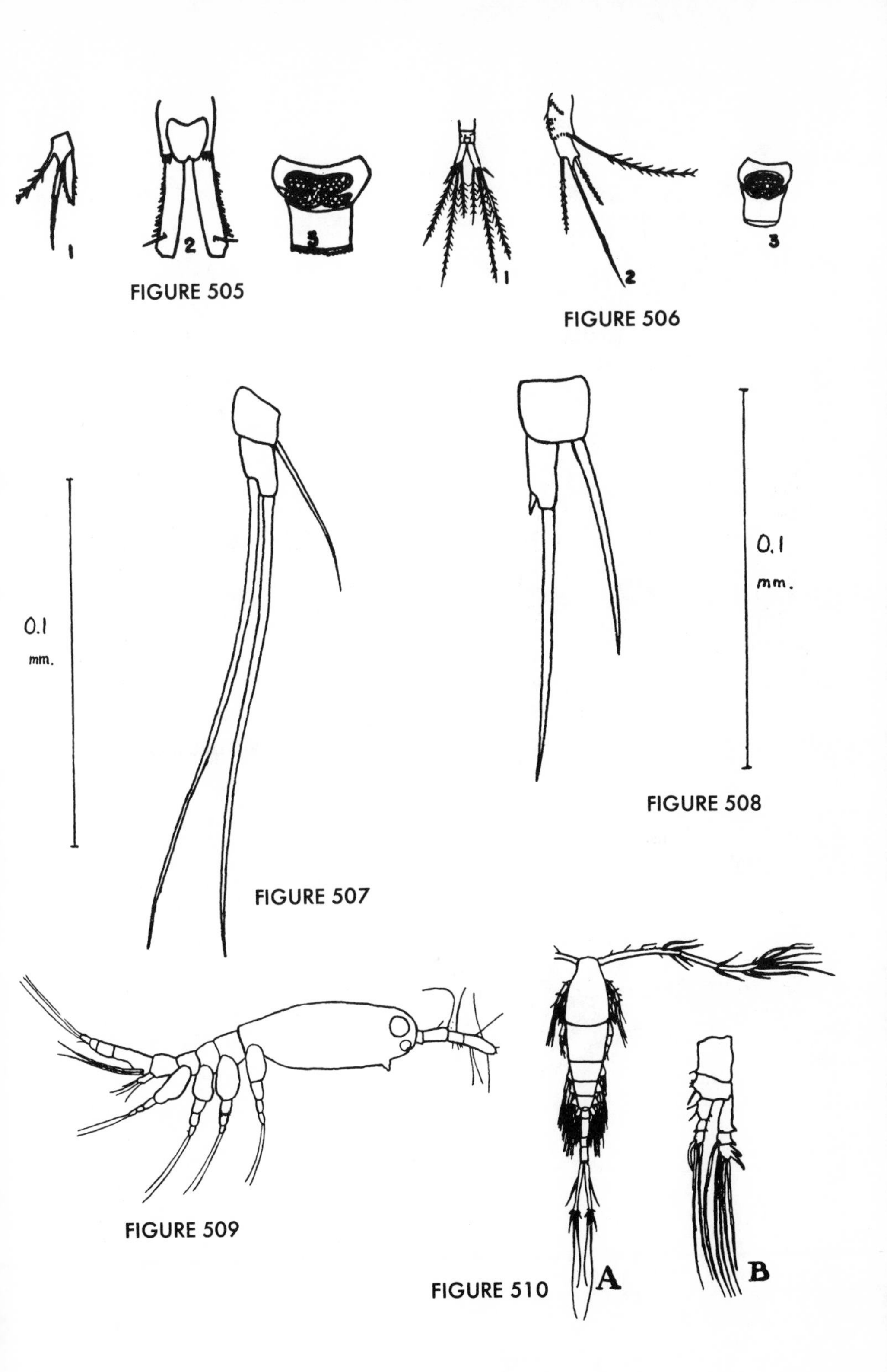

FIGURE 505

FIGURE 506

FIGURE 507

FIGURE 508

FIGURE 509

FIGURE 510

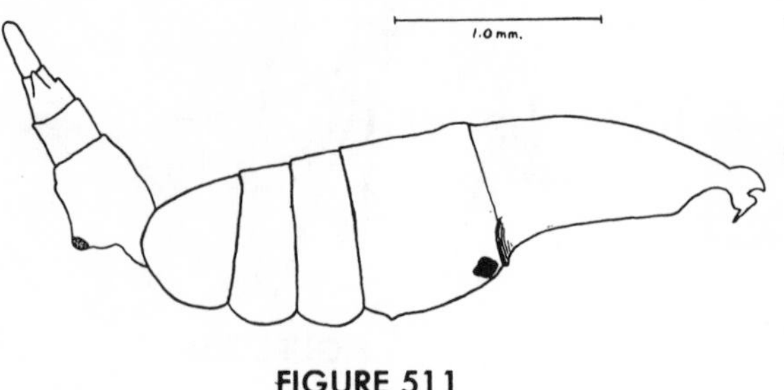

FIGURE 511

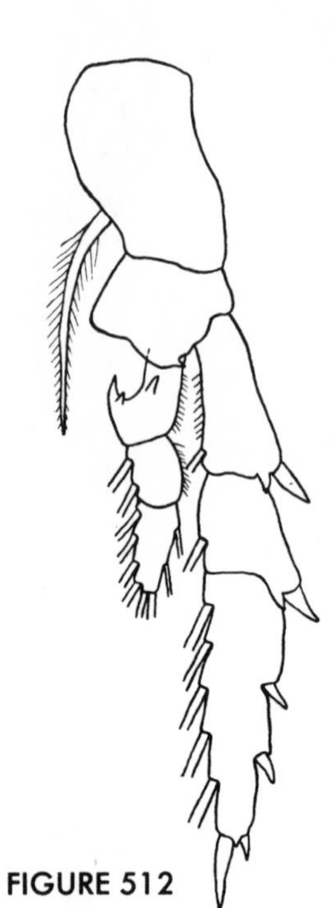

FIGURE 512

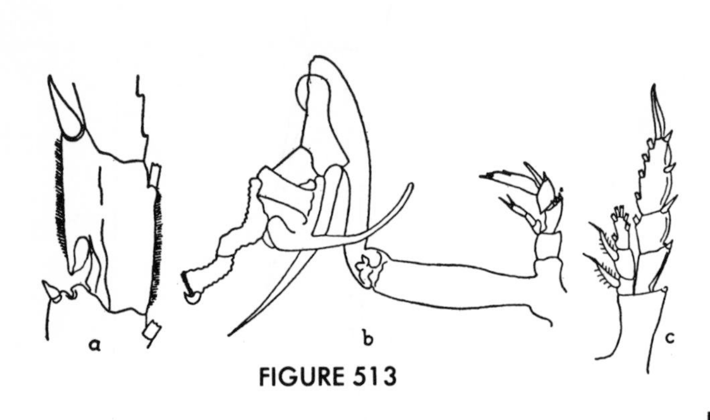

FIGURE 513

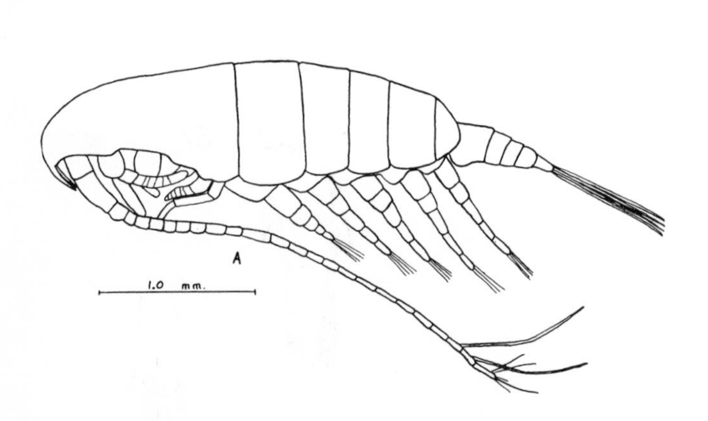

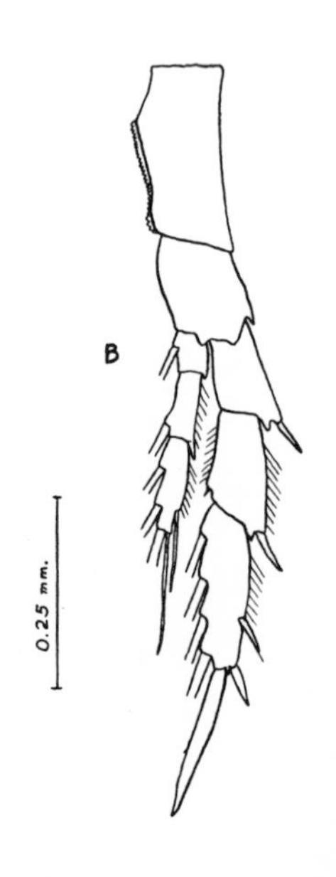

FIGURE 514

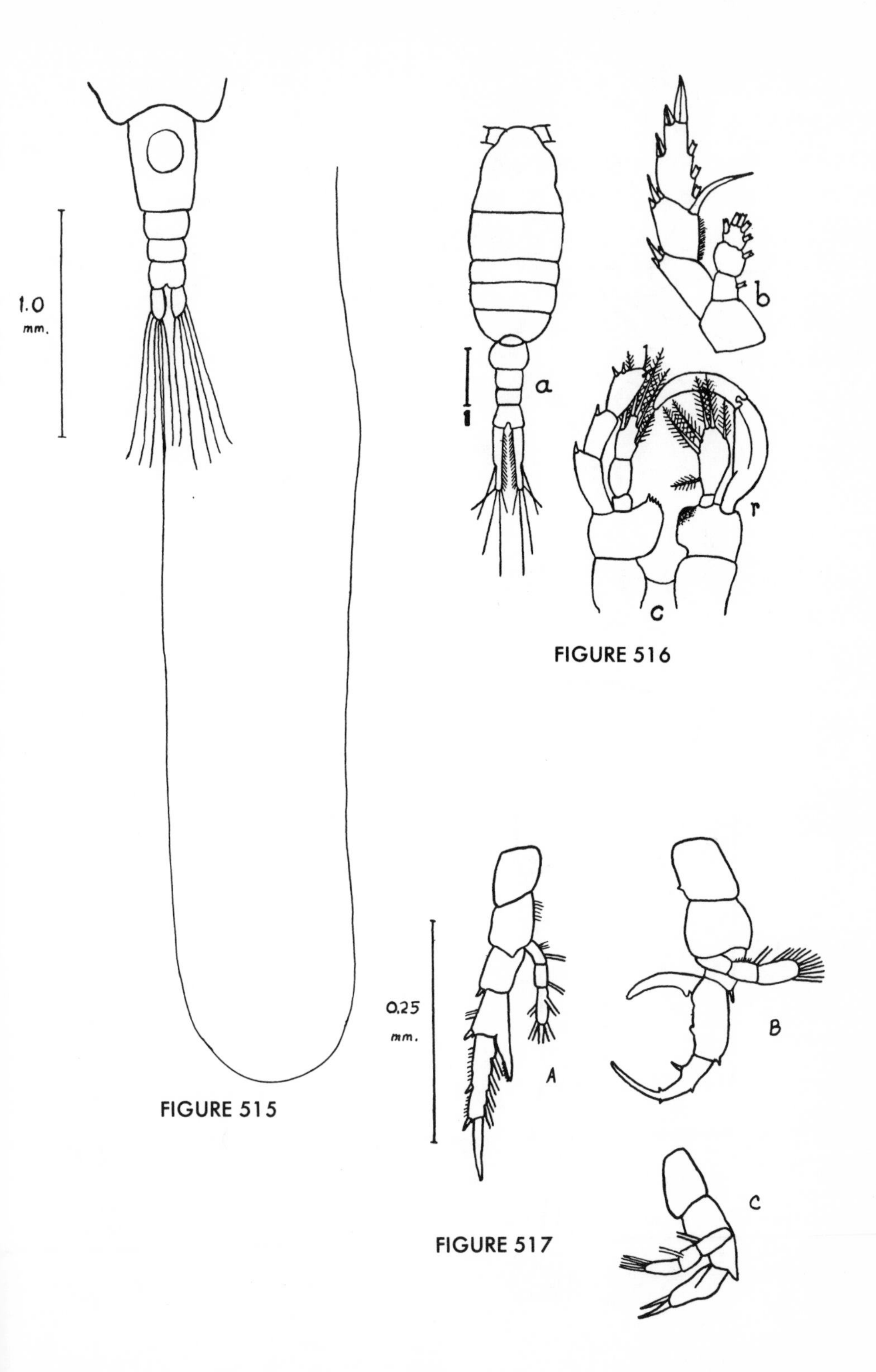

FIGURE 515

FIGURE 516

FIGURE 517

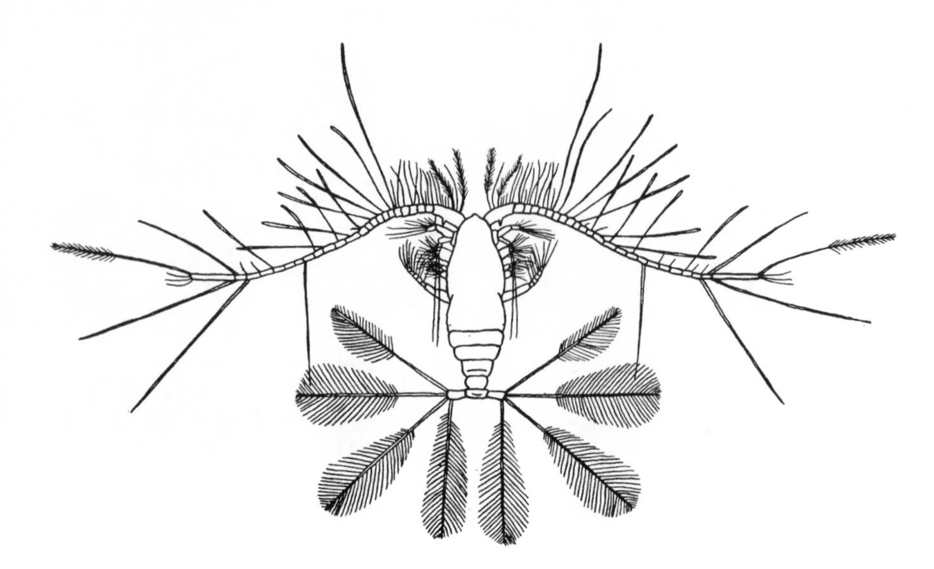

FIGURE 519

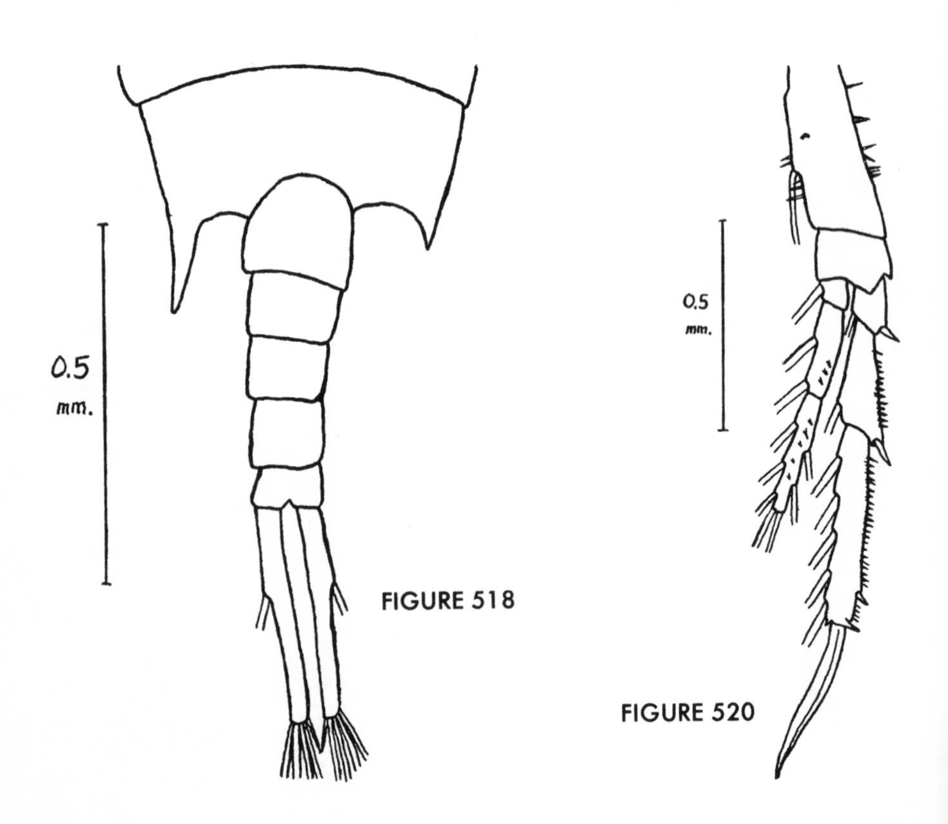

FIGURE 518

FIGURE 520

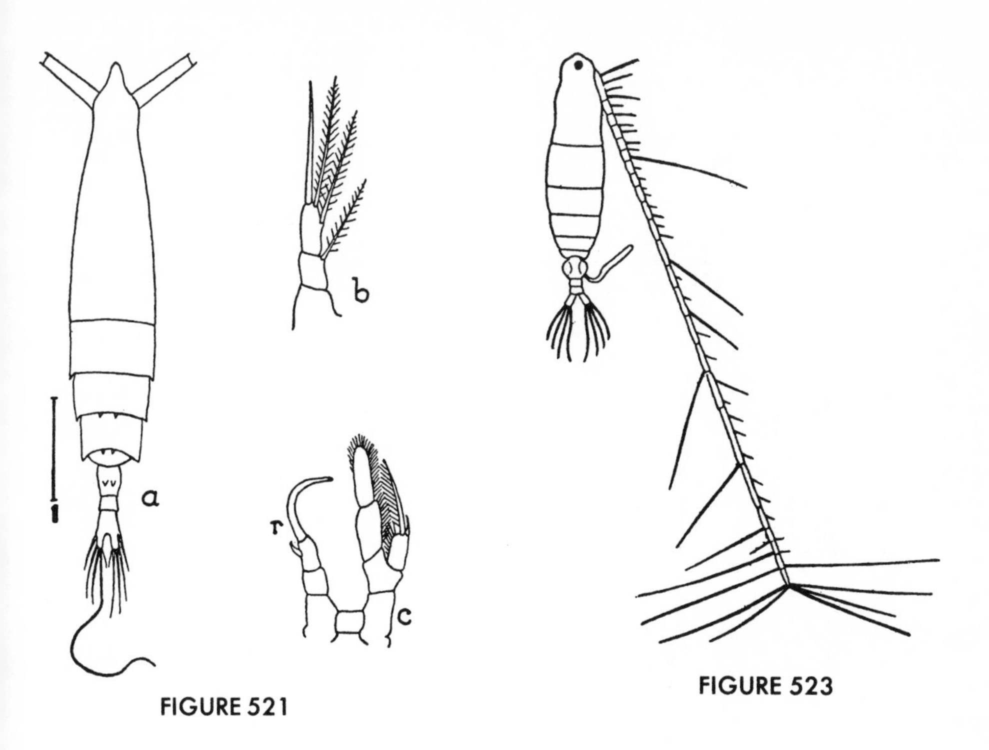

FIGURE 521

FIGURE 523

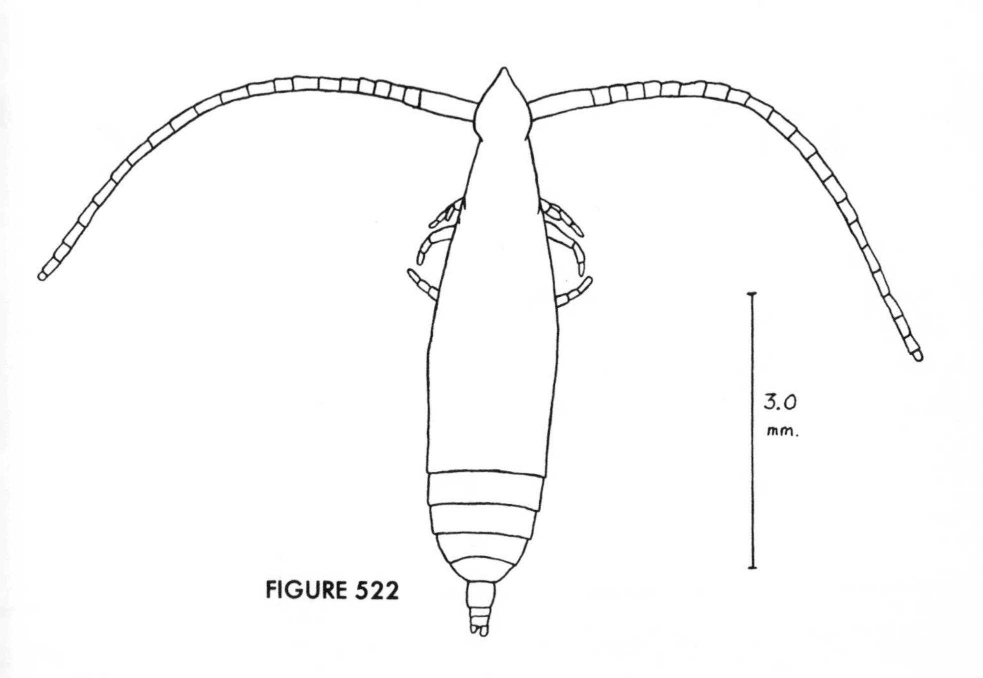

FIGURE 522

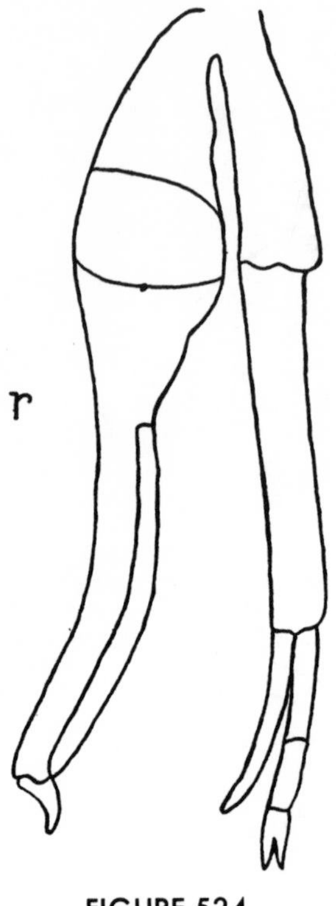

FIGURE 524

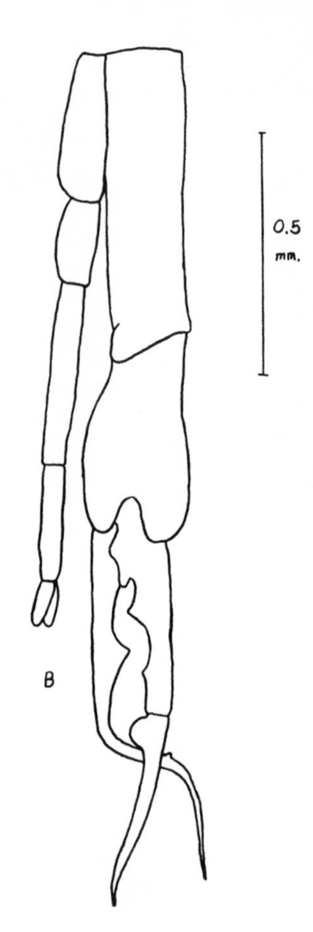

FIGURE 525 B

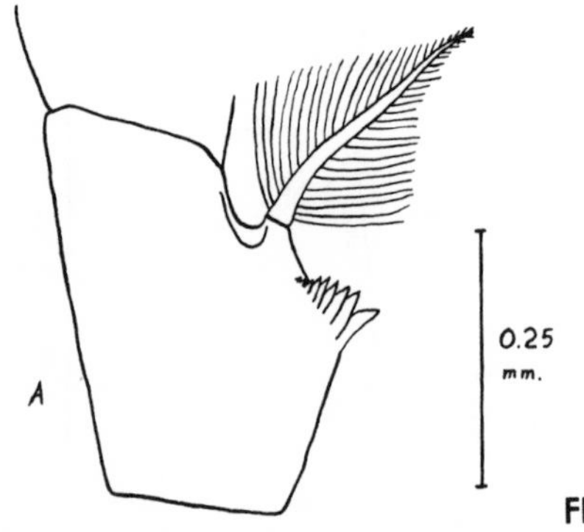

FIGURE 525 A

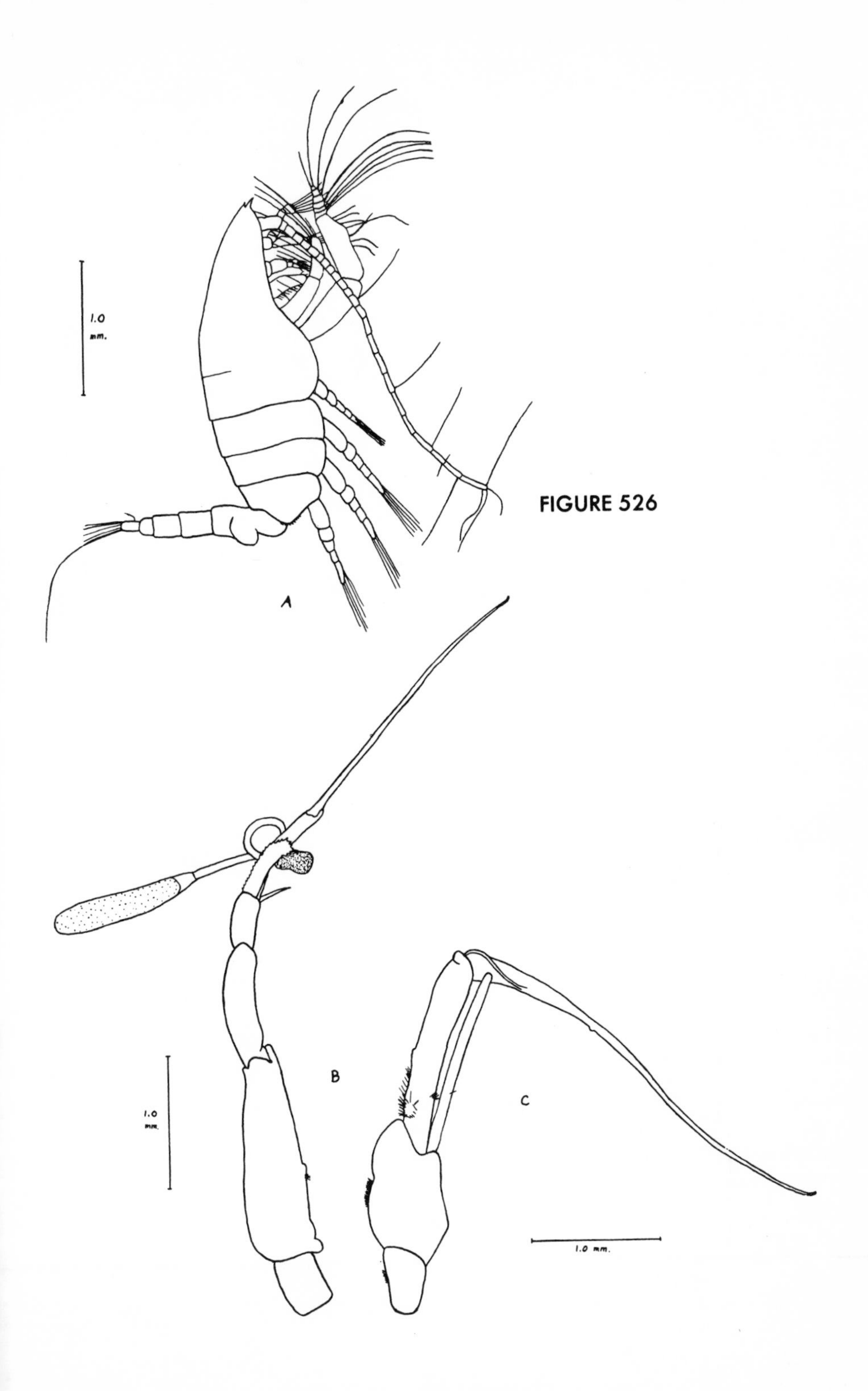

FIGURE 526

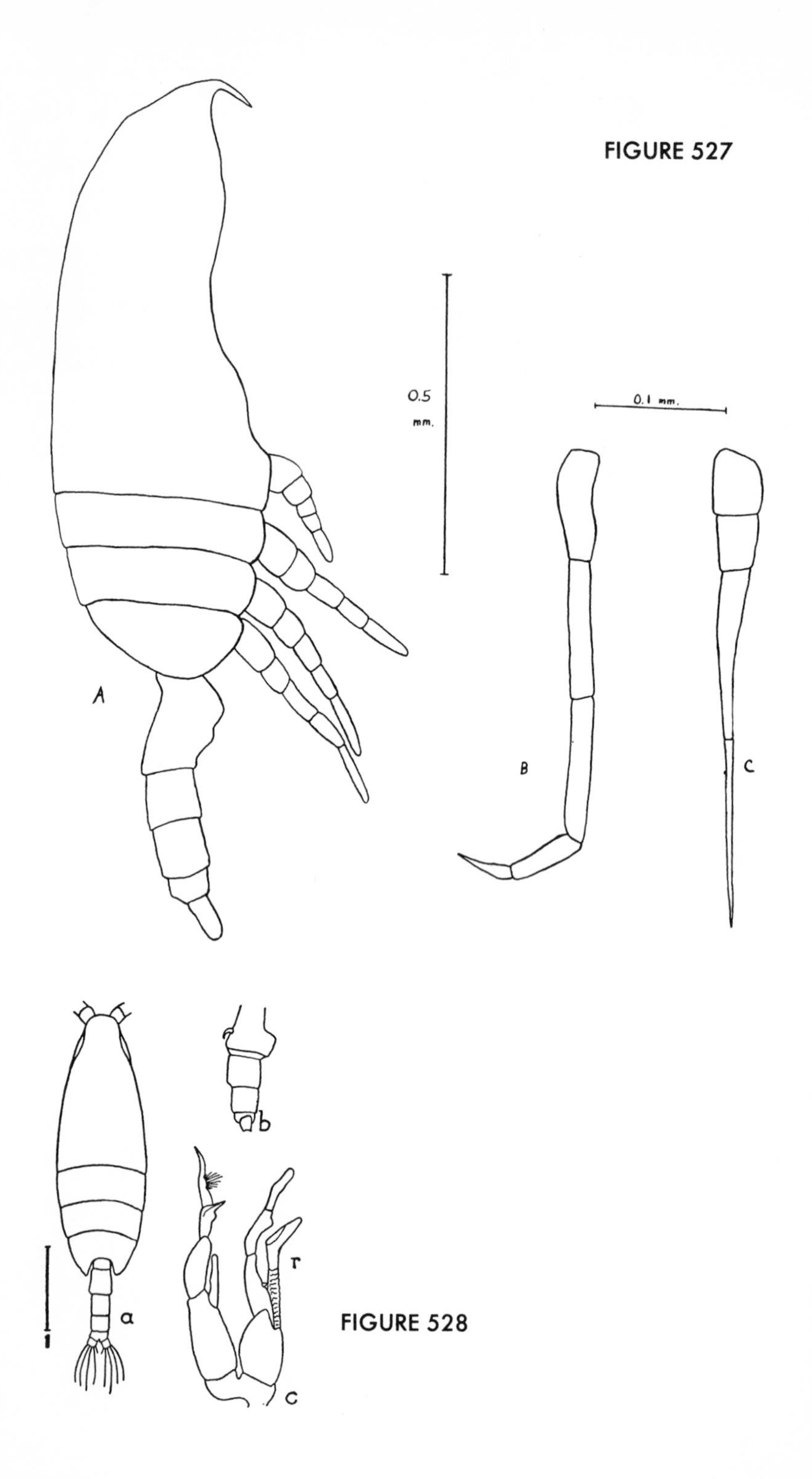
FIGURE 527
0.5
mm.
0.1 mm.
A
B
C
a
b
c
r
FIGURE 528

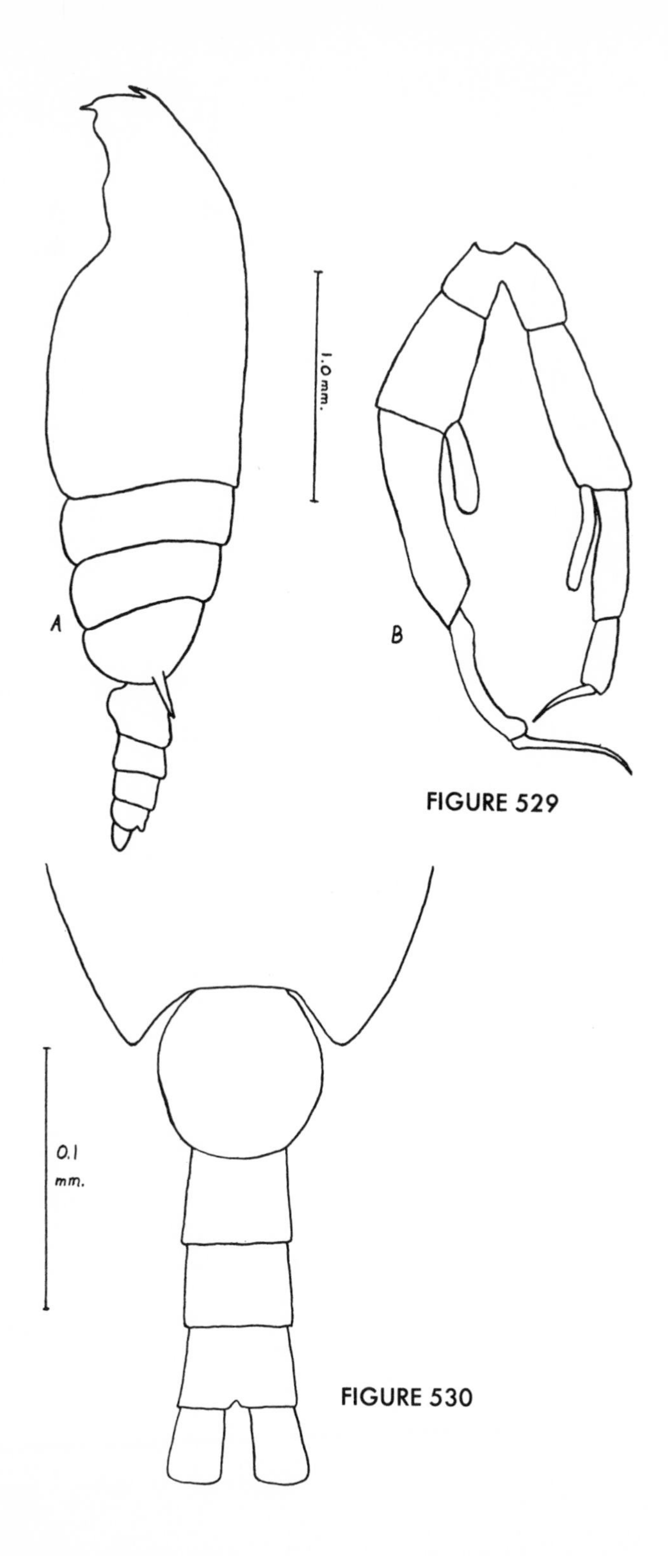

FIGURE 529

FIGURE 530

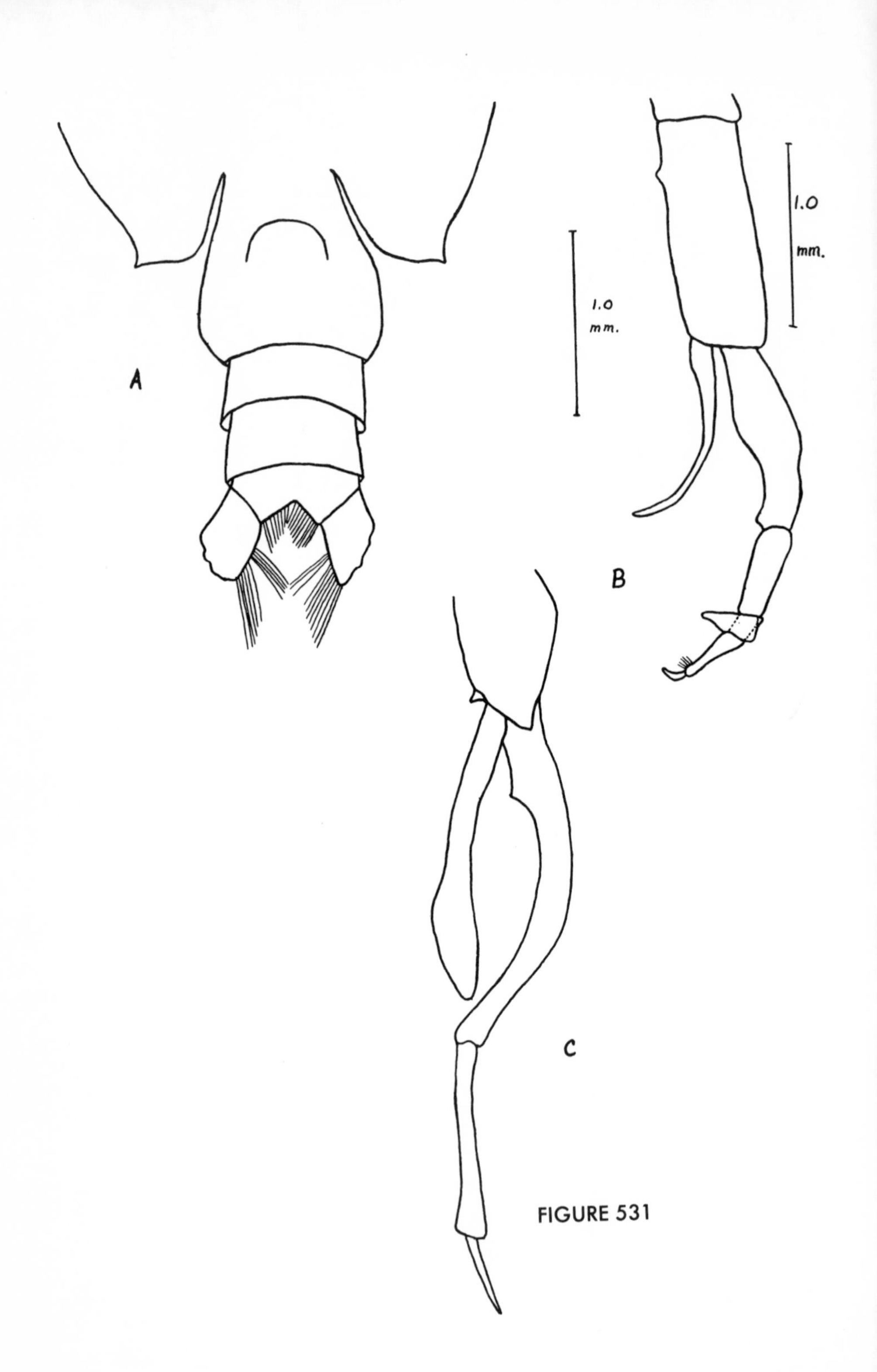

FIGURE 531

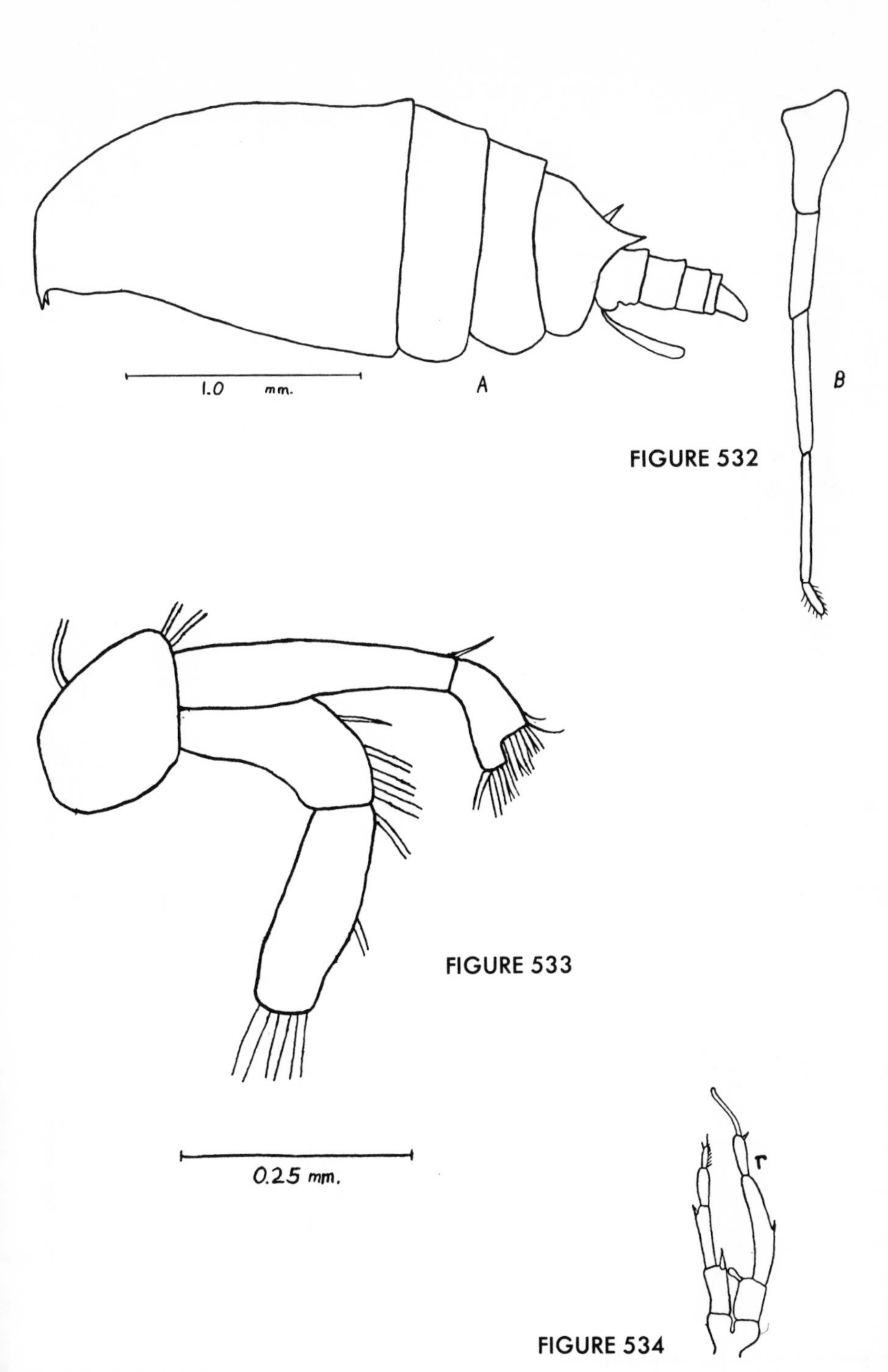

FIGURE 532

FIGURE 533

FIGURE 534

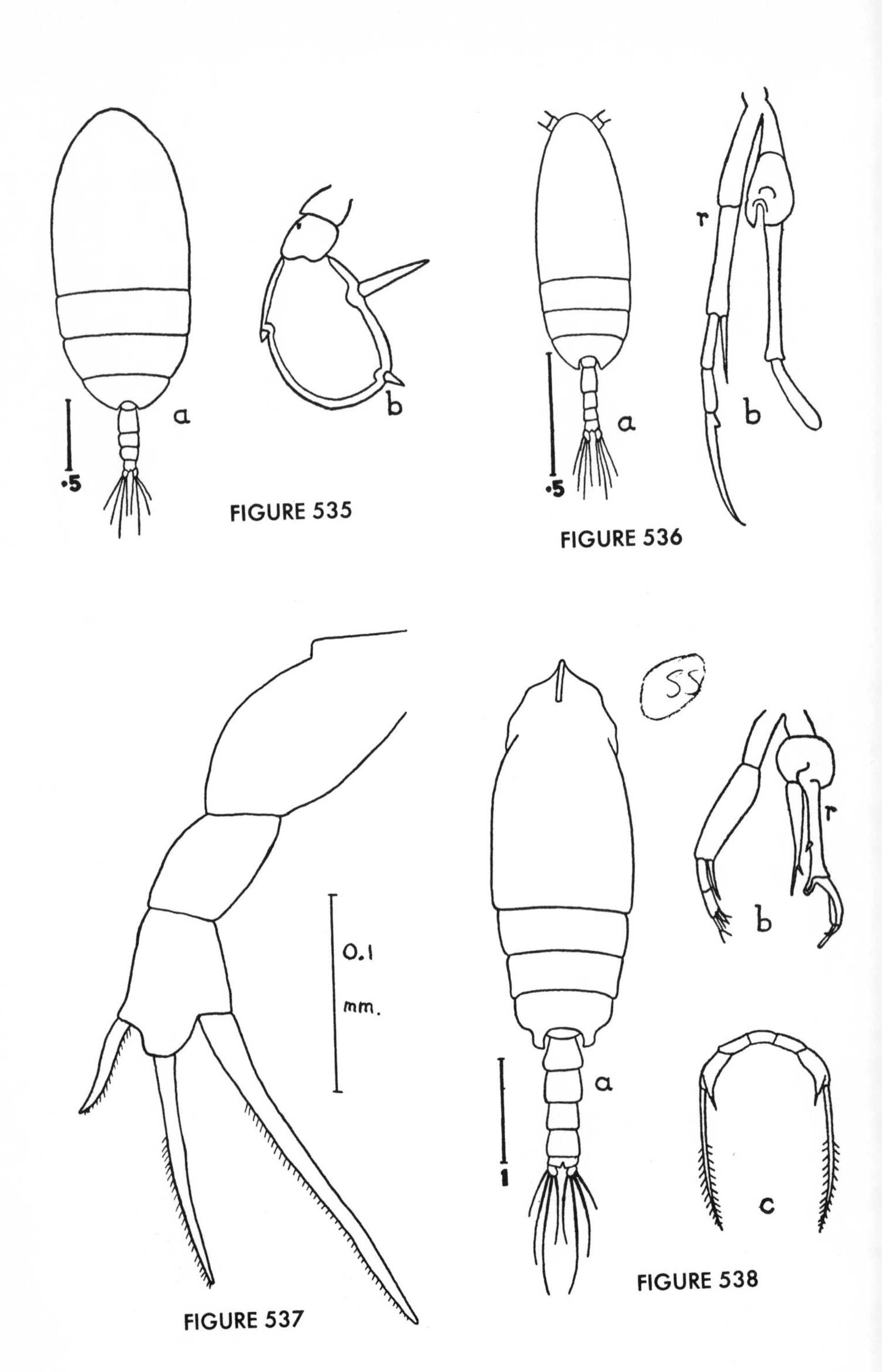

FIGURE 535

FIGURE 536

FIGURE 537

FIGURE 538

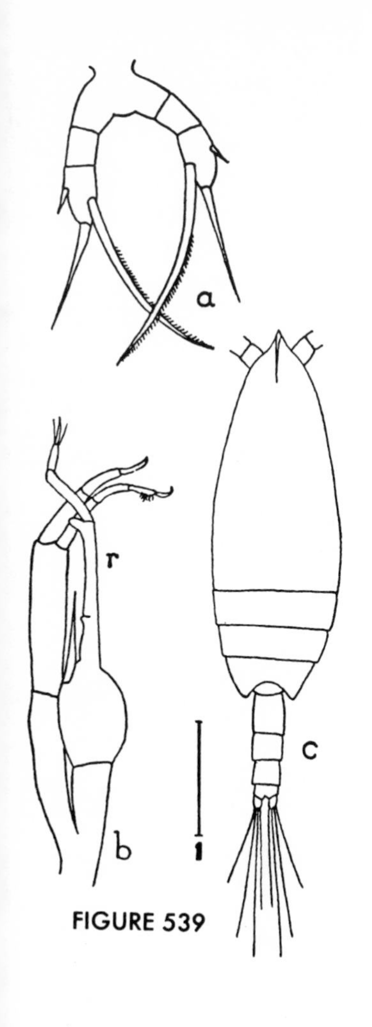

FIGURE 539

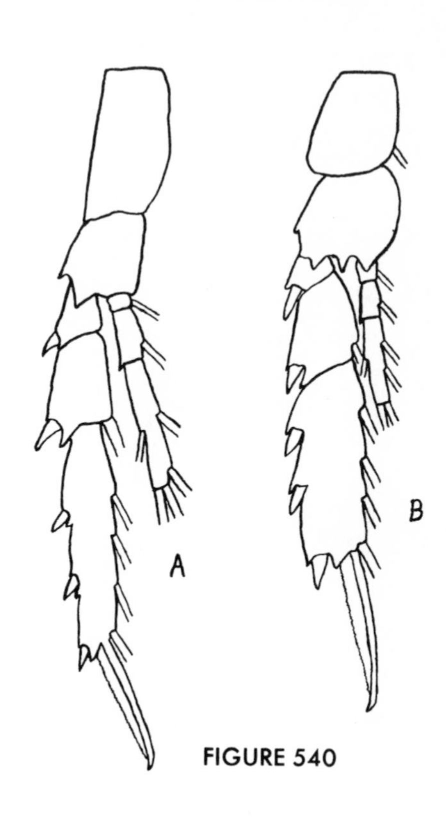

FIGURE 540

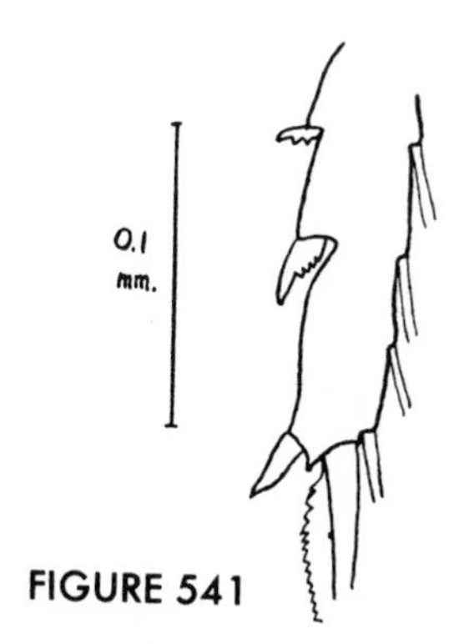

FIGURE 541

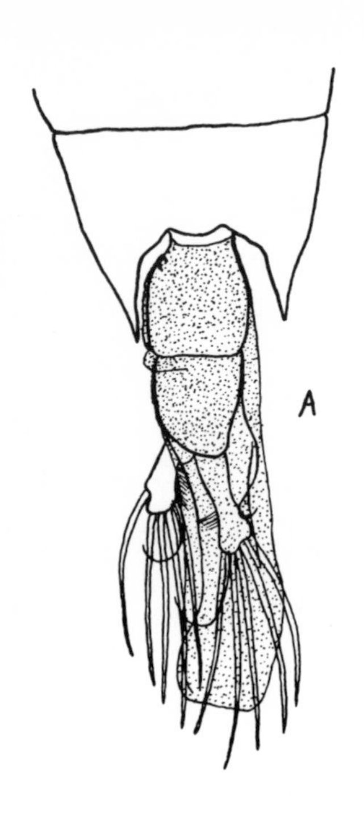

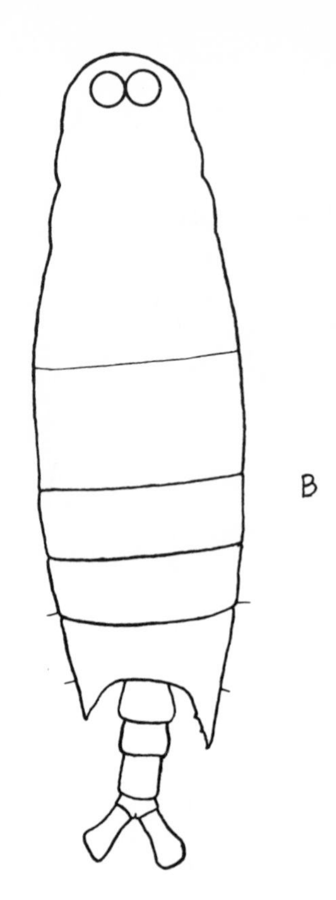

FIGURE 542

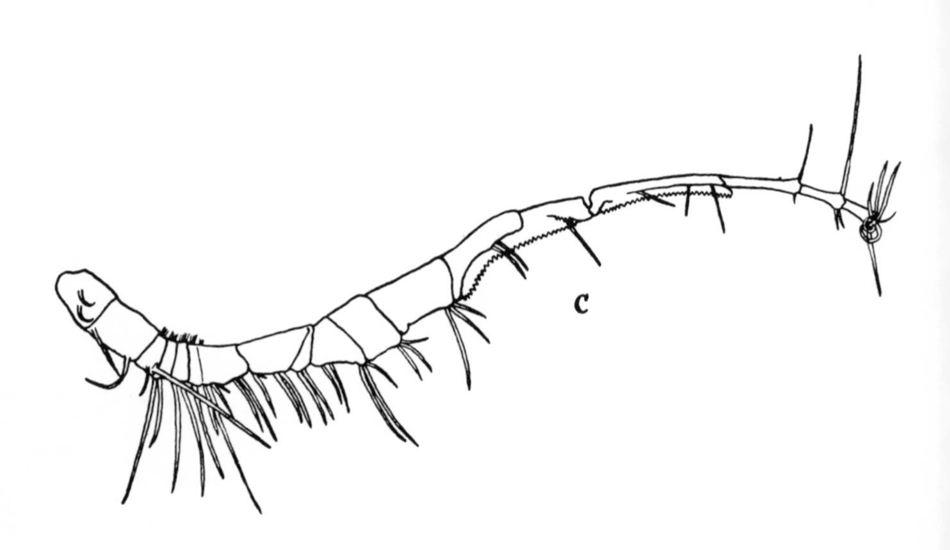

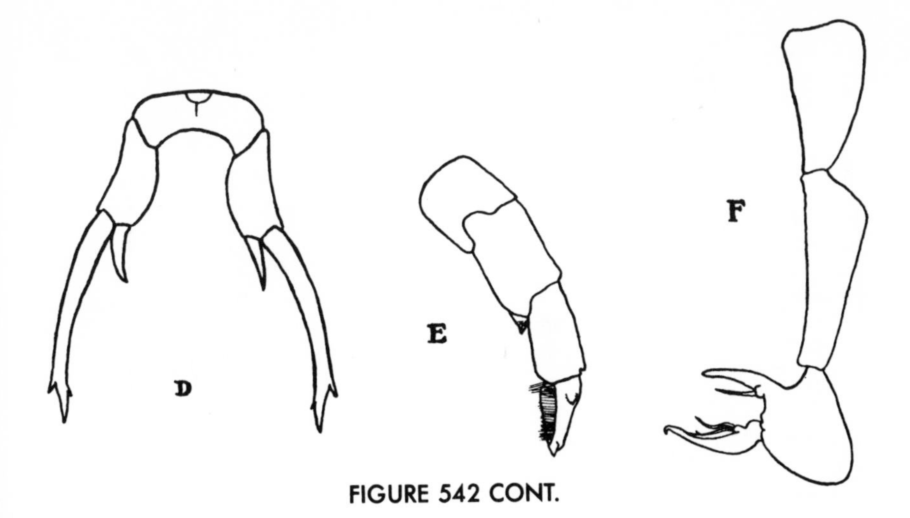

FIGURE 542 CONT.

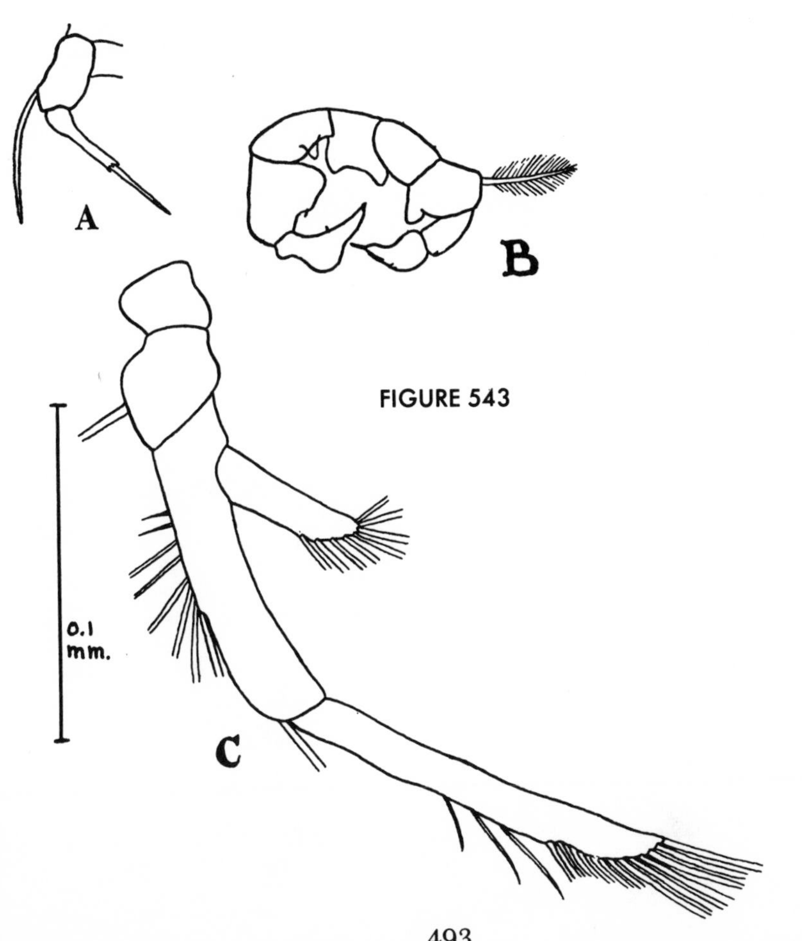

FIGURE 543

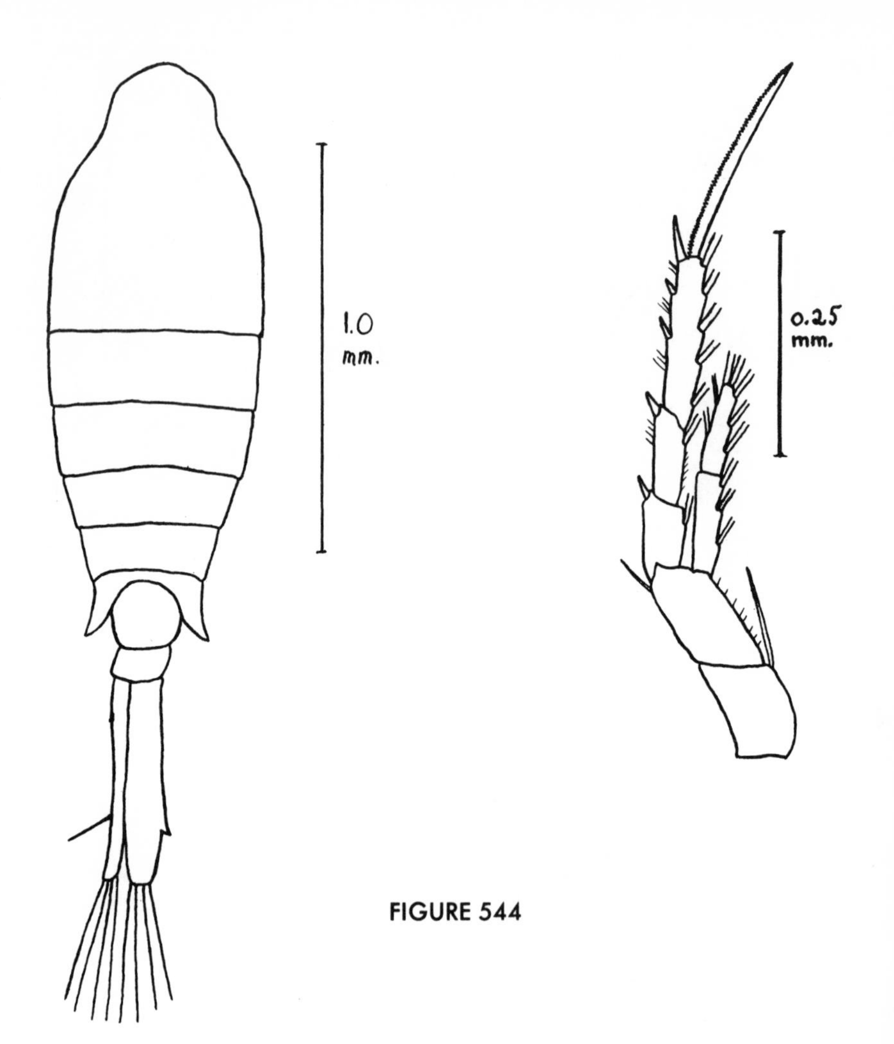

FIGURE 544

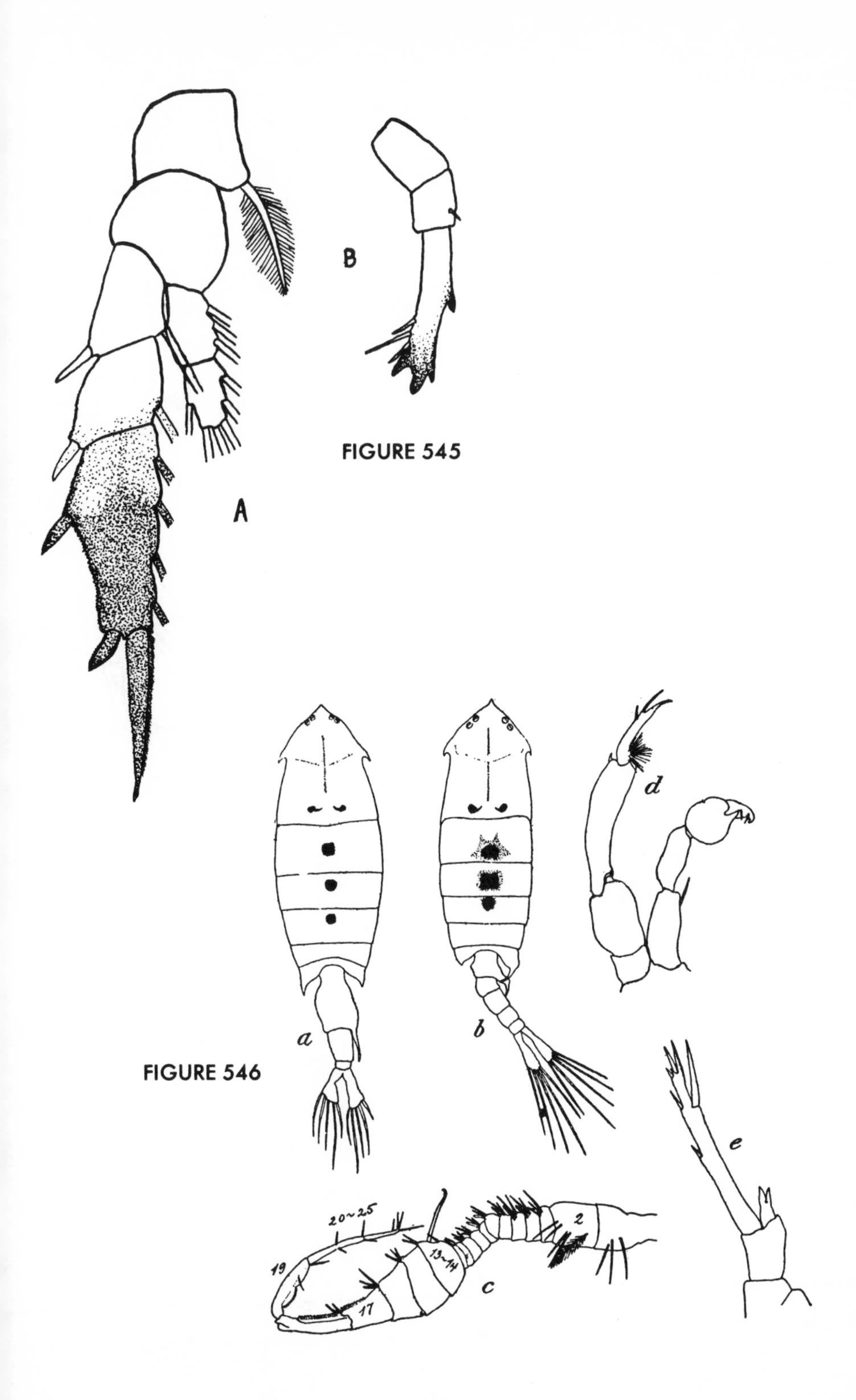

FIGURE 545

FIGURE 546

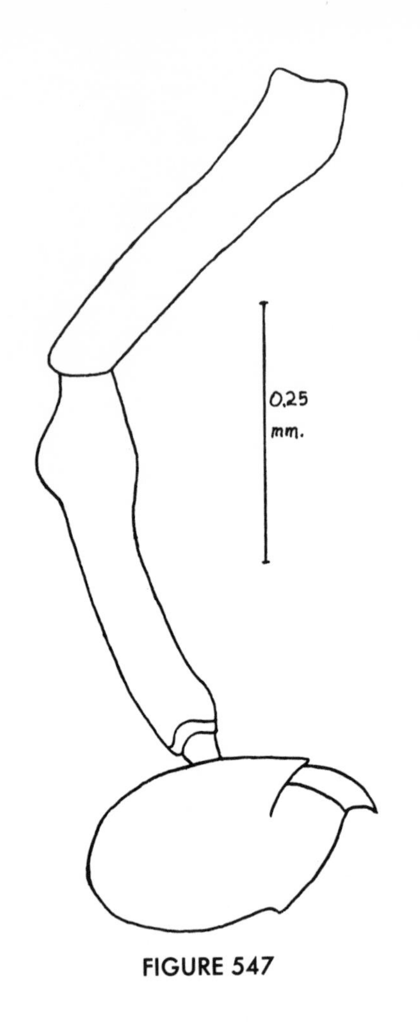

FIGURE 547

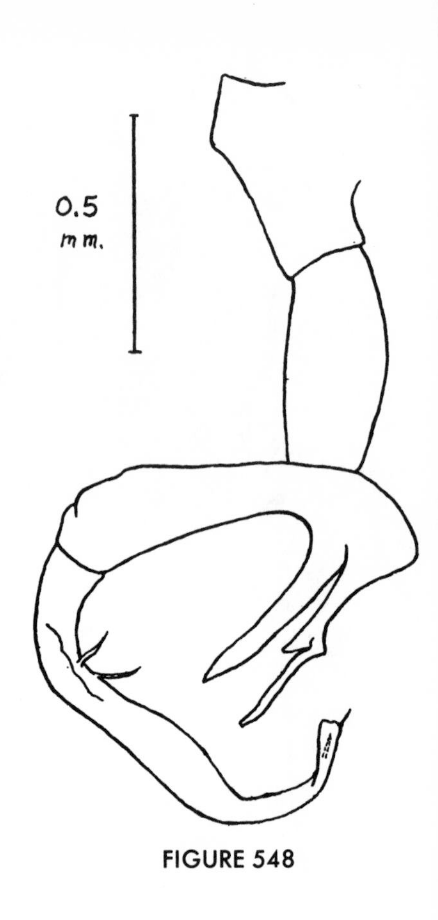

FIGURE 548

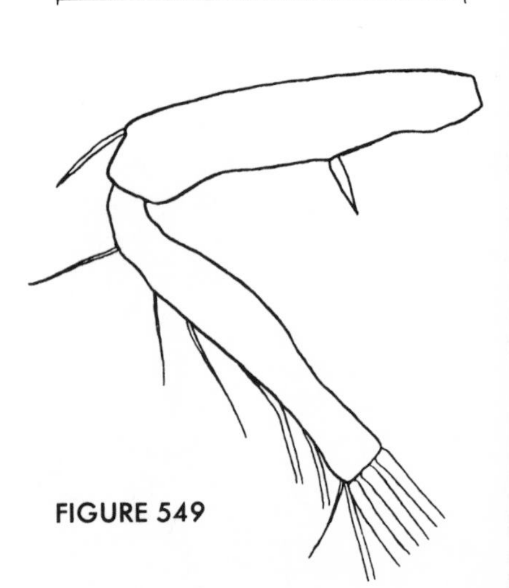

FIGURE 549

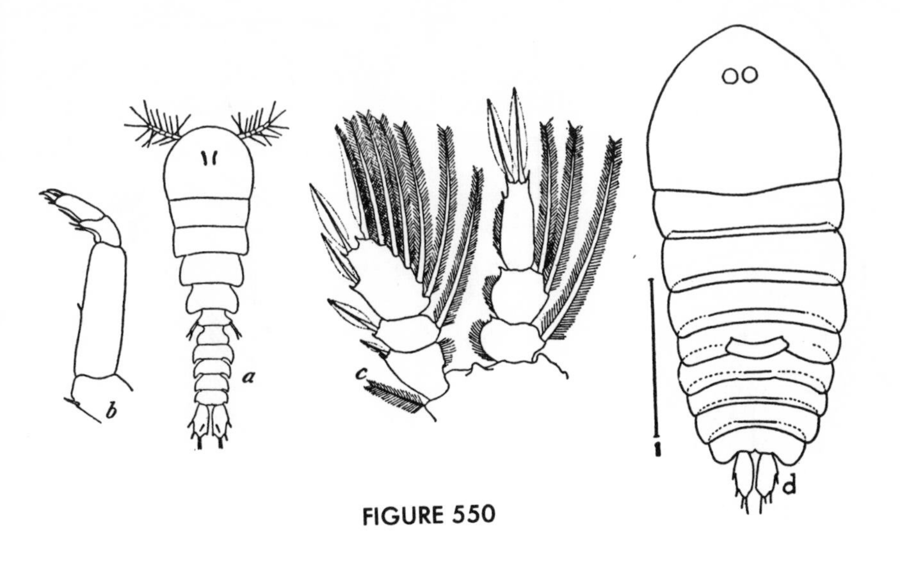

FIGURE 550

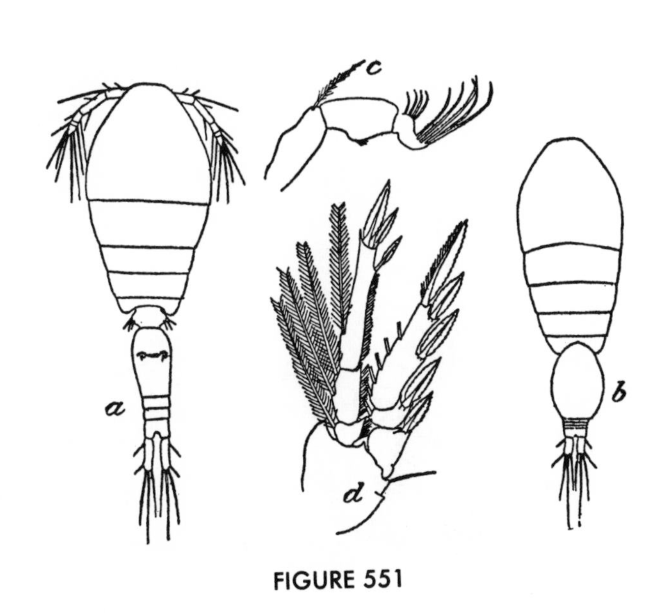

FIGURE 551

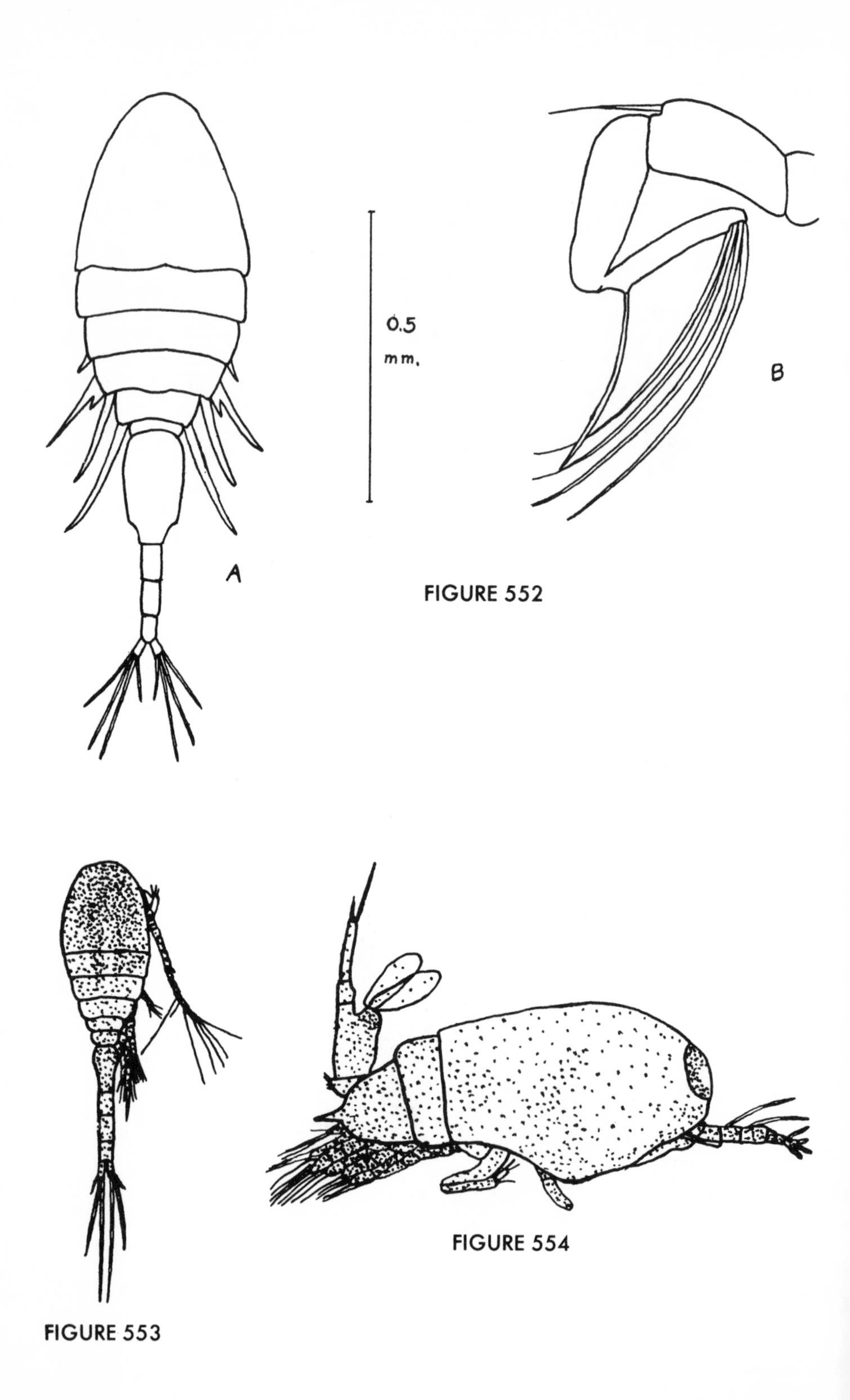

FIGURE 552

FIGURE 553

FIGURE 554

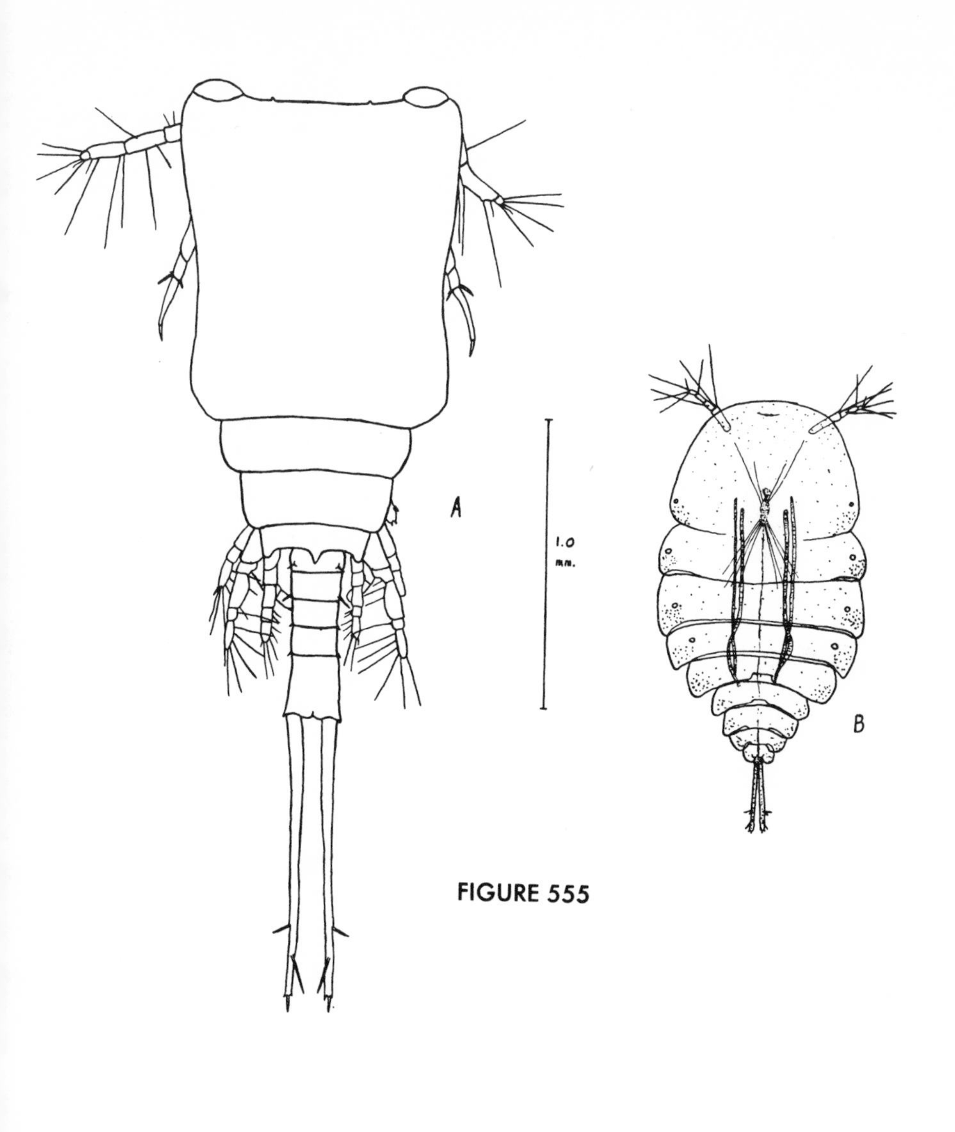

FIGURE 555

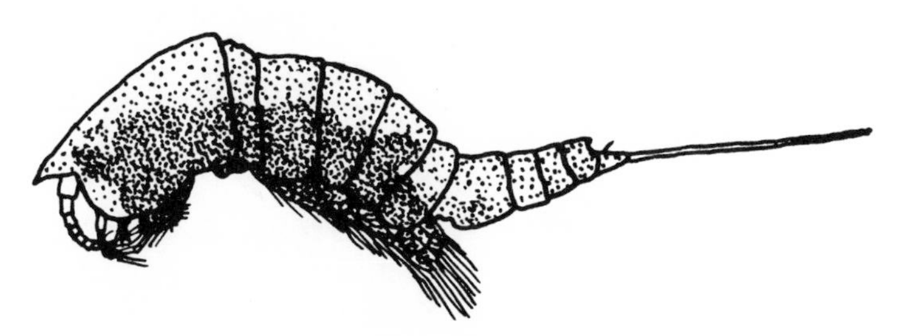

FIGURE 556

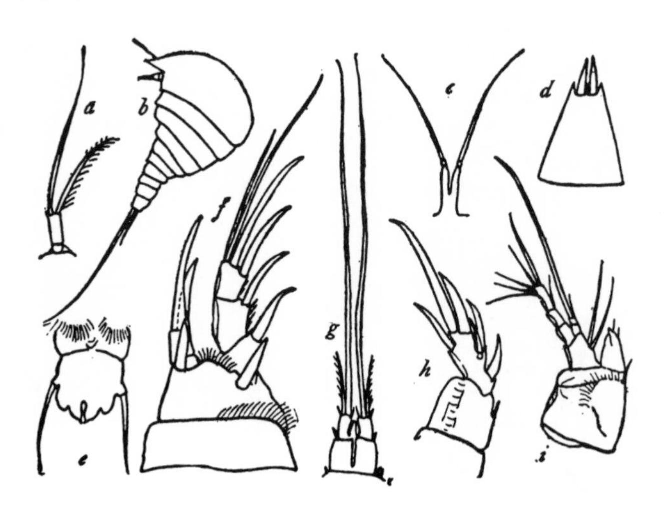

FIGURE 557

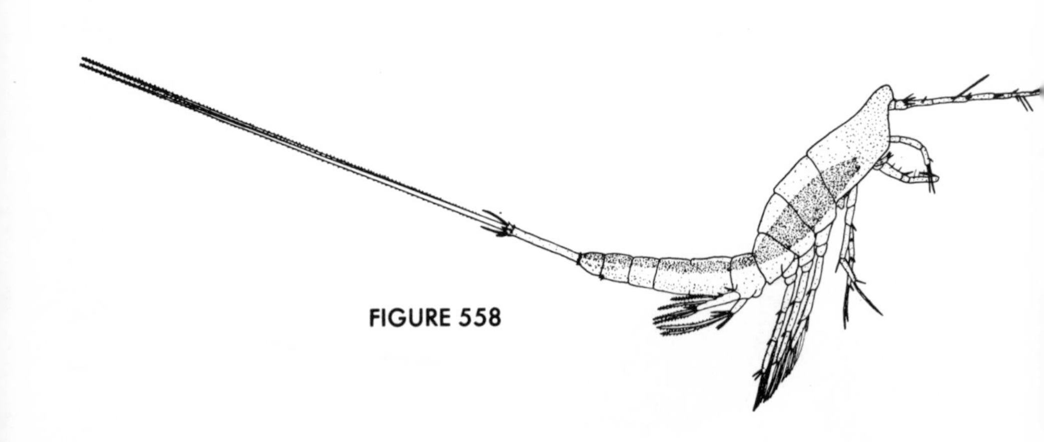

FIGURE 558

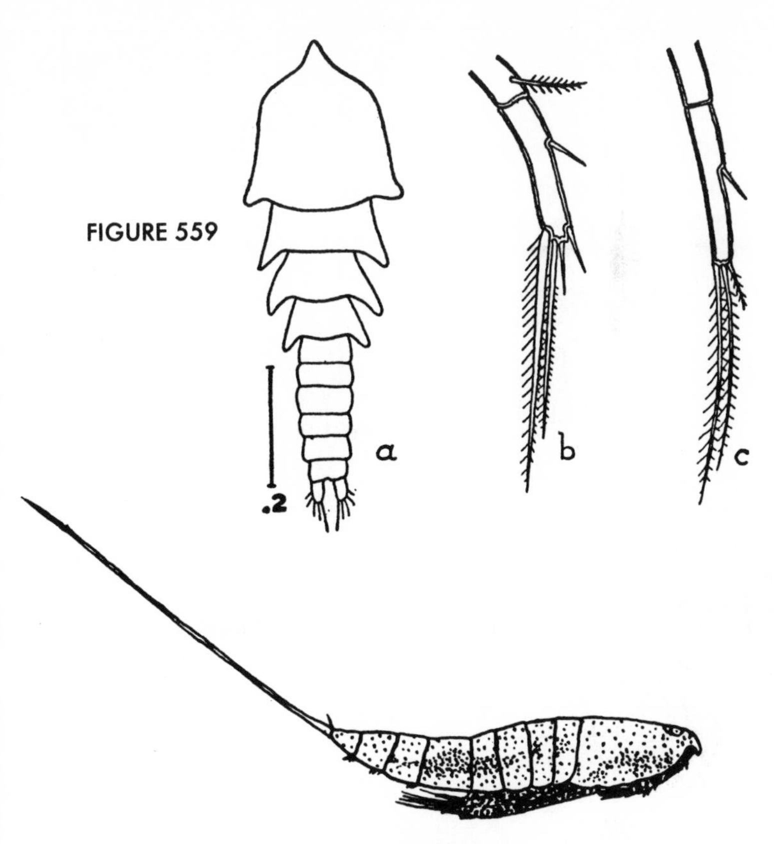

FIGURE 559

FIGURE 560

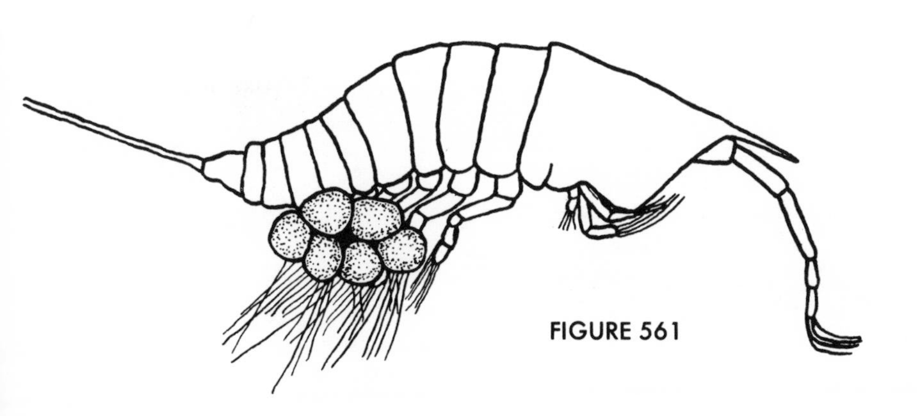
FIGURE 561

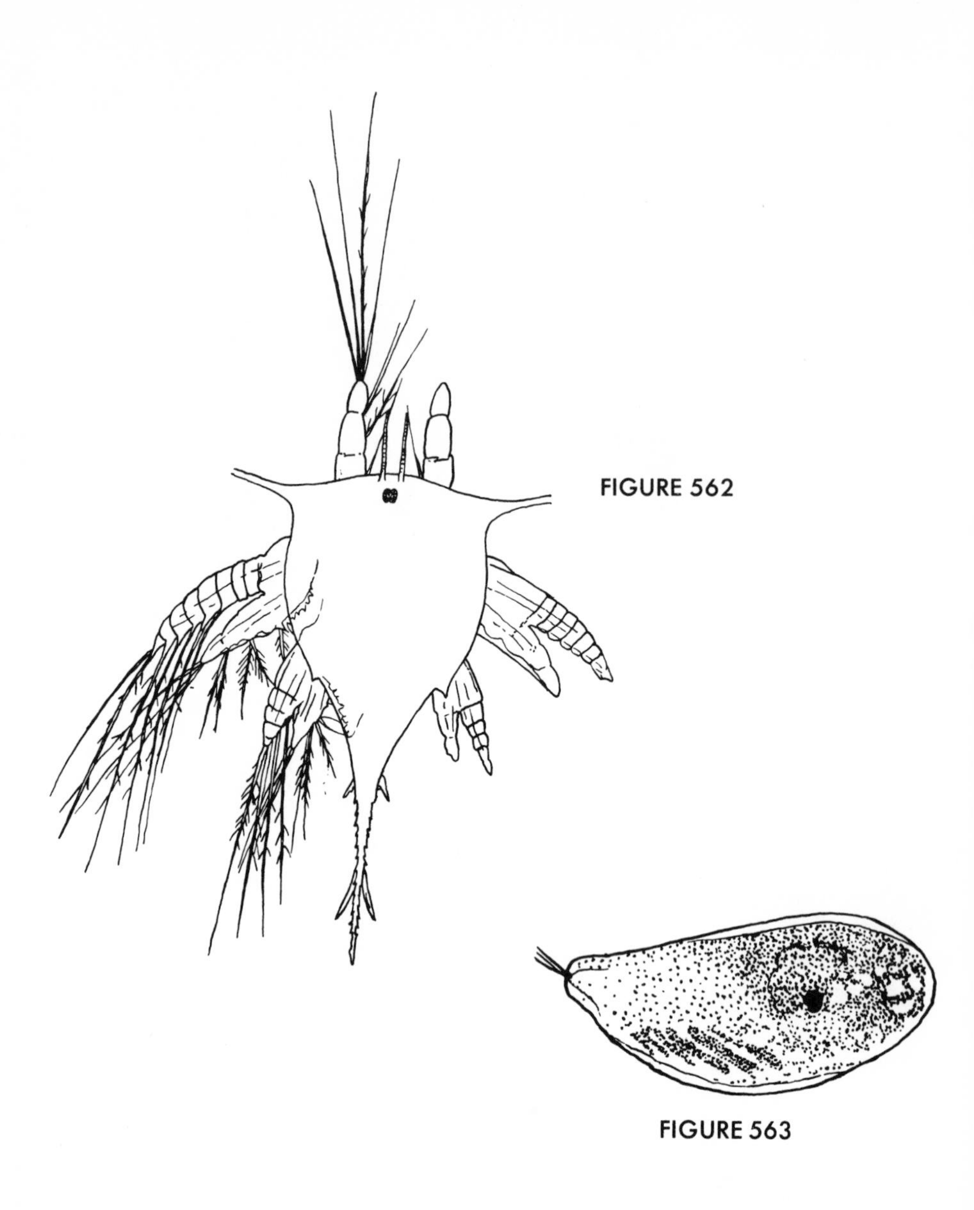

FIGURE 562

FIGURE 563

FIGURE 564

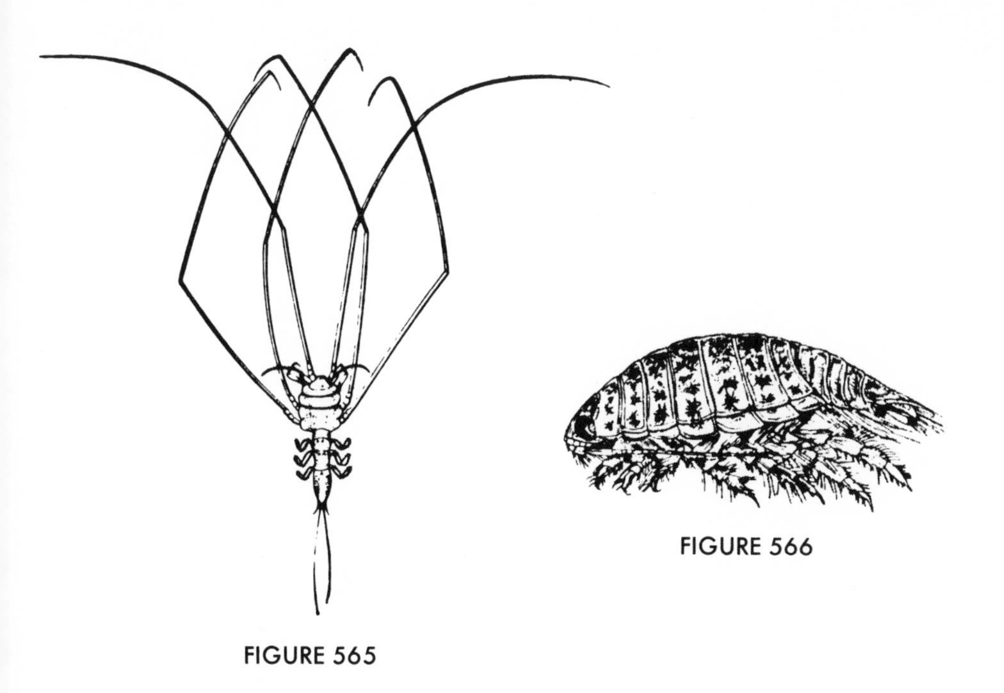

FIGURE 565

FIGURE 566

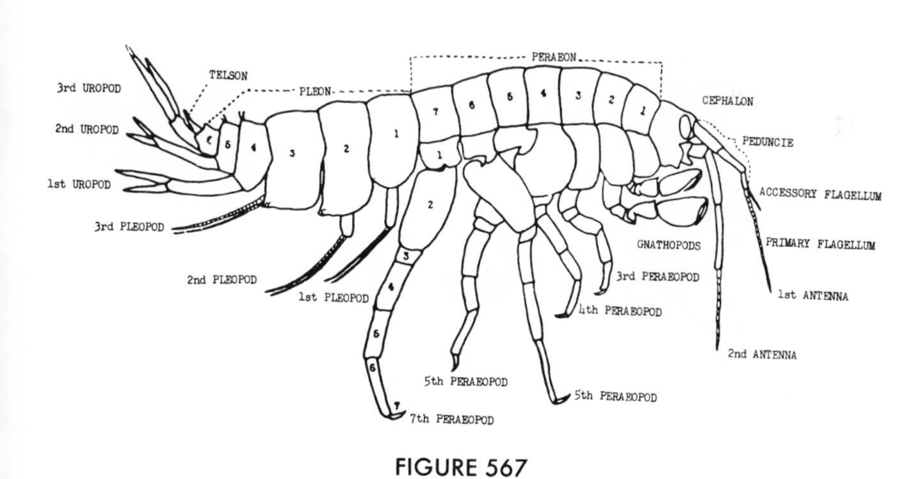

FIGURE 567

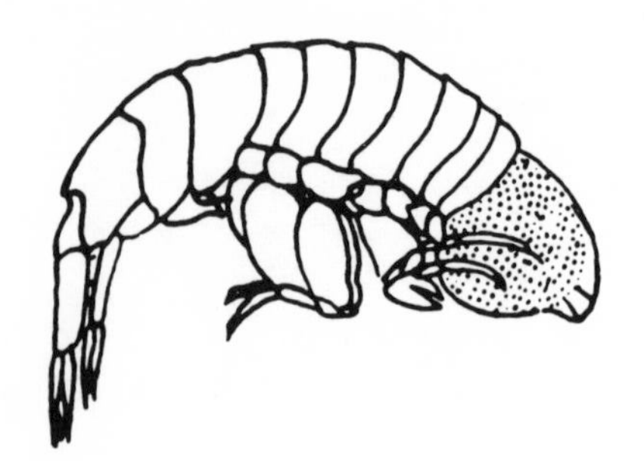

FIGURE 568

FIGURE 569

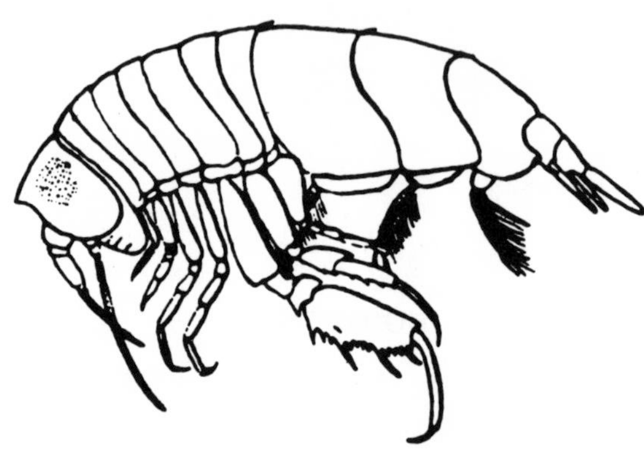

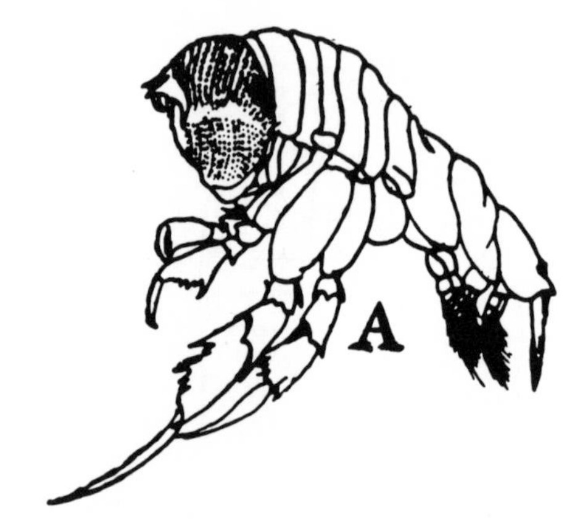

FIGURE 570

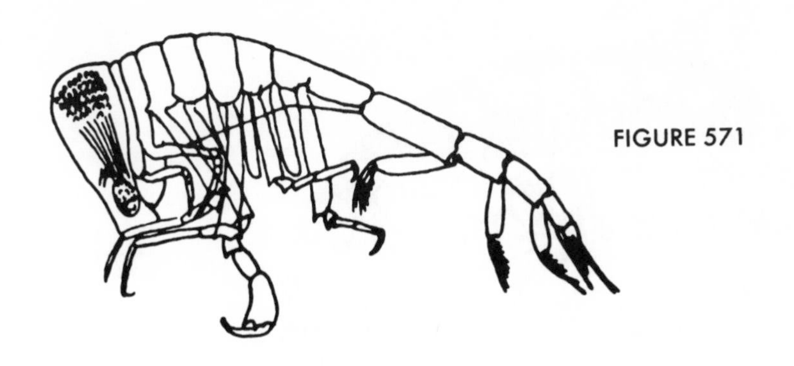

FIGURE 571

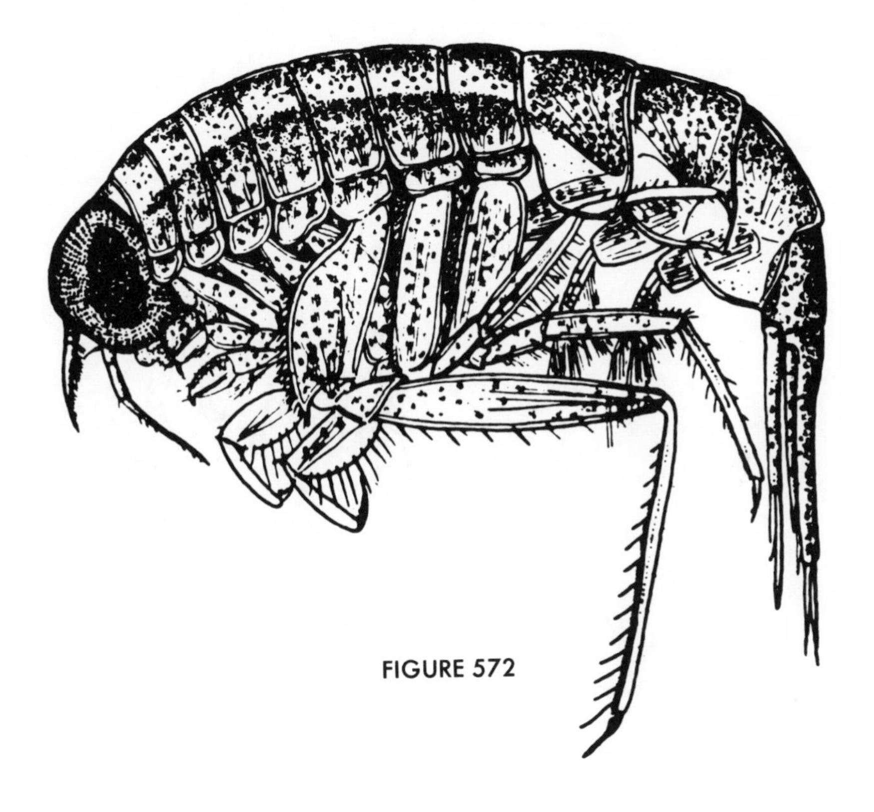

FIGURE 572

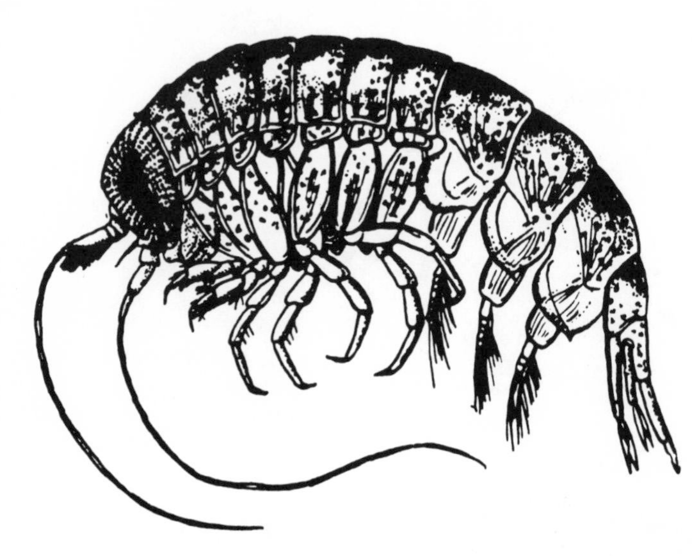

FIGURE 573

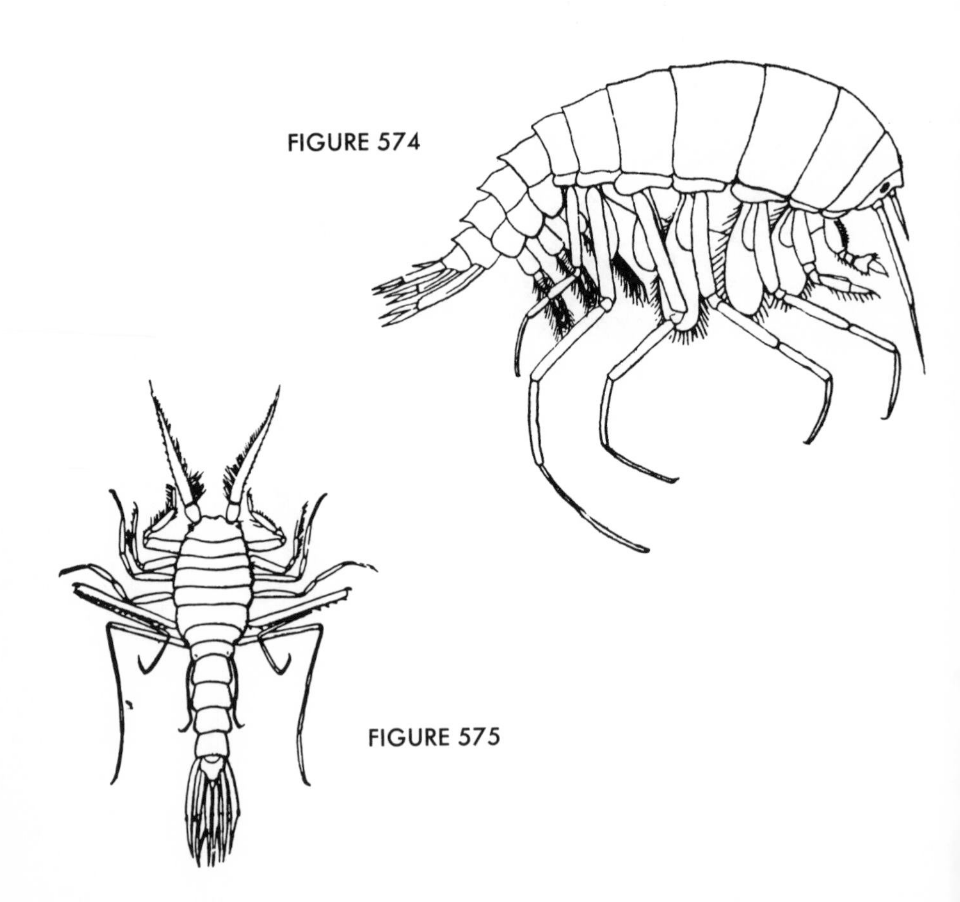

FIGURE 574

FIGURE 575

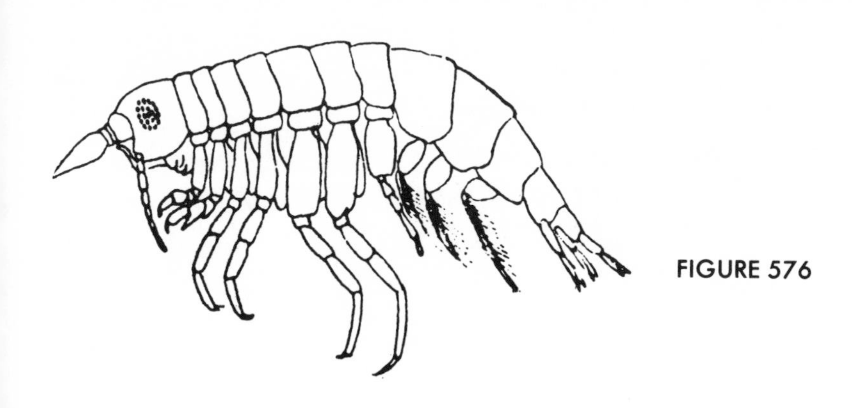

FIGURE 576

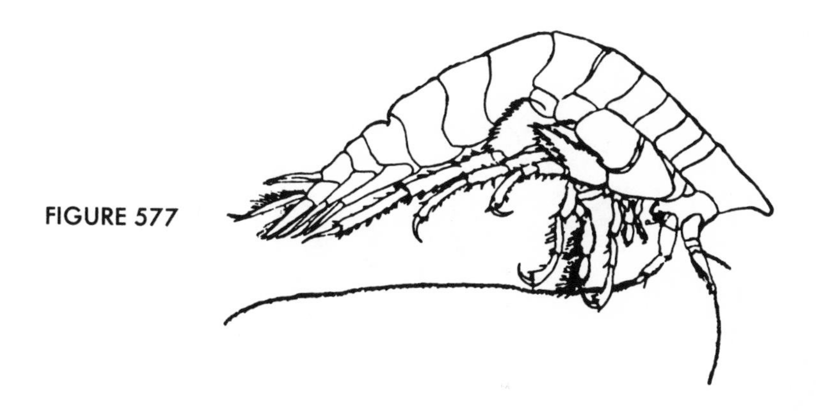

FIGURE 577

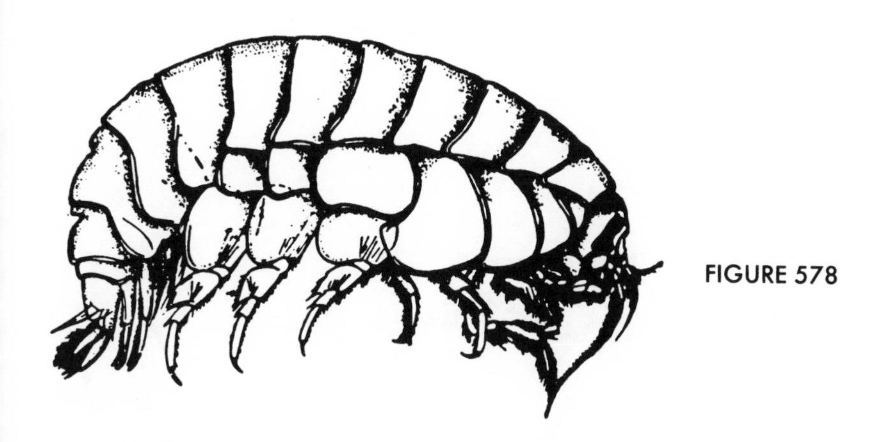

FIGURE 578

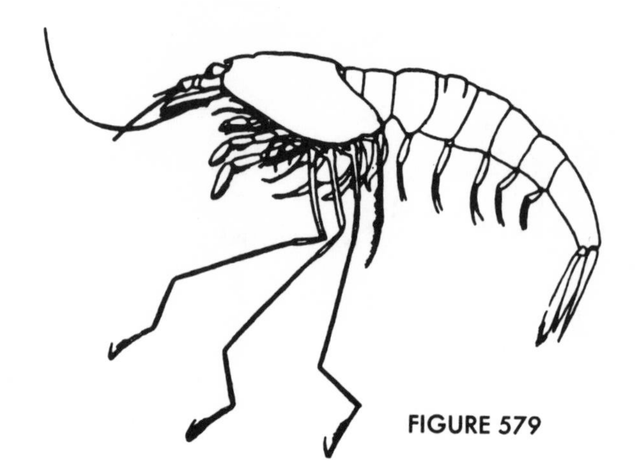

FIGURE 579

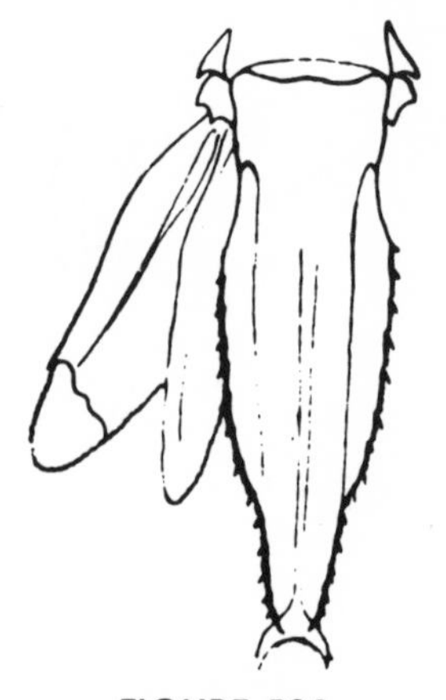

FIGURE 581

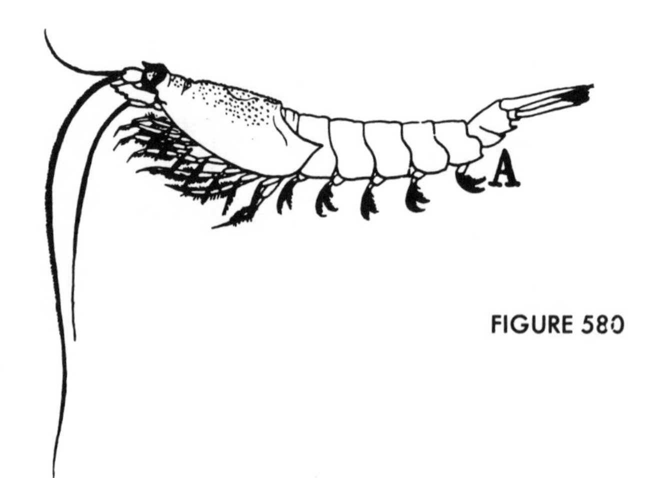

FIGURE 580

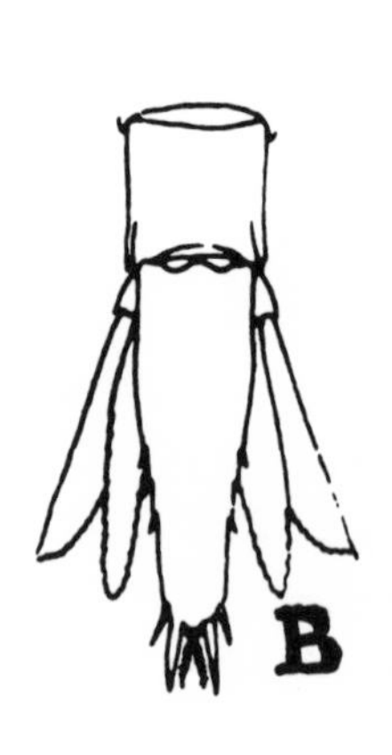

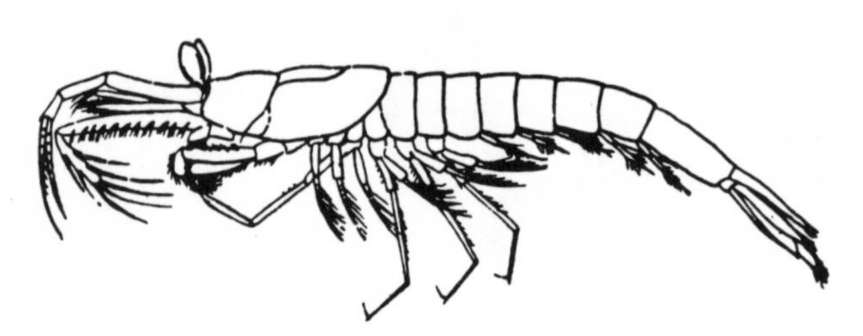

FIGURE 582

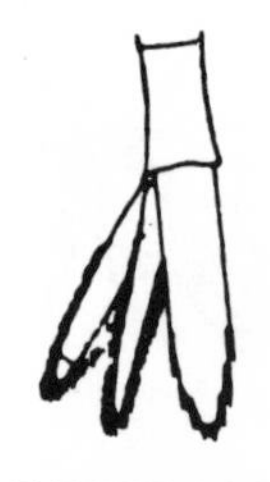

FIGURE 583

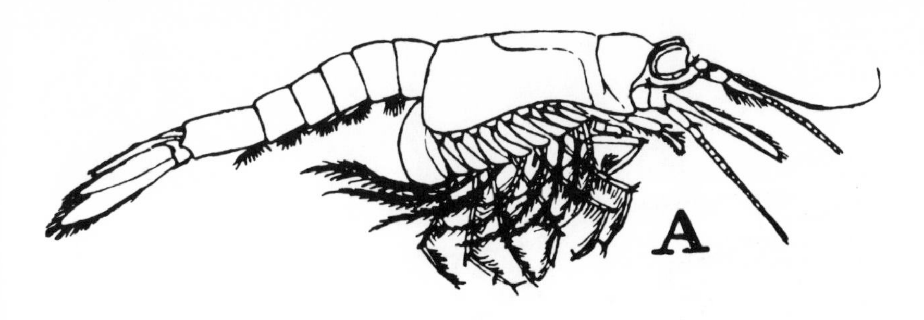

FIGURE 584

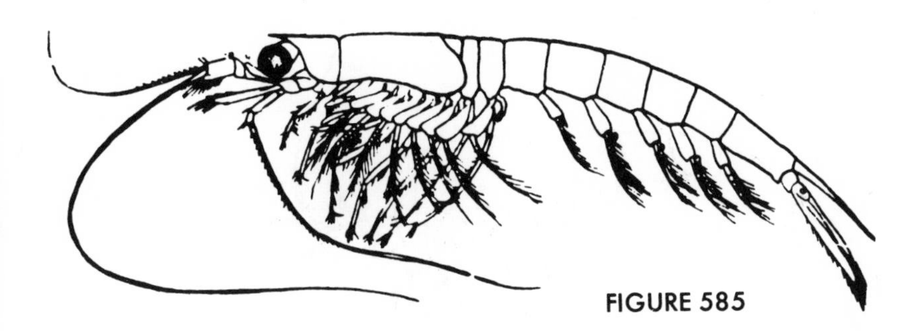

FIGURE 585

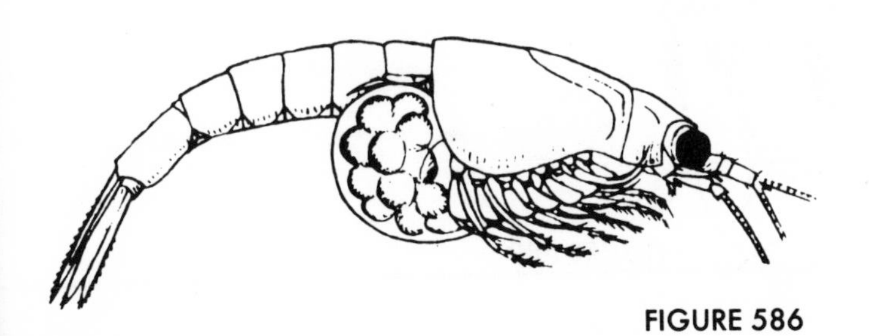

FIGURE 586

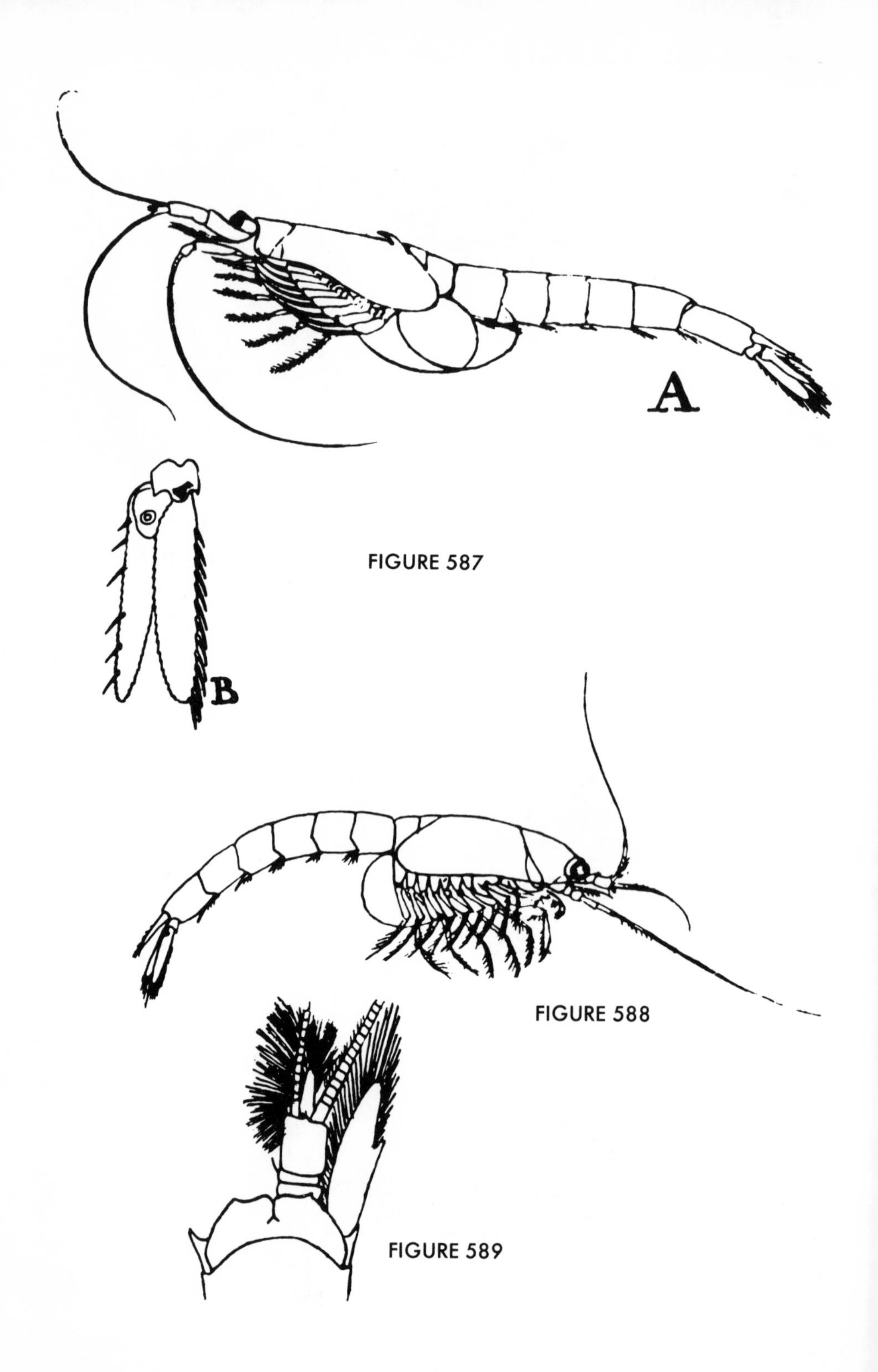

FIGURE 587

FIGURE 588

FIGURE 589

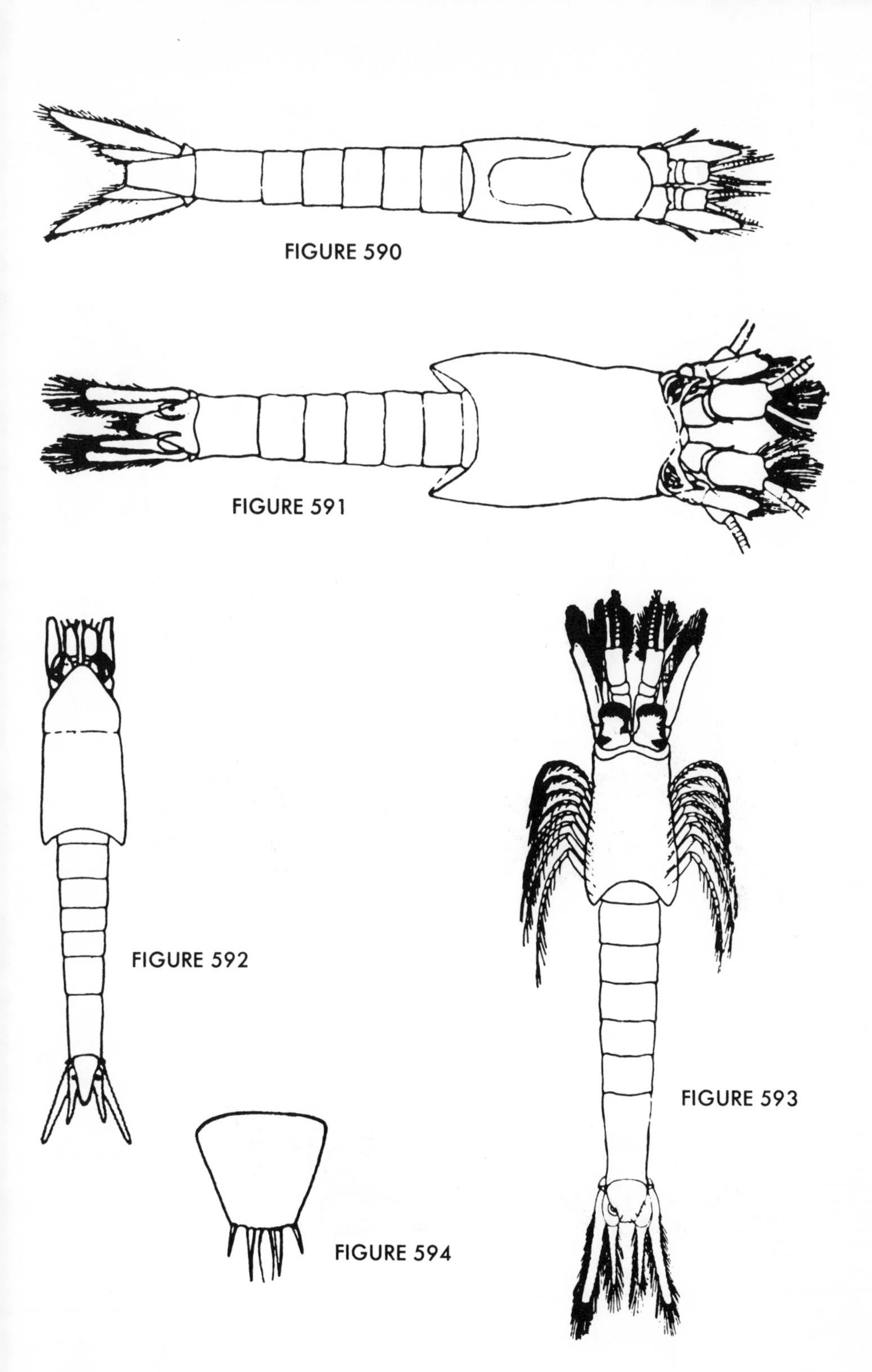
FIGURE 590
FIGURE 591
FIGURE 592
FIGURE 593
FIGURE 594

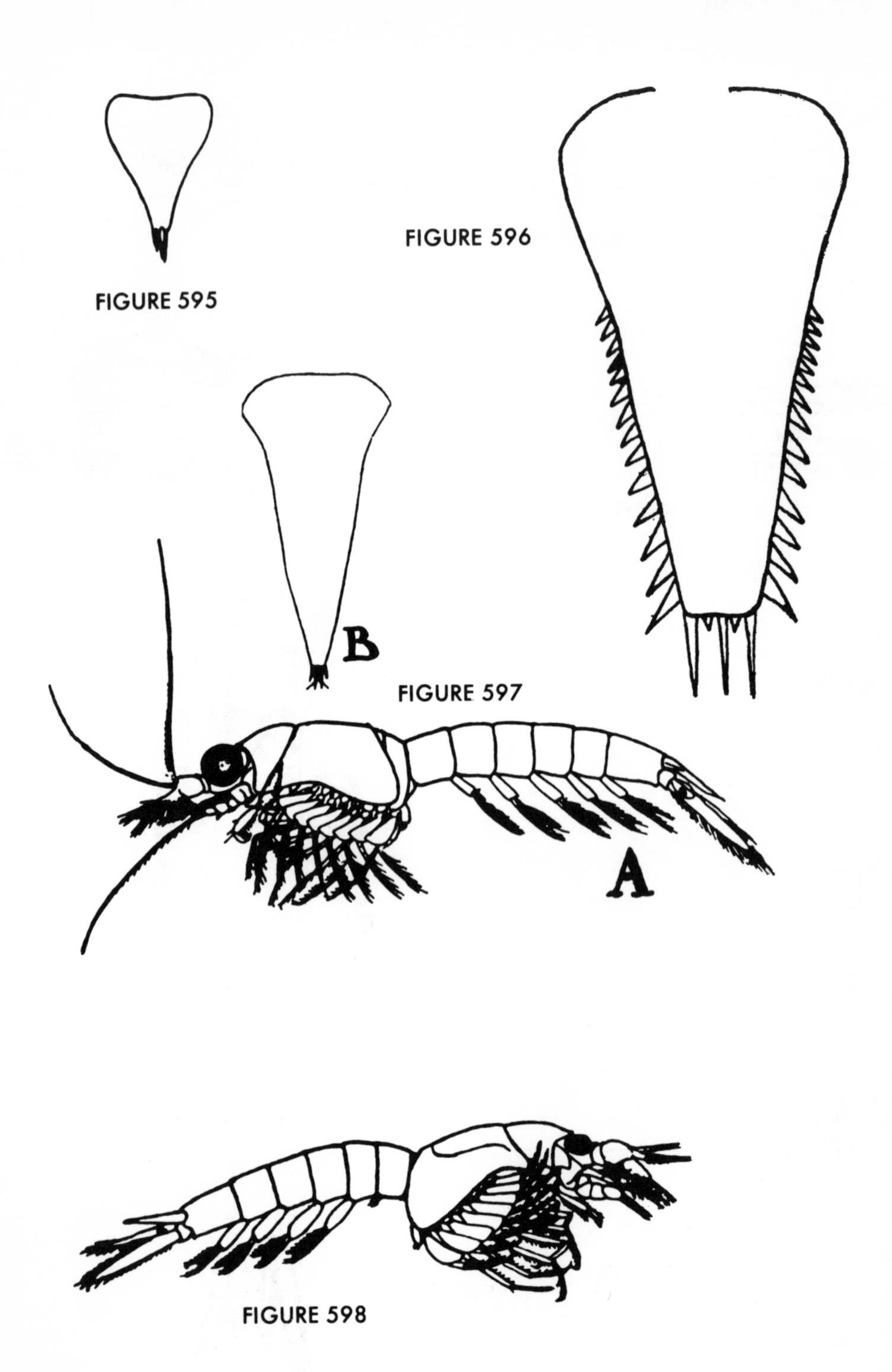

FIGURE 595

FIGURE 596

FIGURE 597

FIGURE 598

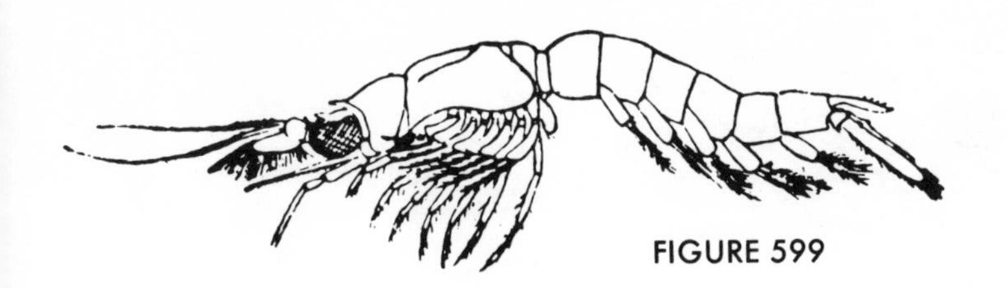

FIGURE 599

FIGURE 600

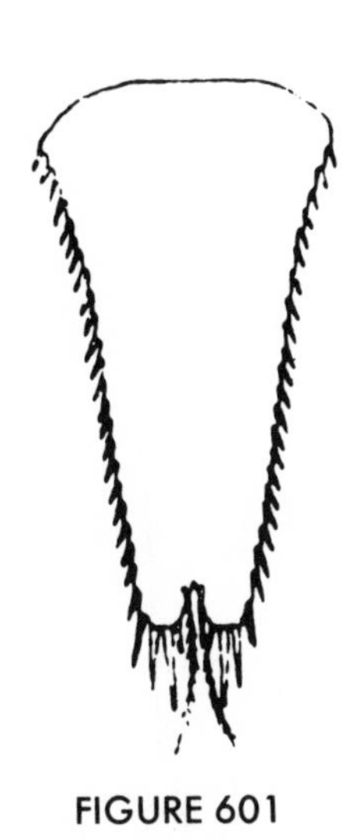

FIGURE 601

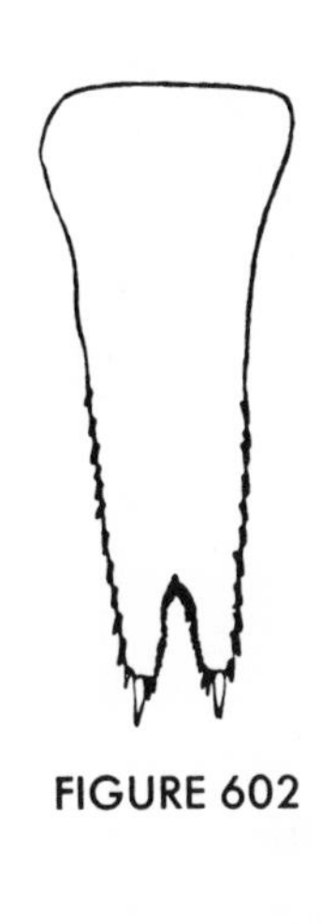

FIGURE 602

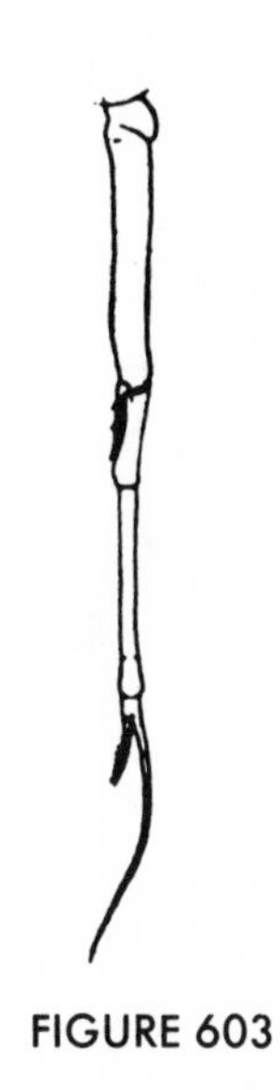

FIGURE 603

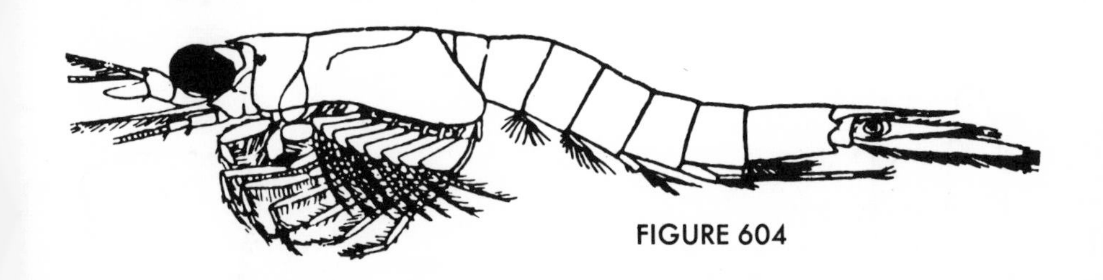

FIGURE 604

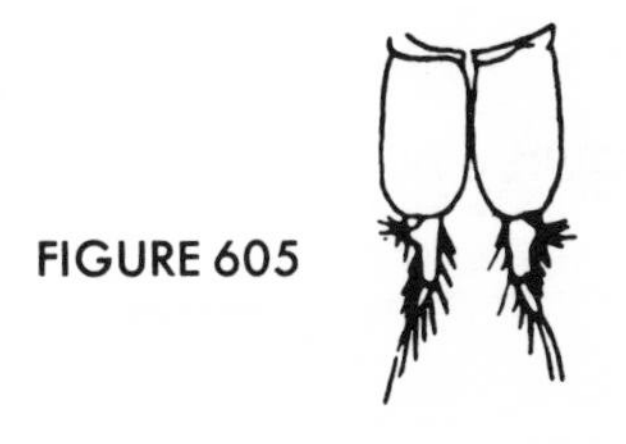

FIGURE 605

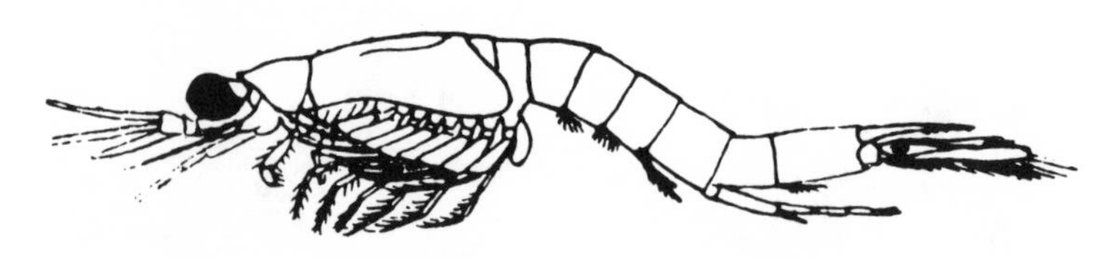

FIGURE 606

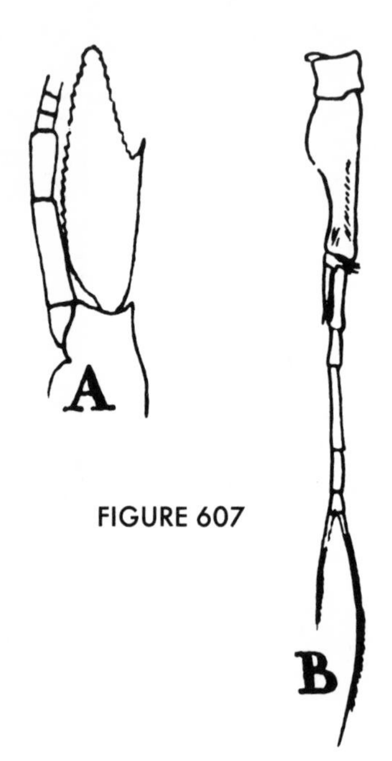

FIGURE 607

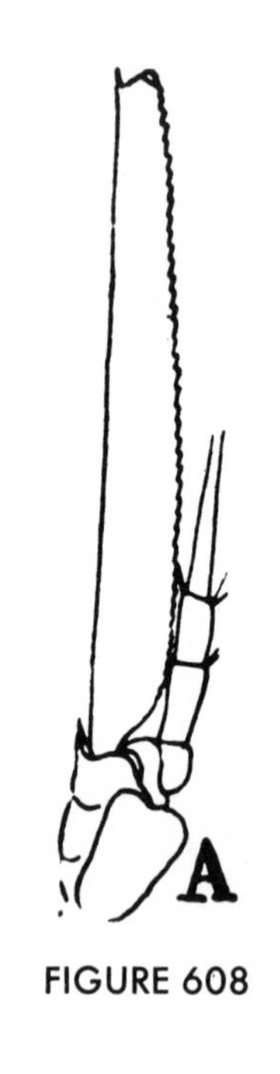

FIGURE 608

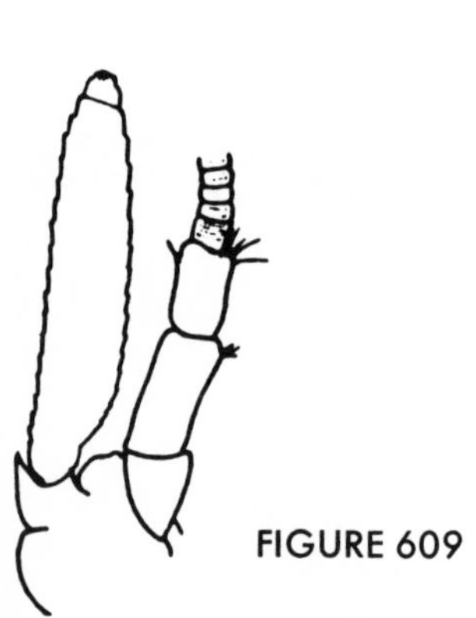

FIGURE 609

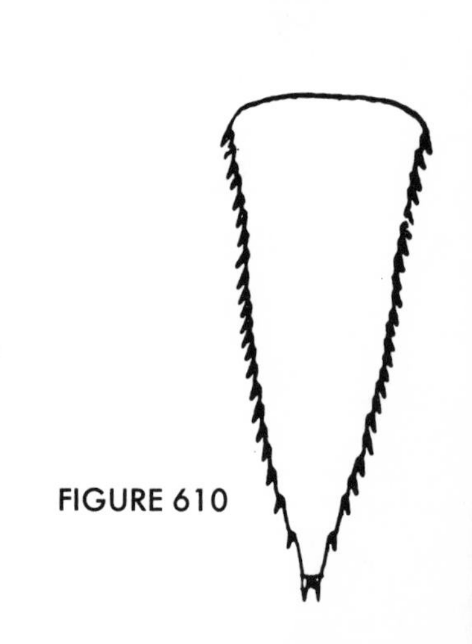

FIGURE 610

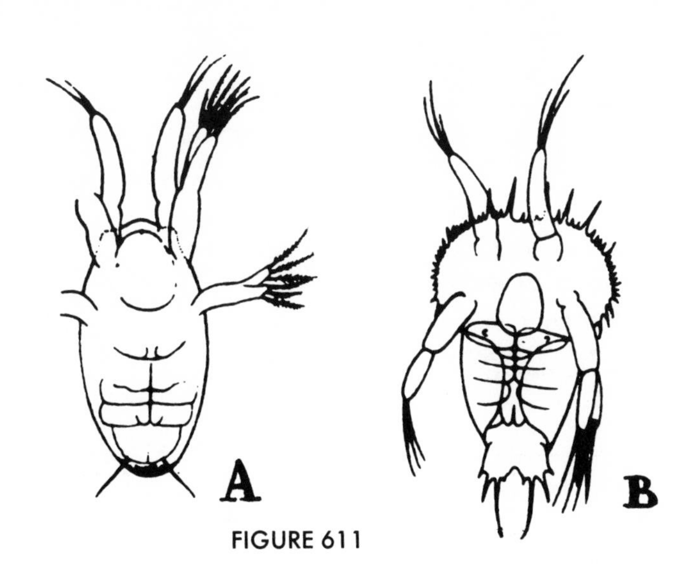

FIGURE 611

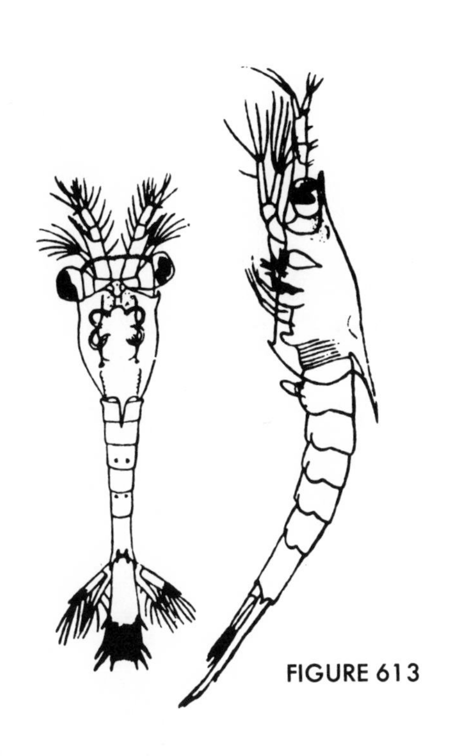
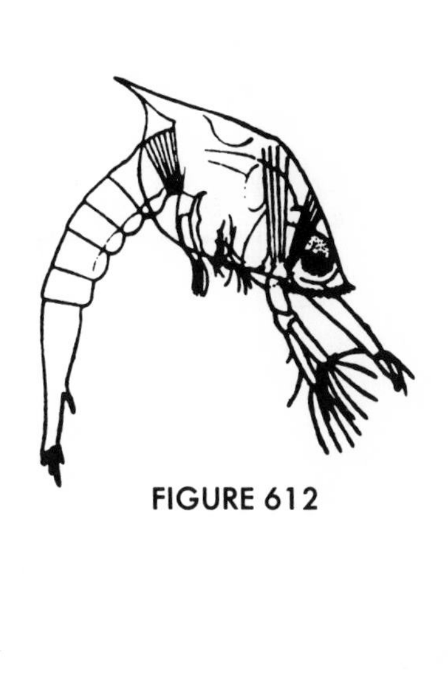
FIGURE 612

FIGURE 613

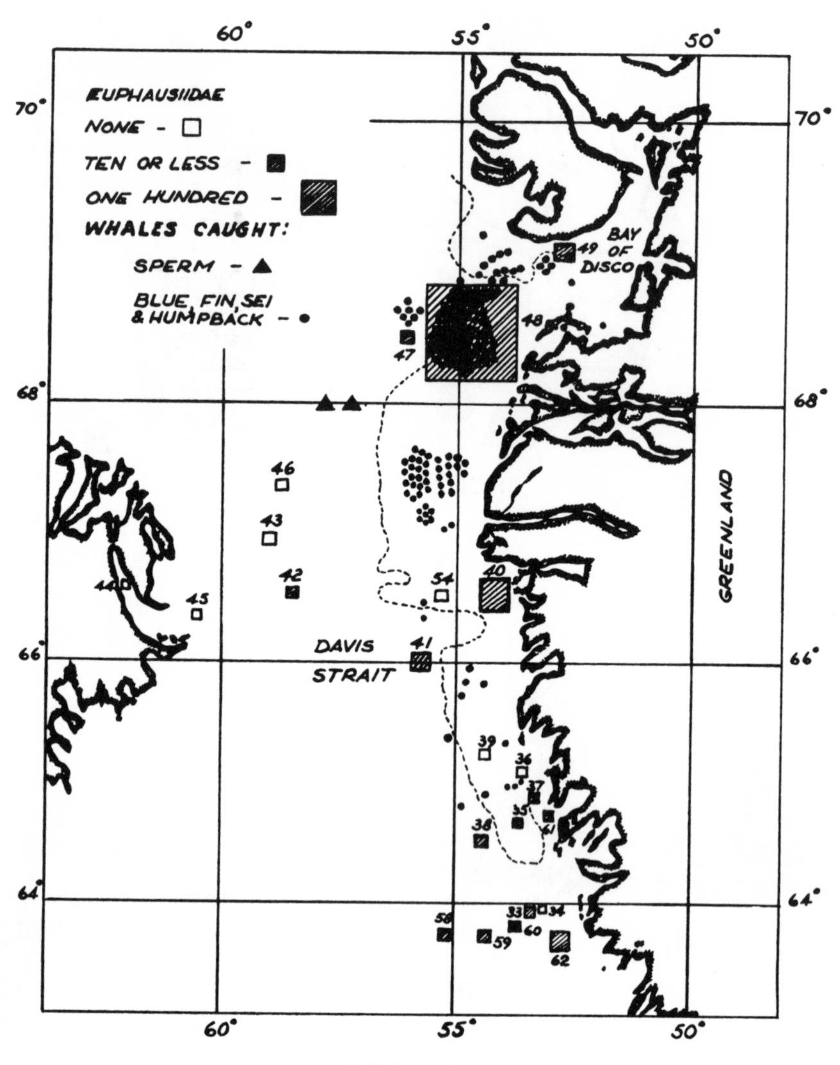
60°
55°
50°
70°
68°
66°
64°
EUPHAUSIIDAE
NONE -
TEN OR LESS -
ONE HUNDRED -
WHALES CAUGHT:
SPERM -
BLUE, FIN, SEI & HUMPBACK -
BAY OF DISCO
DAVIS STRAIT
GREENLAND
49
48
47
46
43
42
44
45
54
40
41
39
36
37
35
61
38
33
34
58
60
59
62

FIGURE 614

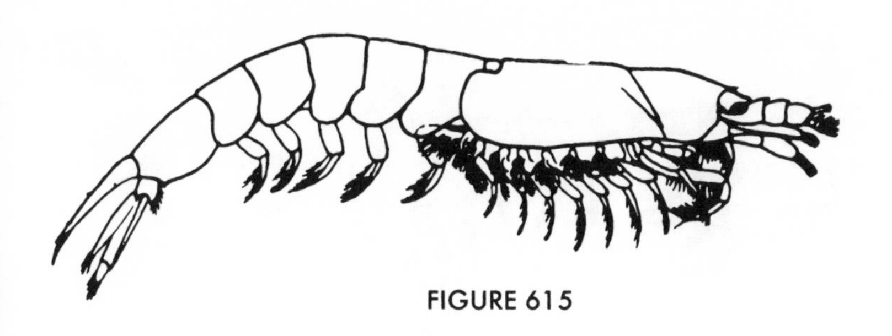

FIGURE 615

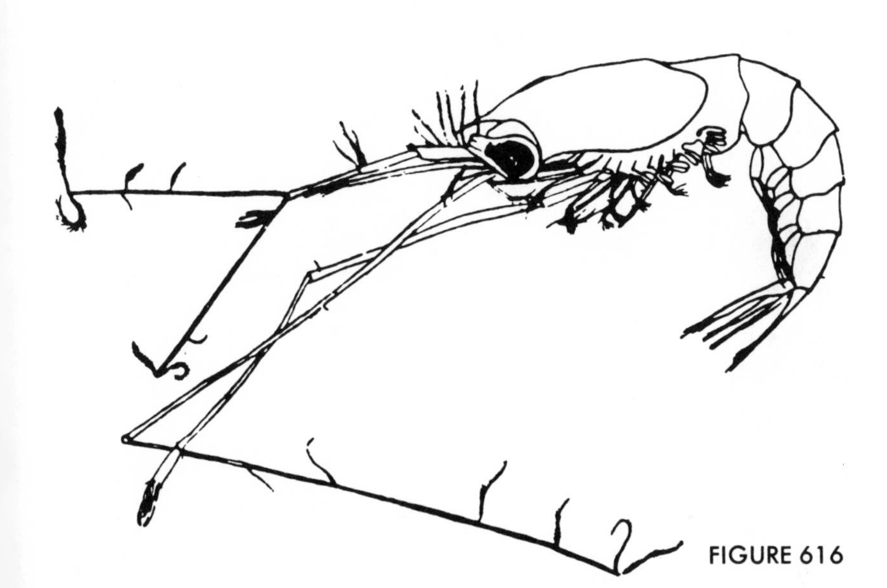

FIGURE 616

FIGURE 617

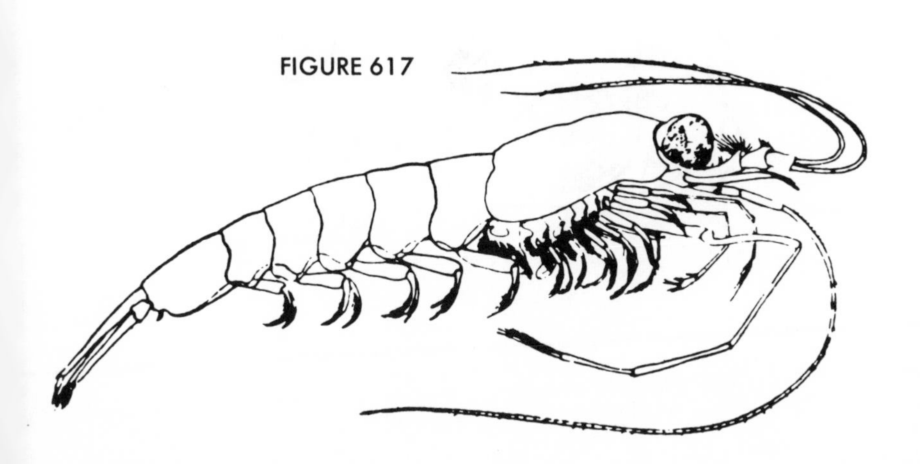

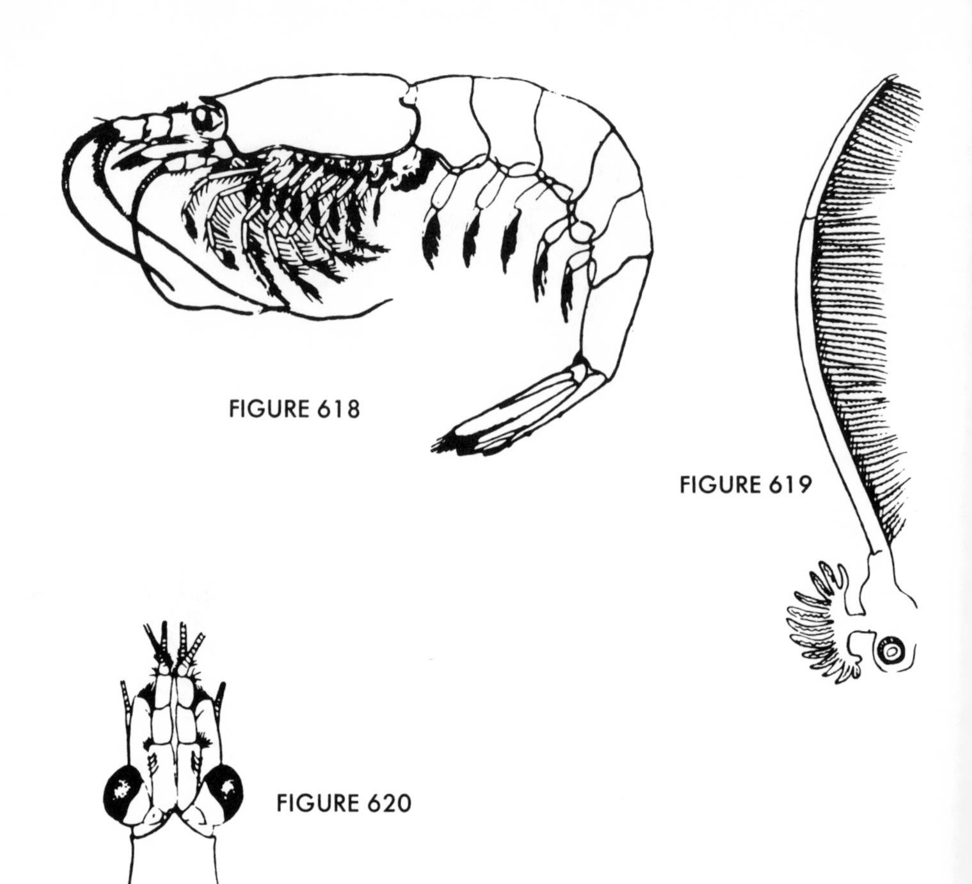

FIGURE 618

FIGURE 619

FIGURE 620

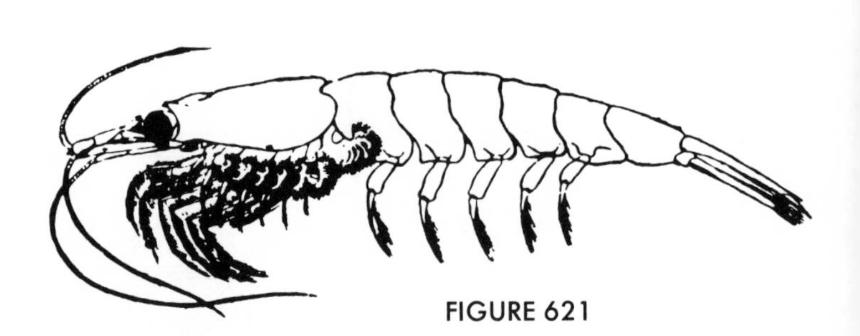

FIGURE 621

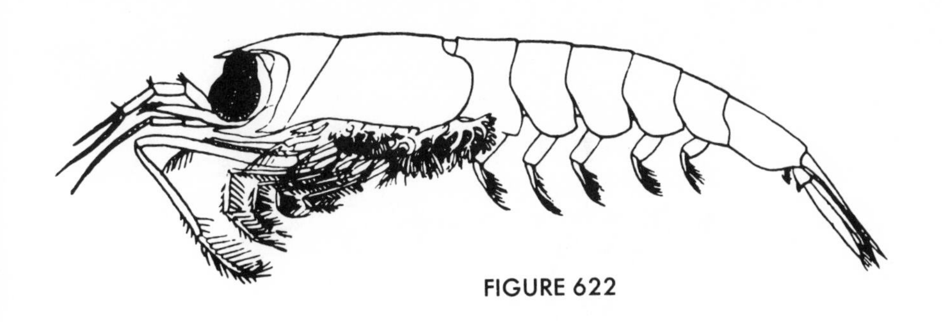

FIGURE 622

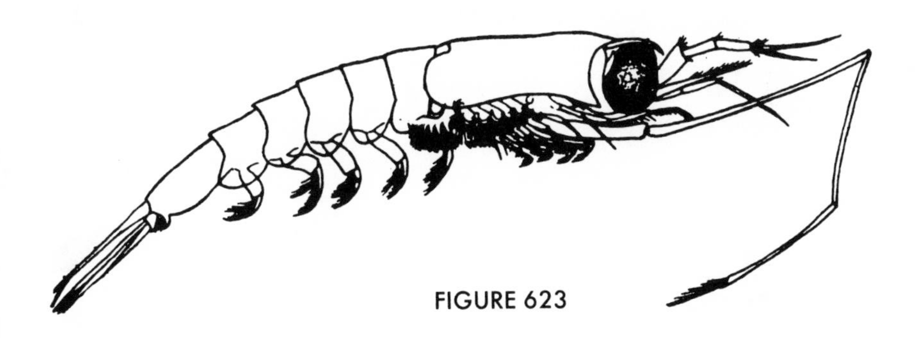

FIGURE 623

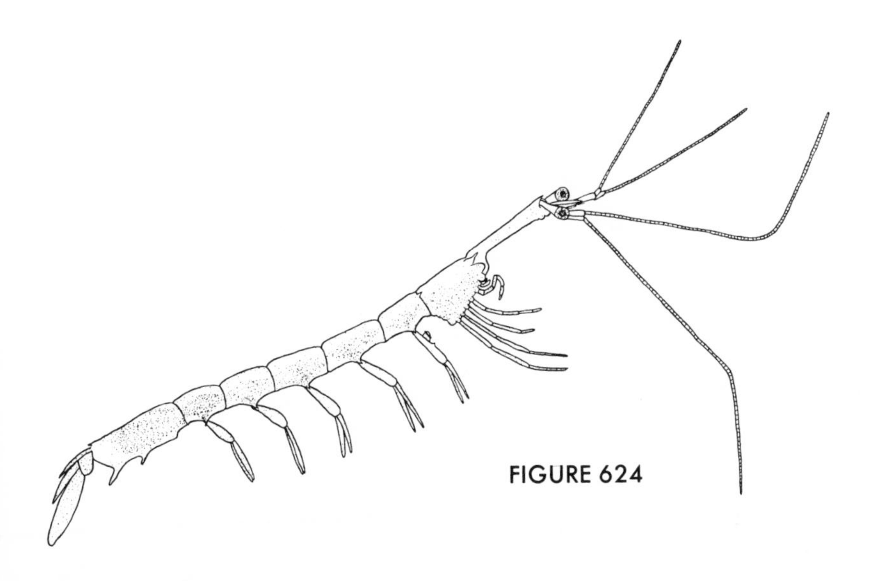

FIGURE 624

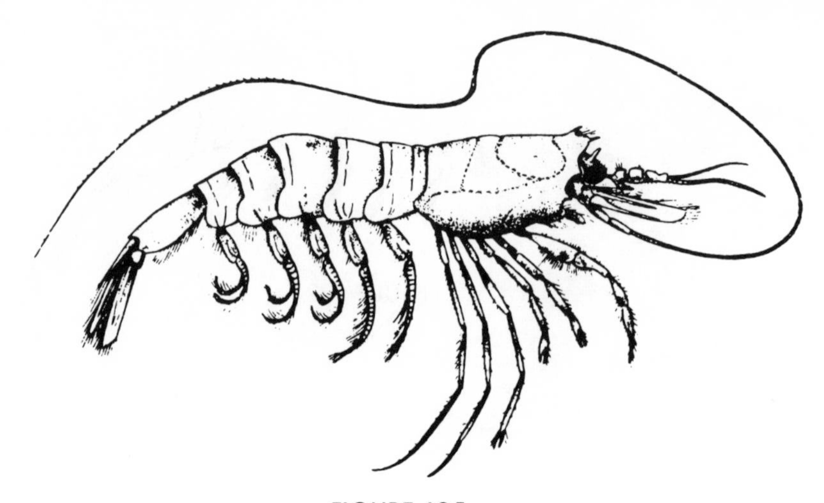

FIGURE 625

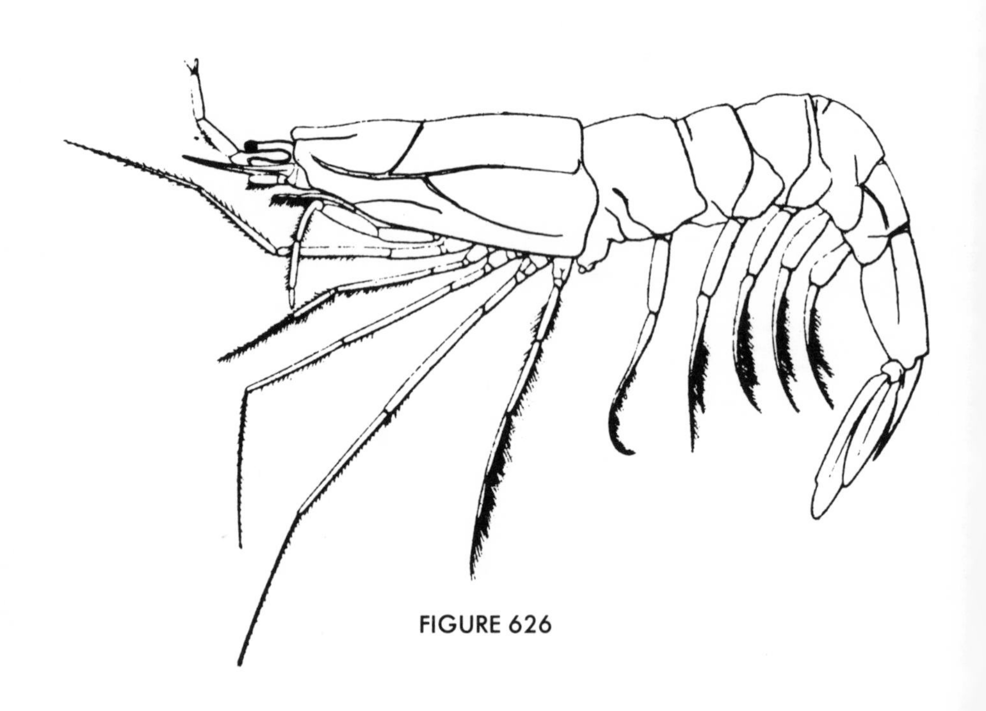

FIGURE 626

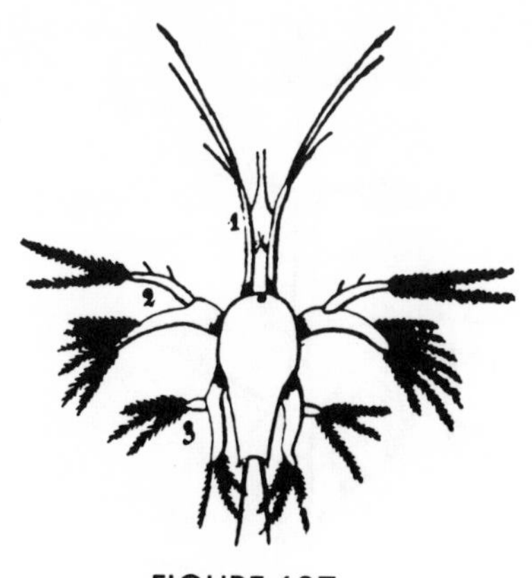

FIGURE 627

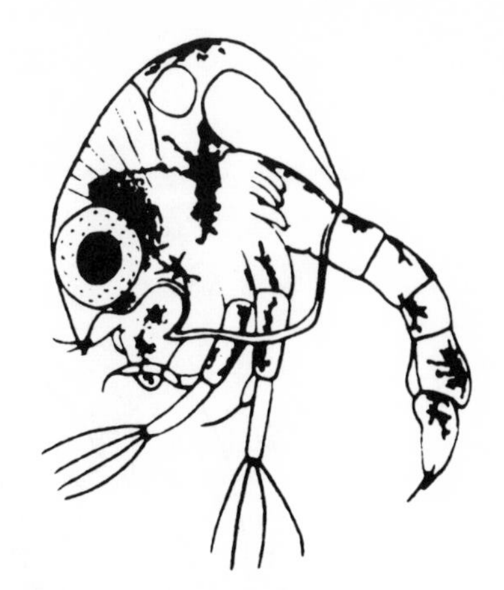
FIGURE 628

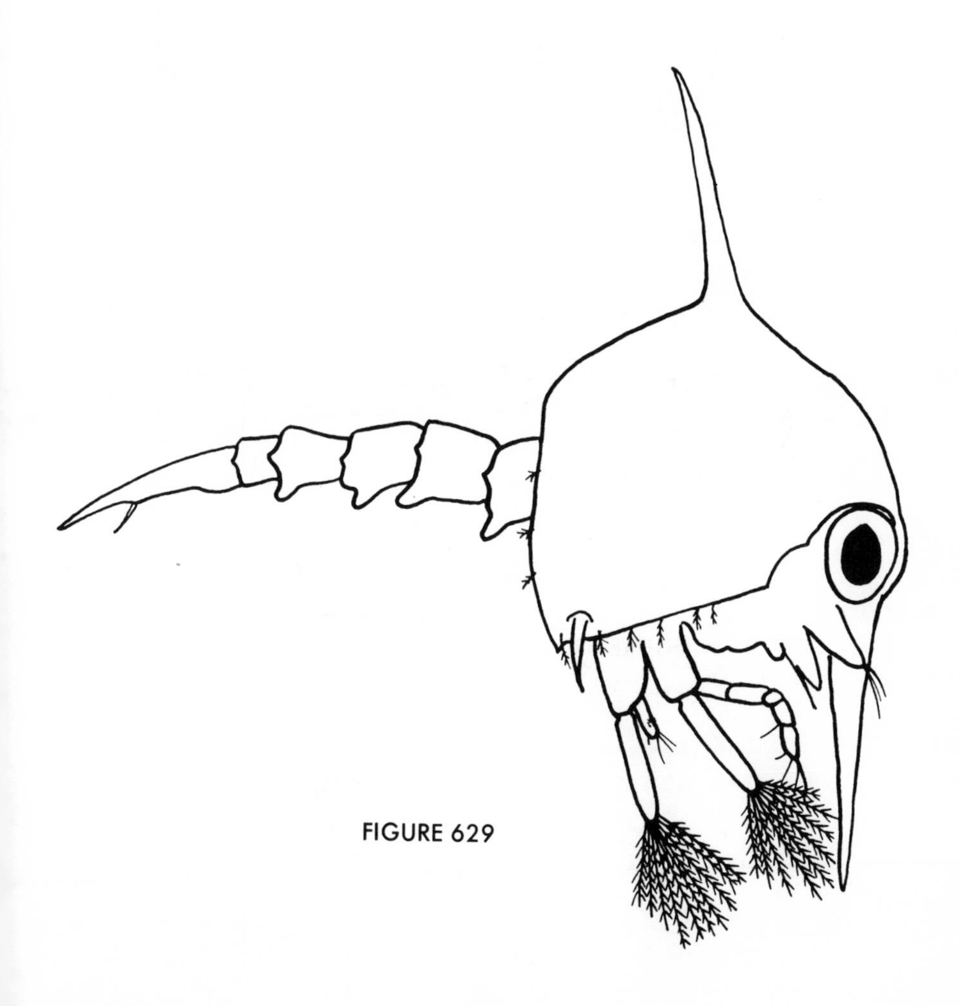
FIGURE 629

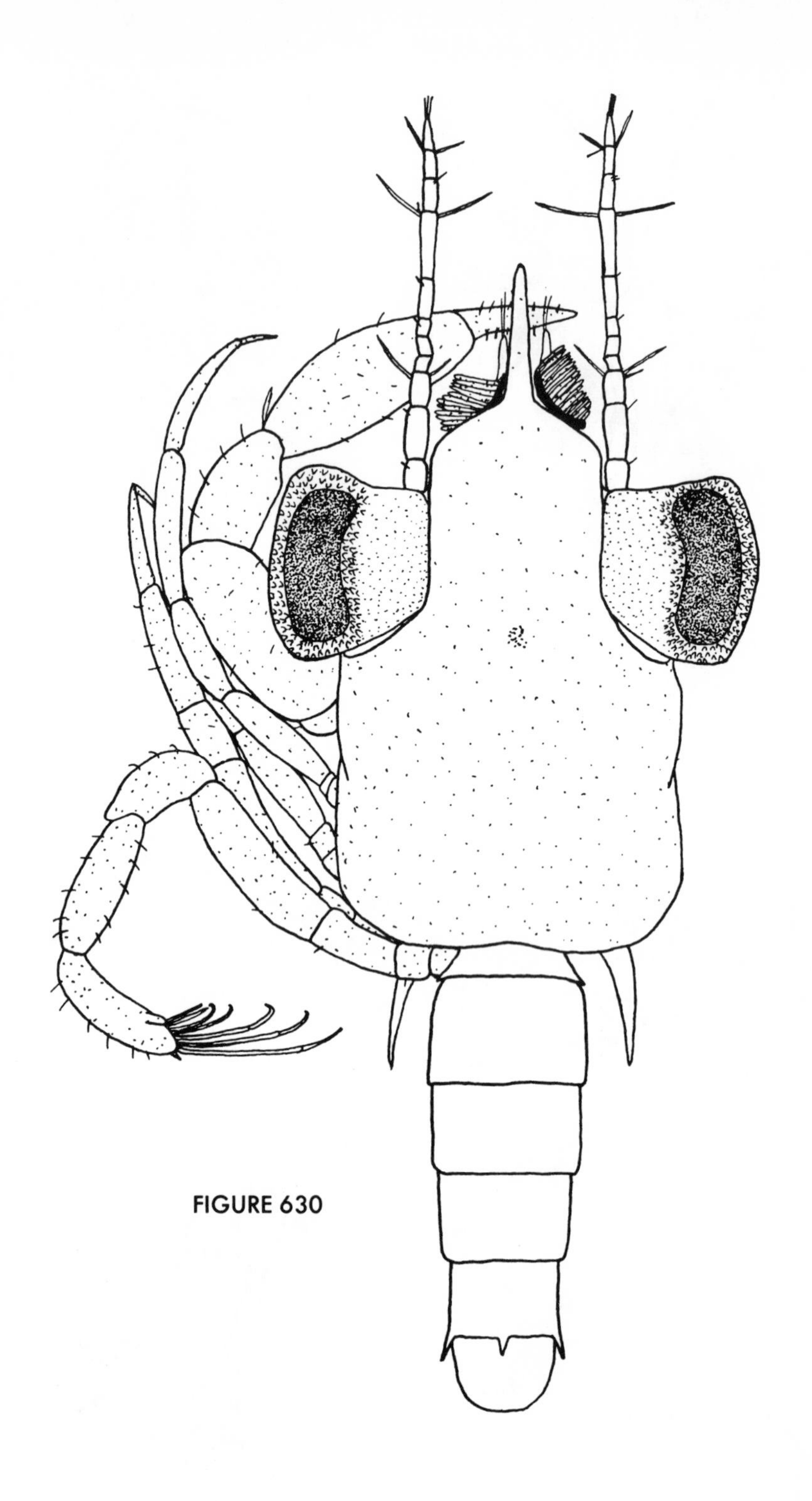

FIGURE 630

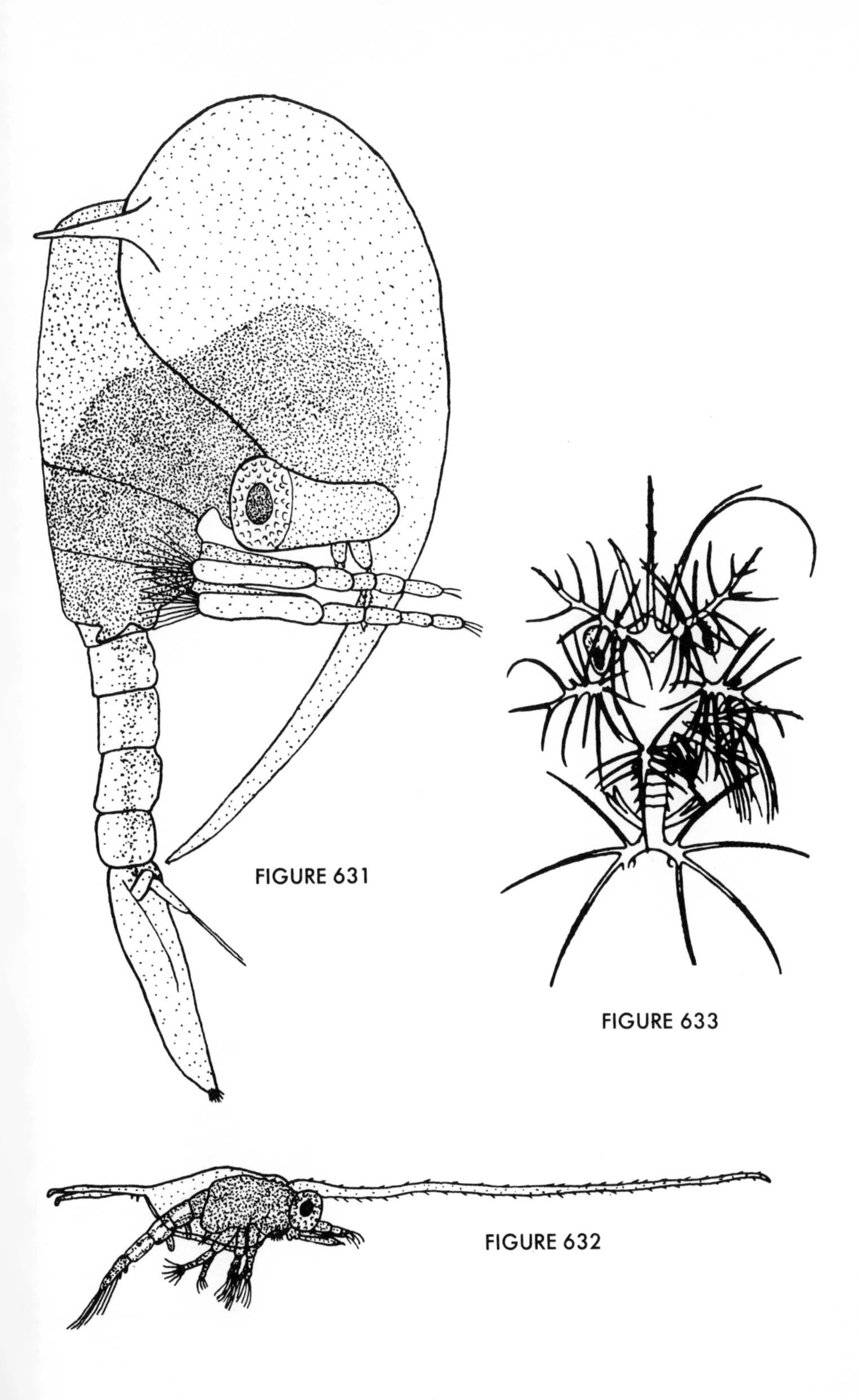
FIGURE 631

FIGURE 633

FIGURE 632

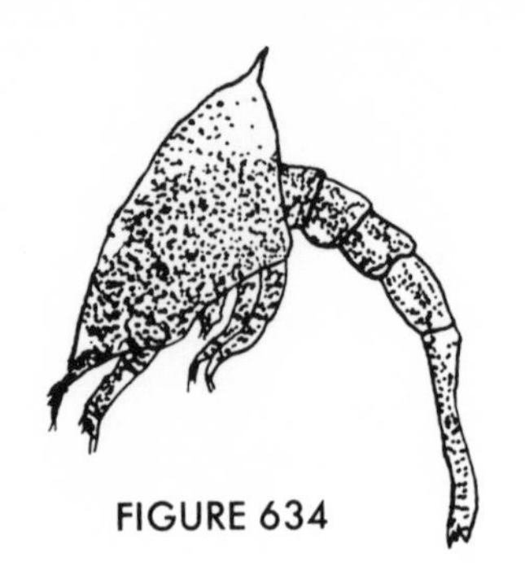

FIGURE 634

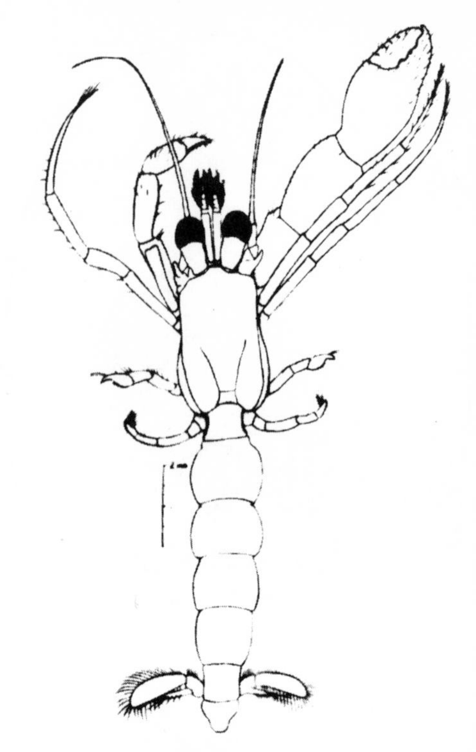

FIGURE 635

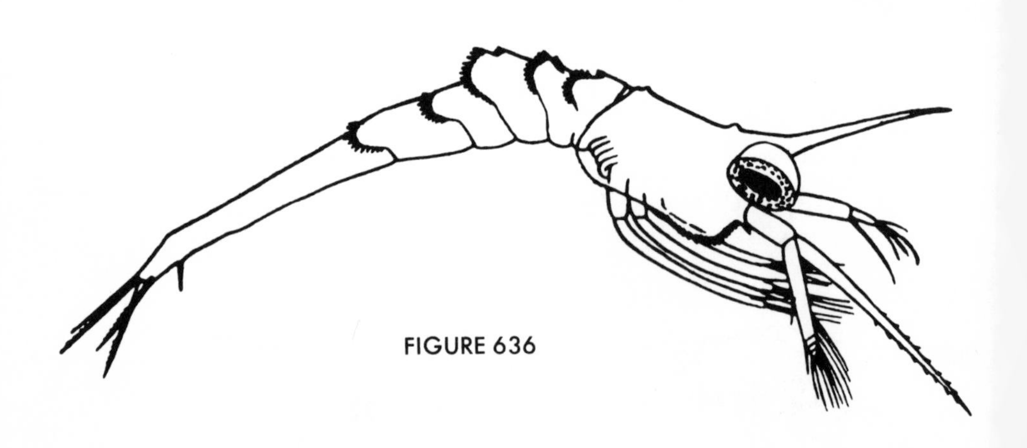

FIGURE 636

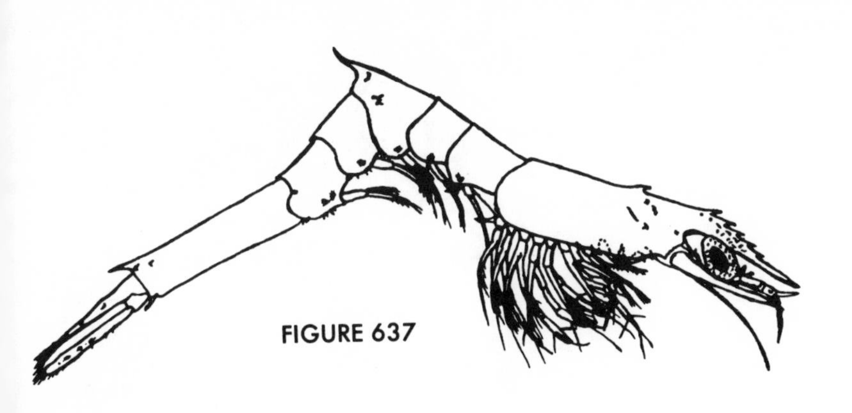
FIGURE 637

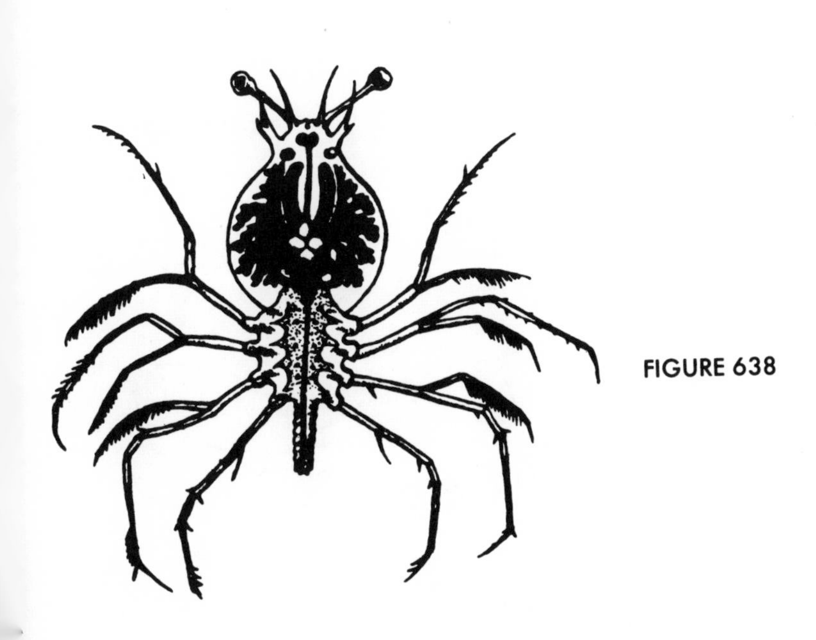
FIGURE 638

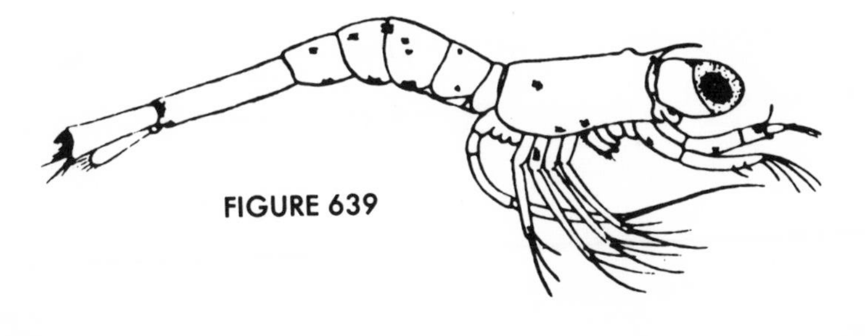
FIGURE 639

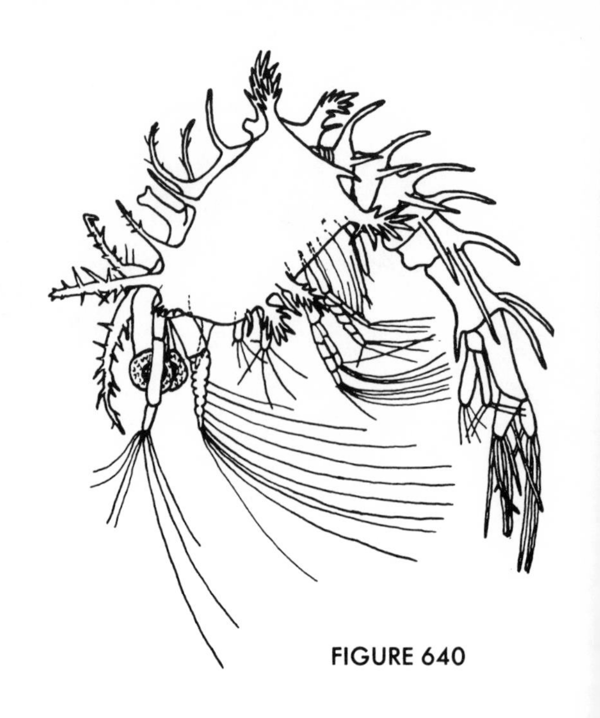

FIGURE 640

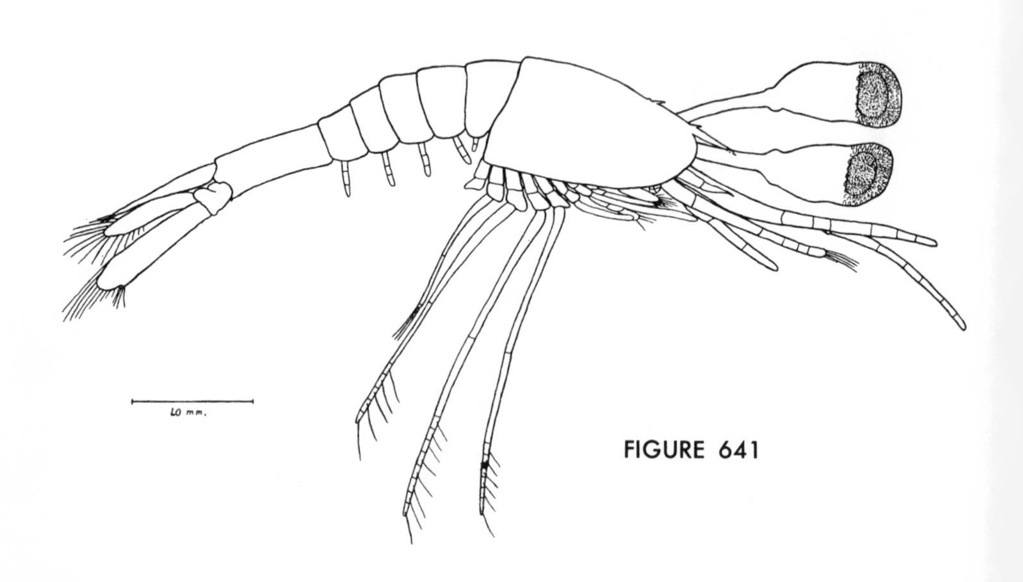

FIGURE 641

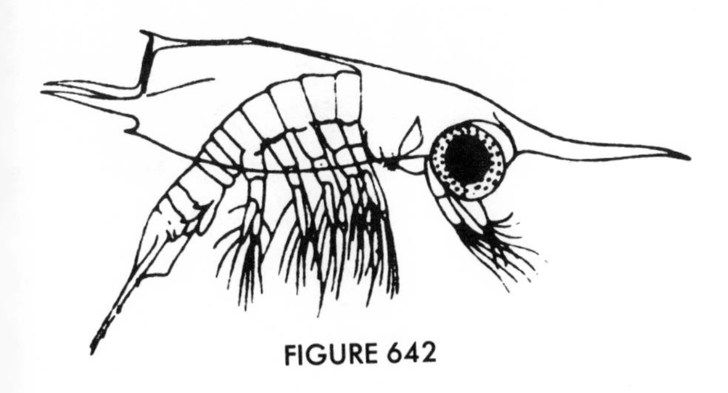

FIGURE 642

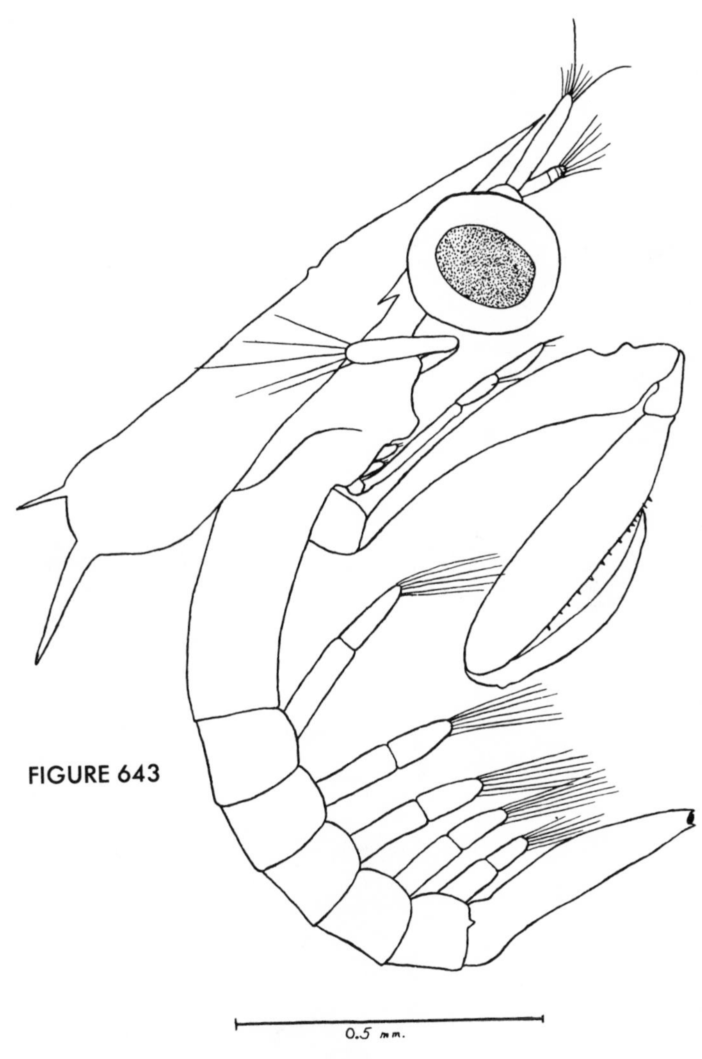

FIGURE 643

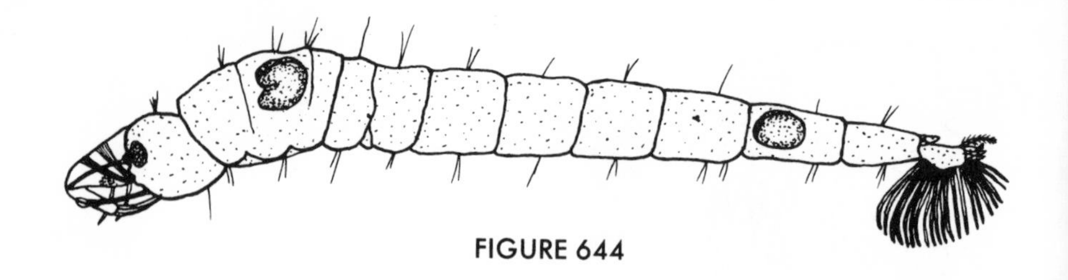

FIGURE 644

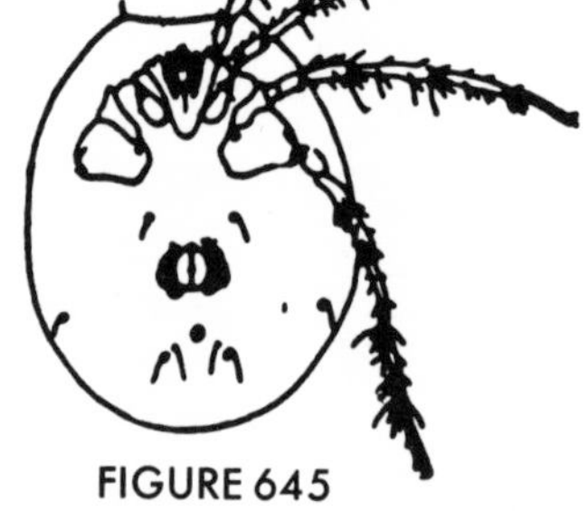

FIGURE 645

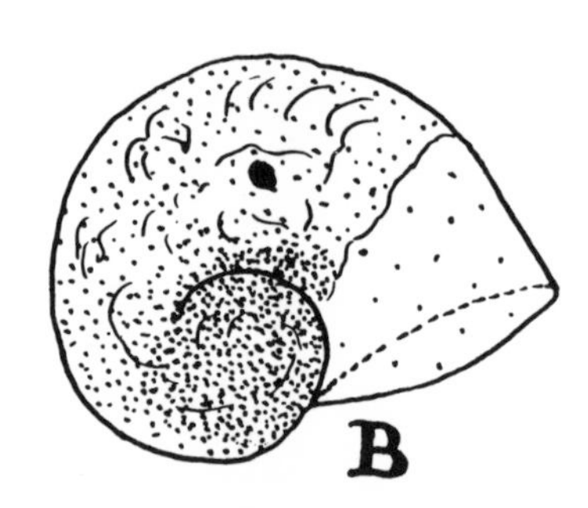

FIGURE 646

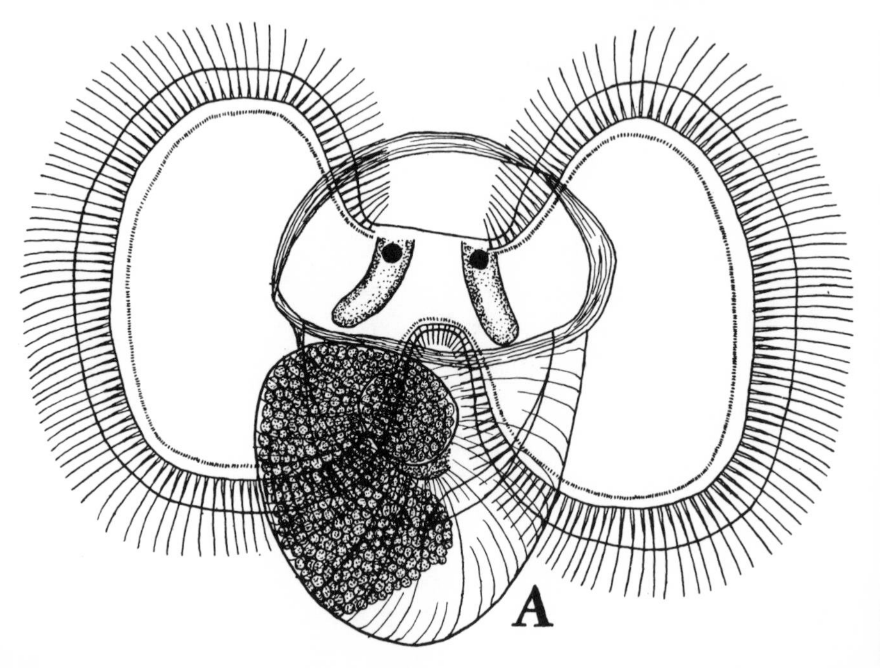

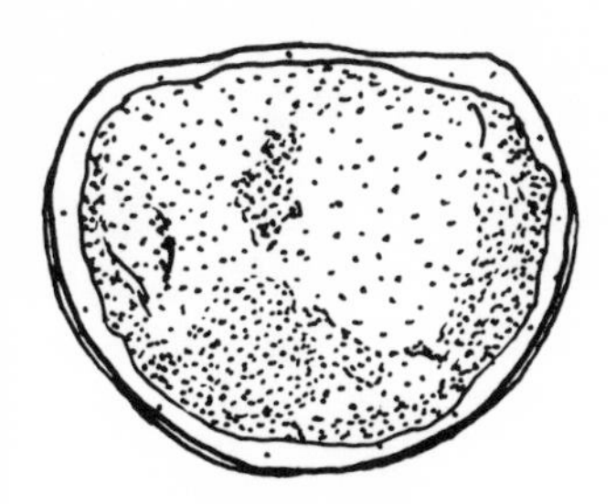

FIGURE 647

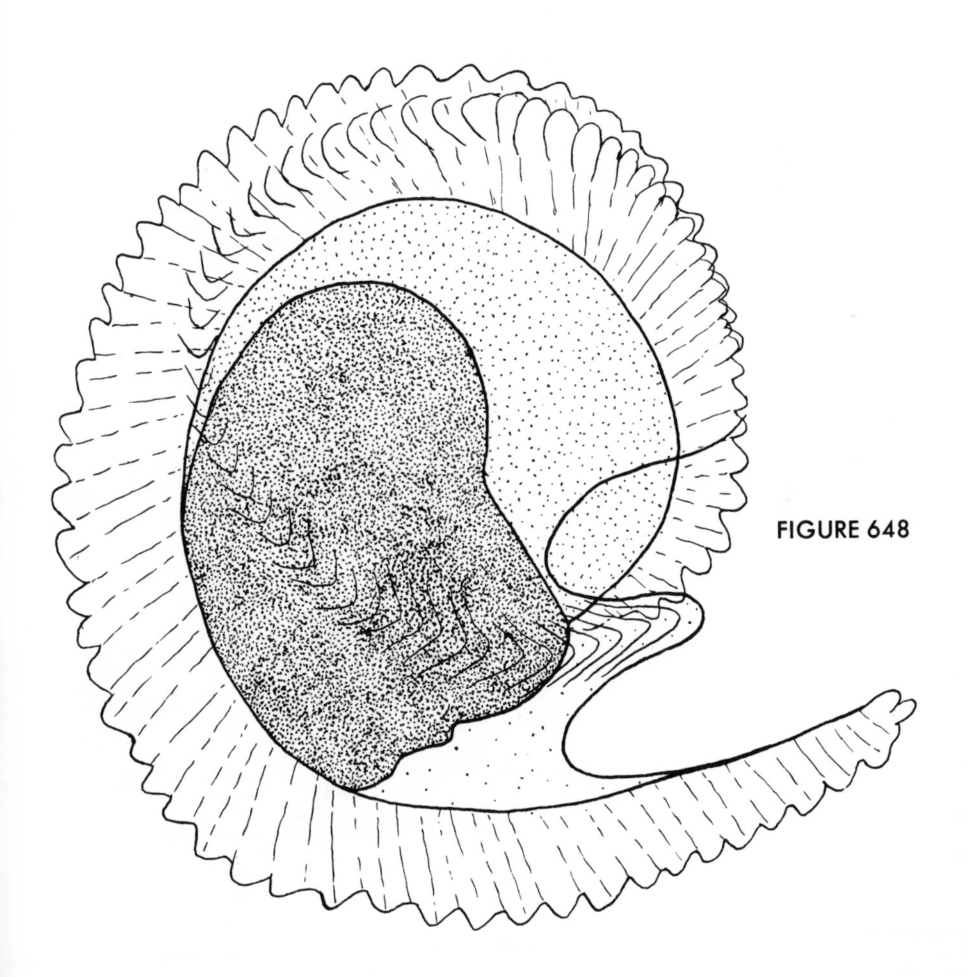

FIGURE 648

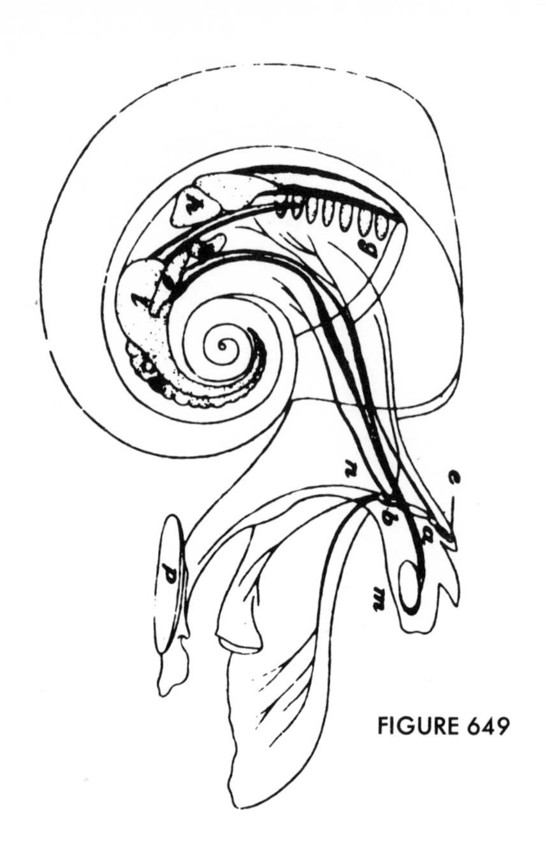

FIGURE 649

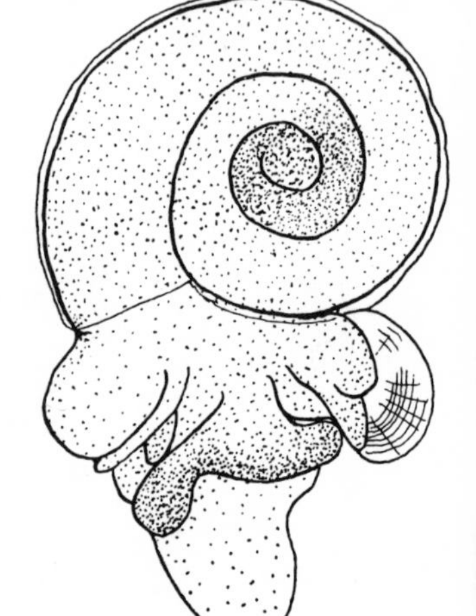

FIGURE 650

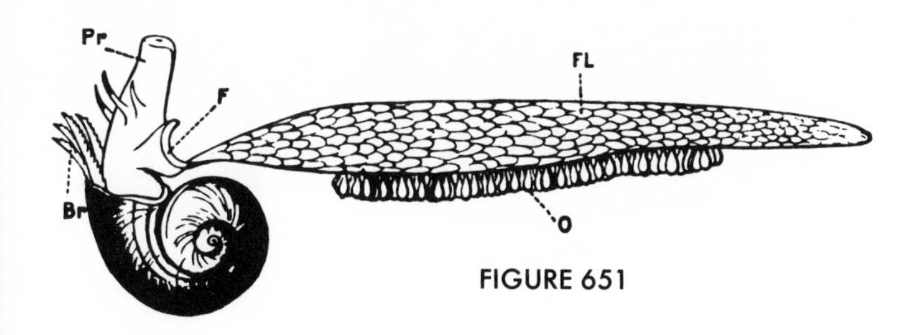

FIGURE 651

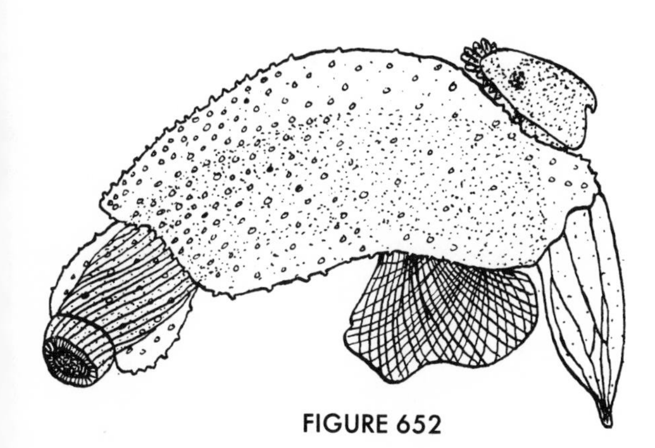

FIGURE 652

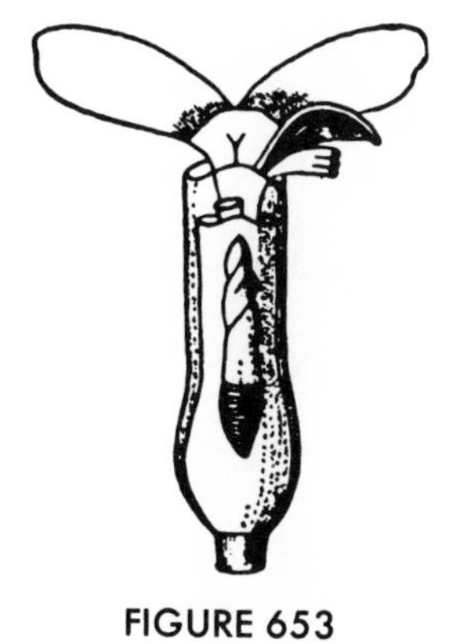

FIGURE 653

FIGURE 654

FIGURE 655

FIGURE 656

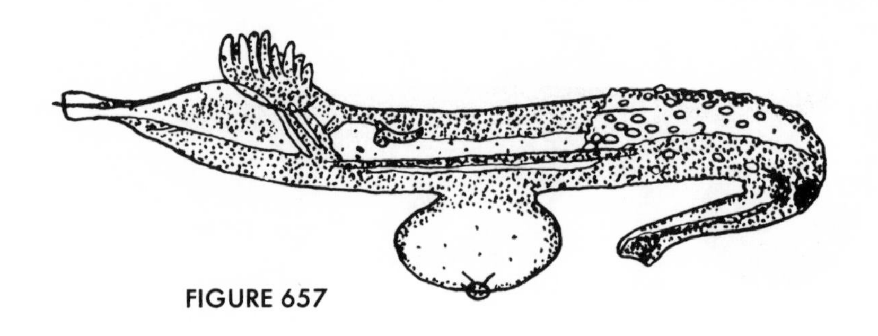

FIGURE 657

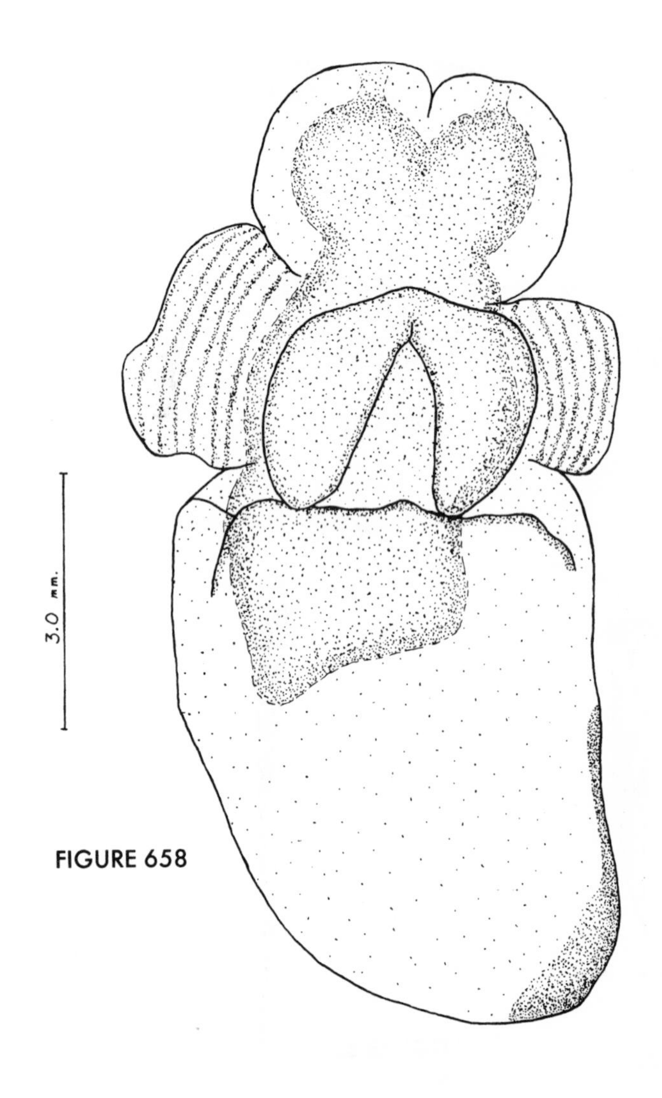

FIGURE 658

FIGURE 659

FIGURE 660

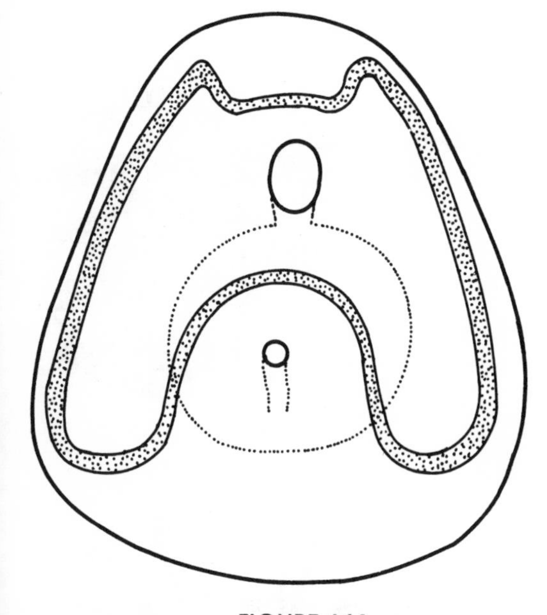

FIGURE 661

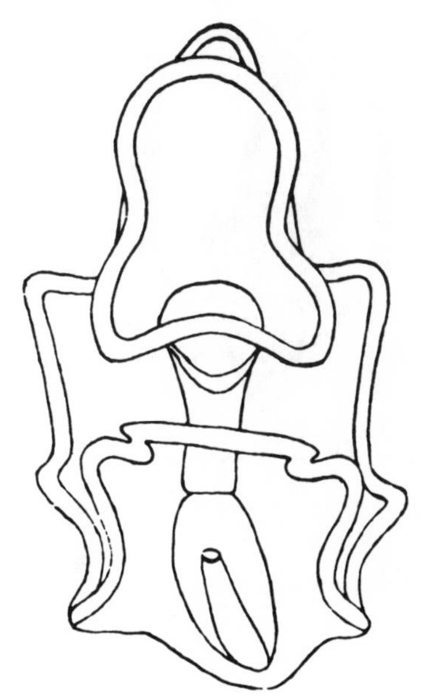

FIGURE 662

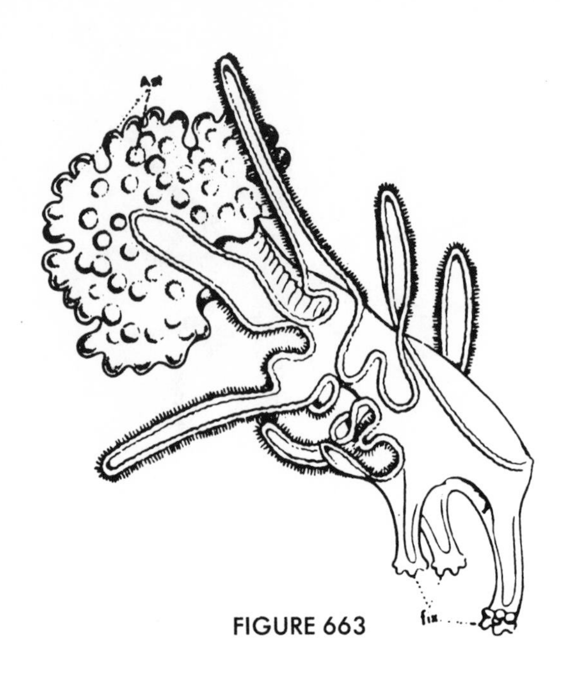

FIGURE 663

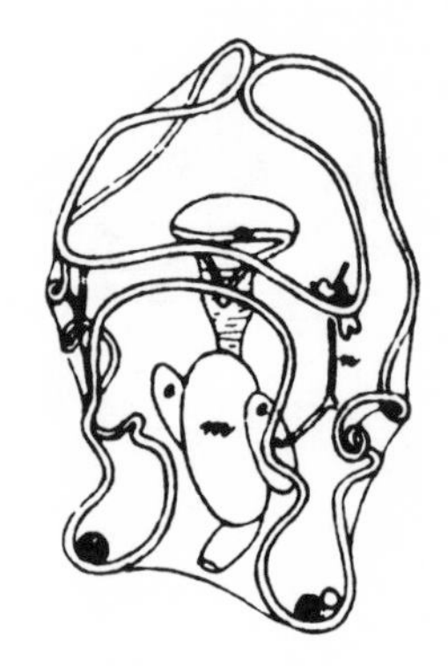

FIGURE 664

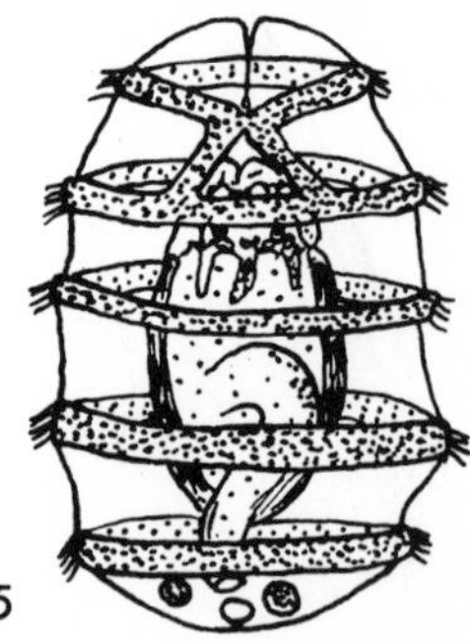

FIGURE 665

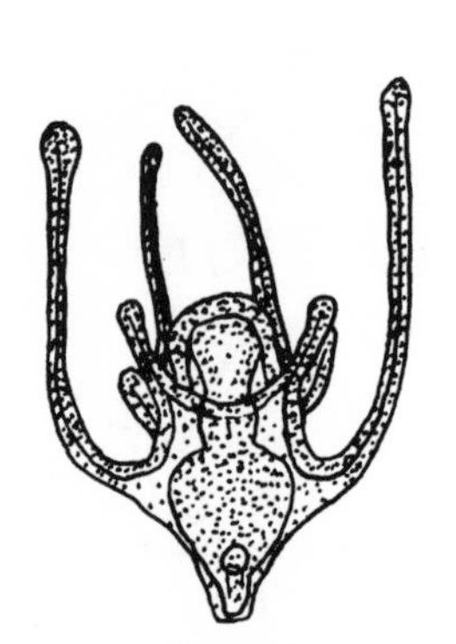

FIGURE 666

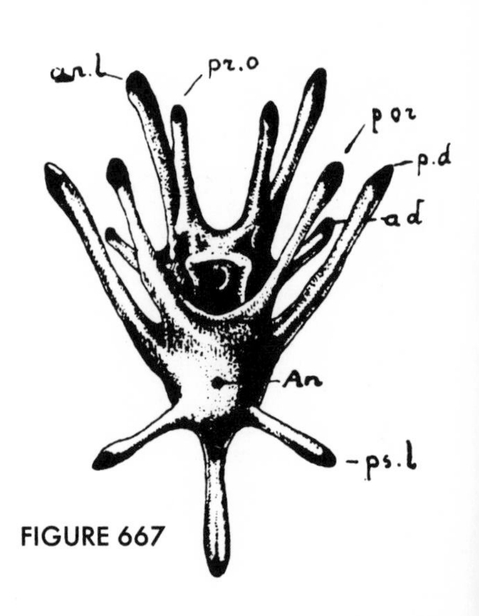

FIGURE 667

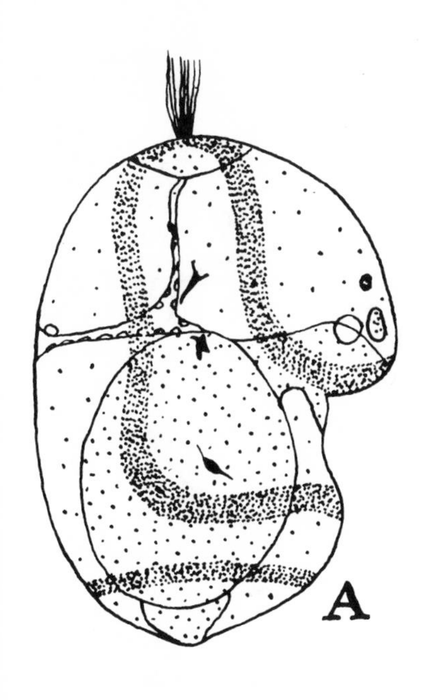

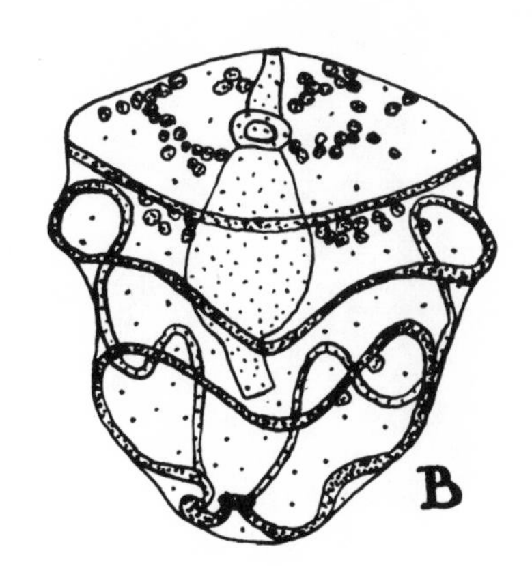

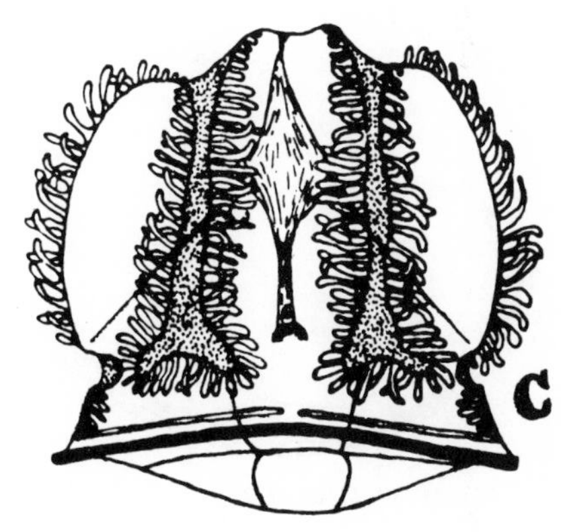

FIGURE 668

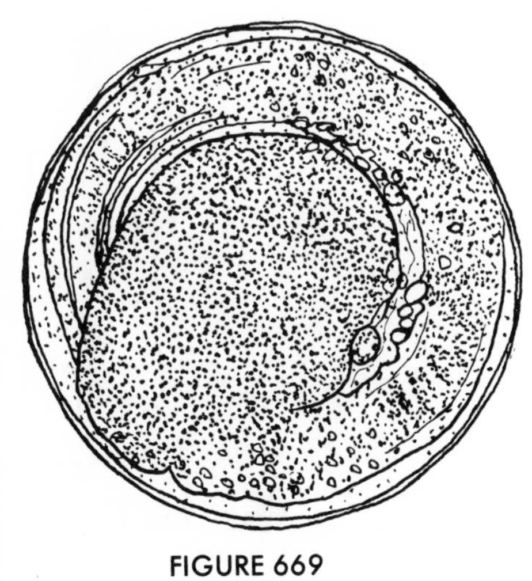

FIGURE 669

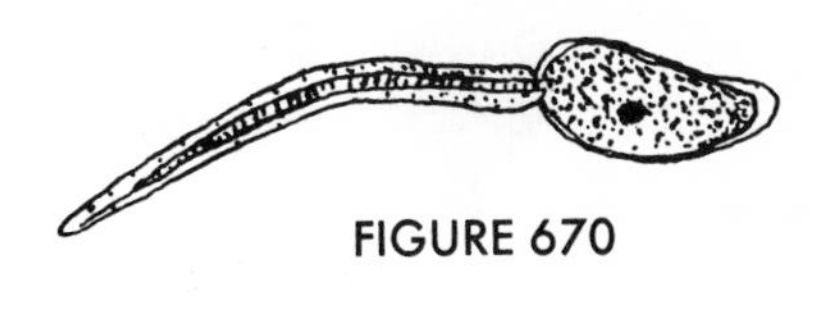

FIGURE 670

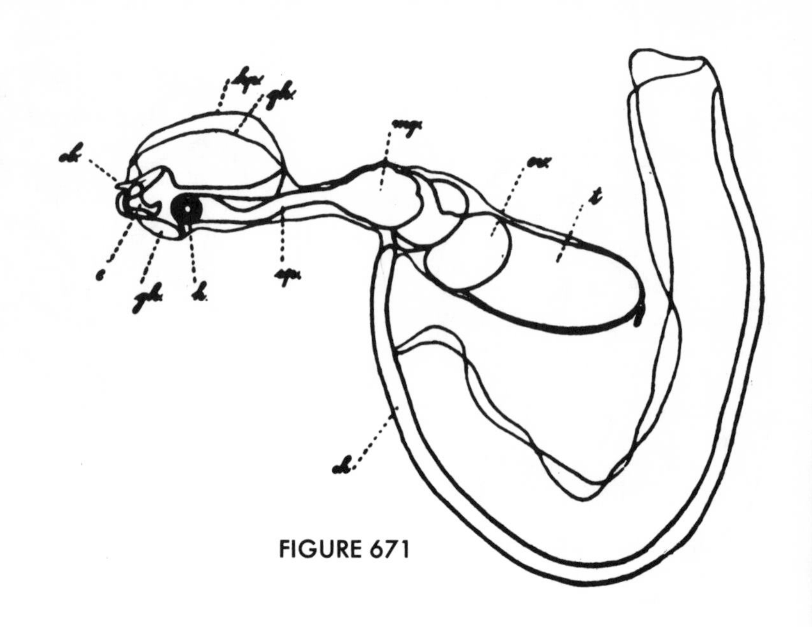

FIGURE 671

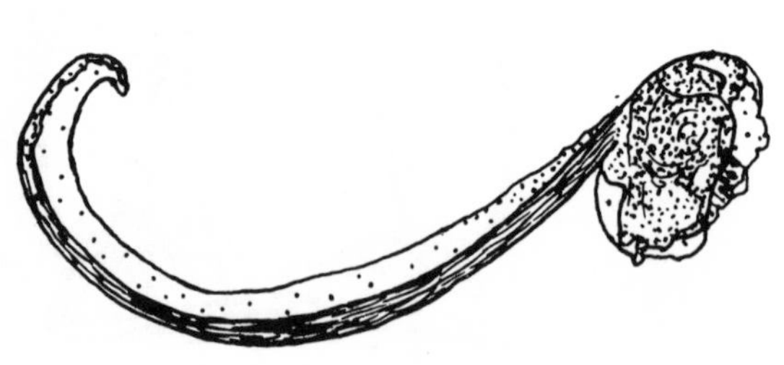

FIGURE 672

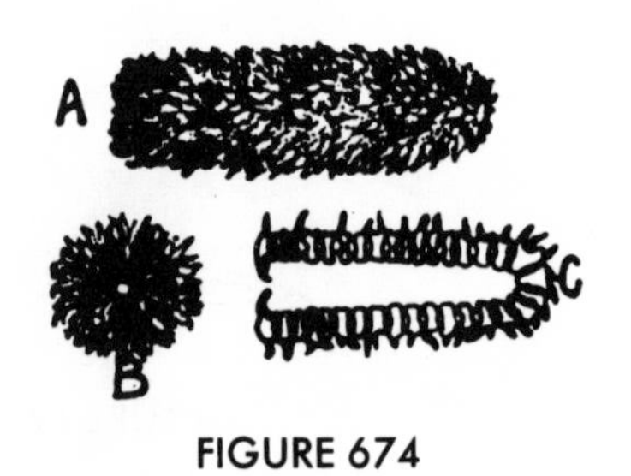

FIGURE 674

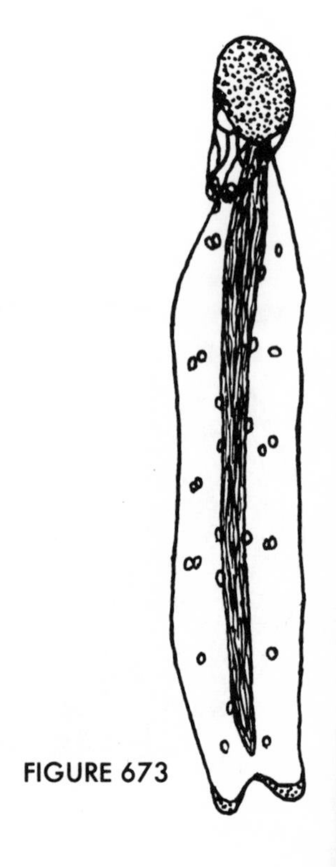

FIGURE 673

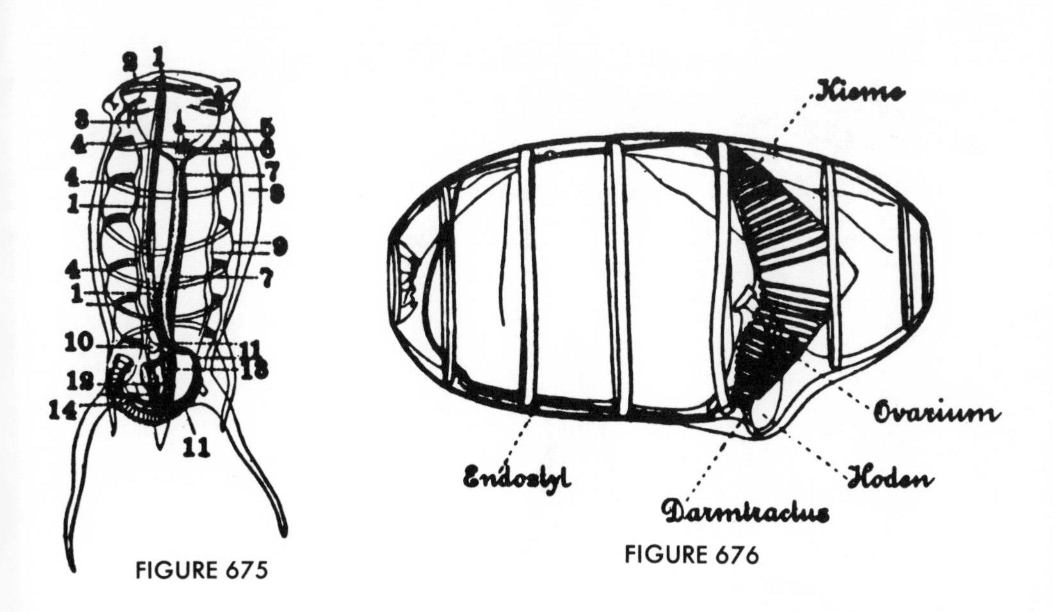

FIGURE 675

FIGURE 676

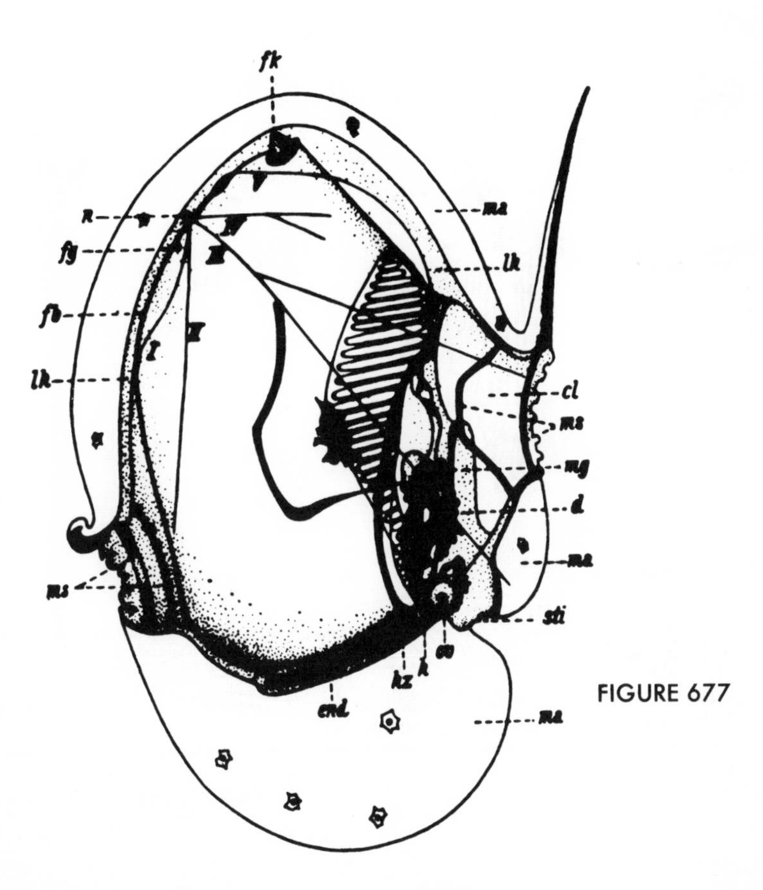

FIGURE 677

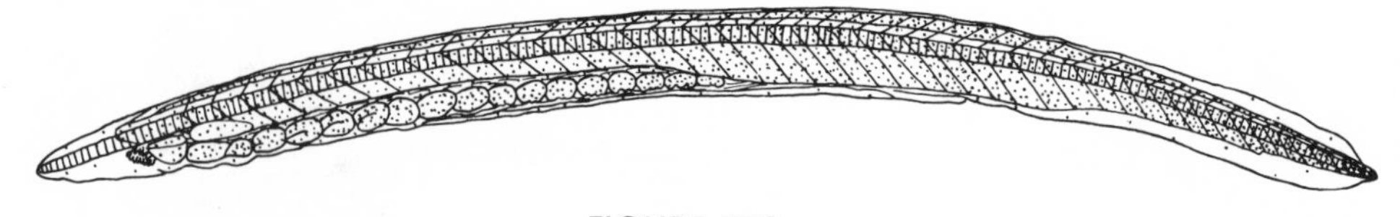

FIGURE 678

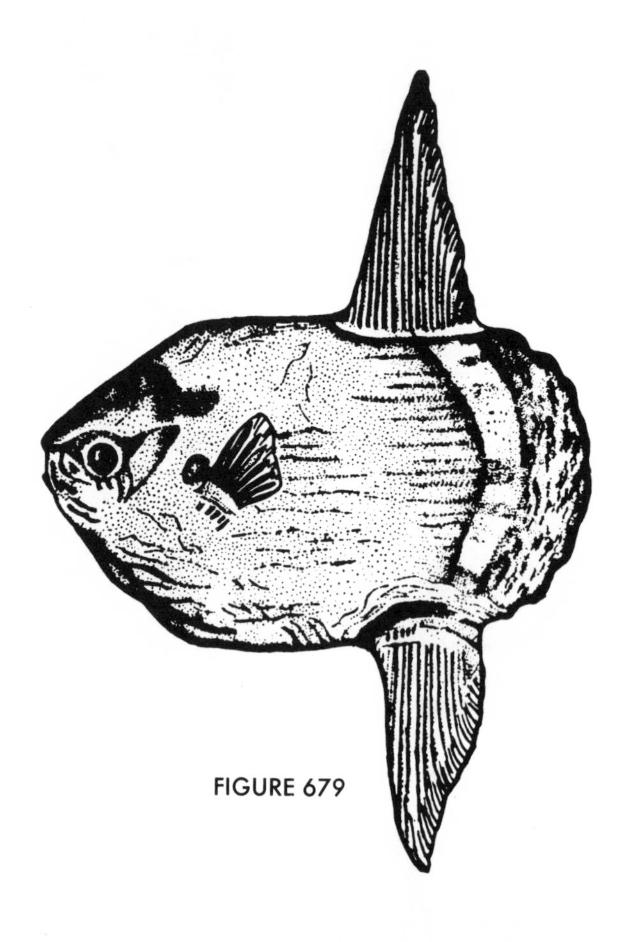

FIGURE 679

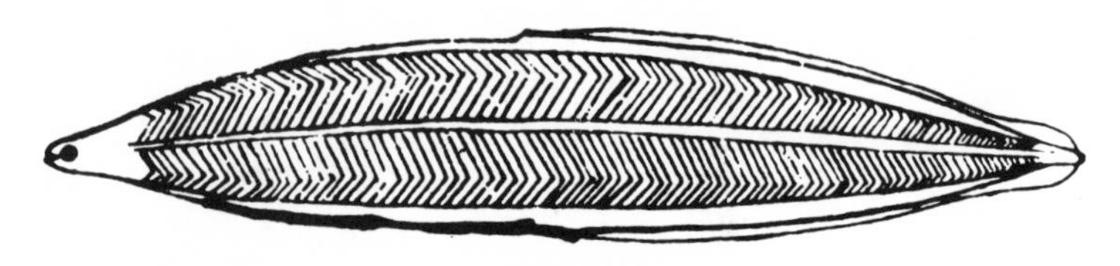

FIGURE 680

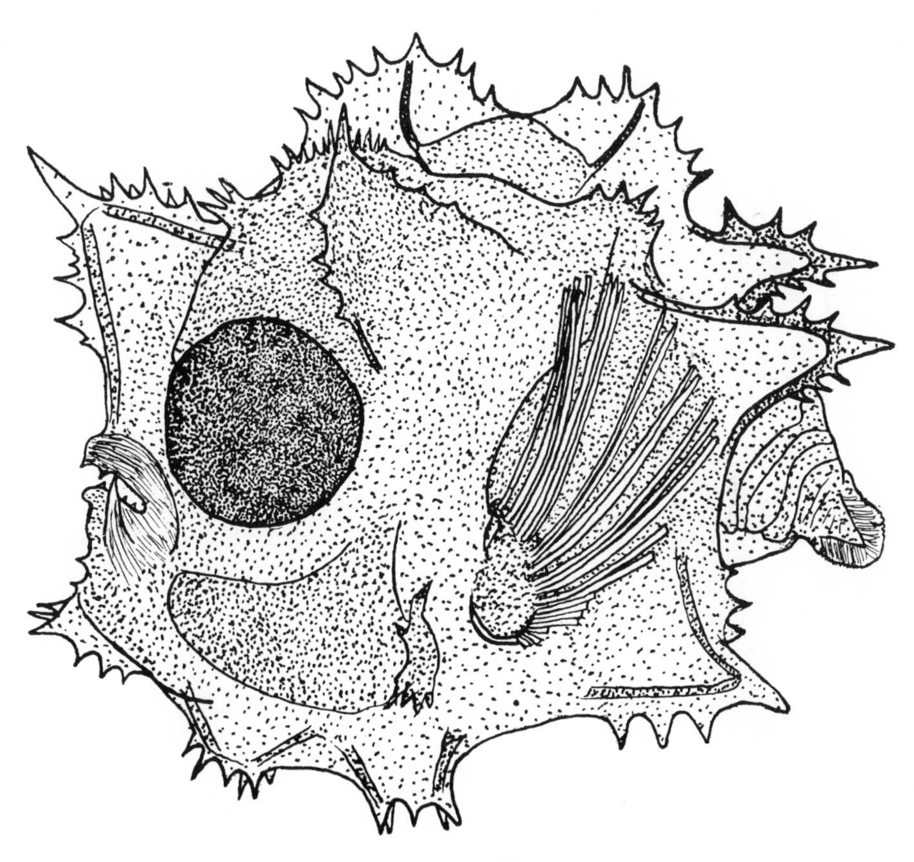

FIGURE 681

INDEX